Parallelism in Matrix Computations

Scientific Computation

More information about this series at http://www.springer.com/series/718

Efstratios Gallopoulos · Bernard Philippe
Ahmed H. Sameh

Parallelism in Matrix Computations

 Springer

Efstratios Gallopoulos
Computer Engineering
  and Informatics Department
University of Patras
Patras
Greece

Bernard Philippe
Campus de Beaulieu
INRIA/IRISA
Rennes Cedex
France

Ahmed H. Sameh
Department of Computer Science
Purdue University
West Lafayette, IN
USA

ISSN 1434-8322   ISSN 2198-2589   (electronic)
Scientific Computation
ISBN 978-94-024-0317-6   ISBN 978-94-017-7188-7   (eBook)
DOI 10.1007/978-94-017-7188-7

Springer Dordrecht Heidelberg New York London

Printed on acid-free paper

Springer Science+Business Media B.V. Dordrecht is part of Springer Science+Business Media
(www.springer.com)

*To the memory of Daniel L. Slotnick,
parallel processing pioneer*

Preface

Computing instruments were developed to facilitate fast calculations, especially in computational science and engineering applications. The fact that this is not just a matter of building a hardware device and its system software, was already hinted to by Charles Babbage, when he wrote in the mid-nineteenth century, *As soon as an Analytical Engine exists, it will necessarily guide the future course of science. Whenever any result is sought by its aid, the question will then arise—By what course of calculation can these results be arrived at by the machine in the shortest time?* [1]. This question points to one of the principal challenges for parallel computing. In fact, in the aforementioned reference, Babbage did consider the advantage of parallel processing and the perfect speedup that could be obtained when adding numbers if no carries were generated. He wrote *If this could be accomplished it would render additions and subtractions with numbers having ten, twenty, fifty or any number of figures as rapid as those operations are with single figures.* He was also well aware of the limitations, in this case the dependencies caused by the carries. A little more than half a century after Babbage, in 1922, an extraordinary idea was sketched by Lewis Fry Richardson. In his treatise *Weather Prediction by Numerical Process* he described his "forecast-factory" fantasy to speed up calculations by means of parallel processing performed by humans [3, Chap. 11, p. 219]. Following the development of the first electronic computer, in the early 1950s, scientists and engineers proposed that one way to achieve higher performance was to build a computing platform consisting of many interconnected von Neumann uniprocessors that can cooperate in handling what were the large computational problems of that era. This idea appeared simple and natural, and quickly attracted the attention of university-, government-, and industrial-research laboratories. Forty years after Richardson's treatise, the designers of the first parallel computer prototype ever built, introduced their design in 1962 as follows: *The Simultaneous Operation Linked Ordinal Modular Network (SOLOMON), a parallel network computer, is a new system involving the interconnections and programming, under the supervision of a central control unit, of many identical processing elements (as few or as many as a given problem requires), in an arrangement that can simulate directly the problem being solved.* It is remarkable how this introductory paragraph underlines the generality and adaptive character of the design, despite the fact that neither the prototype nor subsequent designs went as far. These authors stated further that this architecture shows great promise in aiding progress in certain critical applications that rely on common mathematical denominators that are dominated by matrix computations.

Soon after that, the field of parallel processing came into existence starting with the development of the ILLIAC-IV at the University of Illinois at Urbana-Champaign led by Daniel L. Slotnick (1931–1985), who was one of the principal designers of the SOLOMON computer. The design and building of parallel computing platforms, together with developing the underlying system software as well as the associated numerical libraries, emerged as important research topics. Now, four decades after the introduction of the ILLIAC-IV, parallel computing resources ranging from multicore systems (which are found in most modern desktops and laptops) to massively parallel platforms are within easy reach of most computational scientists and engineers. In essence, parallel computing has evolved from an exotic technology to a widely available commodity. Harnessing this power to the maximum level possible, however, remains the subject of ongoing research efforts.

Massively parallel computing platforms now consist of thousands of nodes cooperating via sophisticated interconnection networks with several layers of hierarchical memories. Each node in such platforms is often a multicore architecture. Peak performance of these platforms has reached the persetascale level, in terms of the number of floating point operations completed in one second, and will soon reach the exascale level. These rapid hardware technological advances, however, have not been matched by system or application software developments. Since the late 1960s different parallel architectures have come and gone in a relatively short time resulting in lack of stable and sustainable parallel software infrastructure. In fact, present day researchers involved in the design of parallel algorithms and development of system software for a given parallel architecture often rediscover work that has been done by others decades earlier. Such lack of stability in parallel software and algorithm development has been pointed out by George Cybenko and David Kuck as early as 1992 in [2].

Libraries of efficient parallel algorithms and their underlying kernels are needed for enhancing the realizable performance of various computational science and engineering (CSE) applications on current multicore and petascale computing platforms. Developing robust parallel algorithms, together with their theoretical underpinnings is the focus of this book. More specifically, we focus exclusively on those algorithms relevant to dense and sparse matrix computations which govern the performance of many CSE applications. The important role of matrix computations was recognized in the early days of digital computers. In fact, after the introduction of the Automatic Computing Engine (ACE), Alan Turing included solving linear systems and matrix multiplication as two of the computational challenges for this computing platform. Also, in what must be one of the first references to sparse and structured matrix computations, he observed that even though the storage capacities available then could not handle dense linear systems of order larger than 50, in practice one can handle much larger systems: *The majority of problems have very degenerate matrices and we do not need to store anything like as much as ... since the coefficients in these equations are very systematic and mostly zero.* The computational challenges we face today are certainly different in scale than those above but they are surprisingly similar in their dependence on matrix computations and numerical linear algebra. In the early 1980s, during building the experimental parallel computing platform "Cedar", led by David Kuck, at the University of Illinois at Urbana-Champaign, a table was compiled that identifies the common computational bottlenecks of major science and engineering applications, and the parallel algorithms that need to be designed, together with their underlying kernels, in order to achieve high performance. Among the algorithms listed, matrix computations are the most prominent. A similar list was created by UC Berkeley in 2009. Among Berkeley's 13 parallel algorithmic methods that capture patterns of computation and communication, which are called "dwarfs", the top two are matrix computation-based. Not only are matrix computations, and especially sparse matrix computations, essential in advancing science and engineering disciplines such as computational mechanics, electromagnetics, nanoelectronics among others, but they are also essential for manipulation of the large graphs that arise in social networks, sensor networks, data mining, and machine learning just to list a few. Thus, we conclude that realizing high performance in dense and sparse matrix computations on parallel computing platforms is central to many applications and hence justify our focus.

Our goal in this book is therefore to provide researchers and practitioners with the basic principles necessary to design efficient parallel algorithms for dense and sparse matrix computations. In fact, for each fundamental matrix computation problem such as solving banded linear systems, for example, we present a family of algorithms. The "optimal" choice of a member of this family will depend on the linear system and the architecture of the parallel computing platform under consideration. Clearly, however, executing a computation on a parallel platform requires the combination of many steps ranging from: (i) the search for an "optimal" parallel algorithm that minimizes the required arithmetic operations, memory references and interprocessor communications, to (ii) its implementation on the underlying platform. The latter step depends on the specific architectural characteristics of the parallel computing platform. Since these architectural characteristics are still evolving rapidly, we will refrain in this book from exposing fine implementation details for each parallel algorithm. Rather, we focus on algorithm robustness and opportunities for parallelism in general. In other words, even though our approach is geared towards numerically reliable algorithms that lend themselves to practical implementation on parallel computing platforms that are currently available, we will also present classes of algorithms that expose the theoretical limitations of parallelism if one were not constrained by the number of cores/ processors, or the cost of memory references or interprocessor communications.

In summary, this book is intended to be both a research monograph as well as an advanced graduate textbook for a course dealing with parallel algorithms in matrix computations or numerical linear algebra. It is assumed that the reader has general, but not extensive, knowledge of: numerical linear algebra, parallel architectures, and parallel programming paradigms. This book consists of four parts for a total of 13 chapters. Part I is an introduction to parallel programming paradigms and primitives for dense and sparse matrix computations. Part II is devoted to dense matrix computations such as solving linear systems, linear least squares and algebraic eigenvalue problems. Part II also deals with parallel algorithms for special matrices such as banded, Vandermonde, Toeplitz, and block Toeplitz. Part III deals with sparse matrix computations: (a) iterative parallel linear system solvers with emphasis on scalable preconditioners, (b) schemes for obtaining few of the extreme or interior eigenpairs of symmetric eigenvalue problems, (c) schemes for obtaining few of the singular triplets. Finally, Part IV discusses parallel algorithms for computing matrix functions and the matrix pseudospectrum.

Acknowledgments

We wish to thank all of our current and previous collaborators who have been, directly or indirectly, involved with topics discussed in this book. We thank especially: Guy-Antoine Atenekeng-Kahou, Costas Bekas, Michael Berry, Olivier Bertrand, Randy Bramley, Daniela Calvetti, Peter Cappello, Philippe Chartier, Michel Crouzeix, George Cybenko, Ömer Eğecioğlu, Jocelyne Erhel, Roland Freund, Kyle Gallivan, Ananth Grama, Joseph Grcar, Elias Houstis, William Jalby, Vassilis Kalantzis, Emmanuel Kamgnia, Alicia Klinvex, Çetin Koç, Efi Kokiopoulou, George Kollias, Erricos Kontoghiorghes, Alex Kouris, Ioannis Koutis, David Kuck, Jacques Lenfant, Murat Manguoğlu, Dani Mezher, Carl Christian Mikkelsen, Maxim Naumov, Antonio Navarra, Louis Bernard Nguenang, Nikos Nikoloutsakos, David Padua, Eric Polizzi, Lothar Reichel, Yousef Saad, Miloud Sadkane, Vivek Sarin, Olaf Schenk, Roger Blaise Sidje, Valeria Simoncini, Aleksandros Sobczyk, Danny Sorensen, Andreas Stathopoulos, Daniel Szyld, Maurice Tchuente, Tayfun Tezduyar, John Tsitsiklis, Marian Vajteršic, Panayot Vassilevski, Ioannis Venetis, Brigitte Vital, Harry Wijshoff, Christos Zaroliagis, Dimitris Zeimpekis, Zahari Zlatev, and Yao Zhu. Any errors and omissions, of course, are entirely our responsibility.

In addition we wish to express our gratitude to Yousuff Hussaini who encouraged us to have our book published by Springer, to Connie Ermel who typed a major part of the first draft, to Eugenia-Maria Kontopoulou for her help in preparing the index, and to our Springer contacts, Kirsten Theunissen and Aldo Rampioni. Finally, we would like to acknowledge the remarkable contributions of the late Gene Golub—a mentor and a friend—from whom we learned a lot about matrix computations. Further, we wish to pay our respect to the memory of our late collaborators and friends: Theodore Papatheodorou, and John Wisniewski.

Last, but not least, we would like to thank our families, especially our spouses, Aristoula, Elisabeth and Marilyn, for their patience during the time it took us to produce this book.

<table>
<tr><td>Patras</td><td>Efstratios Gallopoulos</td></tr>
<tr><td>Rennes</td><td>Bernard Philippe</td></tr>
<tr><td>West Lafayette</td><td>Ahmed H. Sameh</td></tr>
<tr><td>January 2015</td><td></td></tr>
</table>

References

1. Babbage, C.: Passages From the Life of a Philosopher. Longman, Green, Longman, Roberts & Green, London (1864)
2. Cybenko, G., Kuck, D.: Revolution or Evolution, IEEE Spectrum, **29**(9), 39–41 (1992)
3. Richardson, L.F.: Weather Prediction by Numerical Process. Cambridge University Press, Cambridge (1922). (Reprinted by Dover Publications, 1965)

Contents

Part II Dense and Special Matrix Computations

Part IV   Matrix Functions and Characteristics

List of Figures

List of Tables

List of Algorithms

Notations

$i_1 : i_s : i_2$	MATLAB type colon notation
;	MATLAB type start new matrix row
$\lceil \cdot \rceil$ and $\lfloor \cdot \rfloor$	floor and ceiling functions
log	logarithm (base-2) unless mentioned otherwise
$\mathbb{R}, \mathbb{C}$	real number field, complex number field
$\bar{\xi}$	conjugate of ξ
$A^{\mathrm{T}}, A^{\mathrm{H}}$	transpose of A, Hermitian transpose of $A \in \mathbb{C}^{n \times n}$
$\odot, \oslash$	Hadamard (element-by-element) multiplication and division
$\otimes, \oplus$	Kronecker product, Kronecker sum
$0 \in \mathbb{R}^{n \times n}, 0_n$	zero matrix (size implied by the context of $n \times n$)
$0_{m,n}$	zero matrix of size $m \times n$
$I \in \mathbb{R}^{n \times n}, I_n$	identity matrix (size implied by the context of $n \times n$)
e_j or $e_j^{(n)} \in \mathbb{R}^n$	jth column of I or I_n
e or $e^{(n)}$	column vector of 1's, (size implied or n)
J_n or J	order n matrix $(e_2, e_3, \ldots, e_n, 0)$; subscript omitted when implied by context
$\mathrm{diag}(x)$	diagonal matrix with x along the diagonal when it is a vector
$\mathrm{diag}(X)$	diagonal matrix of diagonal elements of X when X is a matrix
$\mathrm{diag}(\xi_1, \ldots, \xi_n)$	same as $\mathrm{diag}(x)$ where $x = (\xi_1, \ldots, \xi_n)^{\mathrm{T}}$
$\mathrm{diag}(A_1, \ldots, A_n)$	the block diagonal matrix with diagonal blocks the matrices A_j
$\det(X)$	determinant of X
$\mathrm{tr}(X)$	trace of X
$\mathrm{tril}(X), \mathrm{triu}(X)$	the strictly lower (resp. upper) triangular section of X
$[\alpha_{i,i-1}, \alpha_{i,i}, \alpha_{i,i+1}]$	tridiagonal matrix with $\alpha_{i,i-1}, \alpha_{i,i}, \alpha_{i,i+1}$ in row i

or $[\alpha_{i,i-1}, \alpha_{i,i}, \alpha_{i,i+1}]_{1:n}$ — order implied or n

$[\beta, \alpha, \gamma]$
or $[\beta, \alpha, \gamma]_n$ — tridiagonal Toeplitz matrix with β, α, γ along the subdiagonal, diagonal and supradiagonal respectively (size implied or $n \times n$)

$[\beta_i, \alpha_i, \gamma_{i+1}]$, $[\beta_i, \alpha_i, \gamma_{i+1}]_{1:n}$
or $[\beta_i, \alpha_i, \gamma_{i+1}]_n$ — tridiagonal matrix (implied order or n) with $\beta_i, \alpha_i, \gamma_{i+1}$ in row i

$\mathscr{T}_n$ — Chebyshev polynomial of the 1st kind and degree n

$\hat{\mathscr{U}}_n$ — Chebyshev polynomial of the 1st kind and degree n, scaled

$\hat{\mathscr{U}}_n$ — Chebyshev polynomial of the 2nd kind and degree n, shifted

Based on the above,

$$
J = \begin{pmatrix}
0 & & & & \\
1 & \ddots & & & \\
0 & \ddots & \ddots & & \\
& \ddots & \ddots & \ddots & \\
& & 0 & 1 & 0
\end{pmatrix}
$$

and

$$
[\alpha_{i,i-1}, \alpha_{i,i}, \alpha_{i,i+1}]_{1:n} = \begin{pmatrix}
\alpha_{1,1} & \alpha_{1,2} & & & \\
\alpha_{2,1} & \alpha_{2,2} & \alpha_{2,3} & & \\
& \ddots & \ddots & \ddots & \\
& & \ddots & \ddots & \alpha_{n-1,n} \\
& & & \alpha_{n,n-1} & \alpha_{n,n}
\end{pmatrix}
$$

Algorithms are frequently expressed in pseudocode, using MATLAB constructs. In most cases, the "Householder notation" is followed, that is matrices are denoted by upper case romans, vectors by lower case romans and scalars by greek letters. There are some notable exceptions: Indices and function names are also denoted by lower case romans (e.g. i, j, k and f, p, q). So the following disclaimer is necessary: whenever we felt that by following the Householder notation conventions, readability would be compromised, they were briefly abandoned.

Part I
Basics

Chapter 1
Parallel Programming Paradigms

In this chapter, we briefly present the main concepts in parallel computing. This is an attempt to make more precise some definitions that will be used throughout this book rather than a survey on the topic. Interested readers are referred to one of the many books available on the subject, e.g. [1–8].

1.1 Computational Models

In order to describe algorithms that are general so as to be useful in a variety of generations of parallel computers, we need to define computational models that are abstract enough so as to encapsulate the main features of the underlying architecture, without obscuring the algorithm description. Such models are helpful in the choice of the programming paradigm as well as designing algorithms that enable faithful performance assessment. It is worth mentioning that creating a computational model that helps in designing efficient algorithms can be vastly different from models targeting the design of compilers and other systems software.

In this chapter we present an outline of different models of parallel architectures and the appropriate way of using them.

1.1.1 Performance Metrics

It is often useful to evaluate the potential performance of a parallel algorithm by assuming that there is an unlimited number of processors, and that interprocessor communication is instantaneous. Such an approach is of value, as it informs the algorithm designer about the maximum expected benefit (such as total runtime and resource usage) that can be achieved by parallel programs implementing these algorithms. Moreover, this approach can steer the designer to search for algorithms that achieve better performance by taking alternative approaches.

© Springer Science+Business Media Dordrecht 2016

E. Gallopoulos et al., *Parallelism in Matrix Computations*,

Scientific Computation, DOI 10.1007/978-94-017-7188-7_1

In what follows:

- p denotes the number of processors.
- $\mathbf{T}_p$ denotes the *number of steps* of the parallel algorithm on p processors. Each step is assumed to consist of (i) the interprocessor communication and memory references, and (ii) the arithmetic operations performed immediately after by the active processors. Thus, for example, as will be shown in Sect. 2.1 of Chap. 2, an inner product of two n-sized vectors, requires at least $\mathbf{T}_p = O(\log n)$ steps on $p = n$ processors.
- $\mathbf{O}_p$ denotes the number of arithmetic operations required by the parallel algorithm when run on p processors.
- $\mathbf{R}_p$ denotes *arithmetic redundancy*. This is the ratio of the total number of arithmetic operations, $\mathbf{O}_p$, in the parallel algorithm, over the least number of arithmetic operations, $\mathbf{O}_1$, required by the sequential algorithm. Specifically

$$\mathbf{R}_p = \mathbf{O}_p/\mathbf{O}_1.$$

Since interprocessor communication, or accessing data from various levels of the memory hierarchy in a multiprocessor, is much more time consuming than arithmetic operations, efficient parallel algorithms often attempt to reduce such communication at the cost of introducing more arithmetic operations than those needed by the best sequential scheme. Therefore, the arithmetic rendundancy is often higher than 1. Note also that $\mathbf{O}_1$ and $\mathbf{T}_1$ are the same.

- $\mathbf{S}_p$ and $\mathbf{E}_p$, respectively denote the *speedup* and *efficiency* of the parallel algorithm. These are defined by

$$\mathbf{S}_p = \frac{\mathbf{T}_1}{\mathbf{T}_p}, \text{ and } \mathbf{E}_p = \frac{\mathbf{S}_p}{p}.$$

The study of speedups is central in most performance evaluation studies undertaken for parallel programs. Speedups and efficiencies are also directly related to the notion of scalability, discussed at the end of the current chapter.

- $\mathbf{V}_p$ denotes the *computational rate*, defined by

$$\mathbf{V}_p = \frac{\mathbf{O}_p}{\mathbf{T}_p}.$$

Based on these definitions, $\mathbf{V}_1 = \mathbf{O}_1 = \mathbf{T}_1$.
- Whenever referenced as above, it would be assumed that the size of the problem is readily available. Sometimes, in our discussions, it would be desirable to express these metrics as functions of the problem size. In these cases, we write $\mathbf{T}_p(n)$, $\mathbf{O}_p(n)$, where n is the problem size.

It is important to note that in order to avoid notation clutter, when we present the number of steps, speedup, number of arithmetic operations and number of processors

Fig. 1.1 SIMD architecture

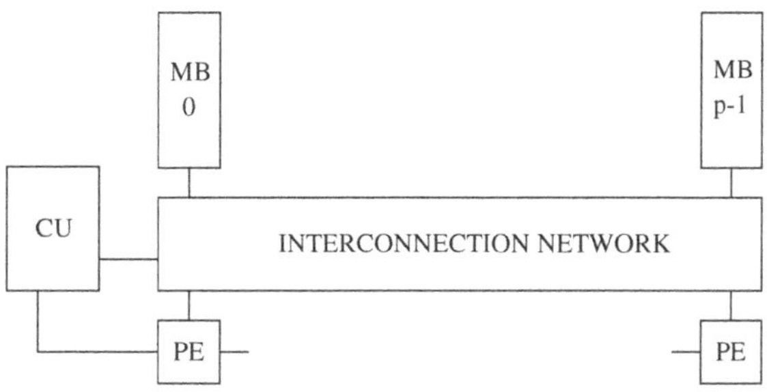

that turn out to be rational functions in the problem size, it must be understood that we are referring to an integer value. Typically this is the ceiling function of the fraction.

1.1.2 Single Instruction Multiple Data Architectures and Pipelining

The earliest computers constructed aiming at high performance via the use of parallelism were of the Single Instruction Multiple Data (SIMD) type, according to the standard classification of parallel architectures [9]. These are characterized by the fact that several processing elements (PE) are controlled by a single control unit (CU); see Fig. 1.1, where MB denotes the memory banks.

For executing a program on such an architecture,

- the CU runs the program and sends: (i) the characteristics of the vectors to the memory, (ii) the permutation to be applied to these vectors by the interconnection network in order to achieve the desired alignment of the operands, and (iii) the instructions to be executed by the PEs,
- the memory delivers the operands which are permuted through the interconnection network,
- the processors perform the operation, and
- the results are sent back via the interconnection network to be correctly stored in the memory.

The computational rate of the operation depends on the vector length. Denoting the duration of the operation on one slice of $N \leq p$ components by t, the parallel steps needed to perform the operation on a vector of length n is given by

$$\mathbf{T}_p = \left\lceil \frac{n}{p} \right\rceil t.$$

Thus, the computational rate realized via the use of this SIMD architecture is given by,

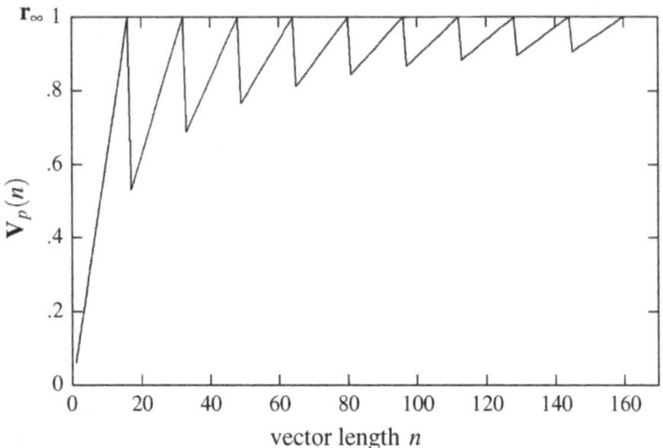

Fig. 1.2 Computational rate and efficiency of a vector operation on $p = 16$ PEs of a SIMD architecture

$$\mathbf{V}_p = \frac{n}{\mathbf{T}_p},$$

with an *asymptotic computational rate* of $\mathbf{r}_\infty = p/t$, which is reached when the vector length is a multiple of the number of processors p (see Fig. 1.2).

Large scale SIMD architectures have totally disappeared. Current CPUs, on the other hand, offer instructions for SIMD processing while graphics processing units (GPUs) utilize related concepts, such as SIMT (Single Instruction stream Multiple Threads); cf. [6, 10].

Vector operations can also be realized using the principles of pipelining. Similar to the assembly line in a factory, in which the total fabrication of an item is split into elementary steps with the items under construction traveling from one post to another until the chain is entirely filled and an item is assembled, we consider the definition of a pipeline for performing a vector floating point addition: As an example, the addition of two vectors of floating-point numbers,

$$c_i = a_i + b_i, \ \text{ for } i = 1, \ldots, n,$$

is typically decomposed into the four following stages:

1. comparing exponents,
2. aligning the mantissas,
3. adding the mantissas,
4. normalizing the resulting floating-point number.

This pipeline scheme is depicted in Fig. 1.3.

In order to describe the performance of a pipeline, we introduce the following auxiliary quantities:

- s denotes the number of stages of the pipeline;
- τ denotes the time consumed by each stage (which is assumed to be uniform);
- $t_0 = (s - 1)\tau$ denotes the start-up time required to fill the pipeline, so that the first result is delivered after time $t_0 + \tau$.

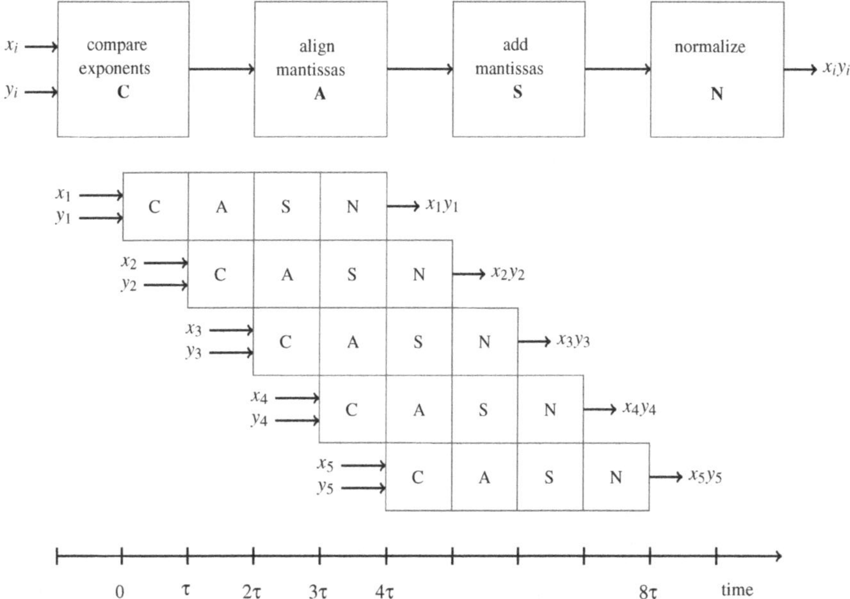

Fig. 1.3 Floating-point adder pipeline

Thus, the elapsed time corresponding to the addition of two vectors of length n is given by,

$$\mathbf{T}_p = t_0 + n\tau$$

which insures a computational rate of,

$$\mathbf{V}_p = \frac{n}{\mathbf{T}_p} = \frac{n}{t_0 + n\tau}. \tag{1.1}$$

Figure 1.4, depicts the computational rate realized with respect to the vector length. The asymptotic computational rate, $\mathbf{r}_\infty = 1/\tau$, is not reached for finite vector lengths. Half of that asymptotic computational rate, however, is reached for a vector of length $\mathbf{n}_{1/2} = t_0/\tau = s - 1$. The two numbers $(\mathbf{r}_\infty, \text{ and } \mathbf{n}_{1/2})$ characterize the pipeline performance since it is easy to see:

$$\mathbf{V}_p = \mathbf{r}_\infty \left(1 - \frac{\mathbf{n}_{1/2}}{n + \mathbf{n}_{1/2}}\right).$$

See [7, 11] for more details on this useful metric.

Most of the pipelines load their operands from vector registers and not directly from the memory. With a vector register of length N (typically $N = 64$ or 128), vector

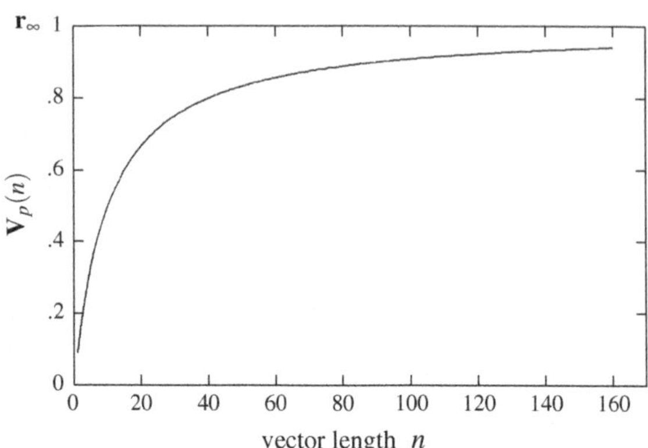

Fig. 1.4 Computational rate of a pipelined operation

operands of length n must be partitioned into slices each of length N. This approach favors operations on short vectors since it decreases the start-up time. By assuming that vector registers are ready with slices of the operands so that an operation is immediately performed on a given slice once the operation on a previous slice is completed, then the rate due to pipelining can be expressed as,

$$\mathbf{V}_p(n) = \frac{n}{kt_0 + n\tau}$$

which is an adaptation of (1.1), where $k = \lceil \frac{n}{N} \rceil$ denotes the number of slices of the operands (the last slice might be shorter than N). The corresponding behavior of the speedup is depicted in Fig. 1.5 in which $\mathbf{r}_\infty$ is equal to $1/\tau$.

Even higher performance can be achieved when the underlying system allows *chaining*, that is letting one pipeline deliver its results directly to another one that performs the subsequent computation; cf. [6].

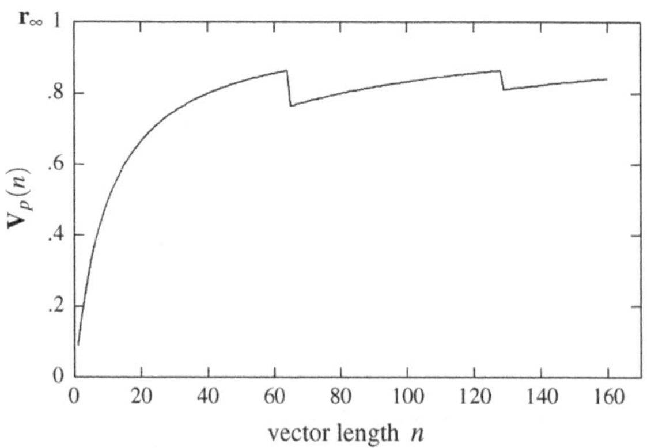

Fig. 1.5 Computational rate of a pipelined operation with vector registers ($N = 64$)

For illustration, let us consider the following vector operation that is quite common in a variety of numerical algorithms (see Sect. 2.1):

$$d_i = a_i + b_i c_i, \text{ for } i = 1, \ldots, n.$$

Assuming that the time for each stage of the two pipelines (which performs the multiplication and the addition) is equal, by chaining the pipelines, the results are still obtained at the same rate. Therefore, the speedup is doubled since two operations are performed at the same time. It is worth noting that in several cases pipelining and chaining are used to increase the performance of elementary operations. In particular, the scalar operation $a + bc$ is implemented as a single instruction, often called Fused Multiply-Add (FMA). This also deliver results with smaller roundoff than if one were to implement it as two separate operations; cf. [12].

Pipelining can also be applied to interconnection networks between memories and PEs to enhance memory bandwidth. In addition, it can be applied to multiprocessors to enhance memory bandwidth.

1.1.3 Multiple Instruction Multiple Data Architectures

Next, we consider architectures belonging to the Multiple Instruction Multiple Data (MIMD) class. In such a class, each of the PEs owns its proper CU and all the PEs are able to run distinct tasks in parallel. This class is subdivided into two major subclasses: the shared memory, and the distributed memory architectures. In the first type, all the PEs exchange their data through a global memory while in the second type, the PEs exchange their data with messages sent through an interconnection network. These two MIMD organizations are depicted in Figs. 1.6 and 1.7, respectively.

By expressing parallelism at the loop level, a shared memory architecture allows a programming style that is closer to that of programming for a uniprocessor. In such a context, exclusion mechanisms are used when instructions for reading and

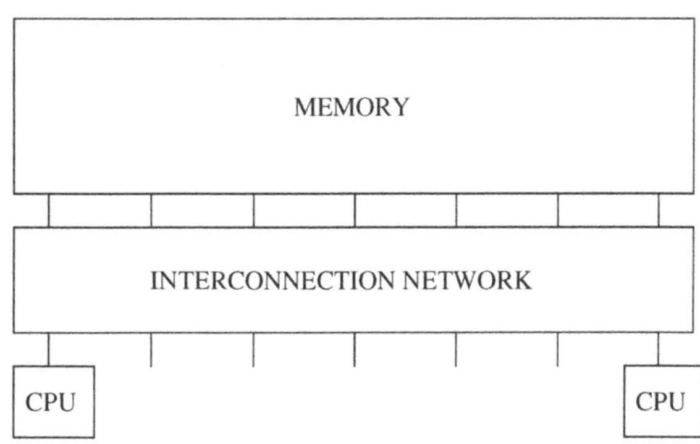

Fig. 1.6 MIMD architecture with shared memory

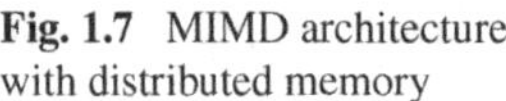

Fig. 1.7 MIMD architecture with distributed memory

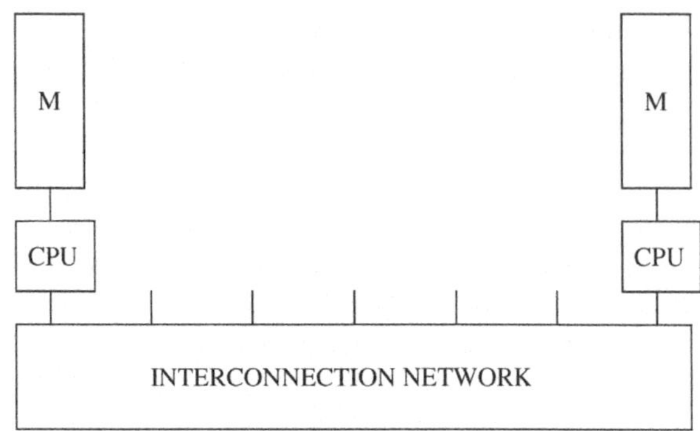

writing must be synchronized to operate in parallel on the same variables. Most of MIMD computing platforms with a massive number of processors belong to the distributed memory class. In such a class, the computing platform consists of many nodes in which each node is a multicore shared-memory architecture with its own proper memory. The nodes are interconnected via physical links. An important issue concerns the topology of the underlying connection network. The most powerful network is the crossbar switch which connects any node to all the rest. Unfortunately, for p nodes, it necessitates $p(p-1)/2$ links, which makes it infeasible for large p.

Other topologies have been investigated for connecting a large number of nodes. Since these topologies do not connect each node to all the rest, they imply routing procedures that permit all possible exchanges of data. An important characteristic of a network is its diameter, i.e. the maximum distance between two nodes (the distance between two nodes is the minimum length of all paths connecting them). For a crossbar network, the diameter is minimum, i.e. $d = 1$, while for a linear array the diameter is maximum, i.e. $d = p - 1$. For other interconnections such as rings, hypercubes and tori, the respective diameters satisfy $1 < d < p - 1$. Clearly, all other things being equal, the smaller the diameter, the faster the interprocessor communication.

The usual way of modeling the time spent in sending a vector of length n between two processors is by $\beta + n\tau_c$, where β is the latency and τ_c the time for sending a single element of the vector, which is independent of the latency.

More elaborate models of parallel architectures can be found in [13–15].

1.1.4 Hierarchical Architectures

Many machines today adopt a hierarchical design in the sense that both the processing and memory systems are organized in layers, each having specific characteristics. Regarding the processing hierarchy, it consists of units capable of scalar processing, vector and SIMD processing, multiple cores each able to process one or multiple threads, the multiple cores organized into interconnected clusters. The memory hierarchy consists of short and long registers, various levels of cache memory local to each processor, and memory shared between processors in a cluster. In many respects, it can be argued that most parallel systems are of this type or simplified versions thereof.

1.2 Principles of Parallel Programming

We present here a simplified picture of parallel programming by introducing two main standards for MIMD programming: MPI and OpenMP. For a more complete description of these two paradigms, the reader is referred to the extensive literature on parallel programming such as [16, 17] and the general books on parallel processing referenced at the beginning of the current chapter. The main difference between these two programming paradigms could be stated simply as follows: MPI is a programming model for MIMD architectures with the necessary interprocessor communications whereas in OpenMP, programs are written using directives (pragmas) in the source code to steer the compiler in the restructuring of loops (a key find in the early 1970s was that loops rather than arithmetic expressions must be the key target for increasing the speedup of parallel programs) and other program constructs.

MIMD Programming

Since an MIMD architecture is a collection of p independent processors, the straightforward MIMD approach would consider p distinct programs to be run on the processors with a mechanism for exchanging information. The most common version of such an approach is to consider the Single Program Multiple Data organization (SPMD) [18]. This implies, in turn, a common but independently run program on each processor. At first sight, this seems to limit the ability of having distinct tasks executed on the processors. If the first instruction of the common program, however, is a switch among all the possible tasks depending on the processor identity, it is then clear that SPMD mimics the original MIMD approach.

In this model it is possible to simulate a parallel architecture of q processes using a smaller number of processors.

The most common software system for message passing programming is the Message Passing Interface (MPI). This is a standard for the syntax and semantics of a set of basic library routines for implementing message passing on the underlying computer platform; cf. [19, 20]. Writing programs in MPI allows program portability between different distributed computing systems. MPI can also be used on a single multicore node. In message passing, there are several types of communications defined, blocking for synchronous and non-blocking for asynchronous; cf. [17] Moreover, the type of communication can be *point-to-point* or *global*.

Expressing Parallelism at the Loop Level

A second option in parallel programming is by means of an application programming interface where parallelism is expressed by directives that mainly apply to loops, in

Table 1.1 Loops considered in this book

do $j = 1 : n,$	**doall** $j = 1 : n,$	**doacross** $j = 1 : n,$
task(j) ;	task(j) ;	**if** $j > 1,$ **then**
end	**end**	wait(j-1) ;
		end if
		task1(j) ; post(j) ;
		task2(j) ;
		end
Sequential.	Parallel.	Pipelined.

which each iteration is independent of the rest. These directives steer the compiler in its restructuring of the program. The most prominent programming paradigm of this category is called OpenMP (Open Multiprocessing Application Program Interface), e.g. see [21, 22]. In such a paradigm, the tasks are implemented via *threads*.

In a given program, directives allow the user to define *parallel* regions which invoke fork and join mechanisms at the beginning and the end of the region, with the ability to define the shared variables of the region. Other directives allow specifying parallelism through loops. For parallel loops, the programmer specifies whether the distribution of the iterations through the tasks (threads) is static or dynamic. Static repartition lowers the overhead cost whereas dynamic allocation adapts better to irregular task loads. Several techniques are provided for synchronizing threads which include locks, barriers, and critical sections.

In this book, when necessary for the sake of illustration, we consider three types of loops which are shown in Table 1.1:

(a) **do** in which the iterations are run sequentially,
(b) **doall** in which the iterations are run independently and possibly simultaneously,
(c) **doacross** in which the iterations are pipelined.

Table 1.1 illustrates these three cases. The **doacross** loop enables the use of pipelining between iterations, so that an iteration can start before the previous one has been completed. For this reason, it also depends on the use of synchronization mechanisms (such as wait and post) to control the execution of the iterations.

It is also worth noting that frequently, it is best to write programs that utilize both of the above paradigms, that is MPI and OpenMP. This hybrid mode is natural in order to take advantage of the hierarchical structure of high performance computer systems.

1.2.1 From Amdahl's Law to Scalability

It was observed quite early in the history of parallel computing that there is a limit to the available parallelism in a given fixed-size computation. The limit is governed by the percentage of the inherently sequential portion of the computation and in turn limits the speedup and efficiency that can be achieved by the program. This observation is known as Amdahl's law and can be expressed as follows:

Proposition 1.1 (Amdahl's law) [23] *Let* $\mathbf{O}_p$ *be the number of operations a parallel program implemented on p processors. If the portion* f_p *(0* $<$ f_p $<$ *1) of the* $\mathbf{O}_p$ *operations is inherently sequential, then the speedup is bounded by* $\mathbf{S}_p < 1/f_p$.

Proof From the definitions of $\mathbf{T}_1$ and $\mathbf{T}_p$ is obvious that $\mathbf{T}_p \geq (f_p + \frac{1-f_p}{p})\mathbf{T}_1$; the above upper bound immediately follows.

This simple result gives a rule-of-thumb for the maximum speedup and efficiency that can be expected of a given algorithm. The limits to parallel processing implied by the law but also the limits of the law's basic assumptions have been discussed extensively in the literature, by designers of parallel systems and algorithms; for example, cf. [24–31]. We make the following remarks.

Remark 1.1 Even for a fully parallel program, there is a loss of efficiency due to the overhead (interprocessor communication time, memory references, and parallel management) which usually implies that $\mathbf{S}_p < p$.

Remark 1.2 An opposite situation may occur when the observed speedup is "super-linear", i.e. $\mathbf{S}_p > p$. This happens, for example, when the data set is too large for the local memory of one processor; whereas storing it across the p processor memories becomes possible. In fact, the ability to manipulate larger datasets is an important advantage of parallel processing that is becoming especially relevant in the context of data intensive computations.

Remark 1.3 The speedup bound in Amdahl's law refers strictly to the performance of a program running in single-user mode on a parallel computer, using one or all of the processors of the system. This was a reasonable assumption at the time of large-scale SIMD systems, but today one can argue that it is too strong of a simplification. For instance, it does not consider the effect of systems that can handle multiple parallel computations across the processors. It also does not capture the possibility of parallel systems consisting of heterogeneous processors with different performance.

Remark 1.4 The aspect of Amdahl's law that has been under criticism by various researchers is the assumption that the program whose speedup is evaluated is solving a problem of fixed size. As was observed by Gustafson in [32], the problem size should be allowed to vary with the number of processors. This meant that the parallel fraction of the algorithm is not constant as the number of processors varies. We discuss this issue in greater detail next.

The performance evaluation of parallel programs usually relies on the graph of $p \to \mathbf{S}_p$, that is that of the "processor to speedup" mapping. A graph close to the straight line $p \to p$ corresponds to a program whose performance (in terms of its speedup and efficiency) is maintained as the number of processors increases.

Assuming that this holds independently of the size of the problem, the program is characterized as *strongly scalable*.

In many cases, the speedup graph exhibits a two-stage behavior: there is some threshold value, say $\tilde{p}$, such that $\mathbf{S}_p$ increases linearly as long as $\tilde{p} \geq p$, whereas for $p > \tilde{p}$, $\mathbf{S}_p$ stagnates or even decreases. This is the sign that for $p > \tilde{p}$, the overhead, i.e. the time spent in managing parallelism, becomes too high when the size of the problem is too small for that number of processors.

The above performance constraint can be partially or fully removed, if we allow the size of the problem to increase with the number of processors. Intuitively, it is not surprising to expect that as the computer system becomes more powerful, so would the size of the problem to be solved. This also means that the fraction of the computations that are computed in parallel does not remain constant as the number of processors increases. The notion of *weak scalability* is then relevant.

Definition 1.1 (*Weak Scalability*) Consider a program capable of solving a family of problems depending on a size parameter n_p associated to the number p of processors on which the program is run. Let $\mathbf{O}_p(n_p) = p\,\mathbf{O}_1(n_1)$, where $\mathbf{O}_p(n_p)$, $\mathbf{O}_1(n_1)$ are the total number of operations necessary for the parallel and sequential programs of size n_1 and n_p respectively, assuming that there is no arithmetic redundancy. Also let $\mathbf{T}_p(n_p)$ denote the steps of the parallel algorithm on p processors for a problem of size n_p. Then a program is characterized as being weakly scalable if the sequence $\mathbf{T}_p(n_p)$ is constant.

Naturally, weak scalability is easier to achieve than strong scalability, which seeks constant efficiency for a problem of fixed size. On the other hand, even by selecting the largest possible size for the problem that can be run on one processor, the problem becomes too small to be run efficiently on a large number of processors.

In conclusion, we note that in order to investigate the scalability potential of a program, it is well worth analyzing the graph of the mapping $p \to (1 - f_p)\mathbf{O}_p$, that is "processors to the total number of operations that can be performed in parallel" (assuming no redundancy). Defining as problem size, $\mathbf{O}_1$, that is the total number of operations that are needed to solve the problem, one question is how fast should the problem size increase in order to keep the efficiency constant as the number of processors increases. An appropriate rate, sometimes called isoefficiency [33], could indeed be neither of the two extremes, namely the constant problem size of Amdahl and the linear increase suggested in [32]; cf. [24] for a discussion.

References

1. Arbenz, P., Petersen, W.: Introduction to Parallel Computing. Oxford University Press (2004)
2. Bertsekas, D.P., Tsitsiklis, J.N.: Parallel and Distributed Computation. Prentice Hall, Englewood Cliffs (1989)
3. Culler, D., Singh, J., Gupta, A.: Parallel Computer Architecture: A Hardware/Software Approach. Morgan Kaufmann, San Francisco (1998)

4. Kumar, V., Grama, A., Gupta, A., Karypis, G.: Introduction to Parallel Computing: Design and Analysis of Algorithms, 2nd edn. Addison-Wesley (2003)
5. Casanova, H., Legrand, A., Robert, Y.: Parallel Algorithms. Chapman & Hall/CRC Press (2008)
6. Hennessy, J., Patterson, D.: Computer Architecture: A Quantitative Approach. Elsevier Science & Technology (2011)
7. Hockney, R., Jesshope, C.: Parallel Computers 2: Architecture, Programming and Algorithms, 2nd edn. Adam Hilger (1988)
8. Tchuente, M.: Parallel Computation on Regular Arrays. Algorithms and Architectures for Advanced Scientific Computing. Manchester University Press (1991)
9. Flynn, M.: Some computer organizations and their effectiveness. IEEE Trans. Comput. **C-21**, 948–960 (1972)
10. Jeffers, J., Reinders, J.: Intel Xeon Phi Coprocessor High Performance Programming, 1st edn. Morgan Kaufmann Publishers Inc., San Francisco (2013)
11. Hockney, R.: The Science of Computer Benchmarking. SIAM, Philadelphia (1996)
12. Higham, N.: Accuracy and Stability of Numerical Algorithms, 2nd edn. SIAM, Philadelphia (2002)
13. Karp, R., Sahay, A., Santos, E., Schauser, K.: Optimal broadcast and summation in the logP model. In: Proceedings of the 5th Annual ACM Symposium on Parallel Algorithms and Architectures SPAA'93, pp. 142–153. ACM Press, Velen (1993). http://doi.acm.org/10.1145/165231.165250
14. Culler, D., Karp, R., Patterson, D., A. Sahay, K.S., Santos, E., Subramonian, R., von Eicken, T.: LogP: towards a realistic model of parallel computation. In: Principles, Practice of Parallel Programming, pp. 1–12 (1993). http://citeseer.ist.psu.edu/culler93logp.html
15. Pjesivac-Grbovic, J., Angskun, T., Bosilca, G., Gabriel, G.F.E., Dongarra, J.: Performance analysis of MPI collective operations. In: Fourth International Workshop on Performance Modeling, Evaluation, and Optimization of Parallel and Distributed Systems (PMEO-PDS'05). Denver (2005). (Submitted)
16. Breshaers, C.: The Art of Concurrency - A Thread Monkey's Guide to Writing Parallel Applications. O'Reilly (2009)
17. Rauber, T., Rünger, G.: Parallel Programming—for Multicore and Cluster Systems. Springer (2010)
18. Darema., F.: The SPMD model : past, present and future. In: Recent Advances in Parallel Virtual Machine and Message Passing Interface. LNCS, vol. 2131/2001, p. 1. Springer, Berlin (2001)
19. Gropp, W., Lusk, E., Skjellum, A.: Using MPI: Portable Parallel Programming with the Message Passing Interface. MIT Press, Cambridge (1994)
20. Snir, M., Otto, S., Huss-Lederman, S., Walker, D., Dongarra, J.: MPI: The Complete Reference (1995). http://www.netlib.org/utk/papers/mpi-book/mpi-book.html
21. Chapman, B., Jost, G., Pas, R.: Using OpenMP: Portable Shared Memory Parallel Programming. The MIT Press, Cambridge (2007)
22. OpenMP Architecture Review Board: OpenMP Application Program Interface (Version 3.1). (2011). http://www.openmp.org/mp-documents/
23. Amdahl, G.M.: Validity of the single processor approach to achieving large scale computing capabilities. Proc. AFIPS Spring Jt. Comput. Conf. **31**, 483–485 (1967)
24. Juurlink, B., Meenderinck, C.: Amdahl's law for predicting the future of multicores considered harmful. SIGARCH Comput. Archit. News **40**(2), 1–9 (2012). doi:10.1145/2234336.2234338. http://doi.acm.org/10.1145/2234336.2234338
25. Hill, M., Marty, M.: Amdahl's law in the multicore era. In: HPCA, p. 187. IEEE Computer Society (2008)
26. Sun, X.H., Chen, Y.: Reevaluating Amdahl's law in the multicore era. J. Parallel Distrib. Comput. **70**(2), 183–188 (2010)
27. Flatt, H., Kennedy, K.: Performance of parallel processors. Parallel Comput. **12**(1), 1–20 (1989). doi:10.1016/0167-8191(89)90003-3. http://www.sciencedirect.com/science/article/pii/0167819189900033

28. Kuck, D.: High Performance Computing: Challenges for Future Systems. Oxford University Press, New York (1996)
29. Kuck, D.: What do users of parallel computer systems really need? Int. J. Parallel Program. **22**(1), 99–127 (1994). doi:10.1007/BF02577794. http://dx.doi.org/10.1007/BF02577794
30. Kumar, V., Gupta, A.: Analyzing scalability of parallel algorithms and architectures. J. Parallel Distrib. Comput. **22**(3), 379–391 (1994)
31. Worley, P.H.: The effect of time constraints on scaled speedup. SIAM J. Sci. Stat. Comput. **11**(5), 838–858 (1990)
32. Gustafson, J.: Reevaluating Amdahl's law. Commun. ACM **31**(5), 532–533 (1988)
33. Grama, A., Gupta, A., Kumar, V.: Isoefficiency: measuring the scalability of parallel algorithms and architectures. IEEE Parallel Distrib. Technol. 12–21 (1993)

Chapter 2
Fundamental Kernels

In this chapter we discuss the fundamental operations, that are the building blocks of dense and sparse matrix computations. They are termed kernels because in most cases they account for most of the computational effort. Because of this, their implementation directly impacts the overall efficiency of the computation. They occur often at the lowest level where parallelism is expressed.

Most basic kernels are of the form $C = C + AB$, where A, B and C can be matrix, vector and possibly scalar operands of appropriate dimensions. For dense matrices, the community has converged into a standard application programming interface, termed *Basic Linear Algebra Subroutines* (BLAS) that have specific syntax and semantics. The set is organized into three separate sets of instructions. The first part of this chapter describes these sets. It then considers several basic sparse matrix operations that are essential for the implementation of algorithms presented in future chapters. In this chapter we frequently make explicit reference to communication costs, on account of the well known growing discrepancy, in the performance characteristics of computer systems, between the rate of performing computations (typically measured by a base unit of the form flops per second) and the rate of moving data (typically measured by a base unit of the form words per second).

2.1 Vector Operations

Operations on vectors are known as Level_1 Basic Linear Algebraic Subroutines (BLAS1) [1]. The two most frequent vector operations are the _AXPY and the _DOT:

  _AXPY: given $x, y \in \mathbb{R}^n$ and $\alpha \in \mathbb{R}$, the instruction updates vector y by: $y = y + \alpha x$.
  _DOT: given $x, y \in \mathbb{R}^n$, the instruction computes the inner product of the two vectors: $s = x^\top y$.

© Springer Science+Business Media Dordrecht 2016
E. Gallopoulos et al., *Parallelism in Matrix Computations*,
Scientific Computation, DOI 10.1007/978-94-017-7188-7_2

A common feature of these instructions is that minimal number of data that needs to be read (loaded) into memory and then stored back in order for the operation to take place is $O(n)$. Moreover, the number of computations required on a uniprocessor is also $O(n)$. Therefore, the ratio of instructions to load from and store to memory relative to purely arithmetic operations is $O(1)$.

With $p = n$ processors, the _AXPY primitive requires 2 steps which yields a perfect speedup, n. The _DOT primitive involves a reduction with the sum of n numbers to obtain a scalar. We assume temporarily, for the sake of clarity, that $n = 2^m$. At the first step, each processor computes the product of two components, and the result can be expressed as the vector $(s_i^{(0)})_{1:n}$. This computation is then followed by m steps such that at each step k the vector $(s_i^{(k-1)})_{1:2^{m-k+1}}$ is transformed into the vector $(s_i^{(k)})_{1:2^{m-k}}$ by computing in parallel $s_i^{(k)} = s_{2i-1}^{(k-1)} + s_{2i}^{(k-1)}$, for $i = 1, \ldots, 2^{m-k}$, with the final result being the scalar $s_1^{(m)}$. Therefore, the inner product consumes $\mathbf{T}_p = m + 1 = (1 + \log n)$ steps, with a speedup of $\mathbf{S}_p = 2n/(1 + \log n)$ and an efficiency of $\mathbf{E}_p = 2/(1 + \log n)$.

On vector processors, these procedures can obtain high performance, especially for the _AXPY primitive which allows chaining of the pipelines for multiplication and addition.

Implementing these instructions on parallel architectures is not a difficult task. It is realized by splitting the vectors in slices of the same length, with each processor performing the operation on its own subvectors. For the _DOT operation, there is an additional summation of all the partial results to obtain the final scalar. Following that, this result has to be broadcast to all the processors. These final steps entail extra costs for data movement and synchronization, especially for computer systems with distributed memory and a large number of processors.

We analyze this issue in greater detail next, departing on this occasion from the assumptions made in Chap. 1 and taking explicitly into account the communication costs in evaluating $\mathbf{T}_p$. The message is that inner products are harmful to parallel performance of many algorithms.

Inner Products Inhibit Parallel Scalability

A major part of this book deals with parallel algorithms for solving large sparse linear systems of equations using preconditioned iterative schemes. The most effective classes of these methods are dominated by a combination of a "global" inner product, that is applied on vectors distributed across all the processors, followed by fan-out operations. As we will show, the overheads involved in such operations cause inefficiency and less than optimal speedup.

To illustrate this point, we consider such a combination in the form of the following primitive for vectors u, v, w of size n that appears often in many computations:

$$w = w - (u^\top v)u. \tag{2.1}$$

We assume that the vectors are stored in a consistent way to perform the operations on the components (each processor stores slices of components of the two vectors with identical indices). The _DOT primitive involves a reduction and therefore an all-to-one (fan-in) communication. Since the result of a dot product is usually needed by all processors in the sequel, the communication actually becomes an all-to-all (fan-out) procedure.

To evaluate the weak scalability on p processors (see Definition 1.1) by taking communication into account (as mentioned earlier, we depart here from our usual definition of $\mathbf{T}_p$), let us assume that $n = pq$. The number of steps required by the primitive (2.1) is $4q - 1$. Assuming no overlap between communication and computation, the cost on p processors, $\mathbf{T}_p(pq)$, is the sum of the computational and communication costs: $\mathbf{T}_p^{\text{cal}}(pq)$ and $\mathbf{T}_p^{\text{com}}(pq)$, respectively. For the all-to-all communication, $\mathbf{T}_p^{\text{com}}(pq)$ is given by: $\mathbf{T}_p^{\text{com}}(pq) = K\ p^\gamma$ in which $1 < \gamma \le 2$, with the constant K depending on the interconnection network technology. The computational load, which is well balanced, is given by $\mathbf{T}_p^{\text{cal}}(pq) = 4q - 1$, resulting in $\mathbf{T}_p(pq) = (4q - 1) + K\ p^\gamma$, which increases with the number of processors. In fact, once $\mathbf{T}_p^{\text{com}}(pq)$ dominates $\mathbf{T}_p^{\text{cal}}(pq)$, the total cost $\mathbf{T}_p(pq)$ increases almost quadratically with the number of processors.

This fact is of crucial importance in parallel implementation of inner products. It makes clear that an important goal of a designer of parallel algorithms on distributed memory architectures is to avoid distributed _DOT primitives as they are detrimental to parallel scalability. Moreover, because of the frequent occurrence and prominence of inner products in most numerical linear algebra algorithms, the aforementioned deleterious effects on the _DOT performance can be far reaching.

2.2 Higher Level BLAS

In order to increase efficiency, vector operations are often packed into a global task of higher level. This occurs for the multiplication of a matrix by a vector which in a code is usually expressed by a doubly nested loop. Classic kernels of this type are gathered into the set known as Level_2 Basic Linear Algebraic Subroutines (BLAS2) [2]. The most common operations of this type, assuming general matrices, are

_GEMV: given $x \in \mathbb{R}^n$, $y \in \mathbb{R}^m$ and $A \in \mathbb{R}^{m \times n}$ this performs the matrix-vector multiplication and accumulate $y = y + Ax$. It is also possible to multiply (row) vector by matrix, scale the result before accumulating.

_TRSV: given $b \in \mathbb{R}^n$ and $A \in \mathbb{R}^{n \times n}$ upper or lower triangular, this solves the triangular system $Ax = b$.

_GER: given scalar α, $x \in \mathbb{R}^n$, $y \in \mathbb{R}^m$ and $A \in \mathbb{R}^{m \times n}$, this performs the rank-one update $A = A + \alpha x y^\top$.

A common feature of these instructions is that the smallest number of data that needs to be read into memory and then stored back in order for the operation to take place when $m = n$ is $O(n^2)$, arithmetic operations is $O(1)$. Moreover, the number of computations required on a uniprocessor is also $O(n^2)$. Therefore, the ratio of instructions to load from and store to memory relative to purely arithmetic ones is

$O(1)$. Typically, the constants involved are a little smaller than those for the BLAS1 instructions. On the other hand, of far more interest in terms of efficiency.

Although of interest, the efficiencies realized by these kernels are easily surpassed by those of (BLAS3) [3], where one additional loop level is considered, e.g. matrix multiplication, and rank-k updates ($k > 1$). The next section of this chapter is devoted to matrix-matrix multiplications.

The set of BLAS is designed for a uniprocessor and used in parallel programs in the sequential mode. Thus, an efficient implementation of the BLAS is of the utmost importance to enable high performance. Versions of the BLAS that are especially fine-tuned for certain types of processors are available (e.g. the Intel Math Kernel Library [4] or the open source set GOTOBLAS [5]). Alternately, one can create a parametrized BLAS set which can be tuned on any processor by an automatic code optimizer, e.g. ATLAS [6, 7]. Yet, it is hard to outperform well designed methods that are based on accurate architectural models and domain expertise; cf. [8, 9].

2.2.1 Dense Matrix Multiplication

Given matrices A, B and C of sizes $n_1 \times n_2$, $n_2 \times n_3$ and $n_1 \times n_3$ respectively, the general operation, denoted by _GEMM, is $C = C + AB$.

Properties of this primitive include:

- The computation involves three nested loops that may be permuted and split; such a feature provides great flexibility in adapting the computation for vector and/or parallel architectures.
- High performance implementations are based on the potential for high data locality that is evident from the relation between the lower bound on the number of data moved ($O(n^2)$) to arithmetic operations, $O(n^3)$ for the classical schemes and $O(n^{2+\mu})$ for some $\mu > 0$ for the "superfast" schemes described later in this section.

Hence, in dense matrix multiplication, it is possible to reuse data stored in cache memory.

Because of these advantages of dense matrix multiplication over lower level BLAS, there has been a concerted effort by researchers for a long time now (see e.g. [10]) to (re)formulate many algorithms in scientific computing to be based on dense matrix multiplications (such as _GEMM and variants).

A Data Management Scheme Dense Matrix Multiplications

We discuss an implementation strategy for _GEMM:

$$C = C + AB. \tag{2.2}$$

We adopt our discussion from [11], where the authors consider the situation of a cluster of p processors with a common cache and count loads and stores in their evaluation. We simplify that discussion and outline the implementation strategy for a uniprocessor equipped with a cache memory that is characterized by fast access. The purpose is to highlight some critical design decisions that will have to be faced by the sequential as well as by the parallel algorithm designer. We assume that reading one floating-point word of the type used in the multiplication from cache can be accomplished in one clock period. Since the storage capacity of a cache memory is limited, the goal of a code developer is to reuse, as much as possible, data stored in the cache memory.

Let M be the storage capacity of the cache memory and let us assume that matrices A, B and C are, respectively, $n_1 \times n_2$, $n_2 \times n_3$ and $n_1 \times n_3$ matrices. Partitioning these matrices into blocks of sizes $m_1 \times m_2$, $m_2 \times m_3$ and $m_1 \times m_3$, respectively, where $n_i = m_i k_i$ for all $i = 1, 2, 3$, our goal is then to estimate the block sizes m_i which maximize data reuse under the cache size constraint.

Instruction (2.2) can be expressed as the nested loop,

```
do i = 1 : k₁,
   do k = 1 : k₂,
      do j = 1 : k₃,
         Cᵢⱼ = Cᵢⱼ + Aᵢₖ × Bₖⱼ;
      end
   end
end
```

where C_{ij}, A_{ik} and B_{kj} are, respectively, blocks of C, A and B, with subscripts denoting here the appropriate block indices. The innermost loop, refers to the identical block A_{ik} in all its iterations. To put it in cache, its dimensions must satisfy $m_1 m_2 \leq M$. Actually, the blocks C_{ij} and B_{kj} must also reside in the cache and the condition becomes

$$m_1 m_3 + m_1 m_2 + m_2 m_3 \leq M. \tag{2.3}$$

Further, since the blocks are obviously smaller than the original matrices, we need the additional constraints:

$$1 \leq m_i \leq n_i \text{ for } i = 1, 2, 3. \tag{2.4}$$

Evaluating the volume of the data moves using the number of data loads necessary for the whole procedure and assuming that the constraints (2.3) and (2.4) are satisfied, we observe that

- all the blocks of the matrix A are loaded only once;
- the blocks of the matrix B are loaded k_1 times;
- the blocks of the matrix C are loaded k_2 times.

Thus the total amount of loads is given by:

$$L = n_1 n_2 + n_1 n_2 n_3 \left(\frac{1}{m_1} + \frac{1}{m_2} \right). \tag{2.5}$$

Choosing $m_3 = 1$, and hence $k_3 = n_3$, (the multiplications of the blocks are performed by columns) and neglecting for simplicity the necessary storage of the columns of the blocks C_{ij} and B_{kj}, the values of m_1 and m_2, which minimize $\frac{1}{m_1} + \frac{1}{m_2}$ under the previous constraints are obtained as follows:

---

if $n_1 n_2 \leq M$ **then**
  $m_1 = n_1$ and $m_2 = n_2$;
else if $n_2 \leq \sqrt{M}$ **then**
  $m_1 = \frac{M}{n_2}$ and $m_2 = n_2$;
else if $n_1 \leq \sqrt{M}$ **then**
  $m_1 = n_1$ and $m_2 = \frac{M}{n_1}$;
else
  $m_1 = \sqrt{M}$ and $m_2 = \sqrt{M}$;
end if

In practice, M should be slightly smaller than the total cache volume to allow for storing the neglected vectors. With this parameter adjustment, at the innermost level, the block multiplication involves $2m_1 m_2$ operations and $m_1 + m_2$ loads as long as A_{ik} resides in the cache. This indicates why the matrix multiplication can be a compute bound program.

The reveal the important decisions that need to be made by the code developer, and leads to a scheme that is very similar to parallel multiplication.

In [11], the authors consider the situation of a cluster of p processors with a common cache and count loads and stores in their evaluation. However, the final decision tree is similar to the one presented here.

2.2.2 Lowering Complexity via the Strassen Algorithm

The classical multiplication algorithms implementing the operation (2.2) for dense matrices use $2n^3$ operations. We next describe the scheme proposed by Strassen which reduces the number of operations in the procedure [12]: assuming that n is even, the operands can be decomposed in 2×2 matrices of $\frac{n}{2} \times \frac{n}{2}$ blocks:

$$\begin{pmatrix} C_{11} & C_{12} \\ C_{21} & C_{22} \end{pmatrix} = \begin{pmatrix} A_{11} & A_{12} \\ A_{21} & A_{22} \end{pmatrix} \begin{pmatrix} B_{11} & B_{12} \\ B_{21} & B_{22} \end{pmatrix}. \tag{2.6}$$

Then, the multiplication can be performed by the following operations on the blocks

$$
\begin{aligned}
P_1 &= (A_{11} + A_{22})(B_{11} + B_{22}), &\quad C_{11} &= P_1 + P_4 - P_5 + P_7, \\
P_2 &= (A_{21} + A_{22})B_{11}, &\quad C_{12} &= P_3 + P_5, \\
P_3 &= A_{11}(B_{12} - B_{22}), &\quad C_{21} &= P_2 + P_4, \\
P_4 &= A_{22}(B_{21} - B_{11}), &\quad C_{22} &= P_1 + P_3 - P_2 + P_6. \\
P_5 &= (A_{11} + A_{12})B_{22}, \\
P_6 &= (A_{21} - A_{11})(B_{11} + B_{12}), \\
P_7 &= (A_{12} - A_{22})(B_{21} + B_{22}),
\end{aligned}
\tag{2.7}
$$

The computation of P_k and C_{ij} is referred as one Strassen step. This procedure involves 7 block multiplications and 18 block additions of blocks, instead of 8 block multiplications and 4 block additions as the case in the classical algorithm. Since the complexity of the multiplications is $O(n^3)$ whereas for an addition it is only $O(n^2)$, the Strassen approach is beneficial for large enough n. This approach was improved in [13] by the following sequence of 7 block multiplications and 15 block additions. It is implemented in the so-called Strassen-Winograd procedure (as expressed in [14]):

$$
\begin{aligned}
T_0 &= A_{11}, &\quad S_0 &= B_{11}, &\quad Q_0 &= T_0 S_0, &\quad U_1 &= Q_0 + Q_3, \\
T_1 &= A_{12}, &\quad S_1 &= B_{21}, &\quad Q_1 &= T_1 S_1, &\quad U_2 &= U_1 + Q_4, \\
T_2 &= A_{21} + A_{22}, &\quad S_2 &= B_{12} + B_{11}, &\quad Q_2 &= T_2 S_2, &\quad U_3 &= U_1 + Q_2, \\
T_3 &= T_2 - A_{12}, &\quad S_3 &= B_{22} - S_2, &\quad Q_3 &= T_3 S_3, &\quad C_{11} &= Q_0 + Q_1, \\
T_4 &= A_{11} - A_{12}, &\quad S_4 &= B_{22} - B_{12}, &\quad Q_4 &= T_4 S_4, &\quad C_{12} &= U_3 + Q_5, \\
T_5 &= A_{12} + T_3, &\quad S_5 &= B_{22}, &\quad Q_5 &= T_5 S_5, &\quad C_{21} &= U_2 - Q_6, \\
T_6 &= A_{22}, &\quad S_6 &= S_3 - B_{21}, &\quad Q_6 &= T_6 S_6, &\quad C_{22} &= U_2 + Q_2.
\end{aligned}
\tag{2.8}
$$

Clearly, (2.7) and (2.8) are still valid for rectangular blocks. If $n = 2^\gamma$, the approach can be repeated for implementing the multiplications of the blocks. If it is recursively applied up to 2×2 blocks, the total complexity of the process becomes $O(n^{\omega_0})$, where $\omega_0 = \log 7$. More generally, if the process is iteratively applied until we get blocks of order $m \leq n_0$, the total number of operations is

$$
T(n) = c_s n^{\omega_0} - 5n^2,
\tag{2.9}
$$

with $c_s = (2n_0 + 4)/n_0^{\omega_0 - 2}$, which achieves its minimum for $n_0 = 8$; cf. [14].

The numerical stability of the above methods has been considered by several authors. In [15], it is shown that the rounding errors in the Strassen algorithm can be worse than those in the classical algorithm for multiplying two matrices, with the situation somewhat worse in Winograd's algorithm. However, ref. [15] indicates that it is possible to get a fast and stable version of _GEMM by incorporating in it steps from the Strassen or Winograd-type algorithms.

Both the Strassen algorithm (2.7), and the Winograd version (2.8), can be implemented on parallel architectures. In particular, the seven block multiplications are independent, as well as most of the block additions. Moreover, each of these operations has yet another inner level of parallelism.

A parallel implementation must allow for the recursive application of several Strassen steps while maintaining good data locality. The Communication-Avoiding Parallel Strassen (CAPS) algorithm, proposed in [14, 16], achieves this aim; cf. [14, 16]. In CAPS, the Strassen steps are implemented by combining two strategies: all the processors cooperate in computing the blocks P_k and C_{ij} whenever the local memories are not large enough to store the blocks. The remaining Strassen steps consist of block operations that are executed independently on seven sets of processors. The latter minimizes the communications but needs extra memory. CAPS is asymptotically optimal with respect to computational cost and data communication.

Theorem 2.1 ([14]) CAPS *has computational cost* $\Theta\left(n^{\omega_0}/p\right)$ *and requires bandwidth* $\Theta\left(\max\left\{\left(n^{\omega_0}/pM^{\omega_0/2}\right)\log p, \log p\right\}\right)$, *assuming p processors, each with local memory of size M words.*

Experiments in [17] show that CAPS uses less communication than some communication optimal classical algorithms and much less than previous implementations of the Strassen algorithm. As a result, it can outperform both classical algorithms for large sized problems, because it requires fewer operations, as well as for small problems, because of lower communication costs.

2.2.3 Accelerating the Multiplication of Complex Matrices

Savings may be realized in multiplying two complex matrices, e.g. see [18]. Let $A = A_1 + iA_2$ and $B = B_1 + iB_2$ two complex matrices where $A_j, B_j \in \mathbb{R}^{n \times n}$ for $j = 1, 2$. The real and imaginary parts C_1 and C_2 of the matrix $C = AB$ can be obtained using only three multiplications of real matrices (and not four as in the classical expression):

$$
\begin{aligned}
T_1 &= A_1 B_1, & C_1 &= T_1 - T_2, \\
T_2 &= A_2 B_2, & C_2 &= (A_1 + A_2)(B_1 + B_2) - T_1 - T_2.
\end{aligned}
\tag{2.10}
$$

The savings are realized through the way the imaginary part C_2 is computed. Unfortunately, the above formulation may suffer from catastrophic cancellations, [18].

For large n, there is a 25 % benefit in arithmetic operations over the conventional approach. Although remarkable, this benefit does not lower the complexity which remains the same, i.e. $O(n^3)$. To push such advantage further, one may use the Strassen's approach in the three matrix multiplications above to realize $O(n^{\omega_0})$ arithmetic operations.

Parallelism is achieved at several levels:

- All the matrix operations are additions and multiplications. They can be implemented with full efficiency. In addition, the multiplication can be realized through the Strassen algorithm as implemented in CAPS, see Sect. 2.2.2.
- The three matrix multiplications are independent, once the two additions are performed.

2.3 General Organization for Dense Matrix Factorizations

In this section, we describe the usual techniques for expressing parallelism in the factorization schemes (i.e. the algorithms that compute any of the well-known decompositions such as LU, Cholesky, or QR). More specific factorizations are included in the ensuing chapters of the book.

2.3.1 Fan-Out and Fan-In Versions

Factorization schemes can be based on one of two basic templates: the *fan-out* template (see Algorithm 2.1) and the *fan-in* version (see Algorithm 2.2). Each of these templates involves two basic procedures which we generically call compute(j) and update(j, k). The two versions, however, differ only by a single loop interchange.

Algorithm 2.1 *Fan-out* version for factorization schemes.

---

do $j = 1 : n$,
  compute(j) ;
  do $k = j + 1 : n$,
    update(j, k) ;
  end
end

Algorithm 2.2 *Fan-in* version for factorization schemes.

---

do $k = 1 : n$,
  do $j = 1 : k - 1$,
    update(j, k) ;
  end
  compute(k) ;
end

The above implementations are also respectively named as the *right-looking* and the *left-looking* versions. The exact definitions of the basic procedures, when applied to a given matrix A, are displayed in Table 2.1 together with their arithmetic complexities on a uniprocessor. They are based on a column oriented organization. For the analysis of loop dependencies, it is important to consider that column j is unchanged by task update(j, k) whereas column k is overwritten by the same task; column j is overwritten by task compute(j).

The two versions are based on vector operations (i.e. BLAS1). It can be seen, however, that for a given j, the inner loop of the *fan-out* algorithm is a rank-one update (i.e. BLAS2), with a special feature for the Cholesky factorization, where only the lower triangular part of A is updated.

Table 2.1 Elementary factorization procedures; MATLAB index notation used for submatrices

Factorization	Procedures	Complexity
Cholesky on $A \in \mathbb{R}^{n \times n}$	C $: A(j:n, j) = A(j:n, j)/\sqrt{A(j, j)}$ U $: A(k:n, k) = A(k:n, j) - A(k:n, j)A(k, j)$	$\frac{1}{3}n^3 + O(n^2)$
LU on $A \in \mathbb{R}^{n \times n}$ (no pivoting)	C $: A(j:n, j) = A(j:n, j)/A(j, j)$ U $: A(j:n, k) = A(j:n, j) - A(j:n, j)A(k, j)$	$\frac{2}{3}n^3 + O(n^2)$
QR on $A \in \mathbb{R}^{m \times n}$ (Householder)	C $: u = \mathrm{house}(A(j:n, j))$ and $\beta = 2/\|u\|^2$ U $: A(j:n, k) = A(j:n, j) - \beta u(u^\top A(j:n, j))$	$2n^2(m - \frac{1}{3}n) +$ $O(mn)$
MGS on $A \in \mathbb{R}^{m \times n}$ (Modified Gram-Schmidt)	C $: A(1:n, j) = A(1:n, j)/\|A(1:n, j)\|$ U $: A(1:n, k) = A(1:n, k) -$ $\qquad A(1:n, k)(A(1:n, k)^\top A(1:n, j))$	$2mn^2 + O(mn)$

Notations C: compute(j); U: update(j, k)

v = house(u): computes the Householder vector (see [19])

2.3.2 Parallelism in the Fan-Out Version

In the *fan-out* version, the inner loop (loop k) of Algorithm 2.1 involves independent iterations whereas in the *fan-in* version, the inner loop (loop j) of Algorithm 2.2 must be sequential because of a recursion on vector k.

The inner loop of Algorithm 2.1 can be expressed as a **doall** loop. The resulting algorithm is referred to as Algorithm 2.3.

Algorithm 2.3 do/doall *fan-out* version for factorization schemes.

```
do j = 1 : n,
  compute(j) ;
  doall k = j + 1 : n,
    update(j, k) ;
  end
end
```

At the outer iteration j, there are $n - j$ independent tasks with identical cost. When the outer loop is regarded as a sequential one, idle processors will result at the end of most of the outer iterations. Let p be the number of processors used, and for the sake of simplicity, let $n = pq + 1$ and assume that the time spent by one processor in executing task compute(j) or task update(j, k) is the same which is taken as the time unit. Note that this last assumption is valid only for the Gram-Schmidt orthogonalization, since for the other algorithms, the cost of task compute(j) and task update(j, k) are proportional to $n - j$ or even smaller for the Cholesky factorization. A simple computation shows that the sequential process consumes $\mathbf{T}_1 = n(n + 1)/2$ steps, whereas the parallel process on p processors consumes $\mathbf{T}_p = 1 + p \sum_{i=2}^{q+1} i = \frac{pq(q+3)}{2} + 1 = \frac{(n-1)(n-1+3p)}{2p} + 1$ steps. For

Table 2.2 Benefit of pipelining the outer loop in MGS (QR factorization)

steps	parallel runs
1	C(1)
2	U(1,2) U(1,3) U(1,4) U(1,5)
3	U(1,6) U(1,7) U(1,8) U(1,9)
4	C(2)
5	U(2,3) U(2,4) U(2,5) U(2,6)
6	U(2,7) U(2,8) U(2,9)
7	C(3)
8	U(3,4) U(3,5) U(3,6) U(3,7)
9	U(3,8) U(3,9)
10	C(4)
11	U(4,5) U(4,6) U(4,7) U(4,8)
12	U(4,9)
13	C(5)
14	U(5,6) U(5,7) U(5,8) U(5,9)
15	C(6)
16	U(6,7) U(6,8) U(6,9)
17	C(7)
18	U(7,8) U(7,9)
19	C(8)
20	U(8,9)
21	C(9)

(a) Sequential outer loop.

steps	parallel runs
1	C(1)
2	U(1,2) U(1,3) U(1,4) U(1,5)
3	C(2) U(1,6) U(1,7) U(1,8)
4	U(1,9) U(2,3) U(2,4) U(2,5)
5	C(3) U(2,6) U(2,7) U(2,8)
6	U(2,9) U(3,4) U(3,5) U(3,6)
7	C(4) U(3,7) U(3,8) U(3,9)
8	U(4,5) U(4,6) U(4,7) U(4,8)
9	C(5) U(4,9)
10	U(5,6) U(5,7) U(5,8) U(5,9)
11	C(6)
12	U(6,7) U(6,8) U(6,9)
13	C(7)
14	U(7,8) U(7,9)
15	C(8)
16	U(8,9)
17	C(9)

Notations : C(j) = compute(j)
U(j,k) = update(j,k)

(b) **doacross** outer loop.

instance, for $n = 9$ and $p = 4$, the parallel calculation is performed in 21 steps (see Table 2.2a) whereas the sequential algorithm requires 45 steps. In Fig. 2.1, the efficiency $\mathbf{E}_p = \frac{n(n+1)}{(n-1)(n-1+3p)+2p}$ is displayed for $p = 4, 8, 16, 32, 64$ processors when dealing with vectors of length n, where $100 \le n \le 1000$.

The efficiency study above is for the Modified Gram-Schmidt (MGS) algorithm. Even though the analysis for other factorizations is more complicated, the general behavior of the corresponding efficiency curves with respect to the vector length, does not change.

The next question to be addressed is whether the iterations of the outer loop can be pipelined so that they can be implemented utilizing the **doacross**.

At step j, for $k = j + 1, \ldots, n$, task update(j, k) may start as soon as task compute(j) is completed but compute(j) may start as soon as all the tasks update(l, j), for $l = 1, \ldots, j - 1$ are completed. Maintaining the serial execution of tasks update(l, j) for $l = 1, \ldots, j - 1$ is equivalent to guaranteeing that any task update(j, k) cannot start before completion of update($j - 1, k$). The resulting scheme is listed as Algorithm 2.4.

In our particular example, scheduling of the elementary tasks is displayed in Table 2.2b. Comparing with the non-pipelined scheme, we clearly see that the processors are fully utilized whenever the number of remaining vectors is large enough. On the other hand, the end of the process is identical for the two strategies. Therefore, pipelining the outer loop is beneficial except when p is much smaller than n.

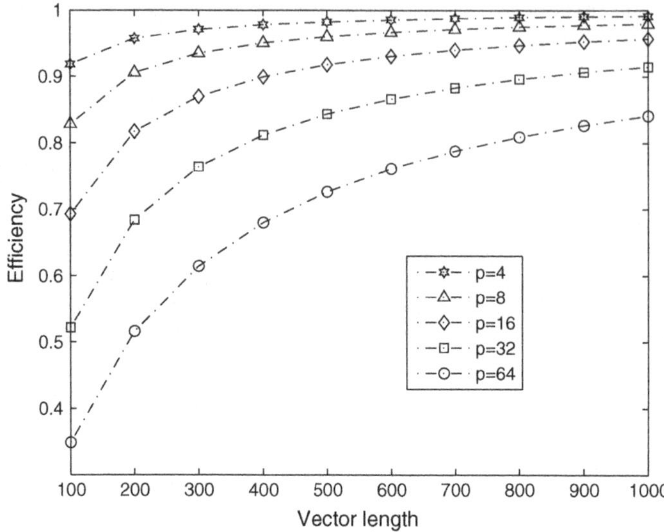

Fig. 2.1 Efficiencies of the **doall** approach with a sequential outer loop in MGS

Algorithm 2.4 doacross/doall version for factorization schemes (*fan-out.*)

```
doacross j = 1 : n,
  if j > 1, then
    wait(j) ;
  end if
  compute(j) ;
  doall k = j + 1 : n,
    update(j, k) ;
    if k = j + 1, then
      post(j + 1) ;
    end if
  end
end
```

2.3.3 *Data Allocation for Distributed Memory.*

The previous analysis is valid for shared or distributed memory architectures. However, for distributed memory systems we need to discuss the data allocation. As an illustration consider a ring of p processors, numbered from 0 to $p - 1$, on which r consecutive columns of A are stored in a round-robin mode. By denoting $\tilde{j} = j - 1$, column j is stored on processor s when $\tilde{j} = r(pv + t) + s$ with $0 \le s < r$ and $0 \le t < p$.

As soon as column j is ready, it is broadcast to the rest of the processors so they can start tasks update(j, k) for the columns k which they own. This implements the **doacross/doall** strategy of the *fan-out* approach, listed as Algorithm 2.5.

To reduce the number of messages, one may transfer only the blocks of r consecutive vectors when they are all ready to be used (i.e. the corresponding compute(j) tasks are completed). The drawback of this option is increasing the periods during which there are idle processors. Therefore, the block size r must be chosen so as to obtain a better trade-off between using as many processors as possible and reducing communication cost. Clearly, the optimum value is architecture dependent as it depends on the smallest efficient task granularity.

Algorithm 2.5 Message passing *fan-out* version for factorization schemes.

Input: processor #q owns the set C_q of columns of A.

    do $j = 1 : n$,
      if $j \in C_q$, **then**
        compute(j) ;
        sendtoall(j) ;
      else
        receive(j) ;
      end if
      do $(k \in C_q)$ & $(k > j)$,
        update(j, k) ;
      end
    end

The discussion above could easily be extended to the case of a torus configuration where each processor of the previous ring is replaced by a ring of q processors. Every column of the matrix A is now distributed into slices on the corresponding ring in a round-robin mode. This, in turn, implies global communication in each ring of q processors.

2.3.4 Block Versions and Numerical Libraries

We have already seen that it is useful to block consecutive columns of A. Actually, there is benefit of doing so, even on a uniprocessor. In Algorithms 2.1 and 2.2, tasks compute(j) and update(j, k) can be redefined to operate on a block of vectors rather than on a single vector. In that case, indices j and k would refer to blocks of r consecutive columns. In Table 2.1 the scalar multiplications correspond now to matrix multiplications involving BLAS3 procedures. It can be shown that for all the above mentioned factorizations, task compute(j) becomes the task performing the original factorization scheme on the corresponding block; cf. [19]. Task update(j, k) remains formally the same but involving blocks of vectors (rank-r update) instead of individual vectors (rank-1 update).

The resulting block algorithms are mathematically equivalent to their vector counterparts but they may have different numerical behavior, especially for the Gram-Schmidt algorithm. This will be discussed in detail in Chap. 7.

Well designed block algorithms for matrix multiplication and rank-k updates for hierarchical machines with multiple levels of memory and parallelism are of critical importance for the design of solvers for the problems considered in this chapter that demonstrate high performance and scalability. The library LAPACK [20], that solves the classic matrix-problems, is a case in point by being based on BLAS3 as well as its parallel version SCALAPACK [21]:

- LAPACK: This is the main reference for a software library for numerical linear algebra. It provides routines for solving systems of linear equations and linear least squares, eigenvalue problems, and singular value decomposition. The involved matrices can be stored as dense matrices or band matrices. The procedures are based on BLAS3 and are proved to be backward stable. LAPACK was originally written in FORTRAN 77, but moved to Fortran 90 in version 3.2 (2008).
- SCALAPACK: This library can be seen as the parallel version of the LAPACK library for distributed memory architectures. It is based on the Message Passing Interface standard MPI [22]. Matrices and vectors are stored on a process grid into a two-dimensional block-cyclic distribution. The library is often chosen as the reference to which compare any new developed procedure.

In fact, many users became fully aware of these gains even when using high-level problem solving environments like MATLAB (cf. [23]). As early works on the subject had shown (we consider it rewarding for the reader to consider the pioneering analyses in [11, 24]), the task of designing primitives is far from simple, if one desires to provide a design that closely resembles the target computer model. The task becomes more difficult as the complexity of the computer architectures increases. It becomes even harder when the target is to build methods that can deliver high performance for a spectrum of computer architectures.

2.4 Sparse Matrix Computations

Most large scale matrix computations in computational science and engineering involve sparse matrices, that is matrices with relatively few nonzero elements, e.g. $n_{nz} = O(n)$, for square matrices of order n. See [25] for instances of matrices from a large variety of applications.

For example, in numerical simulations governed by partial differential equations approximated using finite difference or finite elements, the number of nonzero entries per row is related to the topology of the underlying finite element or finite difference grid. In two-dimensional problems discretized by a 5-point finite difference discretization scheme, the number of nonzeros is about $n_{nz} = 5n$ and the density of the resulting sparse matrix (i.e. the ratio between nonzeros entries and all entries) is $d \approx \frac{5}{n}$, where n is the matrix order.

Methods designed for dense matrix computations are rarely suitable for sparse matrices since they quickly destroy the sparsity of the original matrix leading to the need of storing a much larger number of nonzeros. However, with the availability of large memory capacities in new architectures, factorization methods (LU and QR) exist that control fill-in and manage the needed extra storage. We do not present such algorithms in this book but refer the reader to existing literature, e.g. see [26–28]. Another option is to use *matrix-free* methods in which the sparse matrix is not generated explicitly but used as an operator through the matrix-vector multiplication kernel.

To make feasible large scale computations that involve sparse matrices, they are encoded in some suitable sparse matrix storage format in which only nonzero elements of the matrix are stored together with sufficient information regarding their row and column location to access them in the course of operations.

2.4.1 Sparse Matrix Storage and Matrix-Vector Multiplication Schemes

Let $A = (\alpha_{ij}) \in \mathbb{R}^{n \times n}$ be a sparse matrix, and n_{nz} the number of nonzero entries in A.

Definition 2.1 (*Graph of a sparse matrix*) The *graph* of the matrix is given by the pair of nodes and edges $(< 1 : n >, G)$ where G is characterized by

$$((i, j) \in G) \text{ iff } \alpha_{ij} \neq 0.$$

The adjacency matrix C of the matrix A is $C = (\gamma_{ij}) \in \mathbb{R}^{n \times n}$ such that $\gamma_{ij} = 1$ if $(i, j) \in G$ otherwise $\gamma_{ij} = 0$.

The most common sparse storage schemes are presented below together with their associated kernels: MV for matrix-vector multiplication and MTV for the multiplication by the transpose. For a complete description and some additional storage types see [29].

Compressed Row Sparse Storage (CRS)

All the nonzero entries are successively stored, row by row, in a one-dimensional array a of length n_{nz}. Column indices are stored in the same order in a vector ja of the same length n_{nz}. Since the entries are stored row by row, it is sufficient to define a third vector ia to store the indices of the beginning of each row in a. By convention, the vector is extended by one entry: $ia_{n+1} = n_{nz} + 1$. Therefore, when scanning vector a_k for $k = 1, \ldots, n_{nz}$, the corresponding row index i and column index j are obtained from the following

$$a_k = \alpha_{ij} \Leftrightarrow \begin{cases} j = ja_k, \\ ia_i \leq k < ia_{i+1}. \end{cases} \tag{2.11}$$

The corresponding MV kernel is given by Algorithm 2.6. The inner loop implements a sparse inner product through a so-called gather procedure.

Algorithm 2.6 CRS-type MV.

Input: CRS storage (a, ja, ia) of $A \in \mathbb{R}^{n \times n}$ as defined in (2.11) ; $v, \ w \in \mathbb{R}^n$.
1: **do** $i = 1 : n$,
2:   **do** $k = ia_i : ia_{i+1} - 1$,
3:     $w_i = w_i + a_k v_{ja_k}$; *//Gather*
4:   **end**
5: **end**

Compressed Column Sparse Storage (CCS)

This storage is the dual of CRS: it corresponds to storing $A^\top$ via a CRS format. Therefore, the nonzero entries are successively stored, column by column, in a vector a of length n_{nz}. Row indices are stored in the same order in a vector ia of length n_{nz}. The third vector ja stores the indices of the beginning of each column in a. By convention, the vector is extended by one entry: $ja_{n+1} = n_{nz} + 1$. Thus (2.11) is replaced by,

$$a_k = \alpha_{ij} \Leftrightarrow \begin{cases} j = ia_k, \\ ja_j \leq k < ja_{j+1}. \end{cases} \tag{2.12}$$

The corresponding MV kernel is given by Algorithm 2.7. The inner loop implements a sparse _AXPY through a so-called scatter procedure.

Algorithm 2.7 CCS-type MV.

Input: CCS storage (a, ja, ia) of $A \in \mathbb{R}^{n \times n}$ as defined in (2.12) ; $v, \ w \in \mathbb{R}^n$.
Output: $w = w + Av$.
1: **do** $j = 1 : n$,
2:   **do** $k = ja_j : ja_{j+1} - 1$,
3:     $w_{ia_k} = w_{ia_k} + a_k v_j$; *//Scatter*
4:   **end**
5: **end**

Compressed Storage by Coordinates (COO)

In this storage, no special order of the entries is assumed. Therefore three vectors a, ia and ja of length n_{nz} are used satisfying

$$a_k = \alpha_{ij} \Leftrightarrow \begin{cases} i = ia_k, \\ j = ja_k \end{cases} \tag{2.13}$$

The corresponding MV kernel is given by Algorithm 2.8. It involves both the scatter and gather procedures.

Algorithm 2.8 COO-type MV.

Input: COO storage (a, ja, ia) of $A \in \mathbb{R}^{n \times n}$ as defined in (2.13) ; $v,\ w \in \mathbb{R}^n$.
Output: $w = w + Av$.
1: **do** $k = 1 : n_{nz}$,
2:    $w_{ia_k} = w_{ia_k} + a_k v_{ja_k}$;
3: **end**

MTV Kernel and Other Storages

When A is stored in one of the above mentioned compressed storage formats the MTV kernel

$$w = w + A^\top v,$$

is expressed for a CRS-stored matrix by Algorithm 2.7 and for a CCS-stored one by Algorithm 2.6. For a COO-stored matrix, the algorithm is obtained by inverting the roles of the arrays ia and ja in Algorithm 2.8.

Nowadays, the scatter-gather procedures (see step 3 in Algorithm 2.6 and step 3 in Algorithm 2.7) are pipelined on the architectures allowing vector computations. However, their startup time is often large (i.e. order of magnitude of $\mathbf{n}_{1/2}$—as defined in Sect. 1.1.2—is in the hundreds. If in MV a were a dense matrix $\mathbf{n}_{1/2}$ would be in the tens). The vector lengths in Algorithms 2.6 and 2.7 are determined by the number of nonzero entries per row or per column. They often are so small that the computations are run at sequential computational rates. There have been many attempts to define sparse storage formats that favor larger vector lengths (e.g. see the jagged diagonal format mentioned in [30, 31]).

An efficient storage format which combines the advantages of dense and sparse matrix computations attempts to define a square block structure of a sparse matrix in which most of the blocks are empty. The non empty blocks are stored in any of the above formats, e.g. CSR, or the regular dense storage depending on the sparsity density. Such a sparse storage format is called either *Block Compressed Sparse storage* (BCRS) where the sparse nonempty blocks are stored using the CRS format, or *Block Compressed Column storage* (BCCS) where the sparse nonempty blocks are strored using the CCS format.

Basic Implementation on Distributed Memory Architecture

Let us consider the implementation of $w = w + Av$ and $w = w + A^\top v$ on a distributed memory parallel architecture with p processors where $A \in \mathbb{R}^{n \times n}$ and $v, w \in \mathbb{R}^n$. The first stage consists of partitioning the matrix and allocating respective parts to the local processor memories. Each processor P_q with $q = 1, \ldots, p$, receives a block of rows of A and the corresponding slices of the vectors v and w:

$$
\begin{array}{c}
P_1 : \\ \hline P_2 : \\ \vdots \\ \hline P_p :
\end{array}
\quad
A = \left(\begin{array}{c|c|c|c}
A_{1,1} & A_{1,2} & \cdots & A_{1,p} \\ \hline
A_{2,1} & A_{2,2} & \cdots & A_{2,p} \\ \hline
\vdots & \vdots & & \vdots \\ \hline
A_{p,1} & A_{p,2} & \cdots & A_{p,p}
\end{array} \right)
\quad
v = \begin{pmatrix} v_1 \\ v_2 \\ \vdots \\ v_p \end{pmatrix}
\quad
w = \begin{pmatrix} w_1 \\ w_2 \\ \vdots \\ w_p \end{pmatrix}
$$

With the blocks $A_{q,j}$ $(j = 1, \ldots, p)$ residing on processor P_q, this partition determines the necessary communications for performing the multiplications. All the blocks are sparse matrices with some being empty. To implement the kernels $w = w + Av$ and $w = w + A^\top v$, the communication graph is defined by the sets $\mathscr{R}(q)$ and $\mathscr{C}(q)$ which respectively include the list of indices of the nonempty blocks $A_{q,j}$ of the block row q and $A_{j,q}$ of the block column q $(j = 1, \ldots, p)$. The two implementations are given by Algorithms 2.9 and 2.10, respectively.

Algorithm 2.9 MV: $w = w + Av$

Input: q : processor number.

    $\mathscr{R}(q)$: list of indices of the nonempty blocks of row q.

    $\mathscr{C}(q)$: list of indices of the nonempty blocks of column q.

1: **do** $j \in \mathscr{C}(q)$,
2:    send v_q to processor P_j;
3: **end**
4: compute $w_q = w_q + A_{q,q} v_q$.
5: **do** $j \in \mathscr{R}(q)$,
6:    receive v_j from processor P_j;
7:    compute $w_q = w_q + A_{q,j} v_j$;
8: **end**

Algorithm 2.10 MTV: $w = w + A^\top v$

Input: q : processor number.

    $\mathscr{R}(q)$: list of indices of the nonempty blocks of row q.

    $\mathscr{C}(q)$: list of indices of the nonempty blocks of column q.

1: **do** $j \in \mathscr{R}(q)$,
2:    compute $t_j = A_{q,j}^\top v_q$;
3:    send t_j to processor P_j;
4: **end**
5: compute $w_q = w_q + A_{q,q}^\top v_q$;
6: **do** $j \in \mathscr{C}(q)$,
7:    receive u_j from processor P_j;
8:    compute $w_q = w_q + u_j$;
9: **end**

The efficiency of the two procedures MV and MTV are often quite different, depending on the chosen sparse storage format.

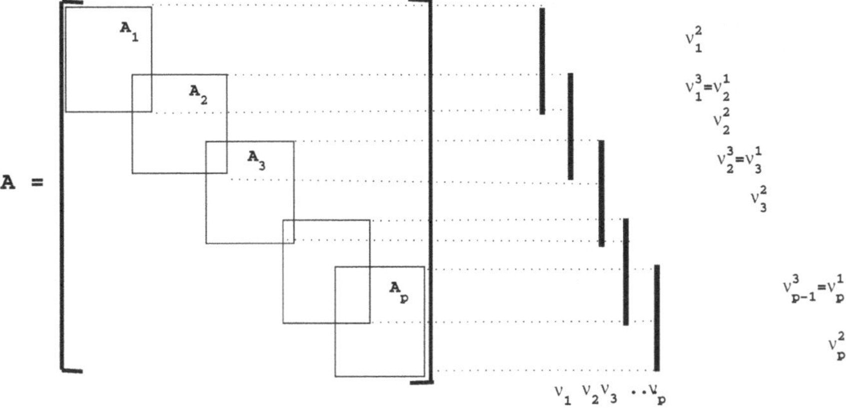

Fig. 2.2 Partition of a block-diagonal matrix with overlapping blocks; vector v is decomposed in overlapping slices

Scalable Implementation for an Overlapped Block-Diagonal Structure

If the sparse matrix $A \in \mathbb{R}^{n \times n}$ can be reordered to result in p overlapping diagonal blocks denoted by A_q, $q = 1, \ldots, p$ as shown in Fig. 2.2, then a matrix-vector multiplication primitive can be designed in high parallel scalability. Let block A_q be stored in the memory of the processor P_q and the vectors $v, w \in \mathbb{R}^n$ stored accordingly. It is therefore necessary to maintain the consistency between the two copies of the components corresponding to the overlaps.

To perform the MV computation $w = w + Av$, the matrix A may be considered as a sum of p blocks B_q ($q = 1, \ldots, p$) where $B_p = A_p$ and all earlier blocks B_q are the same as A_q with the elements of lower right submatrix corresponding to the overlap are replaced by zeros (see Fig. 2.3).

Let vector v_q be the subvector of v corresponding to the q-block row indices. For $2 \leq q \leq p - 1$, the vector v_q is partitioned into $v_q^\top = \left({v_q^1}^\top, {v_q^2}^\top, {v_q^3}^\top \right)$, according to the overlap with the neighboring blocks. The first and the last subvectors are partitioned as $v_1^\top = \left({v_1^2}^\top, {v_1^3}^\top \right)$ and $v_p^\top = \left({v_p^1}^\top, {v_p^2}^\top \right)$.

Denoting $\bar{B}_q$ and $\bar{v}_q$ the prolongation by zeros of B_q and v_q to the full order n, the operation $w + Av = w + \sum_{q=1}^{p} \bar{B}_q \bar{v}_q$ can be performed via Algorithm 2.11. After completion, the vector w is correctly updated and distributed on the processors with consistent subvectors w_q (i.e. $w_{q-1}^3 = w_q^1$ for $q = 2, \ldots, p$). This algorithm

Fig. 2.3 Elementary blocks
for the MV kernel

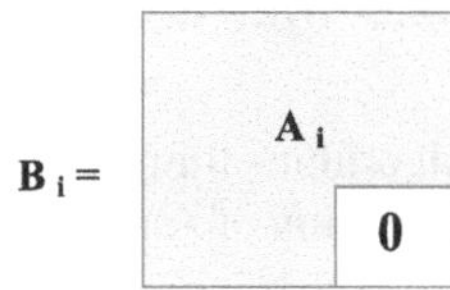

Algorithm 2.11 Scalable MV multiplication $w = w + Av$.

Input: q: processor number.

    In the local memory: B_q, $v_q^\top = [(v_q^1)^\top, (v_q^2)^\top, (v_q^3)^\top]$, and $w_q^\top = [(w_q^1)^\top, (w_q^2)^\top, (w_q^3)^\top]$.

Output: $w = w + Av$.

1: $z_q = B_q v_q$;
2: **if** $q < p$, **then**
3:     send z_q^3 to processor P_{q+1} ;
4: **end if**
5: **if** $q > 1$, **then**
6:     send z_q^1 to processor P_{q-1} ;
7: **end if**
8: $w_q = w_q + z_q$;
9: **if** $q < p$, **then**
10:     receive t from processor P_{q+1} ;
11:     $w_q^3 = w_q^3 + t$;
12: **end if**
13: **if** $q > 1$, **then**
14:     receive t from processor P_{q-1} ;
15:     $w_q^1 = w_q^1 + t$;
16: **end if**

---

does not involve global communications and it can be implemented on a linear array of processors in which every processor only exchanges information with its two immediate neighbors: each processor receives one message from each of its neighbors and it sends back one message to each.

Proposition 2.1 *Algorithm 2.11 which implements the* MV *kernel for a sparse matrix with an overlapped block-diagonal structure on a ring of p processors is weakly scalable as long as the number of nonzeros entries of each block and the overlap sizes are independent of the number of processors p.*

Proof Let T_{BMV} be the bound on the number of steps for the MV kernel of each individual diagonal block, and ℓ being the maximum overlap sizes, then on p processors the number of steps is given by

$$\mathbf{T}_p \le T_{\mathrm{BMV}} + 4(\beta + \ell\tau_c),$$

where β is the latency for a message and τ_c the time for sending a word to an immediate neighbouring node regardless of the latency. Since $\mathbf{T}_p$ is independent of p, weak scalability is assured.

2.4.2 Matrix Reordering Schemes

Algorithms for reordering sparse matrices play a vital role in enhancing the parallel scalability of various sparse matrix algorithms and their underlying primitives, e.g., see [32, 33].

Early reordering algorithms such as minimum-degree and nested dissection have been developed for reducing fill-in in sequential direct methods for solving sparse symmetric positive definite linear systems, e.g., see [26, 34, 35]. Similarly, algorithms such as reverse Cuthill-McKee (RCM), e.g., see [36, 37], have been used for reducing the envelope (variable band or profile) of sparse matrices in order to: (i) enhance the efficiency of uniprocessor direct factorization schemes, (ii) reduce the cost of sparse matrix-vector multiplications in iterative methods such as the conjugate gradients method (CG), for example, and (iii) obtain preconditioners for the PCG scheme based on incomplete factorization [38, 39]. In this section, we here describe a reordering scheme that not only reduces the profile of sparse matrices, but also brings as many of the heaviest (larger magnitude) off-diagonal elements as possible close to the diagonal. For solving sparse linear systems, for example, one aims at realizing an overall cost, with reordering, that is much less than that without reordering. In fact, in many time-dependent computational science and engineering applications, this is possible. In such applications, the relevant nested computational loop occurs as shown in Fig. 2.4.

The outer-most loop deals with time-step t, followed by solving a nonlinear set of equations using a variant of Newton's method, with the inner-most loop dealing with solving a linear system in each Newton iteration to a relatively modest relative residual η_k. Further, it is often the case that it is sufficient to realize the benefits of reordering by keep using the permutation matrices obtained at time step t for several subsequent time steps. This results not only in amortization of the cost of reordering, but also in reducing the total cost of solving all the linear systems arising in such an application.

With such a reordering, we aim to obtain a matrix $C = PAQ$, where $A = (\alpha_{ij})$ is the original sparse matrix, P and Q are permutation matrices, such that C can be split as $C = B + E$, with the most important requirements being that: (i) the sparse matrix E contains far fewer nonzero elements than A, and is of a much lower rank, and (ii) the central band B is a "generalized-banded" matrix with a Frobenius norm that is a substantial fraction of that of A.

Fig. 2.4 Common structure of programs in time-dependent simulations

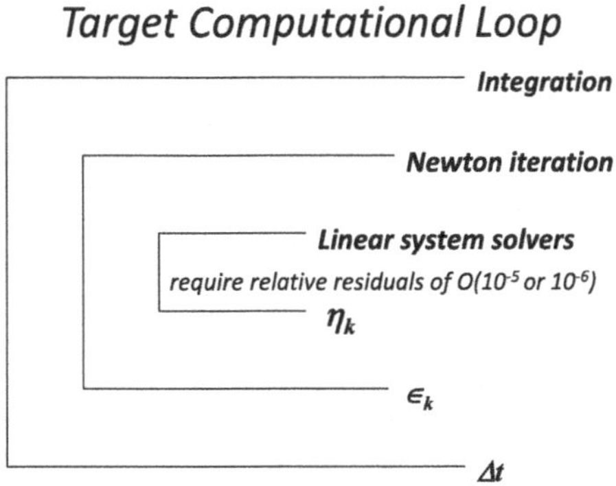

Fig. 2.5 Reordering to a
narrow-banded matrix

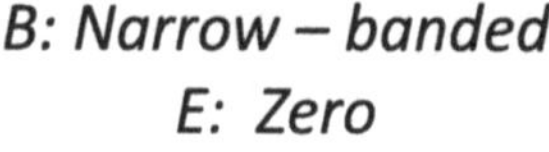

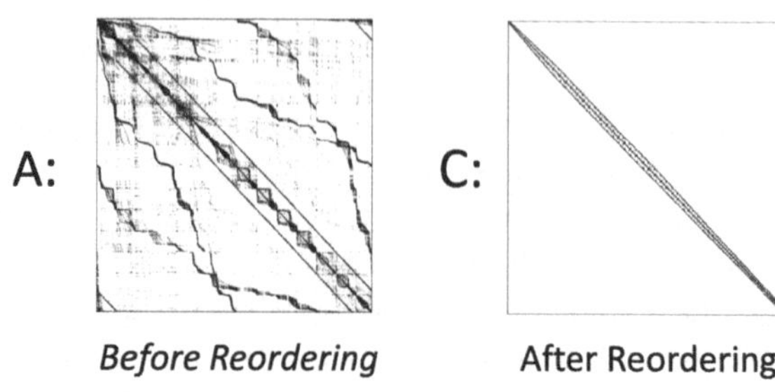

Hence, depending on the original matrix A, the matrix B can be extracted as:

(a) "narrow-banded" of bandwidth β much smaller than the order n of the matrix A, i.e., $\beta = 10^{-4}n$, for $n \geq 10^6$, for example (the most fortunate situation), see Fig. 2.5,

(b) "medium-banded", i.e., of the block-tridiagonal form $[H, G, J]$, in which the elements of the off-diagonal blocks H and J are all zero except for their small upper-right and lower-left corners, respectively, see Fig. 2.6, or

(c) "wide-banded", i.e., consisting of overlapped diagonal blocks, in which each diagonal block is a sparse matrix, see Fig. 2.7.

The motivation for desiring such a reordering scheme is three-fold. First, B can be used as a preconditioner of a Krylov subspace method when solving a linear system $Ax = f$ of order n. Since E is of a rank p much less than n, the preconditioned Krylov subspace scheme will converge quickly. In exact arithmetic, the Krylov subspace method will converge in exactly p iterations. In floating-point arithmetic, however, this translates into the method achieving small relative residuals in less than p iterations. Second, since we require the diagonal of B to be zero-free with the product of its entries maximized, and that the Frobenius norm of B is close to that of A, this will enhance the possibility that B is nonsingular, or close to a nonsingular matrix. Third, multiplying C by a vector can be implemented on a parallel architecture with higher efficiency by splitting the operation into two parts: multiplying the "generalized-banded" matrix B by a vector, and a low-rank sparse matrix E by a vector. The former, e.g. $v = Bu$, can be achieved with high parallel scalability on distributed-memory architectures requiring only nearest neighbor communication, e.g. see Sect. 2.4.1 for the scalable parallel implementation of an overlapped block diagonal matrix-vector multiplication scheme. The latter, e.g. $w = Eu$, however, incurs much less irregular addressing penalty compared to $y = Au$ since E contains far fewer nonzero entries than A.

Since A is nonsymmetric, in general, we could reduce its profile by using RCM (i.e. via symmetric permutations only) applied to $(|A| + |A^\top|)$, [40], or by using the spectral reordering introduced in [41]; see also [42]. However, this will neither

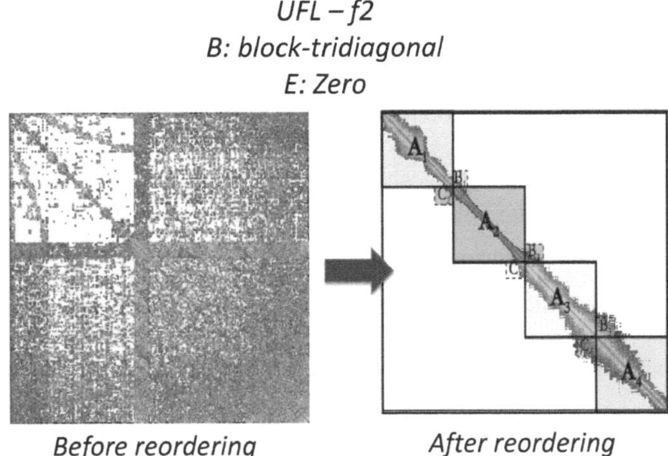

Fig. 2.6 Reordering to a medium-banded matrix

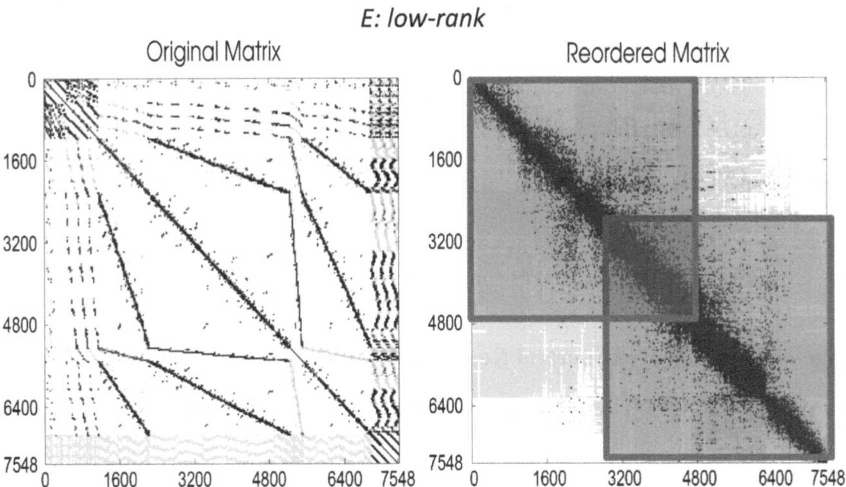

Fig. 2.7 Reordering to a wide-banded matrix

realize a zero-free diagonal, nor insure bringing the heaviest off-diagonal elements close to the diagonal. Consequently, RCM alone will not realize a central "band" B with its Frobenius norm statisfying: $\|B\|_F \geq (1 - \varepsilon)^* \|A\|_F$. In order to solve this *weighted bandwidth reduction* problem, we use a weighted spectral reordering technique which is a generalization of spectral reordering. To alleviate the shortcomings of using only symmetric permutations, and assuming that the matrix A is not structurally singular, this weighted spectral reordering will need to be coupled with a nonsymmetric ordering technique such as the maximum traversal algorithm [43] to guarantee a zero-free diagonal, and to maximize the magnitude of the product of the diagonal elements, via the MPD algorithm (Maximum Product on Diagonal algorithm) [44, 45]. Such a procedure is implemented in the Harwell Subroutine Library [46] as (HSL-MC64).

Thus, the resulting algorithm, which we refer to as WSO (Weighted Spectral Ordering), consists of three stages:

Stage 1: Nonsymmetric Permutations

Here, we obtain a permutation matrix Q that maximizes the product of the absolute values of the diagonal entries of QA, [44, 45]. This is achieved by a maximum traversal search followed by a scaling procedure resulting in diagonal entries of absolute values equal to 1, and all off-diagonal elements with magnitudes less than or equal to 1. After applying this stage, a linear system $Ax = f$, becomes of the form,

$$(QD_2AD_1)(D_1^{-1}x) = (QD_2f) \tag{2.14}$$

in which each D_j, $j = 1, 2$ is a diagonal scaling matrix.

Stage 2: Checking for Irreducibility

In this stage, we need to detect whether the sparse matrix under consideration is irreducible, i.e., whether the corresponding graph has one strongly connected component. This is achieved via Tarjan's strongly connected component algorithm [47], see also related schemes in [48], or [49]. If the matrix is reducible, we apply the weighted spectral reordering simultaneously on each strongly connected component (i.e. on each sparse diagonal block of the resulting upper block triangular matrix). For the rest of this section, we assume that the sparse matrix A under consideration is irreducible, i.e., the corresponding graph has only one strongly connected component.

Stage 3: The Weighted Spectral Reordering Scheme [50]

As in other traditional reordering algorithms, we wish to minimize the half-bandwidth of a matrix A which is given by,

$$BW(A) = \max_{i,j:\alpha_{ij}\neq 0} |i - j|, \tag{2.15}$$

i.e., to minimize the maximum distance of a nonzero entry from the main diagonal. Let us assume for the time being that A is a symmetric matrix, and that we aim at extracting a central band $B = (\beta_{ij})$ of minimum bandwidth such that, for a given tolerance ε,

$$\frac{\sum_{i,j} |\alpha_{ij} - \beta_{ij}|}{\sum_{i,j} |\alpha_{ij}|} \leq \varepsilon, \tag{2.16}$$

and

$$\begin{aligned}
\beta_{ij} &= \alpha_{ij} \text{ if } |i - j| \leq k, \\
\beta_{ij} &= 0
\end{aligned} \tag{2.17}$$

The idea behind this formulation is that if a significant part of the matrix is packed into a central band B, then the rest of the nonzero entries can be dropped to obtain an effective preconditioner. In order to find a heuristic solution to the weighted bandwidth reduction problem, we use a generalization of spectral reordering. Spectral reordering is a linear algebraic technique that is commonly used to obtain approximate solutions to various intractable graph optimization problems [51]. It has also been successfully applied to the bandwidth and envelope reduction problems for sparse matrices [41]. The core idea of spectral reordering is to compute a vector $x = (\xi_i)$ that minimizes

$$\sigma_A(x) = \sum_{i,j:\alpha_{ij} \neq 0} (\xi_i - \xi_j)^2, \tag{2.18}$$

subject to $\|x\|_2 = 1$ and $x^\top e = 0$. As mentioned above we assume that the matrix A is real and symmetric. The vector x that minimizes $\sigma_A(x)$ under these constraints provides a mapping of the rows (and columns) of matrix A to a one-dimensional Euclidean space, such that pairs of rows that correspond to nonzeros are located as close as possible to each other. Consequently, the ordering of the entries of the vector x provides an ordering of the matrix that significantly reduces the bandwidth.

Fiedler [52] first showed that the optimal solution to this problem is given by the eigenvector corresponding to the second smallest eigenvalue of the Laplacian matrix $L = (\lambda_{ij})$ of A,

$$\begin{aligned}
\lambda_{ij} &= -1 && \text{if } i \neq j \wedge \alpha_{ij} \neq 0, \\
\lambda_{ii} &= |\{j : \alpha_{ij} \neq 0\}|.
\end{aligned} \tag{2.19}$$

Note that the matrix L is positive semidefinite, and the smallest eigenvalue of this matrix is equal to zero. The eigenvector x that minimizes $\sigma_A(x) = x^\top L x$, such that $\|x\|_2 = 1$ and $x^\top e = 0$, is the eigenvector corresponding to the second smallest eigenvalue of the Laplacian, i.e. the symmetric eigenvalue problem

$$Lx = \lambda x, \tag{2.20}$$

and is known as the Fiedler vector. The Fiedler vector of a sparse matrix can be computed efficiently using any of the eigensolvers discussed in Chap. 11, see also [53].

While spectral reordering is shown to be effective in bandwidth reduction, the classical approach described above ignores the magnitude of nonzeros in the matrix. Therefore, it is not directly applicable to the weighted bandwidth reduction problem. However, Fiedler's result can be directly generalized to the weighted case [54]. More precisely, the eigenvector x that corresponds to the second smallest eigenvalue of the weighted Laplacian L minimizes

$$\bar{\sigma}_A(x) = x^\top L x = \sum_{i,j} |\alpha_{ij}|(\xi_i - \xi_j)^2, \tag{2.21}$$

where L is defined as

$$\begin{aligned} \lambda_{ij} &= -|\alpha_{ij}| \quad \text{if } i \neq j, \\ \lambda_{ii} &= \sum_j |\alpha_{ij}|. \end{aligned} \tag{2.22}$$

We now show how weighted spectral reordering can be used to obtain a continuous approximation to the weighted bandwidth reduction problem. For this purpose, we first define the relative *bandweight* of a specified band of the matrix as follows:

$$w_k(A) = \frac{\sum_{i,j:|i-j|<k} |\alpha_{ij}|}{\sum_{i,j} |\alpha_{ij}|}. \tag{2.23}$$

In other words, the bandweight of a matrix A, with respect to an integer k, is equal to the fraction of the total magnitude of entries that are encapsulated in a band of half-width k.

For a given α, $0 \leq \alpha \leq 1$, we define α-bandwidth as the smallest half-bandwidth that encapsulates a fraction α of the total matrix weight, i.e.,

$$BW_\alpha(A) = \min_{k:w_k(A)\geq\alpha} k. \tag{2.24}$$

Observe that α-bandwidth is a generalization of half-bandwidth, i.e., when $\alpha = 1$, the α-bandwidth is equal to the half-bandwidth of the matrix. Now, for a given vector $x = (\xi_1, \xi_2, \ldots\ldots, \xi_n)^\top \in \mathbb{R}^n$, define an injective permutation function π: $\{1, 2, \ldots, n\} \to \{1, 2, \ldots, n\}$, such that, for $1 \leq i, j \leq n, \xi_{\pi_i} \leq \xi_{\pi_j}$ iff $i \leq j$. Here, n denotes the number of rows (columns) of the matrix A. Moreover, for a fixed k, define the function $\delta_k(i, j) : \{1, 2, \ldots, n\} \times \{1, 2, \ldots, n\} \to \{0, 1\}$, which quantizes the difference between π_i and π_j with respect to k, i.e.,

$$\delta_k(i, j) = \begin{cases} 0 \text{ if } |\pi_i - \pi_j| \leq k, \\ 1 \text{ else} \end{cases} \tag{2.25}$$

Let $\bar{A}$ be the matrix obtained by reordering the rows and columns of A according to π, i.e.,

$$\bar{A}(\pi_i, \pi_j) = \alpha_{ij} \text{ for } 1 \leq i, j \leq n. \tag{2.26}$$

Then $\delta_k(i, j) = 0$ indicates that α_{ij} is inside a band of half-width k in the matrix $\bar{A}$ while $\delta_k(i, j) = 1$ indicates that it is outside the band. Defining

$$\hat{\sigma}_k(A) = \sum_{i,j} |\alpha_{ij}| \delta_k(i, j), \tag{2.27}$$

then,

$$\hat{\sigma}_k(A) = (1 - w_k(\bar{A})) \sum_{i,j} |\alpha_{ij}|. \tag{2.28}$$

Therefore, for a fixed α, the α-bandwidth of the matrix $\bar{A}$ is equal to the smallest k that satisfies $\hat{\sigma}_A(k)/\sum_{i,j} |\alpha_{ij}| \leq 1 - \alpha$.

Note that the problem of minimizing $\bar{\sigma}_x(A)$ is a continuous relaxation of the problem of minimizing $\hat{\sigma}_k(A)$ for a given k. Therefore, the Fiedler vector of the weighted Laplacian L provides a good basis for reordering A to minimize $\hat{\sigma}_k(A)$. Consequently, for a fixed ε, this vector provides a heuristic solution to the problem of finding a reordered matrix $\bar{A} = (\bar{\alpha}_{ij})$ with minimum $(1 - \varepsilon)$-bandwidth. Once the matrix is obtained, we extract the central band B as follows:

$$B = \{\beta_{ij} = \bar{\alpha}_{ij} \text{ if } |i - j| \leq BW_{1-\epsilon}(\bar{A}), \text{ otherwise } \beta_{ij} = 0\}. \tag{2.29}$$

Clearly, B satisfies (2.16) and is of minimal bandwidth.

Note that spectral reordering is defined specifically for symmetric matrices, and the resulting permutation is symmetric as well. Since our main focus here concerns general nonsymmetric matrices, we apply spectral reordering to nonsymmetric matrices by computing the Laplacian matrix of $|A| + |A^{\top}|$ instead of $|A|$. We note also that this formulation results in a symmetric permutation for a nonsymmetric matrix, which may be considered overconstrained.

Once, the Fiedler vector yields the permutation P, we obtain the matrix C as,

$$C = (PQD_2AD_1P^{\top}), \tag{2.30}$$

and the linear system $Ax = f$ becomes of the final form,

$$(PQD_2AD_1P^{\top})(PD_1^{-1}x) = (PQD_2f). \tag{2.31}$$

References

1. Lawson, C., Hanson, R., Kincaid, D., Krogh, F.: Basic linear algebra subprograms for Fortran usage. ACM Trans. Math. Softw. **5**(3), 308–323 (1979)
2. Dongarra, J., Croz, J.D., Hammarling, S., Hanson, R.: An extended set of FORTRAN basic linear algebra subprograms. ACM Trans. Math. Softw. **14**(1), 1–17 (1988)
3. Dongarra, J., Du Croz, J., Hammarling, S., Duff, I.: A set of level-3 basic linear algebra subprograms. ACM Trans. Math. Softw. **16**(1), 1–17 (1990)
4. Intel company: Intel Math Kernel Library. http://software.intel.com/en-us/intel-mkl

5. Texas advanced computer center, University of Texas: GotoBLAS2. https://www.tacc.utexas.edu/tacc-software/gotoblas2
6. Netlib Repository at UTK and ORNL: Automatically Tuned Linear Algebra Software (ATLAS). http://www.netlib.org/atlas/
7. Whaley, R., Dongarra, J.: Automatically tuned linear algebra software. In: Proceedings of 1998 ACM/IEEE Conference on Supercomputing, Supercomputing'98, pp. 1–27. IEEE Computer Society, Washington (1998). http://dl.acm.org/citation.cfm?id=509058.509096
8. Yotov, K., Li, X., Ren, G., Garzarán, M., Padua, D., Pingali, K., Stodghill, P.: Is search really necessary to generate high-performance BLAS? Proc. IEEE 93(2), 358–386 (2005). doi:10.1109/JPROC.2004.840444
9. Goto, K., van de Geijn, R.: Anatomy of high-performance matrix multiplication. ACM Trans. Math. Softw. 34(3), 12:1–12:25 (2008). doi:10.1145/1356052.1356053. http://doi.acm.org/10.1145/1356052.1356053
10. Gallivan, K.A., Plemmons, R.J., Sameh, A.H.: Parallel algorithms for dense linear algebra computations. SIAM Rev. 32(1), 54–135 (1990). doi:http://dx.doi.org/10.1137/1032002
11. Gallivan, K., Jalby, W., Meier, U.: The use of BLAS3 in linear algebra on a parallel processor with a hierarchical memory. SIAM J. Sci. Stat. Comput. 8(6), 1079–1084 (1987)
12. Strassen, V.: Gaussian elimination is not optimal. Numerische Mathematik 13, 354–356 (1969)
13. Winograd, S.: On multiplication of 2 × 2 matrices. Linear Algebra Appl. 4(4), 381–388 (1971)
14. Ballard, G., Demmel, J., Holtz, O., Lipshitz, B., Schwartz, O.: Communication-optimal parallel algorithm for Strassen matrix multiplication. Technical report UCB/EECS-2012-32, EECS Department, University of California, Berkeley (2012). http://www.eecs.berkeley.edu/Pubs/TechRpts/2012/EECS-2012-32.html
15. Higham, N.J.: Exploiting fast matrix multiplication within the level 3 BLAS. ACM Trans. Math. Softw. 16(4), 352–368 (1990)
16. Ballard, G., Demmel, J., Holtz, O., Schwartz, O.: Graph expansion and communication costs of fast matrix multiplication. J. ACM 59(6), 32:1–32:23 (2012). doi:10.1145/2395116.2395121. http://doi.acm.org/10.1145/2395116.2395121
17. Lipshitz, B., Ballard, G., Demmel, J., Schwartz, O.: Communication-avoiding parallel Strassen: implementation and performance. In: Proceedings of the International Conference on High Performance Computing, Networking, Storage and Analysis, SC'12, pp. 101:1–101:11. IEEE Computer Society Press, Los Alamitos (2012). http://dl.acm.org/citation.cfm?id=2388996.2389133
18. Higham, N.J.: Stability of a method for multiplying complex matrices with three real matrix multiplications. SIAM J. Matrix Anal. Appl. 13(3), 681–687 (1992)
19. Golub, G., Van Loan, C.: Matrix Computations, 4th edn. Johns Hopkins (2013)
20. Anderson, E., Bai, Z., Bischof, C., Blackford, S., Demmel, J., Dongarra, J., Du Croz, J., Greenbaum, A., Hammarling, S., McKenney, A., Sorensen, D.: LAPACK Users' Guide, 3rd edn. Society for Industrial and Applied Mathematics, Philadelphia (1999)
21. Blackford, L., Choi, J., Cleary, A., D'Azevedo, E., Demmel, J., Dhillon, I., Dongarra, J., Hammarling, S., Henry, G., Petitet, A., Stanley, K., Walker, D., Whaley, R.: ScaLAPACK User's Guide. SIAM, Philadelphia (1997). http://www.netlib.org/scalapack
22. Gropp, W., Lusk, E., Skjellum, A.: Using MPI: Portable Parallel Programming with the Message Passing Interface. MIT Press, Cambridge (1994)
23. Moler, C.: MATLAB incorporates LAPACK. Mathworks Newsletter (2000). http://www.mathworks.com/company/newsletters/articles/matlab-incorporates-lapack.html
24. Gallivan, K., Jalby, W., Meier, U., Sameh, A.: The impact of hierarchical memory systems on linear algebra algorithm design. Int. J. Supercomput. Appl. 2(1) (1988)
25. Davis, T., Hu, Y.: The University of Florida Sparse Matrix Collection. ACM Trans. Math. Softw. 38(1), 1:1–1:25 (2011). http://doi.acm.org/10.1145/2049662.2049663
26. Duff, I., Erisman, A., Reid, J.: Direct Methods for Sparse Matrices. Oxford University Press Inc., New York (1989)
27. Davis, T.: Direct Methods for Sparse Linear Systems. SIAM, Philadelphia (2006)

28. Zlatev, Z.: Computational Methods for General Sparse Matrices, vol. 65. Kluwer Academic Publishers, Dordrecht (1991)
29. Bai, Z., Demmel, J., Dongarra, J., Ruhe, A., van der Vorst, H.: Templates for the Solution of Algebraic Eigenvalue Problems: A Practical Guide. SIAM, Philadelphia (2000)
30. Melhem, R.: Toward efficient implementation of preconditioned conjugate gradient methods on vector supercomputers. Int. J. Supercomput. Appl. 1(1), 70–98 (1987)
31. Philippe, B., Saad, Y.: Solving large sparse eigenvalue problems on supercomputers. Technical report RIACS TR 88.38, NASA Ames Research Center (1988)
32. Schenk, O.: Combinatorial Scientific Computing. CRC Press, Switzerland (2012)
33. Kepner, J., Gilbert, J.: Graph Algorithms in the Language of Linear Algebra. SIAM, Philadelphia (2011)
34. George, J., Liu, J.: Computer Solutions of Large Sparse Positive Definite Systems. Prentice Hall (1981)
35. Pissanetzky, S.: Sparse Matrix Technology. Academic Press, New York (1984)
36. Cuthill, E., McKee, J.: Reducing the bandwidth of sparse symmetric matrices. In: Proceedings of 24th National Conference Association Computer Machinery, pp. 157–172. ACM Publications, New York (1969)
37. Liu, W., Sherman, A.: Comparative analysis of the Cuthill-McKee and the reverse Cuthill-McKee ordering algorithms for sparse matrices. SIAM J. Numer. Anal. 13, 198–213 (1976)
38. D'Azevedo, E.F., Forsyth, P.A., Tang, W.P.: Ordering methods for preconditioned conjugate gradient methods applied to unstructured grid problems. SIAM J. Matrix Anal. 13(3), 944–961 (1992)
39. Duff, I., Meurant, G.: The effect of ordering on preconditioned conjugate gradients. BIT 29, 635–657 (1989)
40. Reid, J., Scott, J.: Reducing the total bandwidth of a sparse unsymmetric matrix. SIAM J. Matrix Anal. Appl. 28(3), 805–821 (2005)
41. Barnard, S., Pothen, A., Simon, H.: A spectral algorithm for envelope reduction of sparse matrices. Numer. Linear Algebra Appl. 2, 317–334 (1995)
42. Spielman, D., Teng, S.: Spectral partitioning works: planar graphs and finite element meshes. Numer. Linear Algebra Appl. 421, 284–305 (2007)
43. Duff, I.: On algorithms for obtaining a maximum transversal. ACM Trans. Math. Softw. 7, 315–330 (1981)
44. Duff, I., Koster, J.: On algorithms for permuting large entries to the diagonal of a sparse matrix. SIAM J. Matrix Anal. Appl. 22, 973–966 (2001)
45. Duff, I., Koster, J.: The design and use of algorithms for permuting large entries to the diagonal of sparse matrices. SIAM J. Matrix Anal. Appl. 20, 889–901 (1999)
46. The HSL mathematical software library. See http://www.hsl.rl.ac.uk/index.html
47. Tarjan, R.: Depth-first search and linear graph algorithms. SIAM J. Comput. 1(2), 146–160 (1972)
48. Cheriyan, J., Mehlhorn, K.: Algorithms for dense graphs and networks on the random access computer. Algorithmica 15, 521–549 (1996)
49. Dijkstra, E.: A Discipline of Programming, Chapter 25. Prentice Hall, Englewood Cliffs (1976)
50. Manguoğlu, M., Mehmet, K., Sameh, A., Grama, A.: Weighted matrix ordering and parallel banded preconditioners for iterative linear system solvers. SIAM J. Sci. Comput. 32(3), 1201–1206 (2010)
51. Hendrickson, B., Leland, R.: An improved spectral graph partitioning algorithm for mapping parallel computations. SIAM J. Sci. Comput. 16(2), 452–469 (1995). http://citeseer.nj.nec.com/hendrickson95improved.html
52. Fiedler, M.: Algebraic connectivity of graphs. Czechoslovak Math. J. 23, 298–305 (1973)
53. Kruyt, N.: A conjugate gradient method for the spectral partitioning of graphs. Parallel Comput. 22, 1493–1502 (1997)
54. Chan, P., Schlag, M., Zien, J.: Spectral k-way ratio-cut partitioning and clustering. IEEE Trans. CAD-Integr. Circuits Syst. 13, 1088–1096 (1994)

Part II
Dense and Special Matrix Computations

Chapter 3
Recurrences and Triangular Systems

A recurrence relation is a rule that defines each element of a sequence in terms of the preceding elements; it forms one of the most basic tools in discrete and computational mathematics with myriad applications in scientific computing, ranging from numerical linear algebra, the numerical solution of ordinary and partial differential equations and orthogonal polynomials, to dependence analysis in restructuring compilers and logic design. In fact, almost every computational task relies on recursive techniques and hence involves recurrence relations. Therefore, recurrence solvers are computational kernels and need to be implemented as efficient primitives on various architectures. Throughout this book we will encounter many linear and nonlinear recurrences that can be solved using techniques from this chapter. Linear recurrences can usually be expressed in the form of mostly banded lower triangular linear systems, so much of the discussion is devoted to parallel algorithms for solving $Lx = f$, where L is a lower triangular matrix.

3.1 Definitions and Examples

For illustration, we list a few simple examples from computational mathematics; cf. [1].

Example 3.1 Consider the problem of evaluating the definite integral

$$\psi_k = \int_0^1 \frac{\xi^k}{\xi + 8} d\xi \quad \text{for} \ k = 0, 1, 2, \ldots, n$$

for some integer n. It is not difficult to verify that $\psi_k + 8\psi_{k-1} = \frac{1}{k}$, which is a linear recurrence for evaluating ψ_k, $k \geq 1$, with $\psi_0 = \log_e(9/8)$.

© Springer Science+Business Media Dordrecht 2016
E. Gallopoulos et al., *Parallelism in Matrix Computations*,
Scientific Computation, DOI 10.1007/978-94-017-7188-7_3

Example 3.2 We wish to use a finite-difference method for solving the second-order differential equation $\psi'' = \phi(\xi, \psi)$ with the initial conditions $\psi(0) = \alpha$ and $\psi'(0) = \gamma$. Replacing the derivatives ψ' and ψ'' by the differences

$$
\psi'(\xi_k) \simeq \frac{\psi_{k+1} - \psi_{k-1}}{2h},
$$
$$
\psi''(\xi_k) \simeq \frac{\psi_{k+1} - 2\psi_k + \psi_{k-1}}{h^2},
$$

where $h = \xi_{k+1} - \xi_k$, $k \geq 0$, we obtain the linear recurrence relation

$$
\psi_{k+1} - 2\psi_k + \psi_{k-1} = h^2 \phi(\xi_k, \psi_k) \tag{3.1}
$$

with the initial values

$$
\psi_0 = \alpha, \tag{3.2}
$$

$$
\psi_1 - \psi_{-1} = 2h\gamma. \tag{3.3}
$$

The undefined value ψ_{-1} can be eliminated using (3.1) (with $k = 0$), and (3.2). Hence, the linear recurrence (3.1) can be started with $\psi_0 = \alpha$ and $\psi_1 = \psi_0 + h\gamma + \frac{1}{2}h^2\phi(\xi_0, \psi_0)$.

Example 3.3 ([2]) Chebyshev polynomials of the first kind

$$
\mathscr{T}_k(\xi) = \cos(k \cos^{-1} \xi), \quad k = 0, 1, 2, \ldots,
$$

satisfy the important three-term recurrence relation

$$
\mathscr{T}_{k+1}(\xi) - 2\xi\,\mathscr{T}_k(\xi) + \mathscr{T}_{k-1}(\xi) = 0
$$

for $k \geq 1$, with the starting values $\mathscr{T}_0(\xi) = 1$ and $\mathscr{T}_1(\xi) = \xi$.

Example 3.4 Newton's method for obtaining a root α of a single nonlinear equation $\phi(\xi) = 0$ is given by the nonlinear recurrence involving the first derivative

$$
\xi_{k+1} = \xi_k - (\phi(\xi_k)/\phi'(\xi_k))
$$

with a given initial approximation

$$
\xi_0 = \beta.
$$

See also [3] for examples of recurrences in logic design and restructuring compilers.

Stability Issues

It is well known that computations with recurrence relations are prone to error growth; at each step, computations are performed and new errors generated by operating on past data that maybe be already contaminated with errors [1]. This error propagation and gradual accumulation could be catastrophic. For example, consider the linear recurrence from Example 3.1,

$$\psi_k + 8\psi_{k-1} = \frac{1}{k}$$

with $\psi_0 = \log_e(9/8)$. Using three decimals (with rounding) throughout the evaluation of ψ_i, $i \geq 0$, and taking $\psi_0 \simeq 0.118$, the recurrence yields $\psi_1 \simeq 0.056$, $\psi_2 \simeq 0.052$, and $\psi_3 \simeq -0.083$. The reason for obtaining a negative ψ_3 (note that $\psi_k > 0$ for all values of k) is that the initial error δ_0 in ψ_0 has been highly magnified in ψ_3. In fact, even if we use exact arithmetic in evaluating ψ_1, ψ_2, ψ_3, $\ldots$ the initial rounding error δ_0 propagates such that the error δ_k in ψ_k is given by $(-8)^k\delta_0$. Therefore, since $|\delta_0| \leq 0.5 \times 10^{-3}$ we get $|\delta_3| \leq 0.256$; a very high error, given that the true value of ψ_3 (to three decimals) is 0.028. Note that this numerical instability cannot be eliminated by using higher precision (six decimals, say). Such wrong results will only be postponed, and will show up at a later stage. The relation between the available precision and the highest subscript k of a reasonably accurate ψ_k, however, will play an important role in constructing stable parallel algorithms for handling recurrence relations in general. Such numerical instability may be avoided by observing that the definite integral of Example 3.1 decreases as the value of k increases. Hence, if we assume that $\psi_5 \simeq 0$ (say) and evaluate the recurrence $\psi_k + 8\psi_{k-1} = \frac{1}{k}$ backwards, we should obtain reasonably accurate values for at least ψ_0 and ψ_1, since in this case $\delta_{k-1} = (-1/8)\delta_k$. Performing the calculations to three decimals, we obtain $\psi_4 \simeq 0.025$, $\psi_3 \simeq 0.028$, $\psi_2 \simeq 0.038$, $\psi_1 \simeq 0.058$, and $\psi_0 \simeq 0.118$.

In the remainder of this section we discuss algorithms that are particularly suitable for evaluating linear and certain nonlinear recurrence relations on parallel architectures.

3.2 Linear Recurrences

Consider the linear recurrence system,

$$\xi_1 = \phi_1,$$
$$\xi_i = \phi_i - \sum_{j=k}^{i-1}\lambda_{ij}\xi_j, \tag{3.4}$$

where $i = 2, \ldots, n$ and $k = \max\{1, i - m\}$. In matrix notation, (3.4) may be written as

$$x = f - \hat{L}x, \tag{3.5}$$

where
$$x = (\xi_1, \xi_2, \ldots, \xi_n)^\top,$$
$$f = (\phi_1, \phi_2, \ldots, \phi_n)^\top, \quad \text{and}$$

for $m = 3$ (say) $\hat{L}$ is of the form

$$\hat{L} = \begin{pmatrix} 0 & & & & & & & \\ \lambda_{21} & 0 & & & & & & \\ \lambda_{31} & \lambda_{32} & 0 & & & & & \\ \lambda_{41} & \lambda_{42} & \lambda_{43} & 0 & & & & \\ & \lambda_{52} & \lambda_{53} & \lambda_{54} & 0 & & & \\ & & \ddots & \ddots & \ddots & \ddots & & \\ & & & \lambda_{n,n-3} & \lambda_{n,n-2} & \lambda_{n,n-1} & 0 \end{pmatrix}.$$

Denoting $(I + \hat{L})$ by L, where I is the identity matrix of order n, (3.5) may be written as

$$Lx = f. \tag{3.6}$$

Here L is unit lower triangular with bandwidth $m + 1$ i.e., $\lambda_{ii} = 1$, and $\lambda_{ij} = 0$ for $i - j > m$. The sequential algorithm given by (3.4), forward substitution, requires $2mn + O(m^2)$ arithmetic operations.

We present parallel algorithms for solving the unit lower triangular system (3.6) for the following cases:

(i)   dense systems ($m = n - 1$), denoted by $R\!<\!n\!>$,
(ii) banded systems ($m \ll n$), denoted by $R\!<\!n,m\!>$, and
(iii) Toeplitz systems, denoted by $\hat{R}\!<\!n\!>$, and $\hat{R}\!<\!n,m\!>$, where $\lambda_{ij} = \tilde{\lambda}_{i-j}$.

The kernels used as the basic building blocks of these algorithms will be those dense BLAS primitives discussed in Chap. 2.

3.2.1 Dense Triangular Systems

Letting $m = n - 1$, then (3.6) may be written as

$$\begin{pmatrix} 1 & & & & & \\ \lambda_{21} & 1 & & & & \\ \lambda_{31} & \lambda_{32} & 1 & & & \\ \lambda_{41} & \lambda_{42} & \lambda_{43} & 1 & & \\ \vdots & \vdots & \vdots & \vdots & & \\ \lambda_{n1} & \lambda_{n2} & \lambda_{n3} & \lambda_{n4} & \cdots & \lambda_{n,n-1} & 1 \end{pmatrix} \begin{pmatrix} \xi_1 \\ \xi_2 \\ \xi_3 \\ \xi_4 \\ \vdots \\ \xi_n \end{pmatrix} = \begin{pmatrix} \phi_1 \\ \phi_2 \\ \phi_3 \\ \phi_4 \\ \vdots \\ \phi_n \end{pmatrix}. \tag{3.7}$$

Clearly, $\xi_1 = \phi_1$ so the right hand-side of (3.7) is purified by $f - \xi_1 L e_1$, creating a new right hand-side for the following lower triangular system of order $(n-1)$:

$$\begin{pmatrix} 1 & & & & \\ \lambda_{32} & 1 & & & \\ \lambda_{42} & \lambda_{43} & 1 & & \\ \vdots & \vdots & \vdots & & \\ \lambda_{n2} & \lambda_{n3} & \lambda_{n4} & \cdots & 1 \end{pmatrix} \begin{pmatrix} \xi_2 \\ \xi_3 \\ \xi_4 \\ \vdots \\ \xi_n \end{pmatrix} = \begin{pmatrix} \phi_2^{(1)} \\ \phi_3^{(1)} \\ \phi_4^{(1)} \\ \vdots \\ \phi_n^{(1)} \end{pmatrix},$$

Here $\phi_i^{(1)} = \phi_i - \phi_1 \lambda_{i1}, i = 2, 3, \ldots, n$. The process may be repeated to obtain the rest of the components of the solution vector. Assuming we have $(n-1)$ processors, this algorithm requires $2(n-1)$ parallel steps with no arithmetic redundancy. This method is often referred to as the column-sweep algorithm; we list it as Algorithm 3.1 (CSWEEP). It is straightforward to show that the cost becomes $3(n-1)$ parallel operations for non-unit triangular systems. The column-sweep algorithm can be

---

Algorithm 3.1 CSWEEP: Column-sweep method for unit lower triangular system

---

Input: Lower triangular matrix L of order n with unit diagonal, right-hand side f
Output: Solution of $Lx = f$
1: set $\phi_j^{(0)} = \phi_j$ $j = 1, \ldots, n$ //that is $f^{(0)} = f$
2: **do** $i = 1 : n$
3:    $\xi_i = \phi_i^{(i-1)}$
4:    **doall** $j = i + 1 : n$
5:       $\phi_j^{(i)} = \phi_j^{(i-1)} - \phi_i^{(i-1)} \lambda_{j,i}$ //compute $f^{(i)} = N_i^{-1} f^{(i-1)}$
6:    **end**
7: **end**

---

modified, however, to solve (3.6) in fewer parallel steps but with higher arithmetic redundancy.

Theorem 3.1 ([4, 5]) *The triangular system of equations $Lx = f$, where L is a unit lower triangular matrix of order n, can be solved in $\mathbf{T}_p = \frac{1}{2} \log^2 n + \frac{3}{2} \log n$ using no more than $p = \frac{15}{1024} n^3 + O(n^2)$ processors, yielding an arithmetic redundancy of $\mathbf{R}_p = O(n)$.*

Proof To motivate the proof, we consider the column-sweep algorithm observing that the coefficient matrix in (3.7), L, may be factored as the product

$$L = N_1 N_2 N_3 \ldots N_{n-1},$$

where

$$
N_j = \begin{pmatrix}
1 & & & & & & \\
 & 1 & & & & & \\
 & & \ddots & & & & \\
 & & & 1 & & & \\
 & & & \lambda_{j+1,j} & 1 & & \\
 & & & \vdots & & \ddots & \\
 & & & \lambda_{n,j} & & & 1
\end{pmatrix}
\tag{3.8}
$$

Hence, the solution x of $Lx = f$, is given by

$$
x = N_{n-1}^{-1} N_{n-2}^{-1} \dots N_2^{-1} N_1^{-1} f,
\tag{3.9}
$$

where the inverse of N_j is trivially obtained by reversing the signs of λ_{ij} in (3.8). Forming the product (3.9) as shown in Fig. 3.1, we obtain the column-sweep algorithm which requires $2(n-1)$ parallel steps to compute the solution vector x, given $(n-1)$ processors.

Assuming that n is power of 2 and utilizing a *fan-in* approach to compute

$$
M_{n-1}^{(0)} \cdots M_1^{(0)} f
$$

in parallel, the number of stages in Fig. 3.1 can be reduced from $n-1$ to $\log n$, as shown in Fig. 3.2, where $M_i^{(0)} = -N_i^{-1}$. This is the approach used in Algorithm 3.2 (DTS). To derive the total costs, it is important to consider the structure of the terms in the tree of Fig. 3.2. The initial terms $M_i^{(0)}, i = 1, \dots, n-1$ can each be computed

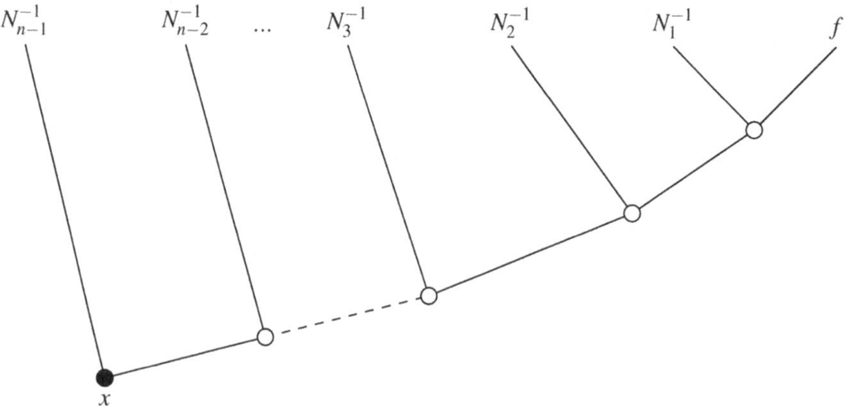

Fig. 3.1 Sequential solution of lower triangular system $Lx = f$ using CSweep (column-sweep Algorithm 3.1)

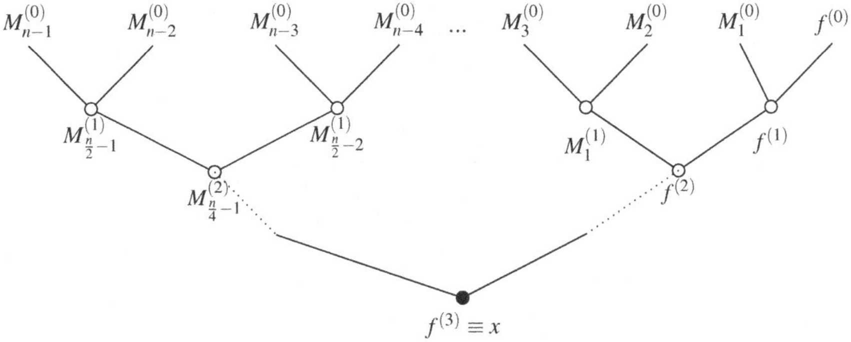

Fig. 3.2 Computation of solution of lower triangular system $Lx = f$ using the fan-in approach of DTS (Algorithm 3.2)

using one parallel division of length $n - i$ and a sign reversal. This requires at most 2 steps using at most $n(n + 1)/2$ processors.

The next important observation, that we show below, is that each $M_k^{(j)}$ has a maximum of $1 + 2^j$ elements in any given row. Therefore,

$$M_k^{(j+1)} = M_{2k+1}^{(j)} M_{2k}^{(j)} \tag{3.10}$$

$$f^{(j+1)} = M_1^{(j)} f^{(j)} \tag{3.11}$$

can be computed using independent inner products of vectors each of $1 + 2^j$ components at most. Therefore, using enough processors (more on that later), all the products for this stage can be accomplished in $j + 2$ parallel steps. Thus the total is approximately

$$\mathbf{T}_p = \sum_{j=0}^{\log n - 1} (j + 2) + 2 = \frac{1}{2}\log^2 n + \frac{3}{2}\log n.$$

We now show our claim about the number of independent inner products in the pairwise products occurring at each stage. It is convenient at each stage $j = 1, \ldots, \log n$, to partition each $M_k^{(j)}$ as follows:

$$M_k^{(j)} = \begin{pmatrix} I_q^{(j)} & \\ & L_k^{(j)} \\ & W_k^{(j)} \; I_r^{(j)} \end{pmatrix}, \tag{3.12}$$

where $L_k^{(j)}$ is unit lower triangular of order $s = 2^j$, $I_q^{(j)}$ and $I_r^{(j)}$ are the identities of order $q = ks - 1$, and $r = (n + 1) - (k + 1)s$, and $W_k^{(j)}$ is of order r-by-s. For $j = 0$,

$$L_s^{(0)} = 1, \text{ and}$$
$$W_k^{(0)} = -(\lambda_{k+1,k}, \ldots, \lambda_{n,k})^\top.$$

Observe that the number of nonzeros at the first stage $j = 0$, in each row of matrices $M_k^{(0)}$ is at most $2 = 2^0 + 1$. Assume the result to be true at stage j. Partitioning $W_k^{(j)}$ as,

$$W_k^{(j)} = \begin{pmatrix} U_k^{(j)} \\ V_k^{(j)} \end{pmatrix},$$

where $U_k^{(j)}$ is a square matrix of order s, then from

$$M_k^{(j+1)} = M_{2k+1}^{(j)} M_{2k}^{(j)},$$

we obtain

$$L_k^{(j+1)} = \begin{pmatrix} L_{2k}^{(j)} & 0 \\ L_{2k+1}^{(j)} U_{2k}^{(j)} & L_{2k+1}^{(j)} \end{pmatrix}, \tag{3.13}$$

and

$$W_k^{(j+1)} = (W_{2k+1}^{(j)} U_{2k}^{(j)} + V_{2k}^{(j)}, W_{2k+1}^{(j)}). \tag{3.14}$$

From (3.12) to (3.14) it follows that the maximum number of nonzeros in each row of $M_k^{(j+1)}$ is $2^{j+1} + 1$. Also, if we partition $f^{(j)}$ as

$$f^{(j)} = \begin{pmatrix} g_1^{(j)} \\ g_2^{(j)} \\ g_3^{(j)} \end{pmatrix}$$

in which $g_1^{(j)}$ is of order $(s-1)$, and $g_2^{(j)}$ is of order s, then

$$\begin{aligned} g_1^{(j+1)} &= g_1^{(j)}, \\ g_2^{(j+1)} &= L_1^{(j)} g_2^{(j)}, \text{ and} \\ g_3^{(j+1)} &= W_1^{(j)} g_2^{(j)} + g_3^{(j)} \end{aligned}$$

with the first two partitions constituting the first $(2s - 1)$ elements of the solution.

    We next estimate the number of processors to accomplish DTS. Terms $L_{2k+1}^{(j)} U_{2k}^{(j)}$ in (3.13) and $W_{2k+1}^{(j)} U_{2k}^{(j)}$ in (3.14) can be computed simultaneously. Each column of the former requires s inner products of s pairs of vectors of size $1, \ldots, s$. Therefore, each column requires $\sum_{i=1}^{s} i = s(s+1)/2$ processors. Moreover, the term $W_{2k+1}^{(j)} U_{2k}^{(j)}$ necessitates sr inner products of length s, where as noted above, at this stage $r = (n+1)-(k+1)s$. The total number of processors for this product therefore is $s^2 r$. The total becomes $s^2(s+1)/2 + s^2 r$ and substituting for r we obtain that the number of processors necessary is

$$p_M^{(j+1)}(k) = \frac{s^2}{2}(2n+3) - \frac{s^3}{2}(2k+1).$$

The remaining matrix addition in (3.14) can be performed with fewer processors. Similarly, the evaluation of $f^{(j+1)}$ requires $p_f^{(j+1)} = \frac{1}{2}(2n+3)s - \frac{3}{2}s^2$ processors. Therefore the total number of processors necessary for stage $j+1$ where $j = 0, 1, \ldots, \log n - 2$ is

$$p^{(j+1)} = \sum_{k=1}^{n/2s-1} p_M^{(j+1)}(k) + p_f^{(j+1)}$$

so the number of processors necessary for the algorithm are

$$\max_j \{p^{(j+1)}, n(n+1)/2\}$$

which is computed to be the value for p in the statement of the theorem.

Algorithm 3.2 DTS: Triangular solver based on a *fan-in* approach

---

Input: Lower triangular matrix L of order $n = 2^\mu$ and right-hand side f
Output: Solution x of $Lx = f$
1: $f^{(0)} = f$, $M_i^{(0)} = N_i^{-1}$, $i = 1, \ldots, n-1$ where N_j as in Eq. (3.8)
2: **do** $j = 0 : \mu - 1$
3: $\quad f^{(j+1)} = M_1^{(j)} f^{(j)}$
4: $\quad$ **doall** $k = 1 : (n/2^{j+1}) - 1$
5: $\quad\quad M_k^{(j+1)} = M_{2k+1}^{(j)} M_{2k}^{(j)}$
6: $\quad$ **end**
7: **end**

3.2.2 Banded Triangular Systems

In many practical applications, the order m of the linear recurrence (3.6) is much less than n. For such a case, algorithm DTS for dense triangular systems can be modified to yield the result in fewer parallel steps.

Theorem 3.2 ([5]) *Let L be a banded unit lower triangular matrix of order n and bandwidth $m+1$, where $m \leq n/2$, and $\lambda_{ik} = 0$ for $i - k > m$. Then the system*

$Lx = f$ can be solved in less than $\mathbf{T}_p = (2 + \log m) \log n$ parallel steps using fewer than $p = m(m+1)n/2$ processors.

Proof The matrix L and the vector f can be written in the form

$$
L = \begin{pmatrix} L_1 & & & & \\ R_1 & L_2 & & & \\ & R_2 & L_3 & & \\ & & & \ddots & \ddots & \\ & & & & R_{\frac{n}{m}-1} & L_{\frac{n}{m}} \end{pmatrix}, \quad f = \begin{pmatrix} f_1 \\ f_2 \\ f_3 \\ \vdots \\ f_{\frac{n}{m}} \end{pmatrix},
$$

where L_i and R_i are $m \times m$ unit lower triangular and upper triangular matrices, respectively. Premultiplying both sides of $Lx = f$ by the matrix $D = \mathrm{diag}(L_1^{-1}, \ldots, L_p^{-1})$, we obtain the system $L^{(0)}x = f^{(0)}$, where

$$
L^{(0)} = \begin{pmatrix} I_m & & & & \\ G_1^{(0)} & I_m & & & \\ & G_2^{(0)} & I_m & & \\ & & \ddots & \ddots & \\ & & & G_{\frac{n}{m}-1}^{(0)} & I_m \end{pmatrix},
$$

$$
L_1 f_1^{(0)} = f_1, \quad \text{and}
$$

$$
L_i(G_{i-1}^{(0)}, f_i^{(0)}) = (R_{i-1}, f_i), \quad i = 2, 3, \ldots, \frac{n}{m}.
$$

(3.15)

From Theorem 3.1, we can show that solving the systems in (3.15) requires $T^{(0)} = \frac{1}{2}\log^2 m + \frac{3}{2}\log m$ parallel steps, using $p^{(0)} = \frac{21}{128}m^2 n + O(mn)$ processors. Now we form matrices $D^{(j)}$, $j = 0, 1, \ldots, \log\frac{n}{2m}$ such that if

$$
L^{(j+1)} = D^{(j)}L^{(j)} \quad \text{and} \quad f^{(j+1)} = D^{(j)}f^{(j)},
$$

then $L^{(\mu)} = I$ and $x = f^{(\mu)}$, where $\mu \equiv \log(n/m)$. Each matrix $L^{(j)}$ is of the form

$$
L^{(j)} = \begin{pmatrix} I_r & & & & \\ G_1^{(j)} & I_r & & & \\ & G_2^{(j)} & I_r & & \\ & & \ddots & \ddots & \\ & & & G_{\frac{n}{r}-1}^{(j)} & I_r \end{pmatrix},
$$

where $r = 2^j \cdot m$. Therefore, $D^{(j)} = \mathrm{diag}((L_1^{(j)})^{-1}, \ldots, (L_p^{(j)})^{-1})$ $(i = 1, 2, \ldots, \frac{n}{2r})$ in which

$$L_i^{(j)^{-1}} = \begin{pmatrix} I_r & \\ -G_{2i-1}^{(j)} & I_r \end{pmatrix}.$$

Hence, for the stage $j + 1$, we have

$$G_i^{(j+1)} = \begin{pmatrix} 0 & G_{2i}^{(j)} \\ 0 & -G_{2i+1}^{(j)} \cdot G_{2i}^{(j)} \end{pmatrix}, \quad i = 1, 2, \ldots, \frac{n}{2r} - 1,$$

and

$$f_i^{(j+1)} = \begin{pmatrix} f_{2i-1}^{(j)} \\ -G_{2i-1}^{(j)} \cdot f_{2i-1}^{(j)} + f_{2i}^{(j)} \end{pmatrix}, \quad i = 1, 2, \ldots, \frac{n}{2r}.$$

Observing that all except the last m columns of each matrix $G_i^{(j)}$ are zero, then $G_{2i+1}^{(j)} G_{2i}^{(j)}$ and $G_{2i-1}^{(j)} f_{2i-1}^{(j)}$ for all i, can be evaluated simultaneously in $1 + \log m$ parallel arithmetic operations using $p' = \frac{1}{2}m(m+1)n - rm^2$ processors. In one final subtraction, we evaluate $f_i^{(j+1)}$ and $G_i^{(j+1)}$, for all i, using $p'' = p'/m$ processors. Therefore, $T^{(j+1)} = 2 + \log m$ parallel steps using $p^{(j+1)} = \max\{p', p''\} = \frac{1}{2}m(m+1)n - rm^2$ processors. The total number of parallel steps is thus given by

$$\mathbf{T}_p = \sum_{j=0}^{\log \frac{n}{m}} T^{(j)} = (2 + \log m)\log n - (1/2)(\log m)(1 + \log m) \tag{3.16}$$

with $p = \max_j \{p^{(j)}\}$ processors. For $m = n/2$, $p \equiv p^{(0)}$, otherwise

$$p \equiv p^{(1)} = \frac{1}{2}m(m+1)n - m^3 \tag{3.17}$$

processors.

We call the corresponding scheme BBTS and list it as Algorithm 3.3.

Corollary 3.1 *Algorithm 3.3 (BBTS) as described in Theorem 3.2 requires* $\mathbf{O}_p = m^2 n \log(n/2m) + O(mn \log n)$ *arithmetic operations, resulting in an arithmetic redundancy of* $\mathbf{R}_p = O(m \log n)$ *over the sequential algorithm.*

3.2.3 Stability of Triangular System Solvers

It is well known that substitution algorithms for solving triangular systems, including CSWEEP, are backward stable: in particular, the computed solution $\tilde{x}$, satisfies a relation of the form $\|b - L\tilde{x}\|_\infty = \|L\|_\infty \|\tilde{x}\|_\infty \mathbf{u}$; cf. [6]. Not only that, but in many cases the forward error is much smaller than what is predicted by the usual upper bound involving the condition number (either based on normwise or componentwise analysis) and the backward error. For some classes of matrices, it is even possible to prove that the theoretical forward error bound does not depend on a condition number. This is the case, for example, for the lower triangular matrix arising from an

---

Algorithm 3.3 BBTS: Block banded triangular solver

Input: Banded lower triangular matrix L s.t. $n/m = 2^\nu$ and vector f
Output: Solution x of $Lx = f$
1: solve $L_1 f_1^{(0)} = f_1$;
2: **doall** $1 : (n/2^{j+1}) - 1$
3:    solve $L_i(G_i^{(0)}, f_i^{(0)}) = (R_i, f_i)$
4: **end**
5: **do** $k = 1 : \nu - 1,$

6:    $f_1^{(k)} = \begin{pmatrix} f_1^{(k-1)} \\ f_2^{(k-1)} - G_1^{(k-1)} f_1^{(k-1)} \end{pmatrix}$

7:    **doall** $j = 1 : 2^{\nu-k} - 1,$

8:      $G_j^{(k)} = \begin{pmatrix} 0 & G_{2j}^{(k-1)} \\ 0 & -G_{2j+1}^{(k-1)} G_{2j}^{(k-1)} \end{pmatrix}$

9:      $f_{j+1}^{(k)} = \begin{pmatrix} f_{2j+1}^{(k-1)} \\ f_{2j+2}^{(k-1)} - G_{2j+1}^{(k-1)} f_{2j+1}^{(k-1)} \end{pmatrix}$

10:    **end**
11: **end**

12: $f_1^{(\nu)} = \begin{pmatrix} f_1^{(\nu-1)} \\ f_2^{(\nu-1)} - G_1^{(\nu-1)} f_1^{(\nu-1)} \end{pmatrix}$

13: $x = f_1^{(\nu)}$

LU factorization with partial or complete pivoting strategies, and the upper triangular matrix resulting from the QR factorization with column pivoting [6]. This is also the case for some triangular matrices arising in the course of parallel factorization algorithms; see for example [7].

The upper bounds obtained for the error of the parallel triangular solvers are less satisfactory; cf. [5, 8]. This is due to the anticipated error accumulation in the (logarithmic length) stages consisting of matrix multiplications building the intermediate values in DTS. Bounds for the residual were first obtained in [5]. These were later improved in [8]. In particular, the residual corresponding to the computed solution, $\tilde{x}$, of DTS satisfies

$$\|b - L\tilde{x}\|_\infty \tilde{d}_n \|(|L||L^{-1}|)^2|L||x|\|_\infty \mathbf{u} + O(\mathbf{u}^2), \tag{3.18}$$

and the forward error

$$|x - \tilde{x}| \leq d_n [M(L)]^{-1} |b| \mathbf{u} + O(\mathbf{u}^2),$$

where $M(L)$ is the matrix with values $|\lambda_{i,i}|$ on the diagonal, and $-|\lambda_{i,j}|$ in the off-diagonal positions, with d_n and $\tilde{d}_n$ constants of the order $n \log n$. When L is an M-matrix and $b \geq 0$, DTS can be shown to be componentwise backward stable to first order; cf. [8].

3.2.4 Toeplitz Triangular Systems

Toeplitz triangular systems, in which $\lambda_{ij} = \tilde{\lambda}_{i-j}$, for $i > j$, arise frequently in practice. The algorithms presented in the previous two sections do not take advantage of this special structure of L. Efficient schemes for solving Toeplitz triangular systems require essentially the same number of parallel arithmetic operations as in the general case, but need fewer processors, $O(n^2)$ rather than $O(n^3)$ processors for dense systems, and $O(mn)$ rather than $O(m^2 n)$ processors for banded systems. The solution of more general banded Toeplitz systems is discussed in Sect. 6.2 of Chap. 6.

To pave the way for a concise presentation of the algorithms for Toeplitz systems, we present the following fundamental lemma.

Lemma 3.1 ([9]) *If L is Toeplitz, then L^{-1} is also Toeplitz, where*

$$L = \begin{pmatrix} 1 & & & & \\ \lambda_1 & 1 & & & \\ \lambda_2 & \lambda_1 & 1 & & \\ \vdots & \ddots & \ddots & \ddots & \\ \lambda_{n-1} & \cdots & \lambda_2 & \lambda_1 & 1 \end{pmatrix}.$$

Proof The matrix L may be written as

$$L = I + \lambda_1 J + \lambda_2 J^2 + \ldots + \lambda_{n-1} J^{n-1}, \tag{3.19}$$

where we recall that in our notation in this book, J is the matrix

$$J = \begin{pmatrix} 0 & & & \\ 1 & \ddots & & \\ 0 & \ddots & \ddots & \\ & \ddots & \ddots & \ddots \\ & & 0 & 1 & 0 \end{pmatrix}$$

Observing that $J^2 = (e_3, e_4, \ldots, e_n, 0, 0)$, where e_i is the ith column of the identity, then $J^3 = (e_4, \ldots, e_n, 0, 0, 0), \ldots, J^{n-1} = (e_n, 0, \ldots, 0)$, and $J^n \equiv 0_n$, then from (3.19) we see that $JL = LJ$. Therefore, solving for the ith column of L^{-1}, i.e., solving the system $Lx_i = e_i$, we have $JLx_i = Je_i = e_{i+1}$, or

$$L(Jx_i) = e_{i+1},$$

where we have used the fact that L and J commute. Therefore, $x_{i+1} = Jx_i$, and L^{-1} can be written as

$$L^{-1} = (x_1, Jx_1, J^2x_1, \ldots, J^{n-1}x_1),$$

where $Lx_1 = e_1$. Now, if $x_1 = (\xi_1, \xi_2, \ldots, \xi_n)^\top$, then

$$Jx_1 = \begin{pmatrix} 0 \\ \xi_1 \\ \vdots \\ \\ \vdots \\ \xi_{n-1} \end{pmatrix}, \quad J^2x_1 = \begin{pmatrix} 0 \\ 0 \\ \xi_1 \\ \vdots \\ \\ \vdots \\ \xi_{n-2} \end{pmatrix}, \quad J^3x_1 = \begin{pmatrix} 0 \\ 0 \\ 0 \\ \xi_1 \\ \vdots \\ \\ \vdots \\ \xi_{n-3} \end{pmatrix}, \ldots, \quad \text{etc.,}$$

establishing that L^{-1} is a Toeplitz matrix.

Theorem 3.3 *Let L be a dense Toeplitz unit lower triangular matrix of order n. Then the system $Lx = f$ can be solved in $\mathbf{T}_p = \log^2 n + 2\log n - 1$ parallel steps, using no more than $p = n^2/4$ processors.*

Proof From Lemma 3.1, we see that the first column of L^{-1} determines L^{-1} uniquely. Using this observation, consider a leading principal submatrix of the Toeplitz matrix L,

$$\begin{pmatrix} L_1 & 0 \\ G_1 & L_1 \end{pmatrix},$$

where L_1 (Toeplitz), G_1 are of order $q = 2^j$, and we assume that $L_1^{-1}e_1$ is known. The first column of the inverse of this principal submatrix is given by

$$\begin{pmatrix} L_1^{-1}e_1 \\ -L_1^{-1}G_1L_1^{-1}e_1 \end{pmatrix}$$

and can be computed in $2(1 + \log q) = 2(j + 1)$ parallel steps using q^2 processors. Note that the only computation involved is that of obtaining $-L_1^{-1}G_1L_1^{-1}e_1$, where $L_1^{-1}e_1$ is given. Starting with a leading submatrix of order 4, i.e., for $j = 1$, we have

$$L_1^{-1} = \begin{pmatrix} 1 & 0 \\ -\lambda_1 & 1 \end{pmatrix},$$

and doubling the size every stage, we obtain the inverse of the leading principal submatrix M of order $n/2$ in

$$\sum_{j=1}^{\log \frac{n}{4}} 2(j+1) = \log^2 n - \log n - 2$$

parallel steps with $(n^2/16)$ processors. Thus, the solution of a Toeplitz system $Lx = f$, or

$$\begin{pmatrix} M & 0 \\ N & M \end{pmatrix} \begin{pmatrix} x_1 \\ x_2 \end{pmatrix} = \begin{pmatrix} f_1 \\ f_2 \end{pmatrix},$$

is given by

$$x_1 = M^{-1} f_1,$$

and

$$x_2 = M^{-1} f_2 - M^{-1} N M^{-1} f_1.$$

Since we have already obtained M^{-1} (actually the first column of the Toeplitz matrix M^{-1}), x_1 and x_2 can be computed in $(1 + 3 \log n)$ parallel arithmetic operations using no more than $n^2/4$ processors. Hence, the total parallel arithmetic operations for solving a Toeplitz system of linear equations is $(\log^2 n + 2 \log n - 1)$ using $n^2/4$ processors.

Let now $n = 2^v$, $Le_1 = (1, \lambda_1, \lambda_2, ..., \lambda_{n-1})^\top$, and let G_k be a square Toeplitz of order 2^k with its first and last columns given by

$$G_k e_1 = (\lambda_{2^k}, \lambda_{2^k+1}, \dots, \lambda_{2^{k+1}-1})^\top,$$

and

$$G_k e_{2^k} = (\lambda_1, \lambda_2, \dots, \lambda_{2^k})^\top.$$

Based on this discussion, we construct the triangular Toeplitz solver TTS that is listed as Algorithm 3.4.

Algorithm 3.4 TTS: Triangular Toeplitz solver

Input: $n = 2^v$; $L \in \mathbb{R}^{n \times n}$ dense Toeplitz unit lower triangular with $Le_1 = [1, \lambda_1, \cdots, \lambda_{n-1}]$ and right-hand side $f \in \mathbb{R}^n$.

Output: Solution x of $Lx = f$

1: $g^{(1)} = L_1^{-1} e_1 = \begin{pmatrix} 1 \\ -\lambda_1 \end{pmatrix}$;

2: **do** $k = 1 : v - 2$,

3: $\quad g^{(k+1)} = L_{k+1}^{-1} e_1 = \begin{pmatrix} g^{(k)} \\ -L_k^{-1} G_k g^{(k)} \end{pmatrix}$;

4: **end**

$\quad$ // $g^{(v-1)}$ determine uniquely L_{v-1}^{-1}

5: $x = \begin{pmatrix} L_{v-1}^{-1} f_1 \\ L_{v-1}^{-1} f_2 - L_{v-1}^{-1} G_{v-1} L_{v-1}^{-1} f_1 \end{pmatrix}$

If L is banded, Algorithm 3.4 is slightly altered as illustrated in the proof of the following theorem.

Theorem 3.4 *Let L be a Toeplitz banded unit lower triangular matrix of order n and bandwidth $(m + 1) \geq 3$. Then the system $Lx = f$ can be solved in less than $(3 + 2 \log m) \log n$ parallel steps using no more than $3mn/4$ processors.*

Proof Let L and f be partitioned as

$$
L = \begin{pmatrix} L_0 & & & & \\ R_0 & L_0 & & & \\ & R_0 & L_0 & & \\ & & \ddots & \ddots & \\ & & & R_0 & L_0 \end{pmatrix}, f = \begin{pmatrix} f_1^{(0)} \\ f_2^{(0)} \\ \vdots \\ f_{n/m}^{(0)} \end{pmatrix}. \tag{3.20}
$$

At the jth stage of this algorithm, we consider the leading principal submatrix L_j of order $2r = m2^j$,

$$
L_j = \begin{pmatrix} L_{j-1} & 0 \\ R_{j-1} & L_{j-1} \end{pmatrix},
$$

where L_{j-1} is Toeplitz, and

$$
R_{j-1} = \begin{pmatrix} 0 & R_0 \\ 0 & 0 \end{pmatrix}.
$$

The corresponding $2r$ components of the right-hand side are given by

$$
f_i^{(j)} = \begin{pmatrix} f_{2i-1}^{(j-1)} \\ f_{2i}^{(j-1)} \end{pmatrix},
$$

where $f^\top = (f_1^{(j)^T}, f_2^{(j)^T}, \ldots, f_{(n/2r)}^{(j)T})$. Now, obtain the column vectors

$$
x_i^{(j)} = L_j^{-1} f_i^{(j)}, \quad i = 1, 2, \ldots, n/2r,
$$

or

$$
\begin{pmatrix} x_{2i-1}^{(j-1)} \\ y_{2i}^{(j-1)} \end{pmatrix} = \begin{pmatrix} L_{j-1}^{-1} f_{2i-1}^{(j-1)} \\ L_{j-1}^{-1} f_{2i}^{(j-1)} - L_{j-1}^{-1} R_{j-1} L_{j-1}^{-1} f_{2i-1}^{(j-1)} \end{pmatrix}. \tag{3.21}
$$

Note that the first $2r$ components of the solution vector x are given by $x_1^{(j)}$. Assuming that we have already obtained $L_{j-1}^{-1} f_{2i-1}^{(j-1)}$ and $L_{j-1}^{-1} f_{2i}^{(j-1)}$ from stage $(j - 1)$, then (3.21) may be computed as shown below in Fig. 3.3, in $T' = (3 + 2 \log m)$ parallel steps, using $p' = \left(mr - \frac{1}{2}m^2 - \frac{1}{2}m\right)(n/2r)$ processors. This is possible only if we have L_{j-1}^{-1} explicitly, i.e., the first column of L_{j-1}^{-1}.

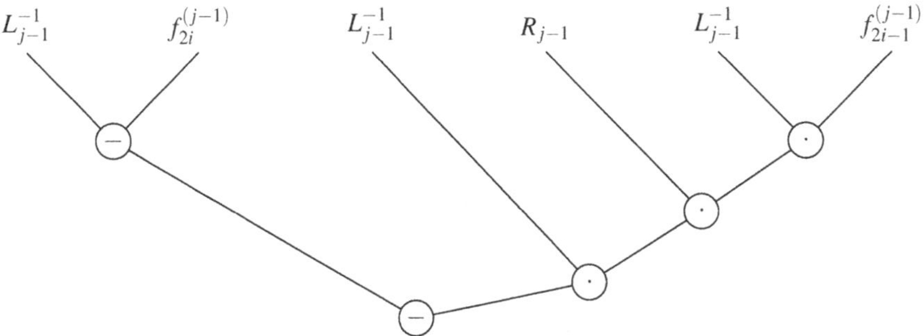

Fig. 3.3 Computation of the terms of the vector (3.21)

Assuming that the first column of L_{j-1}^{-1}, i.e. $L_{j-1}^{-1}e_1$, is available from stage $j-1$, then we can compute the first column of L_j^{-1} simultaneously with (3.21). Since the first column of L_j^{-1} can be written as

$$\begin{pmatrix} L_{j-1}^{-1}e_1 \\ -L_{j-1}^{-1}R_{j-1}L_{j-1}^{-1}e_1 \end{pmatrix},$$

we see that it requires $T'' = (2+2\log m)$ parallel steps using $p'' = mr - \frac{1}{2}m^2 - \frac{1}{2}m$ processors. Hence, stage j requires $T^{(j)} = \max\{T', T''\} = 3 + 2\log m$, parallel steps using $p_j = p' + p''$ processors. Starting with stage 0, we compute $L_0^{-1}f_i^{(0)}$, $i = 1, 2, \ldots, n/m$, and $L_0^{-1}e_1$ in $T^{(0)} = \log^2 m + 2\log m - 1$ parallel steps using $m(n+m)/4$ processors [5]. After stage v, where $v = \log(n/2m)$, the last $n/2$ components of the solution are obtained by solving the system

$$\begin{pmatrix} L_v & 0 \\ R_v & L_v \end{pmatrix}\begin{pmatrix} x_1^{(v)} \\ x_2^{(v)} \end{pmatrix} = \begin{pmatrix} f_1^{(v)} \\ f_2^{(v)} \end{pmatrix}.$$

Again, the solution of this system is obtained in $T^{(v+1)} = (3 + 2\log_2 m)$ parallel steps using $m(n - m - 1)/2$ processors. Consequently, the total number of parallel steps required for this process is given by

$$\mathbf{T}_p = \sum_{k=0}^{\log(n/m)} T^{(k)} = (3 + 2\log m)\log n - (\log^2 m + \log m + 1).$$

The maximum number of processors p used in any given stage, therefore, does not exceed $3mn/4$. For $n \gg m$ the number of parallel steps required is less than twice that of the general-purpose banded solver while the number of processors used is reduced roughly by a factor of $2m/3$.

3.3 Implementations for a Given Number of Processors

In Sects. 3.2.1–3.2.4, we presented algorithms that require the least known number of parallel steps, usually at the expense of a rather large number of processors, especially for dense triangular systems. Throughout this section, we present alternative schemes that achieve the least known number of parallel steps for a given number of processors.

First, we consider banded systems of order n and bandwidth $(m+1)$, i.e., $R\langle n, m\rangle$, and assume that the number of available processors p satisfies the inequality, $m < p \ll n$. Clearly, if $p = m$ we can use the column-sweep algorithm to solve the triangular system in $2(n-1)$ steps. Our main goal here is to develop a more suitable algorithm for $p > m$ that requires $O(m^2 n/p)$ parallel steps given only $p > m$ processors.

Algorithm 3.5 BTS: Banded Toeplitz triangular solver

Input: Banded unit Toeplitz lower triangular matrix $L \in \mathbb{R}^{n \times n}$, integer m s.t. $n/m = 2^\nu$ and
    vector $f \in \mathbb{R}^n$ as defined by (3.20)
Output: Solution x of $Lx = f$
1: $g^{(0)} = L_0^{-1} e_1$ (using Algorithm 3.4) ;
2: **doall** $i = 1 : n/m$
3:    $x_i^{(0)} = L_0^{-1} f_i^{(0)}$;
4: **end**
5: **do** $k = 1 : \nu$,
6:    $g^{(k)} = \begin{pmatrix} g^{(k-1)} \\ -L_{k-1}^{-1} R_{k-1} g^{(k-1)} \end{pmatrix}$;
     // $g^{(k)}$ determines uniquely L_k^{-1}
7:    **doall** $i = 1 : 2^{\nu-k}$,
8:      $x_i^{(k)} = \begin{pmatrix} x_{2i-1}^{(k-1)} \\ x_{2i}^{(k-1)} - L_{k-1}^{-1} R_{k-1} x_{2i-1}^{(k-1)} \end{pmatrix}$
9:    **end**
10: **end**
    //$x_1^{(\nu)}$ is the solution of $Lx = f$, and $g^{(\nu)}$ is the first column of L^{-1}.

In the following, we present two algorithms; one for obtaining all the elements of the solution of $Lx = f$, and one for obtaining only the last m components of x.

Theorem 3.5 *Given p processors $m < p \ll n$, a unit lower triangular system of n equations with bandwidth $(m+1)$ can be solved in*

$$\mathbf{T}_p = 2(m-1) + \tau \frac{n-m}{p(p+m-1)} \tag{3.22}$$

parallel steps, where

$$\tau = \max \begin{cases} (2m^2 + 3m)p - (m/2)(2m^2 + 3m + 5) \\ 2m(m+1)p - 2m \end{cases} \tag{3.23}$$

Proof To motivate the approach, consider the following underdetermined system, $\hat{L}\hat{z} = g$,

$$\begin{pmatrix} R_0\ L_1 & & & & \\ & R_1\ L_2 & & & \\ & & R_2\ L_3 & & \\ & & & \ddots\ \ddots & \\ & & & & R_{p-1}\ L_p \end{pmatrix} \begin{pmatrix} z_0 \\ z_1 \\ z_2 \\ z_3 \\ \vdots \\ z_p \end{pmatrix} = \begin{pmatrix} g_1 \\ g_2 \\ g_3 \\ \vdots \\ g_p \end{pmatrix}, \tag{3.24}$$

of $s = q + p(p - 1)$ equations with $q = mp$, where z_0 is given. Each L_i is unit lower triangular of bandwidth $m + 1$, L_1 is of order q, $L_i\ (i > 1)$ is of order p, and $R_i\ (0 \leq i \leq p - 1)$ is upper triangular containing nonzero elements only on its top m superdiagonals.

System (3.24) can be expressed as

$$\begin{pmatrix} I_{mp} & & & & \\ G_1\ I_p & & & & \\ & G_2\ I_p & & & \\ & & \ddots\ \ddots & & \\ & & & G_{p-1}\ I_p \end{pmatrix} \begin{pmatrix} z_1 \\ z_2 \\ z_3 \\ \vdots \\ z_p \end{pmatrix} = \begin{pmatrix} h_1 \\ h_2 \\ h_3 \\ \vdots \\ h_p \end{pmatrix}, \tag{3.25}$$

where h_i and G_i are obtained by solving the systems

$$L_1 h_1 = g_1 - R_0 z_0, \tag{3.26}$$

$$L_i(G_{i-1}, h_i) = (R_{i-1}, g_i), \quad i = 2, 3, \ldots, p. \tag{3.27}$$

Using one processor, the systems in (3.26) can be solved in

$$\tau_1 = 2m^2 p$$

parallel steps while each of the $(p - 1)$ systems in (3.27) can be solved sequentially in

$$\tau_2 = (2m^2 + m)p - \frac{m}{2}(2m^2 + 3m + 1),$$

parallel steps. Now the reason for choosing $q = mp$ is clear; we need to make the difference in parallel steps between solving (3.26) and any of the $(p - 1)$ systems in (3.27) as small as practically possible. This will minimize the number of parallel steps during which some of the processors remain idle. In fact, $\tau_1 = \tau_2$ for $p = (2m^2 + 3m + 1)/2$. Assigning one processor to each of the p systems in (3.27), they can be solved simultaneously in $\tau_3 = \max\{\tau_1, \tau_2\}$ parallel steps. From (3.25), the solution vector z is given by

$$z_1 = h_1,$$
$$z_i = h_i - G_{i-1}z_{i-1}, \quad i = 2, 3, \ldots, p.$$

Observing that only the last m columns of each G_i are different from zero and using the available p processors, each z_i can be computed in $2m$ parallel steps. Thus, $z_2, z_3, \ldots, z_p$ are obtained in $\tau_4 = 2m(p-1)$ parallel steps, and the system $L\hat{z} = g$ in (3.24) is solved in

$$\tau = \tau_3 + \tau_4$$
$$= \max \begin{cases} (2m^2 + 3m)p - (m/2)(2m^2 + 3m + 5) \\ 2m(m+1)p - 2m \end{cases} \tag{3.28}$$

parallel steps. Now, partitioning the unit lower triangular system $Lx = f$ of n equations and bandwidth $m + 1$, in the form

$$\begin{pmatrix} V_0 & & & & \\ U_1 & V_1 & & & \\ & U_2 & V_2 & & \\ & & \ddots & \ddots & \\ & & & U_k & V_k \end{pmatrix} \begin{pmatrix} x_0 \\ x_1 \\ x_2 \\ \vdots \\ x_k \end{pmatrix} = \begin{pmatrix} f_0 \\ f_1 \\ f_2 \\ \vdots \\ f_k \end{pmatrix},$$

where V_0 is of order m, V_i, $i > 0$, (with the possible exception of V_k) is of order s, i.e., $k = (n-m)/s$, and each U_i is of the same form as R_i. The solution vector x is then given by

$$V_0 x_0 = f_0, \tag{3.29}$$

$$V_i x_i = (f_i - U_i x_{i-1}), \quad i = 1, 2, 3, \ldots, k. \tag{3.30}$$

Solving (3.29) by the column-sweep method in $2(m-1)$ parallel steps (note that $p > m$), the k systems in (3.30) can then be solved one at a time using the algorithm developed for solving (3.24). Consequently, using p processors, $Lx = f$ is solved in

$$\mathbf{T}_p = 2(m-1) + \tau \frac{n-m}{p(p+m-1)}$$

parallel steps in which τ is given by (3.28).

Example 3.5 Let $n = 1024$, $m = 2$, and $p = 16$. From (3.28), we see that $\tau = 205$ parallel steps. Thus, $\mathbf{T}_p = 822$ which is less than that required by the column-sweep method, i.e., $2(n-1) = 2046$.

In general, this algorithm is guaranteed to require fewer parallel steps than the column-sweep method for $p \geq 2m^2$.

Corollary 3.2 *The algorithm in Theorem 3.5 requires* $\mathbf{O}_p = O(m^2 n)$ *arithmetic operations, resulting in an arithmetic redundancy* $\mathbf{R}_p = O(m)$ *over the sequential algorithm.*

In some applications, one is interested only in obtaining the last component of the solution vector, e.g. polynomial evaluation via Horner's rule. The algorithm in Theorem 3.5 can be modified in this case to achieve some savings. In the following theorem, we consider the special case of first-order linear recurrences.

Theorem 3.6 *Consider the unit lower triangular system $Lx = f$ of order n and bandwidth 2. Given p processors, $1 < p \ll n$, we can obtain ξ_n, the last component of x in*

$$\mathbf{T}_p = 3(q - 1) + 2 \log p$$

parallel steps, where $q = (2n + 1)/(2p + 1)$.

Proof Partition the system $Lx = f$ in the form

$$
\begin{pmatrix}
L_1 & & & & \\
R_1 & L_2 & & & \\
& R_2 & L_3 & & \\
& & \ddots & \ddots & \\
& & & R_{p-1} & L_p
\end{pmatrix}
\begin{pmatrix}
x_1 \\ x_2 \\ x_3 \\ \vdots \\ x_p
\end{pmatrix}
=
\begin{pmatrix}
f_1 \\ f_2 \\ f_3 \\ \vdots \\ f_p
\end{pmatrix},
\tag{3.31}
$$

where L_1 is of order r and L_i, $i > 1$, of order q. Now, we would like to choose r and q, so that the number of parallel steps for solving

$$L_1 x_1 = f_1, \tag{3.32}$$

is roughly equal to that needed for solving

$$L_i(G_{i-1}, h_i) = (R_{i-1}, f_i). \tag{3.33}$$

This can be achieved by choosing $q = (2n + 1)/(2p + 1)$, and $r = n - (p - 1)q$. Therefore, the number of parallel steps required for solving the systems (3.32) and (3.33), or reducing (3.31) to the form of (3.25) is given by

$$\sigma' = \max\{2(r - 1),\ 3(q - 1)\} = 3(q - 1).$$

Splitting the unit lower triangular system of order p whose elements are encircled in Fig. 3.4, the p elements $\xi_r, \xi_{r+q}, \xi_{r+2q}, \ldots, \xi_n$ can be obtained using the algorithm of Theorem 3.2 in $2 \log p$ parallel steps using $(p - 1)$ processors. Hence, we can obtain ξ_n in

$$\mathbf{T}_p = \sigma + 2 \log p.$$

Similarly, we consider solving Toeplitz triangular systems using a limited number of processors. The algorithms outlined in Theorems 3.5 and 3.6 can be modified to take advantage of the Toeplitz structure. Such modifications are left as exercises. We

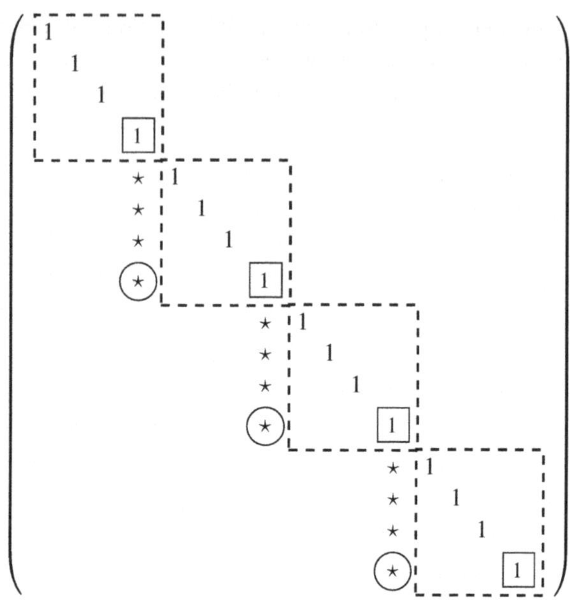

Fig. 3.4 Unit lower triangular matrix in triangular system (3.25) with $n = 16$, $p = 4$ and $m = 1$

will, however, consider an interesting special case; a parallel Horner's rule for the evaluation of polynomials.

Theorem 3.7 *Given p processors, $1 < p \ll n$, we can evaluate a polynomial of degree n in $\mathbf{T}_p = 2(k-1) + 2\log(p-1)$ parallel steps where $k = (n+1)/(p-1)$.*

Proof Consider the evaluation of the polynomial

$$P_n(\xi) = \alpha_1 \xi^n + \alpha_2 \xi^{n-1} + \ldots + \alpha_n \xi + \alpha_{n+1}$$

at $\xi = \theta$. On a uniprocessor this can be achieved via Horner's rule

$$\begin{aligned}
\beta_1 &= \alpha_1, \\
\beta_{i+1} &= \theta\beta_i + \alpha_{i+1} \quad i = 1, 2, \ldots, n,
\end{aligned}$$

where $\beta_{n+1} = P_n(\theta)$. This amounts to obtaining the last component of the solution vector of the triangular system

$$\begin{pmatrix} 1 & & & & \\ -\theta & 1 & & & \\ & -\theta & 1 & & \\ & & \ddots & \ddots & \\ & & & -\theta & 1 \end{pmatrix} \begin{pmatrix} \beta_1 \\ \beta_2 \\ \beta_3 \\ \vdots \\ \beta_{n+1} \end{pmatrix} = \begin{pmatrix} \alpha_1 \\ \alpha_2 \\ \alpha_3 \\ \vdots \\ \alpha_{n+1} \end{pmatrix}. \tag{3.34}$$

Again, partitioning (3.34) as

$$
\begin{pmatrix}
\hat{L} & & & \\
\hat{R} & L & & \\
& R & \ddots & \ddots \\
& & R & L
\end{pmatrix}
\begin{pmatrix}
\beta_1 \\ \beta_2 \\ \beta_3 \\ \vdots \\ \beta_{p-1}
\end{pmatrix}
=
\begin{pmatrix}
\alpha_1 \\ \alpha_2 \\ \alpha_3 \\ \vdots \\ \alpha_{p-1}
\end{pmatrix},
$$

where L is of order $k = (n+1)/(p-1)$, $\hat{L}$ is of order $j = (n+1) - (p-2)k < k$, and $\hat{R}e_j = Re_k = -\theta e_1$. Premultiplying both sides by the block-diagonal matrix, $\mathrm{diag}(\hat{L}^{-1}, \ldots, \hat{L}^{-1})$ we get the system

$$
\begin{pmatrix}
I_j & & & \\
\hat{G} & I_k & & \\
& G & & \\
& & \ddots & \ddots \\
& & & G & I_k
\end{pmatrix}
\begin{pmatrix}
b_1 \\ b_2 \\ b_3 \\ \vdots \\ b_{p-1}
\end{pmatrix}
=
\begin{pmatrix}
h_1 \\ h_2 \\ h_3 \\ \vdots \\ h_{p-1}
\end{pmatrix},
\tag{3.35}
$$

where,

$$
\begin{aligned}
\hat{L}h_1 &= a_1, \\
Lh_i &= a_i, \quad i = 2, 3, \ldots, p-1,
\end{aligned}
\tag{3.36}
$$

and

$$
L(\hat{G}e_j) = L(Ge_k) = -\theta e_1.
\tag{3.37}
$$

Assigning one processor to each of the bidiagonal systems (3.36) and (3.37), we obtain h_i and Ge_k in $2(k-1)$ parallel steps. Since L is Toeplitz, one can easily show that

$$
g = \hat{G}e_j = Ge_k = (-\theta, -\theta^2, \ldots, -\theta^k)^\top.
$$

In a manner similar to the algorithm of Theorem 3.6 we split from (3.35) a smaller linear system of order $(p-1)$

$$
\begin{pmatrix}
1 & & & \\
-\theta^k & 1 & & \\
& -\theta^k & 1 & \\
& & \ddots & \ddots \\
& & & -\theta^k & 1
\end{pmatrix}
\begin{pmatrix}
b_j \\ b_{j+k} \\ b_{j+2k} \\ \vdots \\ b_{n+1}
\end{pmatrix}
=
\begin{pmatrix}
e_j^\top h_1 \\ e_k^\top h_2 \\ e_k^\top h_3 \\ \vdots \\ e_k^\top h_{p-1}
\end{pmatrix}.
\tag{3.38}
$$

From Theorem 3.2 we see that using $(p-2)$ processors, we obtain b_{n+1} in $2\log(p-1)$ parallel steps. Thus the total number of parallel steps for evaluating a polynomial of degree n using $p \ll n$ processors is given by

$$
\mathbf{T}_p = 2(k-1) + 2\log(p-1).
$$

Table 3.1 Summary of bounds for parallel steps, efficiency, redundancy and processor count in algorithms for solving linear recurrences

L		$\mathbf{T}_p$	p	$\mathbf{E}_p$ (proportional to)	$\mathbf{R}_p$
Dense Triangular	$R<n>$	$\frac{1}{2}\log^2 n + \frac{3}{2}\log n$	$15/1024n^3+O(n^2)$	$1/(n\log^2 n)$	$O(n)$
	$\hat{R}<n>$	$\log^2 n + 2\log n - 1$	$n^2/4$	$1/\log^2 n$	$O(1)(=5/4)$
Banded Triangular	$R<n,m>$	$(2+\log m)\log n + O(\log^2 m)$	$\frac{1}{2}m(m+1)n - m^3$	$1/(m\log m \log n)$	$O(m\log n)$
	$\hat{R}<n,m>$	$(3+2\log m)\log n + O(log^2 m)$	$3mn/4$	$1/(\log m \log n)$	$\log n$

Here n is the order of the unit triangular system $Lx = f$ and $m + 1$ is the bandwidth of L

Example 3.6 Let $n = 1024$ and $p = 16$. Hence, $k = 69$ and $\mathbf{T}_p = 144$. This is roughly $(1/14)$ of the number of arithmetic operations required by the sequential scheme, which is very close to $(1/p)$.

The various results presented in this chapter so far are summarized in Tables 3.1 and 3.2. In both tables, we deal with the problem of solving a unit lower triangular system of equations $Lx = f$. Table 3.1 shows upper bounds for the number of parallel steps, processors, and the corresponding arithmetic redundancy for the unlimited number of processors case. In this table, we observe that when the algorithms take advantage of the special Toeplitz structure to reduce the number of processors and arithmetic redundancy, the number of parallel steps is increased. Table 3.2 gives upper bounds on the parallel steps and the redundancy for solving banded unit lower triangular systems given only a limited number of processors, p, $m < p \ll n$. It is also of interest to note that in the general case, it hardly makes a difference in the number of parallel steps whether we are seeking all or only the last m components of the solution. In the Toeplitz case, on the other hand, the parallel steps are cut in half if we seek only the last m components of the solution.

3.4 Nonlinear Recurrences

We have already seen in Sects. 3.1 and 3.2 that parallel algorithms for evaluating linear recurrence relations can achieve significant reduction in the number of required parallel steps compared to sequential schemes. For example, from Table 3.1 we see that the speedup for evaluating short recurrences (small m) as a function of the problem size behaves like $\mathbf{S}_p = O(n/\log n)$. Therefore, if we let the problem size vary, the speedup is unbounded with n. Even discounting the fact that efficiency goes to 0, the result is quite impressive for recurrence computations that appear sequential at first sight. As we see next, and has been known for a long time, speedups are far more restricted for nonlinear recurrences.

$$\xi_{k+r} = f(k, \xi_k, \xi_{k+1}, \ldots, \xi_{k+r-1}). \tag{3.39}$$

Table 3.2 Summary of linear recurrence bounds using a limited number, p, of processors, where $m < p \ll n$ and m and n are as defined in Table 3.1

Problem			$\mathbf{T}_p$	$\mathbf{E}_p$	$\mathbf{R}_p$
Solve the banded unit lower triangular system $Lx = f$	General case	Solve for all or last m components of x	$\frac{2m^2n}{p+m-1/2} + O(\frac{mn}{p})$	proportional to $1/m$	$O(m)$
		($m = 1$) Solve for last component of x	$\frac{3n}{p} + O(\log p)$	$\simeq 2/3$	$\simeq 2$
	Toeplitz case	Solve for all components of x	$\frac{4mn}{p} + O(mp)$	$\simeq 1/2$	$O(m)$
		Solve for last m components of x	$\frac{2mn}{p-m} + O(m^2 \log p)$	$\simeq 1 - \frac{m}{p}$	$\simeq 1 + \frac{1}{p}$
		$m = 1$ (e.g., evaluation of a polynomial of degree $n - 1$)	$\frac{2n}{p-1} + O(\log p)$	$\simeq 1 - \frac{1}{p}$	$\simeq 1 + \frac{1}{p}$

The mathematics literature dealing with the above recurrence is extensive, see for example [10, 11].

The most obvious method for speeding up the evaluation of (3.39) is through linearization, if possible, and then using the algorithms of Sects. 3.2 and 3.3.

Example 3.7 ([10]) Consider the first order rational recurrence of degree one

$$\xi_{k+1} = \frac{\alpha_k \xi_k + \beta_k}{\gamma_k + \xi_k} \quad k \geq 0. \tag{3.40}$$

Using the simple change of variables

$$\xi_k = \frac{\omega_{k+1}}{\omega_k} - \gamma_k, \tag{3.41}$$

we obtain

$$\xi_{k+1}(\gamma_k + \xi_k) = \frac{1}{\omega_k}(\omega_{k+2} - \gamma_{k+1}\omega_{k+1}),$$

and

$$\alpha_k \xi_k + \beta_k = \frac{1}{\omega_k}(\alpha_k \omega_{k+1} + (\beta_k - \alpha_k \gamma_k)\omega_k).$$

Therefore, the nonlinear recurrence (3.40) is reduced to the linear one

$$\omega_{k+2} + \delta_{k+1}\omega_{k+1} + \zeta_k \omega_k = 0, \tag{3.42}$$

where

$$\delta_{k+1} = -(\alpha_k + \gamma_{k+1}), \quad \text{and}$$
$$\zeta_k = \alpha_k \gamma_k - \beta_k.$$

If the initial condition of (3.40) is $\xi_0 = \tau$, where $\tau \neq -\gamma_0$, the corresponding initial conditions of (3.42) are $\omega_0 = 1$ and $\omega_1 = \gamma_0 + \tau$. It is interesting to note that the number of parallel steps required to evaluate ξ_i, $1 \leq i \leq n$, via (3.40) is $O(n)$ whether we use one or more processors. One, however, can obtain $\xi_1, \xi_2, \ldots, \xi_n$ via (3.42) in $O(\log n)$ parallel steps using $O(n)$ processors as we shall illustrate in the following:

  (i) ζ_k, δ_{k+1} ($0 \leq k \leq n - 1$), and ω_1 can be obtained in 2 parallel steps using $2n + 1$ processors,
 (ii) the triangular system of $(n+2)$ equations, corresponding to (3.42) can be solved in $2\log(n + 2)$ parallel steps using no more than $2n - 4$ processors,
(iii) ξ_i, $1 \leq i \leq n$, is obtained from (3.41) in 2 parallel steps employing $2n$ processors.

Hence, given $(2n + 1)$ processors, we evaluate ξ_i, $i = 1, 2, \ldots, n$, in $4 + 2\log(n + 2)$ parallel steps.

If $\gamma_k = 0$, (3.40) is a continued fraction iteration. The number of parallel steps to evaluate $\xi_1, \xi_2, \ldots, \xi_n$ via (3.43) in this case reduces to $1 + 2\log(n + 2)$ assuming that $(2n - 4)$ processors are available.

In general, linearization of nonlinear recurrences should be approached with caution as it may result in an ill-conditioned linear recurrence, or even over- or underflow in the underlying computations. Further, it may not lead to significant reduction in the number of parallel steps. The following theorem and corollary from [12] show that not much can be expected if one allows only algebraic transformations.

Theorem 3.8 ([12]) *Let $\xi_{i+1} = f(\xi_i)$ be a rational recurrence of order 1 and degree $d > 1$. Then the speedup for the parallel evaluation of the final term, ξ_n, of the recurrence on p processors is bounded as follows*

$$S_p \leq \frac{\mathbf{O}_1(f(\xi))}{\log d}.$$

for all n and p, where $\mathbf{O}_1(f(\xi))$ is the number of arithmetic operations required to evaluate $f(\xi)$ on a uniprocessor.

As an example, consider Newton's iteration for approximating the square root of a real number α

$$\xi_{k+1} = (\xi_k^2 + \alpha)/2\xi_k. \tag{3.43}$$

This is a first-order rational recurrence of degree 2. No algebraic change of variables similar to that of Example 3.7 can reduce (3.43) to a linear recurrence. Here, $f(\xi) = \frac{1}{2}\left(\xi + \frac{\alpha}{\xi}\right)$, $\mathbf{T}_1(f) = 3$, and $\log d = 1$, thus yielding 3 as an upper bound for $\mathbf{S}_p$.

Corollary 3.3 ([12]) *By using parallelism, the evaluation of an expression defined by any nonlinear polynomial recurrence, such as $\xi_{i+1} = 2\xi_i^2\xi_{i-1} + \xi_{i-2}$, can be speeded up by at most a constant factor.*

If we are not restricted to algebraic transformations of the variables one can, in certain instances, achieve higher reduction in parallel steps.

Example 3.8 Consider the rational recurrence of order 2 and degree 2

$$\xi_{k+2} = \frac{\alpha(\xi_k + \xi_{k+1})}{(\xi_k\xi_{k+1} - \alpha)}.$$

This is equivalent to

$$\xi_k\xi_{k+1}\xi_{k+2} = \alpha(\xi_k + \xi_{k+1} + \xi_{k+2}).$$

Setting $\xi_k = \sqrt{\alpha}\tan\psi_k$, and using the trigonometric identity

$$\tan(\psi_k + \psi_{k+1}) = \frac{\tan\psi_k + \tan\psi_{k+1}}{1 - \tan\psi_k\tan\psi_{k+1}},$$

we obtain the linear recurrence

$$\psi_k + \psi_{k+1} + \psi_{k+2} = \arcsin 0 = \nu\pi$$

$\nu = 0, 1, 2, \ldots$. If we have efficient sequential algorithms for computing $\tan(\beta)$ or $\arctan(\beta)$, for any parameter β, in τ parallel steps, then computing $\xi_2, \xi_3, \ldots, \xi_n$, from the initial conditions (ξ_0, ξ_1), may be outlined as follows:

(i) compute $\sqrt{\alpha}$ in 2 parallel steps (say),
(ii) compute ψ_0 and ψ_1 using two processors in $(1 + \tau)$ parallel steps,
(iii) the algorithm in Theorem 3.2 may be used to obtain $\psi_2, \psi_3, \ldots, \psi_n$ in $-1 + 3\log(n + 1)$ parallel steps using less than $3n$ processors,
(iv) using $(n - 1)$ processors, compute $\xi_2, \xi_3, \ldots, \xi_n$ in $(1 + \tau)$ parallel steps.

Hence, the total number of parallel steps amounts to

$$\mathbf{T}_p = 2\tau + 3\log 2(n + 1),$$

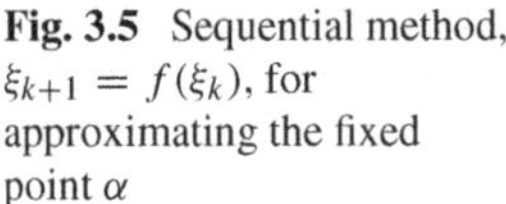

Fig. 3.5 Sequential method, $\xi_{k+1} = f(\xi_k)$, for approximating the fixed point α

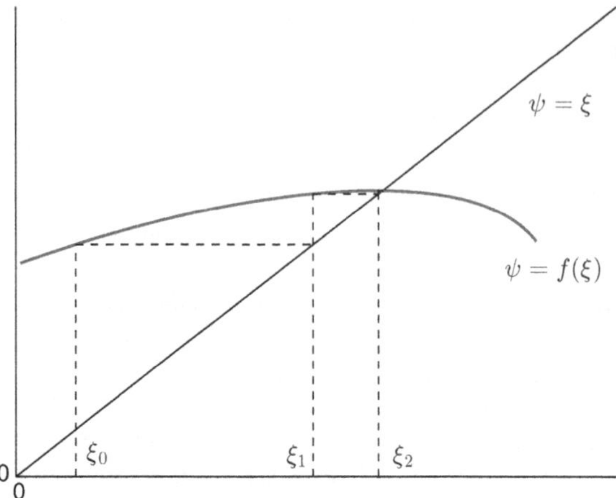

using $p = 3n$ processors, whereas the sequential algorithm for handling the nonlinear recurrence directly requires $\mathbf{O}_1 = 5(n - 1)$ arithmetic operations.

Therefore, provided that $\tau \ll n$, we can achieve speedup $\mathbf{S}_p = O(n/\log n)$. This can be further enhanced if we are seeking only ξ_n and have a parallel algorithm for evaluating the tan and arctan functions. In many cases the first-order nonlinear recurrences $\xi_{k+1} = f(\xi_k)$ arise when one attempts to converge to a fixed point. The goal in this case is not the evaluation of ξ_i for $i \geq 1$ (given the initial iterate ξ_0), but an approximation of $\lim_{i \to \infty} \xi_i$. The classical sequential algorithm is illustrated in Fig. 3.5, where $|f'(\xi)| < 1$ in the neighborhood of the root α. In this case, α is called an attractive fixed point. An efficient parallel algorithm, therefore, should not attempt to linearize the nonlinear recurrence $\xi_{k+1} = f(\xi_k)$, but rather seek an approximation of α by other methods for obtaining roots of a nonlinear function.

We best illustrate this by an example.

Example 3.9 Consider the nonlinear recurrence

$$\xi_{k+1} = \frac{2\xi_k(\xi_k^2 + 6)}{3\xi_k^2 + 2} = f(\xi_k).$$

Starting with $\xi_0 = 10$ as an initial approximation to the attractive fixed point $\alpha = \sqrt{10}$, $f'(\alpha) = 3/8 < 1$, we seek ξ_n such that $|\xi_n - \alpha| \leq 10^{-10}$. By the theory of functional iterations, to approximate α to within the above tolerance we need at least 25 function evaluations of f, each of which requires 6 arithmetic operations, hence $\mathbf{T}_1 \geq 150$.

From the nonlinear recurrence and its initial iterate, it is obvious that we are seeking the positive root of the nonlinear equation $g(\xi) = \xi(\xi^2 - 10)$, see Fig. 3.6. In contrast to Theorem 3.8, we will show that given enough processors we can appreciably accelerate the evaluation of ξ_n. It is not difficult to obtain an interval $(\tau, \tau + \gamma)$

Fig. 3.6 Function g used to compute the fixed point of f

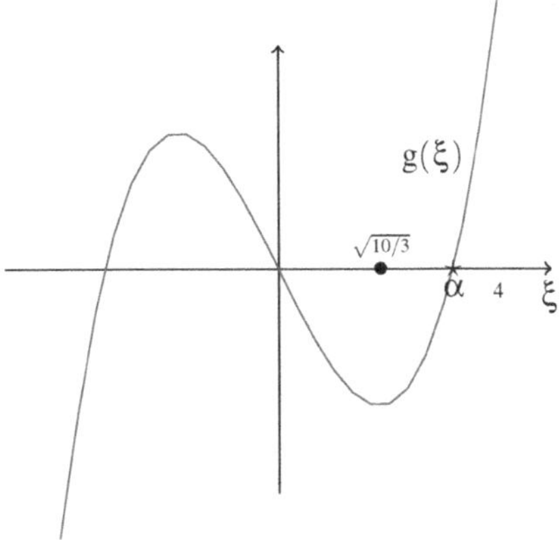

in which $g(\xi)$ changes sign, i.e., $\alpha \in (\tau, \tau + \gamma)$. Assuming we have p processors, we describe a modification of the bisection method for obtaining the single root in $(\tau, \tau + \gamma)$ (see also [13, 14]):

(i)   divide the interval into $(p-1)$ equal subintervals (θ_i, θ_{i+1}), $1 \le i \le p-1$,
(ii) evaluate the function $g(\xi)$ at the points θ_j, $1 \le j \le p$,
(iii) detect that interval with change of sign, i.e., $\text{sign}(g(\theta_i)) \ne \text{sign}(g(\theta_{i+1}))$,
(iv) go back to (i).

This process is repeated until we obtain a subinterval in (iii) $(\theta_\nu, \theta_{\nu+1})$ of width $\le 2\varepsilon$, where $\varepsilon = 10^{-t}$ is some prescribed tolerance. A reasonable approximation to α is therefore given by $(\theta_\nu + \theta_{\nu+1})/2$. Clearly, the process may terminate at an earlier time if for any θ_i, $g(\theta_i) \le \varepsilon$. We can obtain an upper bound on the parallel steps required to locate an approximation to the root. If ν is the number of simultaneous function evaluations such that

$$\frac{\gamma}{(p-1)^\nu} \le 2\varepsilon,$$

then

$$\nu \le \left\lceil \frac{t + \log_{10}(\gamma/2)}{\log_{10}(p-1)} \right\rceil,$$

and

$$\mathbf{T}_p \le 2 + 3\nu.$$

(Note that one function evaluation of g requires 3 arithmetic operations.)

Let $(\tau, \tau + \gamma) = (2,4)$, i.e., $\gamma = 2$, $p = 64$, and $t = 10$ then $\nu \le 6$ and $T_{64} \le 20$. Thus $\mathbf{T}_1/\mathbf{T}_p = (150/20) \simeq 7.5$ which is $O(n/\log n)$ for $n \ge 54$.

References

1. Gautschi, W.: Computational aspects of three-term recurrence relations. SIAM Rev. **9**, 24–82 (1967)
2. Rivlin, T.: The Chebyshev Polynomials. Wiley-Interscience, New York (1974)
3. Kuck, D.: The Structure of Computers and Computations. Wiley, New Yok (1978)
4. Chen, S.C., Kuck, D.: Time and parallel processor bounds for linear recurrence systems. IEEE Trans. Comput. **C-24**(7), 701–717 (1975)
5. Sameh, A., Brent, R.: Solving triangular systems on a parallel computer. SIAM J. Numer. Anal. **14**(6), 1101–1113 (1977)
6. Higham, N.: Accuracy and Stability of Numerical Algorithms, 2nd edn. SIAM, Philadelphia (2002)
7. Sameh, A., Kuck, D.: A parallel QR algorithm for symmetric tridiagonal matrices. IEEE Trans. Comput. **26**(2), 147–153 (1977)
8. Higham, N.: Stability of parallel triangular system solvers. SIAM J. Sci. Comput. **16**(2), 400–413 (1995)
9. Lafon, J.: Base tensorielle des matrices de Hankel (ou de Toeplitz). Appl. Numer. Math. **23**, 249–361 (1975)
10. Boole, G.: Calculus of Finite Differences. Chelsea Publishing Company, New York (1970)
11. Wimp, J.: Computation with Recurrence Relations. Pitman, Boston (1984)
12. Kung, H.: New algorithms and lower bounds for the parallel evaluation of certain rational expressions and recurrences. J. Assoc. Comput. Mach. **23**(2), 252–261 (1976)
13. Miranker, W.: Parallel methods for solving equations. Math. Comput. Simul. **20**(2), 93–101 (1978). doi:10.1016/0378-4754(78)90032-0. http://www.sciencedirect.com/science/article/pii/0378475478900320
14. Gal, S., Miranker, W.: Optimal sequential and parallel seach for finding a root. J. Combinatorial Theory **23**, 1–14 (1977)

Chapter 4
General Linear Systems

One of the most fundamental problems in matrix computations is solving linear systems of the form,

$$Ax = f, \tag{4.1}$$

where A is an n-by-n nonsingular matrix, and $f \in \mathbb{R}^n$ is the corresponding right hand-side. Important related problems are solving for multiple right-hand sides, say $AX = F$ with $F \in \mathbb{R}^{n \times s}$, and computing the inverse A^{-1}.

When A is dense, i.e. when A does not possess special structure and contains relatively few zero elements, the most common method for solving system (4.1) on uniprocessors is some form of Gaussian elimination; cf. [1, 2]. Typically, obtaining a factorization (or decomposition) of A into a product of simpler terms has been the first step in the solution procedure. For example, see the historical remarks in [3, 4], and the seminal early contributions in [5, 6]. It is of interest to note that matrix decompositions are not only needed for solving linear systems, but are also needed in areas such as data analytics; for example in revealing latent information in data, e.g. see [7]. Over the last two decades, a considerable effort has been devoted to the development of parallel algorithms for dense matrix computations. This effort resulted in the development of numerical libraries such as LAPACK , PLASMA and SCALAPACK which are widely used on uniprocessors, multicore architectures, and distributed memory architectures, respectively, e.g. see [8, 9]. A related, but independent, effort has also resulted in the high performance libraries: BLIS, LIBFLAME, and ELEMENTAL e.g. see [10]. As the subject of direct dense linear system solvers has been sufficiently treated elsewhere in the literature, we will briefly summarize basic direct dense linear system solvers, offer variations that enhance their parallel scalability, and propose an approximate factorization scheme that can be used in conjunction with iterative schemes to realize an effective and scalable solver on parallel architectures. Note that such *approximate factorization* approach is frequently justified when:

© Springer Science+Business Media Dordrecht 2016
E. Gallopoulos et al., *Parallelism in Matrix Computations*,
Scientific Computation, DOI 10.1007/978-94-017-7188-7_4

1. An exact factorization might be very expensive or prohibitive; for example if the dense matrix is very large or when communication in the underlying parallel architecture is very expensive.
2. An exact factorization might not exist if certain constraints hold, for example: (i) in order to avoid pivoting, the matrix is subjected to additive modifications of low rank and the factorization corresponds to a modified matrix, or (ii) when the underlying matrix is nonnegative and it is stipulated that the factors are also nonnegative (as in data analytics).
3. Even when an exact factorization exists, the high accuracy obtained in solving the linear system via a direct method is not justified due to uncertainty in the input.

Therefore, in such cases (in large scale applications) the combination of an approximate factorization that lends itself to efficient parallel processing in combination with an iterative technique, ranging from basic *iterative refinement* to *preconditioned projection methods*, can be a preferable alternative. It is also worth noting that recently, researchers have been addressing the issue of designing *communication avoiding (CA) algorithms* for these problems. This consists of building algorithms that achieve as little communication overhead as possible relative to some known lower bounds, without significantly increasing the number of arithmetic (floating-point) operations or negatively affecting the numerical stability, e.g. see for example [11–13], and the survey [14] that discusses many problems of numerical linear algebra (including matrix multiplication, eigenvalue problems and Krylov subspace methods for sparse linear systems) in the CA framework.

4.1 Gaussian Elimination

Given a vector $w \in \mathbb{R}^n$ with $\omega_1 \neq 0$, one can construct an elementary lower triangular matrix M

$$M = \begin{pmatrix} 1 & & & \\ \mu_{21} & 1 & & \\ \vdots & & \ddots & \\ \mu_{n1} & & & 1 \end{pmatrix} = \begin{pmatrix} 1 & 0 \\ m & I_{n-1} \end{pmatrix} \tag{4.2}$$

such that

$$Mw = \omega_1 e_1. \tag{4.3}$$

This is accomplished by choosing

$$\mu_{j1} = -(\omega_j/\omega_1).$$

If $A \equiv A_1 = \begin{pmatrix} \alpha_{11} & a^\top \\ b & B \end{pmatrix}$ is diagonally dominant, then $\alpha_{11} \neq 0$ and B is also diagonally dominant. One can construct an elementary lower triangular matrix M_1 such that $A_2 = M_1 A_1$ is upper-triangular as far as its first column is concerned, i.e.,

$$A_2 = \begin{pmatrix} \alpha_{11} & a^\top \\ 0 & C \end{pmatrix} \tag{4.4}$$

where $C = B + ma^\top$, i.e., C is a rank-1 modification of B. It is also well-known that if A_1 is diagonally dominant, then C will also be diagonally dominant. Thus, the process can be repeated on the matrix C which is of order 1 less than A, and so on in order to produce the factorization,

$$M_{n-1} \cdots M_2 M_1 A = U \tag{4.5}$$

where U is upper triangular, and

$$M_j = \begin{pmatrix} 1 & & & & & \\ & 1 & & \downarrow & & \\ & & \ddots & 1 & & \\ & & & \mu_{j+1,j} & \ddots & \\ & & & \vdots & & \\ & & & \mu_{n,j} & & 1 \end{pmatrix}, \tag{4.6}$$

i.e., $A = LU$, in which $L = M_1^{-1} M_2^{-1} \cdots M_{n-1}^{-1}$, or

$$L = \begin{pmatrix} 1 & & & & \\ -\mu_{21} & 1 & & & \\ -\mu_{31} & -\mu_{32} & 1 & & \\ \vdots & \vdots & \ddots & \ddots & \\ -\mu_{n1} & -\mu_{n,2} & \cdots & -\mu_{n,n-1} & 1 \end{pmatrix}. \tag{4.7}$$

If A is not diagonally dominant, α_{11} is not guaranteed to be nonzero. Hence a pivoting strategy is required to assure that the elements μ_{ij} are less than 1 in magnitude to enhance numerical stability. Adopting *pivoting*, the first step is to choose a permutation P_1 such that the first column of $P_1 A_1$ has as its first element (pivot) $\alpha_{k1}^{(1)}$ for which $|\alpha_{k1}^{(1)}| = \max_{1 \leq i \leq n} |\alpha_{i1}^{(1)}|$. Consequently, the procedure yields,

$$M_{n-1} P_{n-1} \cdots P_2 M_1 P_1 A_1 = U, \tag{4.8}$$

or

$$\hat{M}_{n-1}\hat{M}_{n-2}\cdots\hat{M}_1(P_{n-1}\cdots P_2 P_1)A = U,$$

where $\hat{M}_j$ is identical to M_j, $j = 1, 2, \ldots, n - 2$, but whose elements in the jth column below the diagonal are shuffled, and $\hat{M}_{n-1} = M_{n-1}$. In other words, we obtain the factorization

$$PA = \hat{L}U \tag{4.9}$$

where $\hat{L}$ is unit-lower triangular, U is upper triangular, and $P = P_{n-1}\cdots P_2 P_1$.

In this form of Gaussian elimination, the main operation in each step is the rank-1 update kernel in BLAS2

$$E = E + ma^{\top}. \tag{4.10}$$

Theorem 4.1 ([15]) *Let $A \in \mathbb{R}^{n\times n}$ be nonsingular. Using Gaussian elimination without pivoting, the triangular factorization $A = LU$ may be obtained in $3(n - 1)$ parallel arithmetic operations using no more than n^2 processors.*

Proof Constructing each M_j $(j = 1, \ldots, n - 1)$ using $n - j$ processors can be done in one parallel arithmetic operation. Applying the rank-1 update at step j can be done in two parallel arithmetic operations using $(n - j)^2$ processors. The theorem follows easily.

It is well-known that a rank-k update kernel, for $k > 1$, has higher data-locality than the BLAS2 kernel (4.10). Thus, to realize higher performance, one resorts to using a block form of the Gaussian elimination procedure that utilizes rank-k updates that are BLAS3 operations, see Sect. 2.2.

4.2 Pairwise Pivoting

As outlined above, using a pivoting strategy such as partial pivoting requires obtaining the element of maximum modulus of a vector of order n (say). In a parallel implementation, that vector could be stored across p nodes. Identifying the pivotal row will thus require accessing data along the entire matrix column, potentially giving rise to excessive communication across processors. Not surprisingly, more aggressive forms of pivoting that have the benefit of smaller potential error, are likely to involve even more communication. To reduce these costs, one approach is to utilize a *pairwise pivoting* strategy. Pairwise pivoting leads to a factorization of the form

$$A = S_1^{-1}S_2^{-1}\cdots S_{2n-3}^{-1}U \tag{4.11}$$

where each S_j is a block-diagonal matrix in which each block is of order 2, and U is upper triangular. It is important to note that this is no longer an LU factorization. Specifically, the factorization may be outlined as follows. Let $a^{\top}$ and $b^{\top}$ be two rows of A:

$$\begin{pmatrix} \alpha_1 & \alpha_2 & \alpha_3 & \cdots & \alpha_n \\ \beta_1 & \beta_2 & \beta_3 & \cdots & \beta_n \end{pmatrix},$$

and let G be the 2×2 stabilized elementary transformation $G = \begin{pmatrix} 1 & 0 \\ \gamma & 1 \end{pmatrix}$ if $|\alpha_1| \geq |\beta_1|$ with $\gamma = -\beta_1/\alpha_1$, or $G = \begin{pmatrix} 0 & 1 \\ 1 & \gamma \end{pmatrix}$ if $|\beta_1| > |\alpha_1|$ with $\gamma = -\alpha_1/\beta_1$. Hence, each S_j consists of 2×2 diagonal blocks of the form of G or the identity I_2. For a nonsingular matrix A of order 8, say, the pattern of annihilation of the elements below the diagonal is given in Fig. 4.1 where an '$*$' denotes a diagonal element, and an entry 'k' denotes an element annihilated by S_k. Note that, for n even, S_{n-1} annihilates $(n/2)$ elements simultaneously, with S_1 and S_{2n-3} each annihilating only one element. Note also that S_k^{-1}, for any k, is easily obtained since $(G_j^{(k)})^{-1}$ is either

$$\begin{pmatrix} 1 & 0 \\ -\gamma & 1 \end{pmatrix} \quad \text{or} \quad \begin{pmatrix} -\gamma & 1 \\ 1 & 0 \end{pmatrix}.$$

Consequently, solving a system of the form $Ax = f$ via Gaussian elimination with pairwise pivoting consists of:

(a) factorization:
$$(S_{2n-3} \cdots S_2 S_1)A = \dot{U}, \quad \text{or} \quad A = \dot{S}^{-1}\dot{U},$$

(b) forward sweep:
$$\begin{aligned} v_0 &= f \\ v_j &= S_j v_{j-1}, \quad j = 1, 2, \ldots, 2n - 3 \end{aligned} \tag{4.12}$$

(c) backward sweep: Solve
$$Ux = v_{2n-3} \tag{4.13}$$

via the column sweep scheme.

$*$							
7	$*$						
6	8	$*$					
5	7	9	$*$				
4	6	8	10	$*$			
3	5	7	9	11	$*$		
2	4	6	8	10	12	$*$	
1	3	5	7	9	11	13	$*$

Fig. 4.1 Annihilation in Gaussian elimination with pairwise pivoting

A detailed error analysis of this pivoting strategy was conducted in [16], and in a much refined form in [17], which shows that

$$|A - S_1^{-1} S_2^{-2} \cdots S_{2n-3}^{-1} \dot{U}| \le 2^{n-1}(2^{n-1} - 1)|A|\mathbf{u}. \tag{4.14}$$

This upper bound is 2^{n-1} larger than that obtained for the partial pivoting strategy. Nevertheless, except in very special cases, extensive numerical experiments show that the difference in quality (as measured by relative residuals) of the solutions obtained by these two pivoting strategies is imperceptible.

Another pivoting approach that is related to pairwise pivoting is termed *incremental pivoting* [18]. Incremental pivoting has also been designed so as to reduce the communication costs in Gaussian elimination with partial pivoting.

4.3 Block LU Factorization

Block LU factorization may be described as follows. Let A be partitioned as

$$A = (\dot{A}, \ddot{A}) = \begin{pmatrix} A_{11} & A_{12} \\ A_{21} & A_{22} \end{pmatrix}, \tag{4.15}$$

where $\dot{A}$ consists of v-columns, $v \ll n$. First, we obtain the LU factorization of the rectangular matrix $\dot{A}$ using the Gaussian elimination procedure with partial pivoting described above in which the main operations depend on rank-1 updates,

$$\dot{P}_0 \dot{A} = \begin{pmatrix} L_{11} \\ L_{21} \end{pmatrix} (U_{11}).$$

Thus, $\dot{P}_0 A$ may be written as

$$\dot{P}_0 A = \begin{pmatrix} L_{11} & 0 \\ L_{21} & I_{n-v} \end{pmatrix} \begin{pmatrix} U_{11} & U_{12} \\ 0 & B_0 \end{pmatrix}, \tag{4.16}$$

where U_{12} is obtained by solving

$$L_{11} U_{12} = A_{12} \tag{4.17}$$

for U_{12}, and obtaining

$$B_0 = A_{22} - L_{21} U_{12} \quad (\text{rank}-v \text{ update}).$$

Now, the process is repeated for B_0, i.e., choosing a window $\dot{B}_0$ consisting of the first v columns of B_0, obtaining an LU factorization of $\dot{P}_1 \dot{B}_0$ followed by obtaining v additional rows and columns of the factorization (4.16)

$$\begin{pmatrix} I_v & \\ & \dot{P_1} \end{pmatrix} \dot{P_0} A = \begin{pmatrix} L_{11} & & \\ L'_{21} & L_{22} & \\ L''_{21} & L_{32} & I_{n-2v} \end{pmatrix} \begin{pmatrix} U_{11} & U'_{12} & U''_{12} \\ & U_{22} & U_{23} \\ & & B_1 \end{pmatrix}.$$

The factorization $PA = LU$ is completed in (n/v) stages.

Variants of this basic block LU factorization are presented in Chap. 5 of [19].

4.3.1 Approximate Block Factorization

If A were diagonally dominant or symmetric positive definite, the above block LU factorization could be obtained without the need of a pivoting strategy. Our aim here is to obtain an approximate block LU factorization of general nonsymmetric matrices without incurring the communication and memory reference overhead due to partial pivoting employed in the previous section. The resulting approximate factorization can then be used as an effective preconditioner of a Krylov subspace scheme for solving the system $Ax = f$.

To illustrate this approximate block factorization approach, we first assume that A is diagonally dominant of order mq, i.e., $A \equiv A_0 = [A_{ij}^{(0)}]$, $i, j = 1, 2, \ldots, q$, in which each A_{ij} is of order m. Similar to the pairwise pivoting scheme outlined previously, the block scheme requires $(2q - 3)$ stages to produce the factorization,

$$A = \hat{S}^{-1} \hat{U} \tag{4.18}$$

in which $\hat{S}$ is the product of $(2q - 3)$ matrices each of which consists of a maximum of $\lfloor q/2 \rfloor$ nested independent block-elementary transformation of the form

$$\begin{pmatrix} I_m & \\ -G & I_m \end{pmatrix},$$

and $\hat{U}$ is block-upper-triangular. Given two block rows:

$$\begin{pmatrix} A_{ii} & A_{i,i+1} \cdots A_{i,q} \\ A_{ji} & A_{j,i+1} \cdots A_{j,q} \end{pmatrix},$$

we denote by $[i, j]$ the transformation,

$$\begin{pmatrix} I & \\ -G_{ij} & I \end{pmatrix} \begin{pmatrix} A_{ii} & A_{i,i+1} & \cdots & A_{i,q} \\ A_{j,i} & A_{j,i+1} & \cdots & A_{j,q} \end{pmatrix}$$

$$= \begin{pmatrix} A_{ii} & A_{i,i+1} & \cdots & A_{i,q} \\ 0 & A'_{j,i+1} & \cdots & A'_{j,q} \end{pmatrix} \tag{4.19}$$

in which,

$$G_{ij} = A_{ji} A_{ii}^{-1}, \quad \text{and} \tag{4.20}$$

$$A'_{j,k} = A_{j,k} - A_{ji} A_{ii}^{-1} A_{ik} \tag{4.21}$$

for $k = i + 1, i + 2, \ldots, q$. Thus, the parallel annihilation scheme that produces the factorization (4.18) is given in Fig. 4.2 (here each '$\star$' denotes a matrix of order $m \ll n$) illustrating the available parallelism, with the order of the $[i, j]$ transformations (4.19) shown in Fig. 4.3.

Note that, by assuming that A is diagonally dominant, any A_{ii} or its update A'_{ii} is also diagonally dominant assuring the existence of the factorization $A_{ii} = L_i U_i$ without pivoting, or the factorization $A_{ii} = \hat{S}_i^{-1} \hat{U}_i$.

$\star$							
1	$\star$						
2	3	$\star$					
3	4	5	$\star$				
4	5	6	7	$\star$			
5	6	7	8	9	$\star$		
6	7	8	9	10	11	$\star$	
7	8	9	10	11	12	13	$\star$

Fig. 4.2 Annihilation in block factorization without pivoting

Steps	Nodes $\rightarrow$							
$\downarrow$	1	2	3	4	5	6	7	8
1	fact(A_{11})							
2	$[1,2]$							
3	$[1,3]$	fact(A_{22})						
4	$[1,4]$	$[2,3]$						
5	$[1,5]$	$[2,4]$	fact(A_{33})					
6	$[1,6]$	$[2,5]$	$[3,4]$					
7	$[1,7]$	$[2,6]$	$[3,5]$	fact(A_{44})				
8	$[1,8]$	$[2,7]$	$[3,6]$	$[4,5]$				
9		$[2,8]$	$[3,7]$	$[4,6]$	fact(A_{55})			
10			$[3,8]$	$[4,7]$	$[5,6]$			
11				$[4,8]$	$[5,7]$	fact(A_{66})		
12					$[5,8]$	$[6,7]$		
13						$[6,8]$	fact(A_{77})	
14							$[7,8]$	
15								fact(A_{88})

Fig. 4.3 Concurrency in block factorization

We consider next the general case, where the matrix A is not diagonally dominant. One attractive possibility is to proceed with the factorization of each A_{ii} without partial pivoting but applying, whenever necessary, the procedure of "diagonal boosting" which insures the nonsingularity of each diagonal block. Such a technique was originally proposed for Gaussian elimination in [20].

Diagonal boosting is invoked whenever any diagonal pivot violates the following condition:

$$|\text{pivot}| > 0_{\mathbf{u}} \, \|A_j\|_1,$$

where $0_{\mathbf{u}}$ is a multiple of the unit roundoff, $\mathbf{u}$. If the diagonal pivot does not satisfy the above condition, its value is "boosted" as:

$$\text{pivot} = \text{pivot} + \delta \|A_j\|_1 \ \ \text{if pivot} > 0,$$

$$\text{pivot} = \text{pivot} - \delta \|A_j\|_1 \ \ \text{if pivot} < 0,$$

where δ is often taken as the square root of $0_{\mathbf{u}}$.

In other words, we will obtain a block factorization of a matrix M, rather than the original coefficient matrix A. The matrix $M = \hat{S}^{-1}\hat{U}$ will differ from A in norm by a quantity that depends on δ and differ in rank by the number of times boosting was invoked. A small difference in norm and rank will make M a good preconditioner for a Krylov subspace method (such as those described in Chaps. 9 and 10). When applied to solving $Ax = f$ with this preconditioner, only few iterations will be needed to achieve a reasonable relative residual.

4.4 Remarks

Well designed kernels for matrix multiplication and rank-k updates for hierarchical machines with multiple levels of memory and parallelism are of critical importance for the design of dense linear system solvers. SCALAPACK based on such kernels in BLAS3 is a case in point. The subroutines of this library achieve high performance and parallel scalability. In fact, many users became fully aware of these gains even when using high-level problem solving environments like MATLAB (cf. [21]). As early work on the subject had shown (we consider it rewarding for the reader to consider the pioneering analyses undertaken in [22, 23]), the task of designing kernels with high parallel scalability is far from simple, if one desires to provide a design that closely resembles the target computer model. The task becomes even more difficult as the complexity of the computer architecture increases. It becomes even harder when the target is to build methods that can deliver high performance across a spectrum of computer architectures.

References

1. Golub, G., Van Loan, C.: Matrix Computations, 4th edn. Johns Hopkins (2013)
2. Stewart, G.: Matrix Algorithms. Vol. I. Basic Decompositions. SIAM, Philadelphia (1998)
3. Grcar, J.: John Von Neumann's analysis of Gaussian elimination and the origins of modern numerical analysis. SIAM Rev. **53**(4), 607–682 (2011). doi:10.1137/080734716. http://dx.doi.org/10.1137/080734716
4. Stewart, G.: The decompositional approach in matrix computations. IEEE Comput. Sci. Eng. Mag. 50–59 (2000)
5. von Neumann, J., Goldstine, H.: Numerical inverting of matrices of high order. Bull. Am. Math. Soc. **53**, 1021–1099 (1947)
6. Householder, A.S.: The Theory of Matrices in Numerical Analysis. Dover Publications, New York (1964)
7. Skillicorn, D.: Understanding Complex Datasets: Data Mining using Matrix Decompositions. Chapman Hall/CRC Press, Boca Raton (2007)
8. Dongarra, J., Kurzak, J., Luszczek, P., Tomov, S.: Dense linear algebra on accelerated multicore hardware. In: Berry, M., et al. (eds.) High-Performance Scientific Computing: Algorithms and Applications, pp. 123–146. Springer, New York (2012)
9. Luszczek, P., Kurzak, J., Dongarra, J.: Looking back at dense linear algebrasoftware. J. Parallel Distrib. Comput. **74**(7), 2548–2560 (2014). DOI http://dx.doi.org/10.1016/j.jpdc.2013.10.005
10. Igual, F.D., Chan, E., Quintana-Ortí, E.S., Quintana-Ortí, G., van de Geijn, R.A., Zee, F.G.V.: The FLAME approach: from dense linear algebra algorithms to high-performance multi-accelerator implementations. J. Parallel Distrib. Comput. **72**, 1134–1143 (2012)
11. Demmel, J., Grigori, L., Hoemmen, M., Langou, J.: Communication-optimal parallel and sequential QR and LU factorizations. SIAM J. Sci. Comput. **34**(1), 206–239 (2012). doi:10.1137/080731992
12. Grigori, L., Demmel, J., Xiang, H.: CALU: a communication optimal LU factorization algorithm. SIAM J. Matrix Anal. Appl. **32**(4), 1317–1350 (2011). doi:10.1137/100788926. http://dx.doi.org/10.1137/100788926
13. Irony, D., Toledo, S., Tiskin, A.: Communication lower bounds for distributed-memory matrix multiplication. J. Parallel Distrib. Comput. **64**(9), 1017–1026 (2004). doi:10.1016/j.jpdc.2004.03.021. http://dx.doi.org/10.1016/j.jpdc.2004.03.021
14. Ballard, G., Carson, E., Demmel, J., Hoemmen, M., Knight, N., Schwartz, O.: Communication lower bounds and optimal algorithms for numerical linear algebra. Acta Numerica **23**, 1–155 (2014)
15. Sameh, A.: Numerical parallel algorithms—a survey. In: Kuck, D., Lawrie, D., Sameh, A. (eds.) High Speed Computer and Algorithm Optimization, pp. 207–228. Academic Press, New York (1977)
16. Stern, J.: A fast Gaussian elimination scheme and automated roundoff error analysis on S.I.M.E. machines. Ph.D. thesis, University of Illinois (1979)
17. Sorensen, D.: Analysis of pairwise pivoting in Gaussian elemination. IEEE Trans. Comput. **C-34**(3), 274–278 (1985)
18. Quintana-Ortí, E.S., van de Geijn, R.A.: Updating an LU factorization with pivoting. ACM Trans. Math. Softw. (TOMS) **35**(2), 11 (2008)
19. Dongarra, J., Duff, I., Sorensen, D., van der Vorst, H.: Numerical Linear Algebra for High-Performance Computers. SIAM, Philadelphia (1998)
20. Stewart, G.: Modifying pivot elements in Gaussian elimination. Math. Comput. **28**(126), 537–542 (1974)

21. Moler, C.: MATLAB incorporates LAPACK. Mathworks Newsletter (2000). http://www. mathworks.com/company/newsletters/articles/matlab-incorporates-lapack.html
22. Gallivan, K., Jalby, W., Meier, U.: The use of BLAS3 in linear algebra on a parallel processor with a hierarchical memory. SIAM J. Sci. Statist. Comput. **8**(6), 1079–1084 (1987)
23. Gallivan, K., Jalby, W., Meier, U., Sameh, A.: The impact of hierarchical memory systems on linear algebra algorithm design. Int. J. Supercomput. Appl. **2**(1) (1988)

Chapter 5
Banded Linear Systems

We encounter banded linear systems in many areas of computational science and engineering, including computational mechanics and nanoelectronics, to name but a few. In finite-element analysis, the underlying large sparse linear systems can sometimes be reordered via graph manipulation schemes such as reverse Cuthill-McKee or spectral reordering to result in low-rank perturbations of banded systems that are dense or sparse within the band, and in which the width of the band is often but a small fraction of the size of the overall problem; cf. Sect. 2.4.2. In such cases, a chosen central band can be used as a preconditioner for an outer Krylov subspace iteration. It is thus safe to say that the need to solve banded linear systems arises extremely often in numerical computing. In this chapter, we present parallel algorithms for solving narrow-banded linear systems: (a) direct methods that depend on Gaussian elimination with partial pivoting as well as pairwise pivoting strategies, (b) a class of robust hybrid (direct/iterative) solvers for narrow-banded systems that do not depend on a global LU-factorization of the coefficient matrix and the subsequent global forward and backward sweeps, (c) a generalization of this class of algorithms for solving systems involving wide-banded preconditioners via a tearing method, and, finally, (d) direct methods specially designed for tridiagonal linear systems. Throughout this rather long chapter, we denote the linear system(s) under consideration by $AX = F$, where A is a nonsingular banded matrix of bandwidth $2m + 1$, and F contains $\nu \geq 1$ right hand-sides. However if we want to restrict our discussion to a single right-hand side, we will write $Ax = f$.

5.1 LU-based Schemes with Partial Pivoting

Using the classical Gaussian elimination scheme, with partial pivoting, outlined in Chap. 4, for handling banded systems results in limited opportunity for parallelism. This limitation becomes more pronounced the narrower the system's band. To illustrate this limitation, consider a banded system of bandwidth $(2m + 1)$, shown in Fig. 5.1 for $m = 4$.

© Springer Science+Business Media Dordrecht 2016

E. Gallopoulos et al., *Parallelism in Matrix Computations*,
Scientific Computation, DOI 10.1007/978-94-017-7188-7_5

Fig. 5.1 Illustrating limited parallelism when Gaussian elimination is applied on a system of bandwidth 9

The scheme will consider the triangularization of the leading $(2m+1) \times (2m+1)$ window with partial pivoting, leading to possible additional fill-in. The process is repeated for the following window that slides down the diagonal, one diagonal element at a time.

In view of such limitation, an alternative LU factorization considered in [1] was adopted in the SCALAPACK library; see also [2, 3]. For ease of illustration, we consider the two-partitions case of a tridiagonal system $Ax = f$, of order 18×18, initially setting $A_0 \equiv A$ and $f_0 \equiv f$; see Fig. 5.2 for matrix A. After row permutations, we obtain $A_1 x = f_1$ where $A_1 = P_0 A_0$, with $P_0^\top = [e_{18}, e_1, e_2, \ldots, e_{17}]$ in which e_j is the jth column of the identity I_{18}, see Fig. 5.3. Following this by the column permutations, $A_2 = A_1 P_1$, with

$$P_1 = \begin{pmatrix} I_7 & 0 & 0 & 0 \\ 0 & 0 & I_2 & 0 \\ 0 & I_7 & 0 & 0 \\ 0 & 0 & 0 & I_2 \end{pmatrix}$$

we get,

$$A_2 = \begin{pmatrix} B_1 & 0 & E_1 & F_1 \\ 0 & B_2 & E_2 & F_2 \end{pmatrix},$$

see Fig. 5.4, where $B_1, B_2 \in \mathbb{R}^{9\times7}$, and $E_j, F_j \in \mathbb{R}^{9\times2}$, $j = 1, 2$. Obtaining the LU-factorization of B_1 and B_2 simultaneously, via partial pivoting,

$$P_2 B_1 = L_1 \begin{pmatrix} R_1 \\ 0 \end{pmatrix}; \quad P_3 B_2 = L_2 \begin{pmatrix} R_2 \\ 0 \end{pmatrix},$$

where L_j is unit lower triangular of order 9, and R_j is upper triangular of order 7, then the linear system $Ax = f$ is reduced to solving $A_2 y = f_1$, where $y = P_1^\top x$, which in turn is reduced to solving $A_3 y = g$ where

$$A_3 = \begin{pmatrix} R_1 & 0 & E_1' & F_1' \\ 0 & 0 & E_1'' & F_1'' \\ 0 & R_2 & E_2' & F_2' \\ 0 & 0 & E_2'' & F_2'' \end{pmatrix}$$

Fig. 5.2 Original banded matrix $A \in \mathbb{R}^{18 \times 18}$

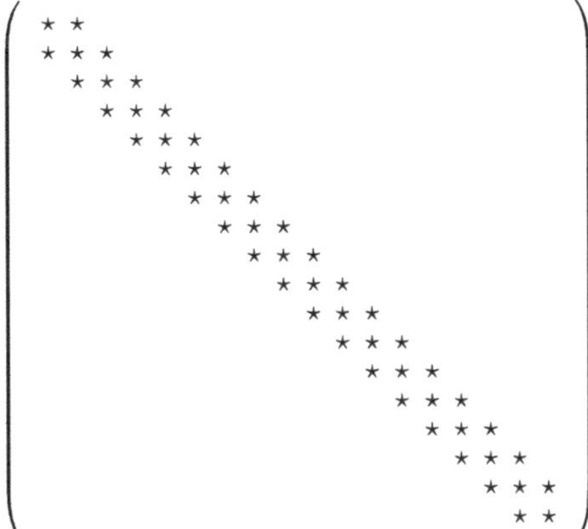

Fig. 5.3 Banded matrix after the row permutations $A_1 = P_0 A_0$, as in SCALAPACK

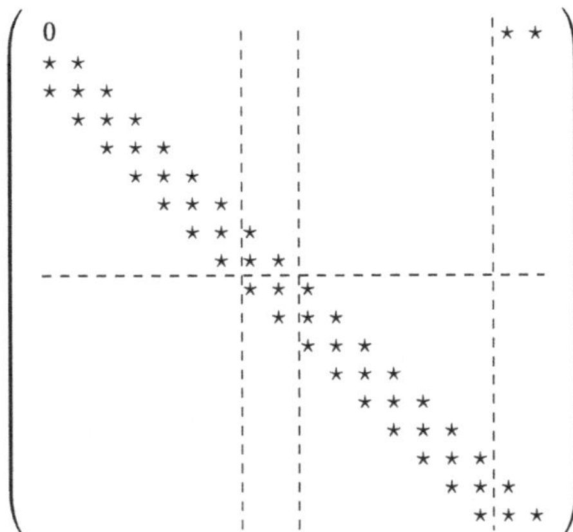

Fig. 5.4 Banded matrix after the row and column permutations $A_2 = A_1 P_1$, as in SCALAPACK

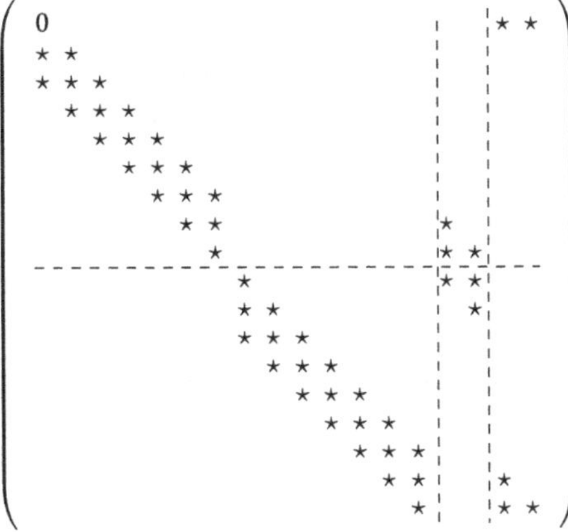

$$g = \begin{pmatrix} L_1^{-1} & 0 \\ 0 & L_2^{-1} \end{pmatrix} \begin{pmatrix} P_2 & 0 \\ 0 & P_3 \end{pmatrix} f_1,$$

and

$$L_j^{-1}(E_j, F_j) = \begin{pmatrix} E_j' & F_j' \\ E_j'' & F_j'' \end{pmatrix},$$

with $E_j', F_j' \in \mathbb{R}^{7 \times 2}$, and $E_j'', F_j'' \in \mathbb{R}^{2 \times 2}$. Using the permutation

$$P_4 = \begin{pmatrix} I_7 & 0 & 0 & 0 & 0 \\ 0 & 0 & I_7 & 0 & 0 \\ 0 & I_2 & 0 & 0 & 0 \\ 0 & 0 & 0 & 0 & I_2 \end{pmatrix},$$

the system $A_3 y = g$ is reduced to the block-upper triangular system $P_4 A_3 y = P_4 g$, where

$$P_4 A_3 = \left(\begin{array}{cc|cc} R_1 & 0 & E_1' & F_1' \\ 0 & R_2 & E_2' & F_2' \\ \hline 0 & 0 & E_1'' & F_1'' \\ 0 & 0 & E_2'' & F_2'' \end{array} \right).$$

Since A is nonsingular, the (4×4) matrix

$$\begin{pmatrix} E_1'' & F_1'' \\ E_2'' & F_2'' \end{pmatrix}$$

is also nonsingular and hence y, and subsequently, the unique solution x can be computed.

In Sects. 5.2 and 5.3, we present alternatives to this parallel direct solver where we present various hybrid (direct/iterative) schemes that possess higher parallel scalability. Clearly, these hybrid schemes are most suitable when the user does not require obtaining solutions whose corresponding residuals have norms of the order of the unit roundoff.

5.2 The Spike Family of Algorithms

Parallel banded linear system solvers have been considered by many authors [1, 4–11]. We focus here on one of these solvers—the Spike algorithm which dates back to the 1970s (the original algorithm created for solving tridiagonal systems on parallel architectures [12] is discussed in detail in Sect. 5.5.5 later in this chapter). Further developments and analysis are given in [13–18]. A distinct feature of the Spike algorithm is that it avoids the poorly scalable parallel scheme of obtaining the classical global LU factorization of the narrow-banded coefficient matrix. The overarching strategy of the Spike scheme consists of two main phases:

(i) the coefficient matrix is reordered or modified so as to consist, for example, of several independent diagonal blocks interconnected somehow so as to allow the extraction of an independent reduced system of a smaller size than that of the original system,

(ii) once the reduced system is solved, the original problem decomposes into several independent smaller problems facilitating almost perfect parallelism in retrieving the rest of the solution vector.

First, we review the Spike algorithm and its features. The Spike algorithm consists of the four stages listed in Algorithm 5.1.

Algorithm 5.1 Basic stages of the Spike algorithm

Input: Banded matrix and right-hand side(s)
Output: Solution vector(s)
   //(a) Pre-processing:
1: partitioning of the original system and distributing it on different nodes, in which each node consists of several or many cores,
2: factorization of each diagonal block and extraction of a reduced system of much smaller size;
   //(b) Post-processing: solving the reduced system,
3: retrieving the overall solution.

Compared to the parallel banded LU-factorization schemes in the existing literature, the Spike algorithm reduces the required memory references and interprocessor communications at the cost of performing more arithmetic operations. Such a trade-off results in the good parallel scalability enjoyed by the Spike family of solvers. The Spike algorithm also enables multilevel parallelism—parallelism across nodes, and parallelism within each multi- or many-core node. Further, it is flexible in the methods used for the pre- and post-processing stages, resulting in a family of solvers. As such, the Spike algorithm has several built-in options that range from using it as a pure direct solver to using it as a solver for banded preconditioners of an outer iterative scheme.

5.2.1 The Spike Algorithm

For ease of illustration, we assume for the time being, that any block-diagonal part of A is nonsingular. Later, however, this assumption will be removed, and the systems $AX = F$ are solved via a Krylov subspace method with preconditioners which are low-rank perturbations of the matrix A. Solving systems involving such banded preconditioners in each outer Krylov iteration will be achieved using a member of the Spike family of algorithms.

Preprocessing Stage

This stage starts with partitioning the banded linear system into a block tridiagonal form with p diagonal blocks A_j ($j = 1, \ldots, p$), each of order $n_j = n/p$ (assuming that n ia an integer multiple of p), with coupling matrices (of order $m \ll n$) $B_j (j = 1, \ldots, p - 1$), and $C_j (j = 2, \ldots, p$), associated with the super- and sub-block diagonals, respectively. Note that it is not a requirement that all the diagonal blocks A_j be of the same order. A straightforward implementation of the Spike algorithm on a distributed memory architecture could be realized by choosing the number of partitions p to be the same as the number of available multicore nodes q. In general, however, $p \leq q$. This stage concludes by computing the LU-factorization of each diagonal block A_j. Figure 5.5 illustrates the partitioning of the matrices A and F for $p = 4$.

The Spike Factorization Stage

Based on our assumption above that each A_j is nonsingular, the matrix A can be factored as $A = DS$, where D is a block-diagonal matrix consisting only of the factored diagonal blocks A_j,

$$D = \mathrm{diag}(A_1, \ldots, A_p),$$

and S is called the "Spike matrix"—shown in Fig. 5.6.

For a given partition j, we call $V_j (j = 1, \ldots, p - 1)$ and $W_j (j = 2, \ldots, p)$, respectively, the right and the left spikes each of order $n_j \times m$.

The spikes V_1 and W_p are given by

$$V_1 = (A_1)^{-1} \begin{pmatrix} 0 \\ I_m \end{pmatrix} B_1, \quad \text{and} \quad W_p = (A_p)^{-1} \begin{pmatrix} I_m \\ 0 \end{pmatrix} C_p. \tag{5.1}$$

The spikes V_j and W_j, $j = 2, \ldots, p - 1$, may be generated by solving,

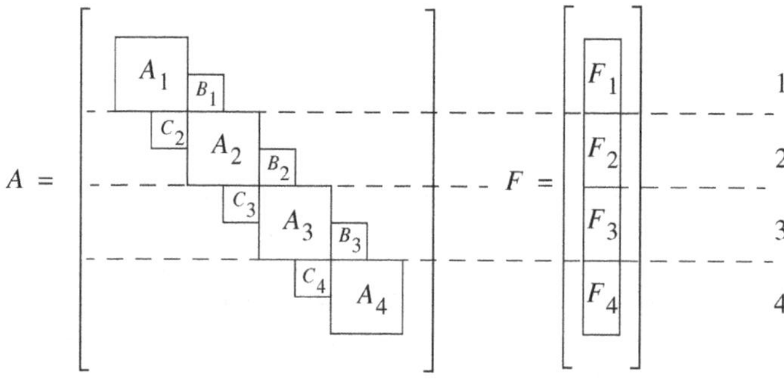

Fig. 5.5 Spike partitioning of the matrix A and block of right-hand sides F with $p = 4$

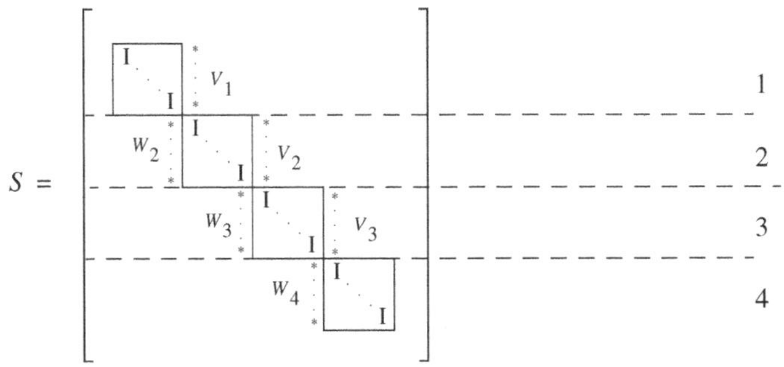

Fig. 5.6 The Spike matrix with 4 partitions

$$
A_j(V_j, W_j) = \begin{pmatrix} 0 & C_j \\ & \vdots & 0 \\ 0 & \vdots \\ B_j & 0 \end{pmatrix}.
\tag{5.2}
$$

Postprocessing Stage

Solving the system $AX = F$ now consists of two steps:

$$\text{(a) solve } DG = F, \tag{5.3}$$

$$\text{(b) solve } SX = G. \tag{5.4}$$

The solution of the linear system $DG = F$ in step (a), yields the modified right hand-side matrix G needed for step (b). Assigning one partition to each node, step (a) is performed with perfect parallelism. If we decouple the pre- and post-processing stages, step (a) may be combined with the generation of the spikes in Eq. (5.2).

Let the spikes V_j and W_j be partitioned as follows:

$$
V_j = \begin{pmatrix} V_j^{(t)} \\ V_j' \\ V_j^{(b)} \end{pmatrix} \text{ and } W_j = \begin{pmatrix} W_j^{(t)} \\ W_j' \\ W_j^{(b)} \end{pmatrix}
\tag{5.5}
$$

where $V_j^{(t)}$, V_j', $V_j^{(b)}$, and $W_j^{(t)}$, W_j', $W_j^{(b)}$, are the top m, the middle $(n_j - 2m)$ and the bottom m rows of V_j and W_j, respectively. In other words,

$$
V_j^{(b)} = [0, \; I_m]V_j; \quad W_j^{(t)} = [I_m \; 0]W_j,
\tag{5.6}
$$

and

$$
V_j^{(t)} = [I_m \; 0]V_j; \quad W_j^{(b)} = [0 \; I_m]W_j.
\tag{5.7}
$$

Similarly, let the jth partitions of X and G, be given by

$$X_j = \begin{pmatrix} X_j^{(t)} \\ X_j' \\ X_j^{(b)} \end{pmatrix} \quad \text{and} \quad G_j = \begin{pmatrix} G_j^{(t)} \\ G_j' \\ G_j^{(b)} \end{pmatrix}. \tag{5.8}$$

Thus, in solving $SX = G$ in step (b), we observe that the problem can be reduced further by solving the reduced system,

$$\hat{S}\hat{X} = \hat{G}, \tag{5.9}$$

where $\hat{S}$ is a block-tridiagonal matrix of order $t = 2m(p - 1)$. While t may not be small, in practice it is much smaller than n (the size of the original system). Each kth diagonal block of $\hat{S}$ is given by,

$$\begin{pmatrix} I_m & V_k^{(b)} \\ W_{k+1}^{(t)} & I_m \end{pmatrix}, \tag{5.10}$$

with the corresponding left and right off-diagonal blocks,

$$\begin{pmatrix} W_k^{(b)} & 0 \\ 0 & 0 \end{pmatrix} \quad \text{and} \quad \begin{pmatrix} 0 & 0 \\ 0 & V_{k+1}^{(t)} \end{pmatrix}, \tag{5.11}$$

and the associated solution and right hand-side matrices

$$\begin{pmatrix} X_k^{(b)} \\ X_{k+1}^{(t)} \end{pmatrix} \quad \text{and} \quad \begin{pmatrix} G_k^{(b)} \\ G_{k+1}^{(t)} \end{pmatrix}.$$

Finally, once the solution $\hat{X}$ of the reduced system (5.9) is obtained, the global solution X is reconstructed with perfect parallelism from $X_k^{(b)}$ ($k = 1, \ldots, p - 1$) and $X_k^{(t)}$ ($k = 2, \ldots, p$) as follows:

$$\begin{cases} X_1' = G_1' - V_1' X_2^{(t)}, \\ X_j' = G_j' - V_j' X_{j+1}^{(t)} - W_j' X_{j-1}^{(b)}, \quad j = 2, \ldots, p - 1, \\ X_p' = G_p' - W_j' X_{p-1}^{(b)}. \end{cases}$$

Remark 5.1 Note that the generation of the spikes is included in the factorization step. In this way, the solver makes use of the spikes stored in memory thus allowing solving the reduced system quickly and efficiently. Since in many applications one has to solve many linear systems with the same coefficient matrix A but with different right hand-sides, optimization of the solver step is crucial for assuring high parallel scalability of the Spike algorithm.

5.2.2 Spike: A Polyalgorithm

Several options are available for efficient implementation of the Spike algorithm on parallel architectures. These choices depend on the properties of the linear system as well as the parallel architecture at hand. More specifically, each of the following three tasks in the Spike algorithm can be handled in several ways:

1. factorization of the diagonal blocks A_j,
2. computation of the spikes,
3. solution of the reduced system.

In the first task, each linear system associated with A_j can be solved via: (i) a direct method making use of the LU-factorization with partial pivoting, or the Cholesky factorization if A_j is symmetric positive definite, (ii) a direct method using an LU-factorization without pivoting but with a diagonal boosting strategy, (iii) an iterative method with a preconditioning strategy, or (iv) via an appropriate approximation of the inverse of A_j. If more than one node is associated with each partition, then linear systems associated with each A_j may be solved using the Spike algorithm creating yet another level of parallelism. In the following, however, we consider only the case in which each partition is associated with only one multicore node.

In the second task, the spikes can be computed either explicitly (fully or partially) using Eq. (5.2), or implicitly—"on-the-fly".

In the third task, the reduced system (5.9) can be solved via (i) a direct method such as LU with partial pivoting, (ii) a "recursive" form of the Spike algorithm, (iii) a preconditioned iterative scheme, or (iv) a "truncated" form of the Spike scheme, which is ideal for diagonally dominant systems, as will be discussed later.

Note that an outer iterative method will be necessary to assure solutions with acceptable relative residuals for the original linear system whenever we do not use numerically stable direct methods for solving systems (5.3) and (5.9). In such a case, the overall hybrid solver consists of an outer Krylov subspace iteration for solving $Ax = f$, in which the Spike algorithm is used as a solver of systems involving a banded preconditioner consisting of an approximate Spike factorization of A in each outer iteration. In the remainder of this section, we describe several variants of the Spike algorithm depending on: (a) whether the whole spikes are obtained explicitly, or (b) whether we obtain only an approximation of the independent reduced system. We describe these variants depending on whether the original banded system is diagonally dominant.

5.2.3 The Non-diagonally Dominant Case

If it is known apriori that all the diagonal blocks A_j are nonsingular, then the LU-factorization of each block A_j with partial pivoting is obtained using either the relevant single core LAPACK routine [19], or its multithreaded counterpart on one multicore node. Solving the resulting banded triangular system to generate the spikes in (5.2), and update the right hand-sides in (5.3), may be realized using a BLAS3 based primitive [16], instead of BLAS2 based LAPACK primitive. In the absence of any knowledge of the nonsingularity of the diagonal blocks A_j, an LU-factorization is performed on each diagonal block without pivoting, but with a diagonal boosting strategy to overcome problems associated with very small pivots. Thus, we either obtain an LU-factorization of a given diagonal block A_j, or the factorization of a slightly perturbed A_j. Such a strategy circumvents difficulties resulting from computationally singular diagonal blocks. If diagonal boosting is required for the factorization of any diagonal block, an outer Krylov subspace iteration is used to solve $Ax = f$ with the preconditioner being $M = \hat{D}\hat{S}$, where $\hat{D}$ and $\hat{S}$ are the Spike factors resulting after diagonal boosting. In this case, the Spike scheme reduces to solving systems involving the preconditioner in each outer Krylov subspace iteration.

One natural way to solve the reduced system (5.9) in parallel is to make use of an inner Krylov subspace iterations with a block Jacobi preconditioner obtained from the diagonal blocks of the reduced system (5.10). For these non-diagonally dominant systems, however, this preconditioner may not be effective in assuring a small relative residual without a large number of outer iterations. This, in turn, will result in high interprocessor communication cost. If the unit cost of interprocessor communication is excessively high, the reduced system may be solved directly on a single node. Such alternative, however, may have memory limitations if the size of the reduced system is large. Instead, we propose the following parallel recursive scheme for solving the reduced system. This "Recursive" scheme involves successive applications of the Spike algorithm resulting in better balance between the computational and communication costs.

First, we dispense with the simple case of two partitions, $p = 2$. In this case, the reduced system consists only of one diagonal block (5.10), extracted from the central part of the system (5.4)

$$\begin{pmatrix} I_m & V_1^{(b)} \\ W_2^{(t)} & I_m \end{pmatrix} \begin{pmatrix} X_1^{(b)} \\ X_2^{(t)} \end{pmatrix} = \begin{pmatrix} G_1^{(b)} \\ G_2^{(t)} \end{pmatrix}, \tag{5.12}$$

that can be solved directly as follows:

- Form $H = I_m - W_2^{(t)} V_1^{(b)}$,
- Solve $H X_2^{(t)} = G_2^{(t)} - W_2^{(t)} G_1^{(b)}$ to obtain $X_2^{(t)}$,
- Compute $X_1^{(b)} = G_1^{(b)} - V_1^{(b)} X_2^{(t)}$.

Note that if we obtain the LU factorization of A_1, and the UL factorization of A_2, then the bottom tip $V_1^{(b)}$ of the right spike, and and the top tip $W_2^{(t)}$ of the left spike can be obtained very economically without obtaining all of the spikes V_1 and W_2. Once $X_1^{(b)}$ and $X_2^{(t)}$ are computed, retrieving the rest of the solution of $AX = F$ cannot be obtained using Eq. (5.12) since the entire spikes are not computed explicitly. Instead, we purify the right hand-side from the contributions of the coupling blocks B and C, thus decoupling the system into two independent subsystems. This is followed by solving these independent systems simultaneously using the previously computed factorizations.

The Recursive Form of the Spike Algorithm

In this variant of the Spike algorithm we wish to solve the reduced system recursively using the Spike algorithm. In order to proceed this way, however, we first need the number of partitions to be a power of 2, i.e. $p = 2^d$, and second we need to work with a slightly modified reduced system. The independent reduced system that is suitable for the first level of this recursion has a coefficient matrix $\tilde{S}_1$ that consists of the matrix $\hat{S}$ in (5.9), augmented at the top by the top m rows of the first partition, and augmented at the bottom by the bottom m rows of the last partition. The structure of this new reduced system, of order $2mp$, remains block-tridiagonal, with each diagonal block being the identity matrix of order $2m$, and the corresponding off-diagonal blocks associated with the kth diagonal block (also of order $2m$) are given by

$$\begin{pmatrix} 0 & W_k^{(t)} \\ 0 & W_k^{(b)} \end{pmatrix} \text{ for } k = 2, \ldots, p, \text{ and} \begin{pmatrix} V_k^{(t)} & 0 \\ V_k^{(b)} & 0 \end{pmatrix} \text{ for } k = 1, \ldots, p - 1. \quad (5.13)$$

Let the spikes of the new reduced system at level 1 of the recursion be denoted by $V_k^{[1]}$ and $W_k^{[1]}$, where

$$V_k^{[1]} = \begin{pmatrix} V_k^{(t)} \\ V_k^{(b)} \end{pmatrix} \text{ and } W_k^{[1]} = \begin{pmatrix} W_k^{(t)} \\ W_k^{(b)} \end{pmatrix}. \quad (5.14)$$

In preparation for level 2 of the recursion of the Spike algorithm, we choose now to partition the matrix $\tilde{S}_1$ using $p/2$ partitions with diagonal blocks each of size $4m$. The matrix can then be factored as

$$\tilde{S}_1 = D_1 \tilde{S}_2,$$

where $\tilde{S}_2$ represents the new Spike matrix at level 2 composed of the spikes $V_k^{[2]}$ and $W_k^{[2]}$. For $p = 4$ partitions, these matrices are of the form

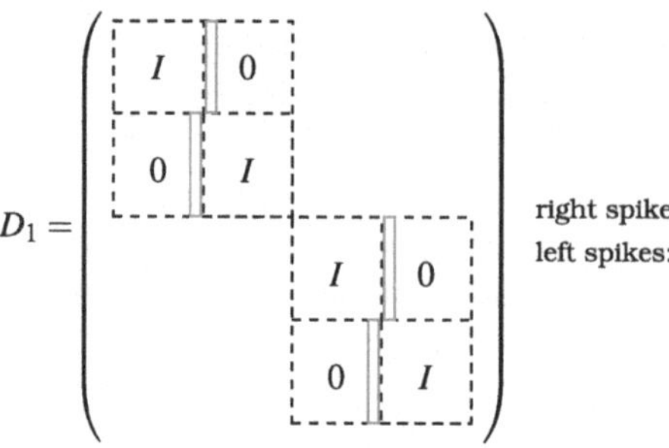

$$D_1 =$$

right spikes: $v_k^{[1]}, k = 1, 3$
left spikes: $\quad w_k^{[1]}, k = 2, 4$

and

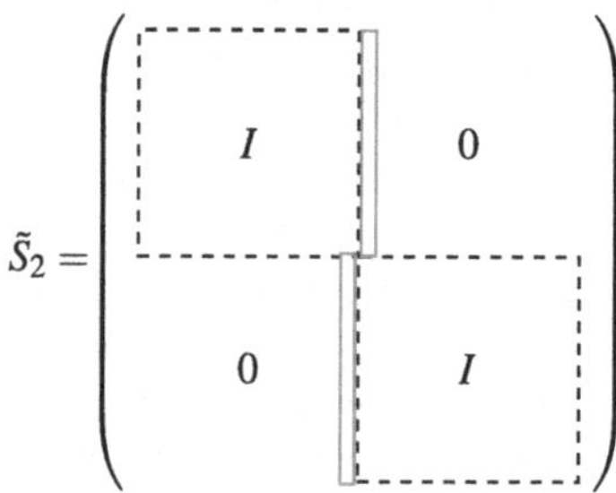

$$\tilde{S}_2 =$$

right spikes: $v_k^{[2]}, k = 2$
left spikes: $\quad w_k^{[2]}, k = 2$

In general, at level i of the recursion, the spikes $V_k^{[i]}$, and $W_k^{[i]}$, with k ranging from 1 to $p/(2^i)$, are of order $2^i m \times m$. Thus, if the number of the original partitions p is equal to 2^d, the total number of recursion levels is $d - 1$ and the matrix $\tilde{S}_1$ is given by the product

$$\tilde{S}_1 = D_1 D_2 \dots D_{d-1} \tilde{S}_d,$$

where the matrix $\tilde{S}_d$ has only two spikes $V_1^{[d]}$ and $W_2^{[d]}$. Thus, the reduced system can be written as,

$$\tilde{S}_d \tilde{X} = B, \tag{5.15}$$

where B is the modified right hand side given by

$$B = D_{d-1}^{-1} \dots D_2^{-1} D_1^{-1} \tilde{G}. \tag{5.16}$$

If we assume that the spikes $V_k^{[i]}$ and $W_k^{[i]}$, for all k, of the matrix $\tilde{S}_i$ are known at a given level i, then we can compute the spikes $V_k^{[i+1]}$ and $W_k^{[i+1]}$ at level $i + 1$ as follows:

STEP 1: Denoting the bottom and the top blocks of the spikes at the level i by

$$V_k^{[i](b)} = [0 \ I_m] V_k^{[i]}; \quad W_k^{[i](t)} = [I_m \ 0] W_k^{[i]},$$

and the middle block of $2m$ rows of the spikes at level $i + 1$ by

$$\begin{pmatrix} \dot{V}_k^{[i+1]} \\ \ddot{V}_k^{[i+1]} \end{pmatrix} = [0 \ \ I_{2m} \ \ 0] V_k^{[i+1]}; \qquad \begin{pmatrix} \dot{W}_k^{[i+1]} \\ \ddot{W}_k^{[i+1]} \end{pmatrix} = [0 \ \ I_{2m} \ \ 0] W_k^{[i+1]},$$

one can form the following reduced systems:

$$\begin{pmatrix} I_m & V_{2k-1}^{[i](b)} \\ W_{2k}^{[i](t)} & I_m \end{pmatrix} \begin{pmatrix} \dot{V}_k^{[i+1]} \\ \ddot{V}_k^{[i+1]} \end{pmatrix} = \begin{pmatrix} 0 \\ V_{2k}^{[i](t)} \end{pmatrix}, \quad k = 1, 2, \ldots, \frac{p}{2^{i-1}} - 1, \quad (5.17)$$

and

$$\begin{pmatrix} I_m & V_{2k-1}^{[i](b)} \\ W_{2k}^{[i](t)} & I_m \end{pmatrix} \begin{pmatrix} \dot{W}_k^{[i+1]} \\ \ddot{W}_k^{[i+1]} \end{pmatrix} = \begin{pmatrix} W_{2k-1}^{[i](b)} \\ 0 \end{pmatrix}, \quad k = 2, 3, \ldots, \frac{p}{2^{i-1}}. \quad (5.18)$$

These reduced systems are solved similarly to (5.12) to obtain the central portion of all the spikes at level $i + 1$.

STEP 2: The rest of the spikes at level $i + 1$ are retrieved as follows:

$$[I_{2^i m} \ \ 0] V_k^{[i+1]} = -V_{2k-1}^{[i]} \ddot{V}_k^{[i+1]}, \quad [0 \ \ I_{2^i m}] V_k^{[i+1]} = V_{2k}^{[i]} - W_{2k}^{[i]} \dot{V}_k^{[i+1]}, \quad (5.19)$$

and

$$[I_{2^i m} \ \ 0] W_k^{[i+1]} = W_{2k-1}^{[i]} - V_{2k-1}^{[i]} \ddot{W}_k^{[i+1]}, \quad [0 \ \ I_{2^i m}] W_k^{[i+1]} = -W_{2k}^{[i]} \dot{W}_k^{[i+1]}. \quad (5.20)$$

In order to compute the modified right hand-side B in (5.16), we need to solve $d - 1$ block-diagonal systems of the form, $D_i \tilde{G}_i = \tilde{G}_{i-1}$, where $\tilde{G}_1 = D_1^{-1} \tilde{G}$). For each diagonal block k, the reduced system is similar to that in (5.12), or in (5.17) and (5.18), but the right hand-side is now defined as a function of $\tilde{G}_{i-1}$. Once we get the central portion of each $\tilde{G}_i$, associated with each block k in D_i, the entire solution is retrieved as in (5.19) and (5.20). Consequently, the linear system (5.15) involves only one reduced system to solve and only one retrieval stage to get the solution $\tilde{X}$, from which the overall solution X is retrieved via (5.12).

The Spike Algorithm on-the-Fly

One can completely avoid explicitly computing and storing all the spikes, and hence the reduced system. Instead, the reduced system is solved using a Krylov subspace method which requires only matrix-vector multiplication. Note that the $2m \times 2m$ diagonal and off-diagonal blocks are given by (5.10) and (5.11), respectively. Since $V_j^{(t)}$, $V_j^{(b)}$, $W_{j+1}^{(t)}$, and $W_{j+1}^{(b)}$, for $j = 1, 2, \ldots, p - 1$ are given by (5.6) and (5.7), in which V_k and W_k are given by (5.1), each matrix-vector product exhibits abundant parallelism in which the most time-consuming computations concerns solving systems of equations involving the diagonal blocks A_j whose LU-factors are available.

5.2.4 The Diagonally Dominant Case

In this section, we describe a variant of the Spike algorithm—the truncated Spike algorithm—that is aimed at handling diagonally dominant systems. This truncated scheme generates an effective preconditioner for such diagonally dominant systems. Further, it yields a modified reduced system that is block-diagonal rather than block-tridiagonal. This, in turn, allows solving the reduced system with higher parallel scalability. In addition, the truncated scheme facilitates different new options for the factorization of each diagonal making it possible to compute the upper and lower tips of each spike rather than the entire spike.

The Truncated Spike Algorithm

If matrix A is diagonally dominant, one can show that the magnitude of the elements of the right spikes V_j decay in magnitude from bottom to top, while the elements of the left spikes W_j decay in magnitude from top to bottom [20]. The decay is more pronounced the higher the degree of diagonal dominance. Since the size of each A_j is much larger than the size of the coupling blocks B_j and C_j, the bottom blocks of the left spikes $W_j^{(b)}$ and the top blocks of the right spikes $V_j^{(t)}$ can be ignored. This leads to an approximate block-diagonal reduced system in which each block is of the form (5.10) leading to enhanced parallelism.

The Truncated Factorization Stage

Since the matrix is diagonally dominant, the LU factorization of each block A_j can be performed without pivoting via the relevant single core LAPACK routine or its multicore counterpart. Using the resulting factors of A_j, the bottom $m \times m$ tip of the right spike V_j in (5.1) is obtained using only the bottom $m \times m$ blocks of L and U. Obtaining the upper $m \times m$ tip of the left spike W_j still requires computing the entire spike with complete forward and backward sweeps. However, if one computes the UL-factorization of A_j, without pivoting, the upper $m \times m$ tip of W_j can be similarly obtained using the top $m \times m$ blocks of the new U and L. Because of the data locality-rich BLAS3, adopting such a LU/UL strategy on many parallel architectures, results in superior performance compared to computing only the LU-factorization for each diagonal block and generating the entire left spikes. Note that, using an appropriate permutation, that reverses the order of the rows and columns of each diagonal block, one can use the same LU-factorization scheme to obtain the corresponding UL-factorization.

Approximation of the Factorization Stage

For high degree of diagonal dominance, an alternative to performing a UL-factorization of the diagonal blocks A_j is to approximate the upper tip of the left spike, $W_j^{(t)}$, of order m, by inverting only the leading $l \times l$ ($l > m$) diagonal block (top left corner) of the banded block A_j. Typically, we choose $l = 2m$ to get a suitable approximation of the $m \times m$ left top corner of the inverse of A_j. With such an approximation strategy, Spike is used as a preconditioner that could yield low relative residuals after few outer iterations of a Krylov subspace method.

As outlined above for the two-partition case, once the reduced system is solved, we purify the right hand-side from the contributions of the coupling blocks B_j and C_j, thus decoupling the system into independent subsystems, one corresponding to each diagonal block. Solving these independent systems simultaneously using the previously computed LU- or UL-factorizations as shown below:

$$\begin{cases} A_1 X_1 = F_1 - \begin{pmatrix} 0 \\ I_m \end{pmatrix} B_j X_2^{(t)}, \\[2em] A_j X_j = F_j - \begin{pmatrix} 0 \\ I_m \end{pmatrix} B_j X_{j+1}^{(t)} - \begin{pmatrix} I_m \\ 0 \end{pmatrix} C_j X_{j-1}^{(b)}, \quad j = 2, \ldots, p-1 \\[2em] A_p X_p = F_p - \begin{pmatrix} I_m \\ 0 \end{pmatrix} C_j X_{p-1}^{(b)}. \end{cases}$$

5.3 The Spike-Balance Scheme

We introduce yet another variant of the Spike scheme for solving banded systems that depends on solving simultaneously several independent underdetermined linear least squares problems under certain constraints [7].

Consider the nonsingular linear system

$$Ax = f, \tag{5.21}$$

where A is an $n \times n$ banded matrix with bandwidth $\beta = 2m+1$. For ease of illustration we consider first the case in which n is even, e.g. $n = 2s$, and A is partitioned into two equal block rows (each of s rows), as shown in Fig. 5.7. The linear system can be represented as

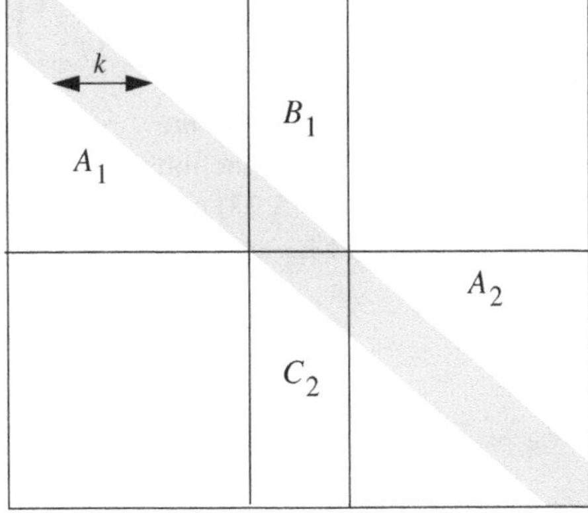

Fig. 5.7 The Spike-balance scheme for *two block rows*

$$\begin{pmatrix} A_1 \ B_1 \\ \ C_2 \ A_2 \end{pmatrix} \begin{pmatrix} x_1 \\ \xi \\ x_2 \end{pmatrix} = \begin{pmatrix} f_1 \\ f_2 \end{pmatrix}, \tag{5.22}$$

where A_1 and A_2 are of order $(s \times (s - m))$, and B_1, C_2 are of order $(s \times 2m)$.

Denoting that portion of the solution vector which represents the unknowns common to both blocks by ξ, the above system gives rise to the following two underdetermined systems,

$$(A_1 , \ B_1) \begin{pmatrix} x_1 \\ \xi \end{pmatrix} = f_1, \tag{5.23}$$

$$(C_2 , \ A_2) \begin{pmatrix} \tilde{\xi} \\ x_2 \end{pmatrix} = f_2. \tag{5.24}$$

While these two underdetermined systems can be solved independently, they will yield the solution of the global system (5.22), only if $\xi = \tilde{\xi}$. Here, the vectors x_1, x_2 are of order $(s - m)$, and ξ, $\tilde{\xi}$ each of order $2m$.

Let the matrices of the underdetermined systems be denoted as follows:

$$E_1 = (A_1, B_1),$$
$$E_2 = (C_2, A_2),$$

then the general form of the solution of the two underdetermined systems (5.23) and (5.24) is given by

$$v_i = p_i + Z_i y_i, \quad i = 1, 2,$$

where p_i is a particular solution, Z_i is a basis for $\mathcal{N}(E_i)$, and y_i is an arbitrary vector. Note the p_i is of order $s + m$ and Z_i is of order $(s + m) \times m$. Thus,

$$\begin{pmatrix} x_1 \\ \xi \end{pmatrix} = \begin{pmatrix} p_{1,1} \\ p_{1,2} \end{pmatrix} + \begin{pmatrix} Z_{1,1} \\ Z_{1,2} \end{pmatrix} y_1, \tag{5.25}$$

$$\begin{pmatrix} \tilde{\xi} \\ x_2 \end{pmatrix} = \begin{pmatrix} p_{2,1} \\ p_{2,2} \end{pmatrix} + \begin{pmatrix} Z_{2,1} \\ Z_{2,2} \end{pmatrix} y_2, \tag{5.26}$$

where y_1 and y_2 are of order m.

Clearly ξ and $\tilde{\xi}$ are functions of y_1 and y_2, respectively. Since the underdetermined linear systems (5.23) and (5.24) are consistent under the assumption that the submatrices E_1 and E_2 are of full row rank, enforcing the condition:

$$\xi(y_1) = \tilde{\xi}(y_2),$$

i.e.

$$p_{1,2} + Z_{1,2} y_1 = p_{2,1} + Z_{2,1} y_2.$$

assures us of obtaining the global solution of (5.21) and yields the *balance* linear system

$$My = g$$

of order $2m$ where $M = (Z_{1,2}, -Z_{2,1})$, $g = p_{2,1} - p_{1,2}$, and whose solution $y = (y_1^\top, y_2^\top)^\top$, satisfies the above condition.

Thus, the *Spike-balance* scheme for the two-block case consists of the following steps:

1. Obtain the orthogonal factorization of E_i, $i = 1, 2$ to determine Z_i, and to allow computing the particular solution $p_i = E_i^\top (E_i E_i^\top)^{-1} f_i$ in a numerically stable way, thus solving the underdetermined systems (5.23) and (5.24).
2. Solve the balance system $My = g$.
3. Back-substitute y in (5.25) and (5.26) to determine x_1, x_2, and ξ.

The first step requires the solution of underdetermined systems of the form

$$Bz = d,$$

where B is an $(s \times (s+m))$ matrix with full row-rank. The general solution is given by

$$z = B^+ d + P_{\mathscr{N}(B)} u,$$

where $B^+ = B^\top (BB^\top)^{-1}$ is the generalized inverse of B, and $P_{\mathscr{N}(B)}$ is the orthogonal projector onto $\mathscr{N}(B)$. Furthermore, the projector $P_{\mathscr{N}(B)}$ can be expressed as,

$$P_{\mathscr{N}(B)} = I - B^+ B.$$

The general solution z can be obtained using a variety of techniques, e.g., see [21, 22]. Using sparse orthogonal factorization, e.g. see [23], we obtain the decomposition,

$$B^\top = (Q, Z)\begin{pmatrix} R \\ 0 \end{pmatrix},$$

where R is an $s \times s$ triangular matrix and Z is an $(s+m) \times m$ matrix with orthonormal columns. Since Z is a basis for $\mathscr{N}(B)$, the general solution is given as

$$z = \bar{z} + Zy,$$
$$\text{where } \bar{z} = QR^{-T}d.$$

Before presenting the solution scheme of the balance system $My = g$, we show first that it is nonsingular. Let,

$$E_i^\top = (Q_i, Z_i)\begin{pmatrix} R_i \\ 0 \end{pmatrix} \quad i = 1, 2,$$

where R_i is upper triangular, nonsingular of order $s = n/2$. Partitioning Q_1 and Q_2 as

$$Q_1 = \begin{pmatrix} Q_{11} \\ Q_{12} \end{pmatrix}, \qquad Q_2 = \begin{pmatrix} Q_{21} \\ Q_{22} \end{pmatrix},$$

where Q_{12}, Q_{21} are of order $2m \times s$, and Q_{11}, Q_{22} are of order $(s - m) \times s$, and similarly for Z_1 and Z_2,

$$Z_1 = \begin{pmatrix} Z_{11} \\ Z_{12} \end{pmatrix}, \qquad Z_2 = \begin{pmatrix} Z_{21} \\ Z_{22} \end{pmatrix}$$

in which Z_{12}, Z_{21} are of order $(2m \times m)$, and Z_{11}, Z_{22} are of order $(s - m) \times m$. Now, consider

$$\tilde{A} = \begin{pmatrix} R_1^{-T} & 0 \\ 0 & R_2^{-T} \end{pmatrix} A = \begin{pmatrix} Q_{11}^\top & Q_{12}^\top & 0 \\ 0 & Q_{21}^\top & Q_{22}^\top \end{pmatrix},$$

and

$$\tilde{A}\tilde{A}^\top = \begin{pmatrix} I_m & Q_{12}^\top Q_{21} \\ Q_{21}^\top Q_{12} & I_m \end{pmatrix}.$$

Let, $H = Q_{12}^\top Q_{21}$. Then, since $(\tilde{A}\tilde{A}^\top)$ is symmetric positive definite,

$$\lambda(\tilde{A}\tilde{A}^\top) = 1 \pm \eta > 0$$

in which $\lambda(H^\top H) = \eta^2$. Consequently, $-1 < \eta < 1$. Consider next the structure of $\tilde{A}^\top \tilde{A}$,

$$\tilde{A}^\top \tilde{A} = \left(\begin{array}{c|c|c} Q_{11}Q_{11}^\top & Q_{11}Q_{12}^\top & 0 \\ \hline Q_{12}Q_{11}^\top & Q_{12}Q_{12}^\top + Q_{21}Q_{21}^\top & Q_{21}Q_{22}^\top \\ \hline 0 & Q_{22}Q_{21}^\top & Q_{22}Q_{22}^\top \end{array} \right).$$

Since $\tilde{A}^\top \tilde{A}$ is positive definite, then the diagonal block $N = (Q_{12}Q_{12}^\top + Q_{21}Q_{21}^\top)$ is also positive definite. Further, since

$$Q_i Q_i^\top = I - Z_i Z_i^\top \quad i = 1, 2,$$

we can easily verify that

$$N = 2I - (Z_{12}Z_{12}^\top + Z_{21}Z_{21}^\top),$$
$$= 2I - MM^\top.$$

But, the eigenvalues of N lie within the spectrum of $\tilde{A}^\top \tilde{A}$, or $\tilde{A}\tilde{A}^\top$, i.e.,

$$1 - \rho \le 2 - \lambda(MM^\top) \le 1 + \rho$$

in which $\rho^2 < 1$ is spectral radius of $(H^\top H)$. Hence,

$$0 < 1 - \rho \le \lambda(MM^\top) \le 1 + \rho.$$

In other words, M is nonsingular. Next, we consider the general case of Spike-balance scheme with p partitions, i.e. $n = s \times p$ (e.g., see Fig. 5.8),

$$\begin{pmatrix} A_1 & B_1 & & & & \\ & C_2 & A_2 & B_2 & & \\ & & C_3 & A_3 & B_3 & \\ & & & \ddots & \ddots & \\ & & & & C_{p-1} & A_{p-1} & B_{p-1} \\ & & & & & C_p & A_p \end{pmatrix} \begin{pmatrix} x_1 \\ \xi_1 \\ x_2 \\ \xi_2 \\ \vdots \\ \xi_{p-1} \\ x_p \end{pmatrix} = \begin{pmatrix} f_1 \\ f_2 \\ f_3 \\ \vdots \\ f_{p-1} \\ f_p \end{pmatrix}. \tag{5.27}$$

With p equal block rows, the columns of A are divided into $2p - 1$ column blocks, where the unknowns ξ_i, $i = 1, \ldots p - 1$, are common to consecutive blocks of the

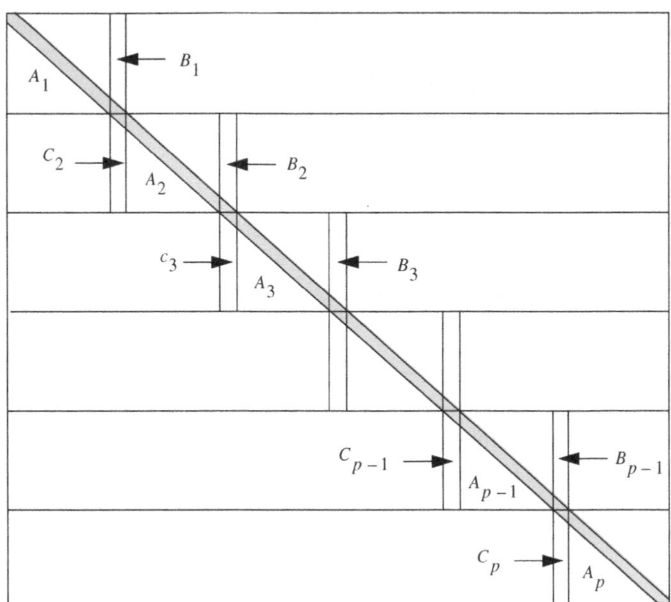

Fig. 5.8 The Spike-balance scheme for p *block rows*

matrix. Then, A_1 and A_p are of order $s \times (s - m)$, and each A_i, $i = 2, 3, \ldots, p - 1$ is of order $s \times (s - 2m)$, and B_i, C_i are of order $s \times 2m$. These p block rows give rise to the following set of underdetermined systems:

$$(A_1 \quad B_1) \begin{pmatrix} x_1 \\ \xi_1 \end{pmatrix}, \tag{5.28}$$

$$(C_i \quad A_i \quad B_i) \begin{pmatrix} \tilde{\xi}_{i-1} \\ x_i \\ \xi_i \end{pmatrix} = f_i, \quad i = 2, \ldots, p - 1, \tag{5.29}$$

$$(C_p \quad A_p) \begin{pmatrix} \tilde{\xi}_{p-1} \\ x_p \end{pmatrix} = f_p. \tag{5.30}$$

Denoting the coefficient matrices of the above underdetermined systems as follows:

$$\begin{aligned} E_1 &= (A_1, B_1), \\ E_i &= (C_i, A_i, B_i), \quad i = 2, \ldots, p - 1, \\ E_p &= (C_p, A_p), \end{aligned}$$

the general solution of the ith underdetermined system is given by

$$v_i = p_i + Z_i y_i,$$

where p_i is the particular solution, Z_i is a basis for $\mathcal{N}(E_i)$, and y_i is an arbitrary vector. The solutions of the systems (5.28) through (5.30) are then given by

$$\begin{pmatrix} x_1 \\ \xi_1 \end{pmatrix} = \begin{pmatrix} p_{1,1} \\ p_{1,2} \end{pmatrix} + \begin{pmatrix} Z_{1,1} \\ Z_{1,2} \end{pmatrix} y_1, \tag{5.31}$$

$$\begin{pmatrix} \tilde{\xi}_{i-1} \\ x_i \\ \xi_i \end{pmatrix} = \begin{pmatrix} p_{i,1} \\ p_{i,2} \\ p_{i,3} \end{pmatrix} + \begin{pmatrix} Z_{i,1} \\ Z_{i,2} \\ Z_{i,3} \end{pmatrix} y_i, \quad i = 2, \ldots, p - 1, \tag{5.32}$$

$$\begin{pmatrix} \tilde{\xi}_{p-1} \\ x_p \end{pmatrix} = \begin{pmatrix} p_{p,1} \\ p_{p,2} \end{pmatrix} + \begin{pmatrix} Z_{p,1} \\ Z_{p,2} \end{pmatrix} y_p. \tag{5.33}$$

Here, each ξ and $\tilde{\xi}$ is of order $2m$, and each y_i, $i = 2, 3, \ldots, p - 1$ is of the same order $2m$, with y_1, y_p being of order m. Since the submatrices E_i are of full row-rank, the above underdetermined linear systems are consistent, and we can enforce the following conditions to insure obtaining the unique solution of the global system:

$$\xi_i(y_i) = \tilde{\xi}_{i-1}(y_{i+1}), \quad i = 1, \ldots, p - 1.$$

This gives rise to the system of equations

$$p_{1,2} + Z_{1,2}y_1 = p_{2,1} + Z_{2,1}y_2,$$
$$p_{i,3} + Z_{i,3}y_i = p_{i+1,1} + Z_{i+1,1}y_{i+1}, \quad i = 2, \ldots, p-1,$$

which yields the balance system

$$My = g.$$

Here, M is of order $v = 2m(p-1)$,

$$M = \begin{pmatrix} Z_{1,2} & -Z_{2,1} & & & \\ & Z_{2,3} & -Z_{3,1} & & \\ & & Z_{3,3} & -Z_{4,1} & \\ & & & \ddots & \ddots \\ & & & & Z_{p-1,3} & -Z_{p,1} \end{pmatrix}, \tag{5.34}$$

with y consisting of the subvectors y_i, $i = 1, 2, \ldots, p$, and the right-hand side g given by

$$g = \begin{pmatrix} p_{2,1} - p_{1,2} \\ p_{3,1} - p_{2,3} \\ \vdots \\ p_{p,1} - p_{p-1,3} \end{pmatrix}.$$

Thus, the banded linear system (5.27) may be solved by simultaneously computing the solutions of the underdetermined systems (5.28)–(5.30), followed by solving the balance system $My = g$. Next, we summarize the Spike-balance scheme for p blocks.

The Spike-Balance Scheme

1. Simultaneously solve the underdetermined systems (5.28)–(5.30) to obtain the particular solution p_i and the bases Z_i, $i = 1, \ldots, p$.
2. Solve the balance system $My = g$.
3. Simultaneously back-substitute the subvectors y_i to determine all the subvectors x_i and ξ_i.

Solving the underdetermined systems, in Step 1, via orthogonal factorization of the matrices E_i has been outlined above. Later, we will consider a solution strategy that does not require the complete orthogonal factorization of the E_i's and storing the relevant parts of the matrices Z_i which constitute the bases of $\mathcal{N}(E_i)$.

In Step 2, we can solve the balance system in a number of ways. The matrix M can be computed explicitly, as outlined above, and the balance system is handled via a direct or preconditioned iterative solver. When using a direct method, it is advantageous to exploit the block-bidiagonal structure of the balance system. Iterative methods may be used when M is sufficiently large and storage is at a premium.

In fact, for large linear systems (5.21), and a relatively large number of partitions p, e.g. when solving (5.21) on a cluster with large number of nodes, it is preferable not to form the resulting large balance matrix M. Instead, we use an iterative scheme in which the major kernels are: (i) matrix–vector multiplication, and (ii) solving systems involving the preconditioner. From the above derivation of the balance system, we observe the following:

$$r(y) = g - My = \tilde{\xi}(y) - \xi(y),$$

and computing the matrix-vector product My is given by

$$My = r(0) - r(y),$$

where $r(0) = g$. In this case, however, one needs to solve the underdetermined systems in each iteration.

Note that conditioning the matrix M is critical for the rapid convergence of any iterative method used for solving the balance system. The following theorem provides an estimate of the condition number of M, $\kappa(M)$.

Theorem 5.1 ([7]) *The coefficient matrix M of the balance system has a condition number which is at most equal to that of the coefficient matrix A of the original banded system (5.21), i.e.,*

$$\kappa(M) \leq \kappa(A).$$

In order to form the coefficient matrix M of the balance system, we need to obtain the matrices Z_i, $i = 1, 2, \ldots, p$. Depending on the size of the system and the parallel architecture at hand, however, the storage requirements could be excessive. Below, we outline an alternate approach that does not form the balance system explicitly.

A Projection-Based Spike-Balance Scheme

In this approach, the matrix M is not computed explicitly, rather, the balance system is available only implicitly in the form of a matrix-vector product, in which the matrix under consideration is given by $MM^\top$. As a result, an iterative scheme such as the conjugate gradient method (CG) can be used to solve the system $MM^\top \hat{w} = g$ instead of the balance system in step 2 of the Spike-balance scheme.

This algorithm is best illustrated by considering the 2-partition case (5.23) and (5.24),

$$E_1 \begin{pmatrix} x_1 \\ \xi \end{pmatrix} = f_1,$$

$$E_2 \begin{pmatrix} \tilde{\xi} \\ x_2 \end{pmatrix} = f_2,$$

where the general solution for these underdetermined systems are given by

$$\begin{pmatrix} x_1 \\ \xi \end{pmatrix} = p_1 + (I - P_1)u_1, \tag{5.35}$$

$$\begin{pmatrix} \tilde{\xi} \\ x_2 \end{pmatrix} = p_2 + (I - P_2)u_2, \tag{5.36}$$

in which the particular solutions p_i are expressed as

$$p_i = E_i^{\top}(E_i E_i^{\top})^{-1} f_i, \quad i = 1, 2,$$

and the projectors P_i are given by

$$P_i = E_i^{\top}(E_i E_i^{\top})^{-1} E_i, \quad i = 1, 2. \tag{5.37}$$

Imposing the condition

$$\xi = \tilde{\xi},$$

we obtain the identical balance system, but expressed differently as

$$(N_1, N_2) \begin{pmatrix} u_1 \\ u_2 \end{pmatrix} = g$$

in which,

$$\begin{aligned}
g &= (0, \hat{I})p_1 - (\hat{I}, 0)p_2, \\
N_1 &= -(0, \hat{I})(I - P_1), \\
N_2 &= (\hat{I}, 0)(I - P_2).
\end{aligned}$$

Here, $\hat{I} \equiv I_{2\beta}$, and $I \equiv I_{(\frac{n}{2}+\beta)}$.

Let, $u_1 = N_1^{\top} w$ and $u_2 = N_2^{\top} w$ to obtain the symmetric positive definite system

$$MM^{\top} w = g, \tag{5.38}$$

where

$$MM^{\top} = (0, \hat{I})(I - P_1) \begin{pmatrix} 0 \\ \hat{I} \end{pmatrix} + (\hat{I}, 0)(I - P_2) \begin{pmatrix} \hat{I} \\ 0 \end{pmatrix}. \tag{5.39}$$

The system (5.38) is solved using the CG scheme with a preconditioning strategy. In each CG iteration, the multiplication of $MM^{\top}$ by a vector requires the simultaneous multiplications (or projections),

$$c_j = (I - P_j)b_j, \quad j = 1, 2$$

forming an outer level of parallelism, e.g., using two nodes. Such a projection corresponds to computing the residuals of the least squares problems,

$$\min_{c_j} \|b_j - E_j c_j\|_2, \quad j = 1, 2.$$

These residuals, in turn, can be computed using inner CG iterations with each iteration involving matrix-vector multiplication of the form,

$$h_i = E_i E_i^\top v_i, \quad i = 1, 2$$

forming an inner level of parallelism, e.g., taking advantage of the multicore architecture of each node. Once the solution w of the balance system in (5.38) is obtained, the right-hand sides of (5.35) and (5.36) are updated as follows:

$$\begin{pmatrix} x_1 \\ \xi \end{pmatrix} = p_1 - (I - P_1) \begin{pmatrix} 0 \\ w \end{pmatrix},$$

and

$$\begin{pmatrix} \hat{\xi} \\ x_2 \end{pmatrix} = p_2 + (I - P_2) \begin{pmatrix} w \\ 0 \end{pmatrix}.$$

For the general p-partition case, the matrix $MM^\top$ in (5.38) becomes of the form,

$$MM^\top = \sum_{i=1}^{p} M_i M_i^\top, \tag{5.40}$$

in which each $M_i M_i^\top$ is actually a section of the projector

$$I - P_i = I - E_i^+ E_i. \tag{5.41}$$

More specifically,

$$\tilde{M}_i \tilde{M}_i^\top = \begin{pmatrix} -\hat{I} \, 0 \, 0 \\ 0 \, 0 \, \hat{I} \end{pmatrix} (I - P_i) \begin{pmatrix} -\hat{I} \, 0 \, 0 \\ 0 \, 0 \, \hat{I} \end{pmatrix}^\top.$$

Hence, it can be seen that $MM^\top$ can be expressed as the sum of sections of the projectors $(I - P_i), i = 1, \ldots, p$. As outlined above, for the 2-partition case, such a form of $MM^\top$ allows for the exploitation of multilevel parallelism when performing matrix-vector product in the conjugate gradient algorithm for solving the balance system. Once the solution of the balance system is obtained, the individual particular solutions for the p partitions are updated simultaneously as outlined above for the 2-partition case.

While the projection-based approach has the advantage of replacing the orthogonal factorization of the block rows E_i with projections onto the null spaces of E_i, leading to significant savings in storage and computation, we are now faced with a problem of solving balance systems in the form of the normal equations. Hence, it is essential to adopt a preconditioning strategy (e.g. using the block-diagonal of $MM^\top$ as a preconditioner) to achieve a solution for the balance system in few CG iterations.

5.4 A Tearing-Based Banded Solver

5.4.1 Introduction

Here we consider solving wide-banded linear systems that can be expressed as overlapping diagonal blocks in which each is a block-tridiagonal matrix. Our approach in this tearing-based scheme is different from the Spike algorithm variants discussed above. This scheme was first outlined in [24], where the study was restricted to diagonally dominant symmetric positive definite systems . This was later generalized to nonsymmetric linear systems without the requirement of diagonal dominance, e.g. see [25]. Later, in Chap. 10, we extend this tearing scheme for the case when we strive to obtain a central band preconditioner that encapsulates as many nonzero elements as possible.

First, we introduce the algorithm by showing how it *tears* the original block-tridiagonal system and extract a smaller balance system to solve. Note that the extracted balance system here is not identical to that described in the Spike-balance scheme. Second, we analyze the conditions that guarantee the nonsingularity of the balance system. Further, we show that if the original system is symmetric positive definite and diagonally dominant then the smaller balance system is also symmetric positive definite as well. Third, we discuss preconditioned iterative methods for solving the balance system.

5.4.2 Partitioning

Consider the linear system

$$Ax = f, \tag{5.42}$$

where $A \in \mathbb{R}^{n \times n}$ is nonsingular and $x, f \in \mathbb{R}^n$. Let $A = [a_{ij}]$ be a banded matrix of bandwidth $2\tau + 1$, i.e. $a_{ij} = 0$ for $|i - j| \geq \tau \ll n$. Hence we can cast our banded linear system (5.42) in the block-tridiagonal form. For clarity of the presentation, we illustrate the partitioning and tearing scheme using three overlapped partitions ($p = 3$). Also, without loss of generality, we assume that all the partitions, or overlapped block-diagonal matrices, are of equal size s, and that all the overlaps are of identical size τ. Generalization to the case of $p > 3$ partitions of different sizes is straightforward.

Let the banded matrix A and vectors x and f be partitioned as

$$A = \begin{pmatrix} A_{11} & A_{12} & & & & & \\ A_{21} & A_{22} & A_{23} & & & & \\ & A_{32} & A_{33} & A_{34} & & & \\ & & A_{43} & A_{44} & A_{45} & & \\ & & & A_{54} & A_{55} & A_{56} & \\ & & & & A_{65} & A_{66} & A_{67} \\ & & & & & A_{76} & A_{77} \end{pmatrix} , \quad x = \begin{pmatrix} x_1 \\ x_2 \\ x_3 \\ x_4 \\ x_5 \\ x_6 \\ x_7 \end{pmatrix} , \quad \text{and } f = \begin{pmatrix} f_1 \\ f_2 \\ f_3 \\ f_4 \\ f_5 \\ f_6 \\ f_7 \end{pmatrix} .$$

$$(5.43)$$

Here, A_{ij}, x_i and f_i for $i, j = 1, \ldots, 7$ are blocks of appropriate size. Let the partitions of A, delineated by lines in the illustration of A in (5.43), be of the form

$$A_k = \begin{pmatrix} A_{11}^{(k)} & A_{12}^{(k)} & \\ A_{21}^{(k)} & A_{22}^{(k)} & A_{23}^{(k)} \\ & A_{32}^{(k)} & A_{33}^{(k)} \end{pmatrix} , \quad \text{for } k = 1, 2, 3. \tag{5.44}$$

The blocks $A_{\mu\nu}^{(k)} = A_{\eta+\mu,\eta+\nu}$ for $\eta = 2(k-1)$ and $\mu, \nu = 1, 2, 3$, except for the overlaps between partitions. The overlaps consist of the top left blocks of the last and middle partitions, and bottom right blocks of the first and middle partitions. For these blocks the following equality holds $A_{33}^{(k-1)} + A_{11}^{(k)} = A_{\eta+1,\eta+1}$. The exact choice of splitting A_{33} into $A_{33}^{(k-1)}$ and $A_{11}^{(k)}$ will be discussed below.

Thus, we can rewrite the original system as a set of smaller linear systems, $k = 1, 2, 3$,

$$\begin{pmatrix} A_{11}^{(k)} & A_{12}^{(k)} & \\ A_{21}^{(k)} & A_{22}^{(k)} & A_{23}^{(k)} \\ & A_{32}^{(k)} & A_{33}^{(k)} \end{pmatrix} \begin{pmatrix} x_1^{(k)} \\ x_2^{(k)} \\ x_3^{(k)} \end{pmatrix} = \begin{pmatrix} (1 - \alpha_{k-1}) f_{\eta+1} - y_{k-1} \\ f_{\eta+2} \\ \alpha_k f_{\eta+3} + y_k \end{pmatrix} , \tag{5.45}$$

where $\alpha_0 = 0$, $\alpha_3 = 1$, $y_0 = y_3 = 0$, and y_ζ, α_ζ for $\zeta = 1, 2$ are yet to be specified. Now, we need to choose the scaling parameters $0 \le \alpha_1, \alpha_2 \le 1$ and determine the *adjustment vector* $y^\top = (y_1^\top, y_2^\top)$.

Notice that the parameters α_ζ, can be used for example to give more weight to a part of the original right hand-side corresponding to a particular partition. For simplicity, however, we set all the α_ζ's to 0.5.

The adjustment vector y is used to modify the right hand-sides of the smaller systems in (5.45), such that their solutions match at the portions corresponding to the overlaps. This is equivalent to determining the adjustment vector y so that

$$x_3^{(\zeta)} = x_1^{(\zeta+1)} \quad \text{for } \zeta = 1, 2. \tag{5.46}$$

If we assume that the partition A_k in (5.44) is selected to be nonsingular with,

$$A_k^{-1} = \begin{pmatrix} B_{11}^{(k)} & B_{12}^{(k)} & B_{13}^{(k)} \\ B_{21}^{(k)} & B_{22}^{(k)} & B_{23}^{(k)} \\ B_{31}^{(k)} & B_{32}^{(k)} & B_{33}^{(k)} \end{pmatrix},$$

then, we from (5.45) we have

$$\begin{pmatrix} x_1^{(k)} \\ x_2^{(k)} \\ x_3^{(k)} \end{pmatrix} = \begin{pmatrix} B_{11}^{(k)} & B_{12}^{(k)} & B_{13}^{(k)} \\ B_{21}^{(k)} & B_{22}^{(k)} & B_{23}^{(k)} \\ B_{31}^{(k)} & B_{32}^{(k)} & B_{33}^{(k)} \end{pmatrix} \begin{pmatrix} (1 - \alpha_{k-1})f_{\eta+1} - y_{k-1} \\ f_{\eta+2} \\ \alpha_k f_{\eta+3} + y_k \end{pmatrix}, \tag{5.47}$$

i.e.

$$\begin{cases} x_1^{(k)} = B_{11}^{(k)}((1 - \alpha_{k-1})f_{\eta+1} - y_{k-1}) + B_{12}^{(k)} f_{\eta+2} + B_{13}^{(k)}(\alpha_k f_{\eta+3} + y_k), \\ x_3^{(k)} = B_{31}^{(k)}((1 - \alpha_{k-1})f_{\eta+1} - y_{k-1}) + B_{32}^{(k)} f_{\eta+2} + B_{33}^{(k)}(\alpha_k f_{\eta+3} + y_k). \end{cases} \tag{5.48}$$

Using (5.46) and (5.48) we obtain

$$(B_{33}^{(\zeta)} + B_{11}^{(\zeta+1)})y_\zeta = g_\zeta + B_{31}^{(\zeta)}y_{\zeta-1} + B_{13}^{(\zeta+1)}y_{\zeta+1} \tag{5.49}$$

for $\zeta = 1, 2$, where

$$g_\zeta = ((\alpha_{\zeta-1}-1)B_{31}^{(\zeta)}, -B_{32}^{(\zeta)}, (1-\alpha_\zeta)B_{11}^{(\zeta+1)}-\alpha_\zeta B_{33}^{(\zeta)}, B_{12}^{(\zeta+1)}, \alpha_{\zeta+1}B_{13}^{(\zeta+1)}) \begin{pmatrix} f_{\eta-1} \\ f_\eta \\ f_{\eta+1} \\ f_{\eta+2} \\ f_{\eta+3} \end{pmatrix}. \tag{5.50}$$

Finally, letting $g^\top = (g_1^\top, g_2^\top)$, the adjustment vector y can be found by solving the balance system

$$My = g, \tag{5.51}$$

where

$$M = \begin{pmatrix} B_{33}^{(1)} + B_{11}^{(2)} & -B_{13}^{(2)} \\ -B_{31}^{(2)} & B_{33}^{(2)} + B_{11}^{(3)} \end{pmatrix}. \tag{5.52}$$

Once y is obtained, we can solve the linear systems in (5.45) independently for each $k = 1, 2, 3$. Next we focus our attention on solving (5.51). First we note that the matrix M is not available explicitly, thus using a direct method to solve the balance system (5.52) is not possible and we need to resort to iterative schemes that utilize M implicitly for performing matrix-vector multiplications of the form $z = M * v$. For example, one can use Krylov subspace methods, e.g. CG or BiCGstab for the symmetric positive definite and nonsymmetric cases, respectively. However, to use such solvers we must be able to compute the initial residual $r_{\text{init}} = g - M y_{\text{init}}$, and matrix-vector products of the form $q = Mp$. Further, we need to determine those conditions that A and its partitions A_k must satisfy in order to yield a nonsingular *balance system*.

Note that for an arbitrary number p of partitions, i.e. p overlapped block-diagonal matrices, the balance system is block-tridiagonal of the form

$$
\begin{pmatrix}
B_{33}^{(1)} + B_{11}^{(2)} & -B_{13}^{(2)} & & & \\
-B_{31}^{(2)} & B_{33}^{(2)} + B_{11}^{(3)} & -B_{13}^{(3)} & & \\
& & \ddots & & \\
& & -B_{31}^{(p-2)} & B_{33}^{(p-2)} + B_{11}^{(p-1)} & -B_{13}^{(p-1)} \\
& & & -B_{31}^{(p-1)} & B_{33}^{(p-1)} + B_{11}^{(p)}
\end{pmatrix}. \tag{5.53}
$$

5.4.3 The Balance System

Since the tearing approach that enables parallel scalability depends critically on how effectively we can solve the balance system via an iterative method, we explore further the characteristics of its coefficient matrix M.

The Symmetric Positive Definite Case

First, we assume that A is symmetric positive definite (SPD) and investigate the conditions under which the balance system is also SPD.

Theorem 5.2 ([25]) *If the partitions A_k in (5.44) are SPD for $k = 1, \ldots, p$ then the balance system in (5.51) is also* SPD.

Proof Without loss of generality, let $p = 3$. Since each A_k is SPD then A_k^{-1} is also SPD. Next, let

$$
Q^\top = \begin{pmatrix} I & 0 & 0 \\ 0 & 0 & -I \end{pmatrix}, \tag{5.54}
$$

then from (5.47), we obtain

$$
Q^\top A_k^{-1} Q = \begin{pmatrix} B_{11}^{(k)} & -B_{13}^{(k)} \\ -B_{31}^{(k)} & B_{33}^{(k)} \end{pmatrix}, \tag{5.55}
$$

which is also symmetric positive definite. But, since the *balance system* coefficient matrix M can be written as the sum

$$
M = M_1 + M_2 + M_3 = \begin{pmatrix} B_{33}^{(1)} & 0 \\ 0 & 0 \end{pmatrix} + \begin{pmatrix} B_{11}^{(2)} & -B_{13}^{(2)} \\ -B_{31}^{(2)} & B_{33}^{(2)} \end{pmatrix} + \begin{pmatrix} 0 & 0 \\ 0 & B_{11}^{(3)} \end{pmatrix}, \tag{5.56}
$$

then for any nonzero vector z, we have $z^\top M z > 0$.

If, in addition, A is also diagonally dominant (d.d.), i.e., $\sum_{j=1,j\neq i}^{n} |a_{ij}| < |a_{ii}|$ for $i = 1, \ldots, n$, then we can find a splitting that results in each A_k that is also SPD and d.d., which, in turn, guarantees that M is SPD but not necessarily diagonally dominant [25].

Theorem 5.3 ([25]) *If A in (5.42) is* SPD *and d.d., then the partitions A_k in (5.44) can be chosen such that they inherit the same properties. Further, the coefficient matrix M, of the resulting balance system in (5.51), is also* SPD.

Proof Since A is d.d., we only need to obtain a splitting that ensures the diagonal dominance of the overlapping parts. Let e_i be the ith column of the identity, $e = [1, \ldots, 1]^\top$, diag(C) and offdiag(C) denote the diagonal and off-diagonal elements of the matrix C respectively. Now let the elements of the diagonal matrices $H_\zeta^{(1)} = [h_{ii}^{(\zeta,1)}]$ and $H_{\zeta+1}^{(2)} = [h_{ii}^{(\zeta+1,2)}]$ of appropriate sizes, be given by

$$h_{ii}^{(\zeta,1)} = e_i^\top |A_{32}^{(\zeta)}| e + \frac{1}{2} e_i^\top |\text{ offdiag } (A_{33}^{(\zeta)} + A_{11}^{(\zeta+1)})| e, \text{ and} \tag{5.57}$$

$$h_{ii}^{(\zeta+1,2)} = e_i^\top |A_{12}^{(\zeta+1)}| e + \frac{1}{2} e_i^\top |\text{ offdiag } (A_{33}^{(\zeta)} + A_{11}^{(\zeta+1)})| e, \tag{5.58}$$

respectively. Note that $h_{ii}^{(\zeta,1)}$ and $h_{ii}^{(\zeta+1,2)}$ are the sum of absolute values of all the off-diagonal elements, with elements in the overlap being halved, in the ith row to the left and right of the diagonal, respectively. Next, let the difference between the positive diagonal elements and the sums of absolute values of all off-diagonal elements in the same row be given by

$$D_\zeta = \text{diag}(A_{33}^{(\zeta)} + A_{11}^{(\zeta+1)}) - H_\zeta^{(1)} - H_{\zeta+1}^{(2)}. \tag{5.59}$$

Now, if

$$A_{33}^{(\zeta)} = H_\zeta^{(1)} + \frac{1}{2} D_\zeta + \frac{1}{2} \text{ offdiag } (A_{33}^{(\zeta)} + A_{11}^{(\zeta+1)}),$$

$$A_{11}^{(\zeta+1)} = H_{\zeta+1}^{(2)} + \frac{1}{2} D_\zeta + \frac{1}{2} \text{ offdiag } (A_{33}^{(\zeta)} + A_{11}^{(\zeta+1)}), \tag{5.60}$$

it is easy to verify that $A_{33}^{(\zeta)} + A_{11}^{(\zeta+1)} = A_{2\zeta+1,2\zeta+1}$ and each A_k, for $k = 1, \ldots, p$, is SPDd.d. Consequently, if (5.43) is SPD and d.d., so are the partitions A_k and by Theorem 5.2, the balance system is guaranteed to be SPD.

The Nonsymmetric Case

Next, if A is nonsymmetric, we explore under which conditions will the balance system (5.51) become nonsingular.

Theorem 5.4 ([25]) *Let the matrix A in (5.42) be nonsingular with partitions A_k, $k = 1, 2, \ldots, p$, in (5.44) that are also nonsingular. Then the coefficient matrix M of the balance system in (5.51), is nonsingular.*

Proof Again, let $p = 3$ and write A as

$$
A = \begin{pmatrix} A_1 & & \\ & 0_{s-2\tau} & \\ & & A_3 \end{pmatrix} + \begin{pmatrix} 0_{s-\tau} & & \\ & A_2 & \\ & & 0_{s-\tau} \end{pmatrix}. \tag{5.61}
$$

Next, let

$$
A_L = \begin{pmatrix} A_1 & & \\ & I_{s-2\tau} & \\ & & A_3 \end{pmatrix}, \quad A_R = \begin{pmatrix} I_{s-\tau} & & \\ & A_2 & \\ & & I_{s-\tau} \end{pmatrix}, \tag{5.62}
$$

and consider the nonsingular matrix C given by

$$
\begin{aligned}
C &= A_L^{-1} A A_R^{-1} \\
&= \begin{pmatrix} I_s & & \\ & 0_{s-2\tau} & \\ & & I_s \end{pmatrix} \begin{pmatrix} I_{s-\tau} & & \\ & A_2^{-1} & \\ & & I_{s-\tau} \end{pmatrix} + \begin{pmatrix} A_1^{-1} & & \\ & I_{s-2\tau} & \\ & & A_3^{-1} \end{pmatrix} \begin{pmatrix} 0_{s-\tau} & & \\ & I_s & \\ & & 0_{s-\tau} \end{pmatrix}.
\end{aligned} \tag{5.63}
$$

where I_s and 0_s are the identity and zero matrices of order s, respectively.
   Using (5.61), (5.63) and (5.47), we obtain

$$
C = \begin{pmatrix} I_\tau & & & & & \\ & I_{s-2\tau} & & & & \\ & & B_{11}^{(2)}\ B_{12}^{(2)}\ B_{13}^{(2)} & & & \\ & & 0\ \ 0_{s-2\tau}\ \ 0 & & & \\ & & B_{31}^{(2)}\ B_{32}^{(2)}\ B_{33}^{(2)} & & & \\ & & & I_{s-2\tau} & & \\ & & & & I_\tau & \end{pmatrix} + \begin{pmatrix} 0_\tau & 0 & B_{13}^{(1)} & & & \\ 0 & 0_{s-2\tau} & B_{23}^{(1)} & & & \\ 0 & 0 & B_{33}^{(1)} & & & \\ & & & I_{s-2\tau} & & \\ & & & & B_{11}^{(3)}\ \ 0\ \ 0 & \\ & & & & B_{21}^{(3)}\ 0_{s-2\tau}\ 0 & \\ & & & & B_{31}^{(3)}\ \ 0\ \ 0_\tau & \end{pmatrix},
$$

$$
= \begin{pmatrix} I_\tau & 0 & B_{13}^{(1)} & & & & \\ 0 & I_{s-2\tau} & B_{23}^{(1)} & & & & \\ 0 & 0 & B_{33}^{(1)} + B_{11}^{(2)} & B_{12}^{(2)} & B_{13}^{(2)} & & \\ & & 0 & I_{s-2\tau} & 0 & & \\ & & B_{31}^{(2)} & B_{32}^{(2)} & B_{33}^{(2)} + B_{11}^{(3)} & 0 & 0 \\ & & & & B_{21}^{(3)} & I_{s-2\tau} & 0 \\ & & & & B_{31}^{(3)} & 0 & I_\tau \end{pmatrix}, \tag{5.64}
$$

where the zero matrices denoted by 0 without subscripts, are considered to be of the appropriate sizes. Using the orthogonal matrix P, of order $3s - 2\tau$, given by

$$
P^\top = \begin{pmatrix}
I_\tau & & & & & \\
 & I_{s-2\tau} & 0 & & & \\
 & 0 & I_{s-2\tau} & 0 & & \\
 & 0 & & 0 & I_{s-2\tau} & \\
 & 0 & & 0 & & I_\tau \\
 & I_\tau & & 0 & & \\
 & & & -I_\tau & &
\end{pmatrix}, \tag{5.65}
$$

we obtain

$$
P^\top C P = \left(\begin{array}{ccccc|cc}
I_\tau & & & & & B_{13}^{(1)} & \\
 & I_{s-2\tau} & & & & B_{23}^{(1)} & \\
 & & I_{s-2\tau} & & & & \\
 & & & I_{s-2\tau} & & & -B_{21}^{(3)} \\
 & & & & I_\tau & & B_{31}^{(3)} \\
\hline
B_{12}^{(2)} & & & & & B_{33}^{(1)} + B_{11}^{(2)} & -B_{13}^{(2)} \\
-B_{32}^{(2)} & & & & & -B_{31}^{(2)} & B_{33}^{(2)} + B_{11}^{(3)}
\end{array}\right). \tag{5.66}
$$

Rewriting (5.66) as the block 2×2 matrix,

$$
P^\top C P = \left(\begin{array}{c|c}
I_{3s-4\tau} & Z_1 \\
\hline
Z_2^\top & M
\end{array}\right), \tag{5.67}
$$

and considering the eigenvalue problem

$$
\left(\begin{array}{c|c}
I_{3s-4\tau} & Z_1 \\
\hline
Z_2^\top & M
\end{array}\right)
\begin{pmatrix} u_1 \\ u_2 \end{pmatrix} = \lambda \begin{pmatrix} u_1 \\ u_2 \end{pmatrix}, \tag{5.68}
$$

we see that premultiplying the first block row of (5.68) by $Z_2^\top$ and noticing that $Z_2^\top Z_1 = 0$, we obtain

$$
(1 - \lambda) Z_2^\top u_1 = 0. \tag{5.69}
$$

Hence, either $\lambda = 1$ or $Z_2^\top u_1 = 0$. If $\lambda \neq 1$, then the second block row of (5.68) yields

$$
M u_2 = \lambda u_2. \tag{5.70}
$$

Thus, the eigenvalues of C are either 1 or identical to those of the balance system, i.e.,

$$
\lambda(C) = \lambda(P^\top C P) \subseteq \{1, \lambda(M)\}. \tag{5.71}
$$

Hence, since C is nonsingular, the balance system is also nonsingular.

Since the size of the coefficient matrix M of the balance system is much smaller than n, the above theorem indicates that A_L and A_R are effective left and right preconditioners of system (5.42).

Next, we explore those conditions which guarantee that there exists a splitting of the coefficient matrix in (5.42) resulting in nonsingular partitions A_k, and provide a scheme for computing such a splitting.

Theorem 5.5 ([25]) *Assume that the matrix A in (5.42) is nonsymmetric with a positive definite symmetric part $H = \frac{1}{2}(A + A^\top)$. Also let*

$$
B_1 = \begin{pmatrix}
0_{(s-\tau)\times\tau} & & & & \\
A_{11}^{(2)} & & & & \\
A_{21}^{(2)} & & & & \\
& A_{11}^{(3)} & & & \\
& A_{21}^{(3)} & & & \\
& & \ddots & & \\
& & & A_{11}^{(p)} & \\
& & & A_{21}^{(p)} & \\
& & & & 0_\tau
\end{pmatrix}
\quad \text{and} \quad
B_2 = \begin{pmatrix}
0_{(s-\tau)\times\tau} & & & & \\
A_{11}^{(2)^T} & & & & \\
A_{12}^{(2)^T} & & & & \\
& A_{11}^{(3)^T} & & & \\
& A_{12}^{(3)^T} & & & \\
& & \ddots & & \\
& & & A_{11}^{(p)^T} & \\
& & & A_{12}^{(p)^T} & \\
& & & & 0_\tau
\end{pmatrix}.
$$

$$(5.72)$$

Assuming that partial symmetry holds, i.e., $B_1 = B_2 = B$, then there exists a splitting such that the partitions A_k in (5.44) for $k = 1, \ldots, p$ are nonsingular.

Proof Let $p = 3$, and let $\tilde{A}$ be the block-diagonal matrix in which the blocks are the partitions A_k.

$$
\tilde{A} = \begin{pmatrix}
A_{11}^{(1)} & A_{12}^{(1)} & & & & & & \\
A_{21}^{(1)} & A_{22}^{(1)} & A_{23}^{(1)} & & & & & \\
& A_{32}^{(1)} & A_{33}^{(1)} & & & & & \\
& & & A_{11}^{(2)} & A_{12}^{(2)} & & & \\
& & & A_{21}^{(2)} & A_{22}^{(2)} & A_{23}^{(2)} & & \\
& & & & A_{32}^{(2)} & A_{33}^{(2)} & & \\
& & & & & & A_{11}^{(3)} & A_{12}^{(3)} \\
& & & & & & A_{21}^{(3)} & A_{22}^{(3)} & A_{23}^{(3)} \\
& & & & & & & A_{32}^{(3)} & A_{33}^{(3)}
\end{pmatrix}.
$$

$$(5.73)$$

Let $V^\top$ be the nonsingular matrix of order $3m$ given by

$$
V^\top =
\begin{pmatrix}
I_\tau & & & & & & \\
I_{s-2\tau} & & & & & & \\
& I_\tau & I_\tau & & & & \\
& & I_{s-2\tau} & & & & \\
& & & I_\tau & I_\tau & & \\
& & & & I_{s-2\tau} & & \\
& & & & & I_\tau & \\
0_\tau & I_\tau & & & & & \\
& & & 0_\tau & I_\tau & &
\end{pmatrix},
\tag{5.74}
$$

then,

$$
J^\top \tilde{A} J =
\left(
\begin{array}{ccccccc|c}
A_{11}^{(1)} & A_{12}^{(1)} & & & & & & \\
A_{21}^{(1)} & A_{22}^{(1)} & A_{23}^{(1)} & & & & & \\
& A_{32}^{(1)} & A_{33}^{(1)} + A_{11}^{(2)} & A_{12}^{(2)} & & & & A_{11}^{(2)} \\
& & A_{21}^{(2)} & A_{22}^{(2)} & A_{23}^{(2)} & & & A_{21}^{(2)} \\
& & & A_{32}^{(2)} & A_{33}^{(2)} + A_{11}^{(3)} & A_{12}^{(3)} & & A_{11}^{(3)} \\
& & & & A_{21}^{(3)} & A_{22}^{(3)} & A_{23}^{(3)} & A_{21}^{(3)} \\
& & & & & A_{32}^{(3)} & A_{33}^{(3)} & \\
\hline
A_{11}^{(2)} & A_{12}^{(2)} & & & & & & A_{11}^{(2)} \\
& & & A_{11}^{(3)} & A_{12}^{(3)} & & & A_{11}^{(3)}
\end{array}
\right).
\tag{5.75}
$$

Writing (5.75) as a block 2×2 matrix, we have

$$
J^\top \tilde{A} J =
\left(
\begin{array}{c|c}
A & B_1 \\
\hline
B_2^\top & K
\end{array}
\right).
\tag{5.76}
$$

Using the splitting $A_{11}^{(k)} = \frac{1}{2}(A_{\eta+1,\eta+1} + A_{\eta+1,\eta+1}^\top) + \beta I$ and $A_{33}^{(k-1)} = \frac{1}{2}(A_{\eta+1,\eta+1} - A_{\eta+1,\eta+1}^\top) - \beta I$, we can choose β so as to ensure that B_1 and B_2 are of full rank and K is SPD Thus, using Theorem 3.4 on p. 17 of [26], we conclude that $J^\top \tilde{A} J$ is of full rank, hence $\tilde{A}$ has full rank and consequently the partitions A_k are nonsingular.

Note that the partial symmetry assumption $B_1 = B_2$ in the theorem above is not as restrictive as it seems. Recalling that the original matrix is banded, it is easy to see that the matrices $A_{12}^{(k)}$ and $A_{21}^{(k)}$ are almost completely zero except for small parts in their respective corners, which are of size no larger than the overlap. This condition then can be viewed as a requirement of symmetry surrounding the overlaps.

Let us now focus on two special cases. First, if the matrix A is SPD the conditions of Theorem 5.5 are immediately satisfied and we obtain the following.

Corollary 5.1 *If the matrix A in (5.42) is* SPD *then there is a splitting (as described in Theorem 5.5) such that the partitions A_k in (5.44) for $k = 1, \ldots, p$ are nonsingular and consequently the coefficient matrix M of the balance system in (5.51), is nonsingular.*

Second, note that Theorem 5.3 still holds even if the symmetry requirement is dropped. Combining the results of Theorems 5.3 and 5.4, without any requirement of symmetry, we obtain the following.

Corollary 5.2 *If the matrix A in (5.42) is d.d., then the partitions A_k in (5.44) can be chosen such that they are also nonsingular and d.d. for $k = 1, \ldots, p$ and consequently the coefficient matrix M, of the balance system in (5.51), is nonsingular.*

5.4.4 The Hybrid Solver of the Balance System

Next, we show how one can compute the residual r_{init} needed to start a Krylov subspace scheme for solving the balance system. Rewriting (5.47) as

$$\begin{pmatrix} x_1^{(k)} \\ x_2^{(k)} \\ x_3^{(k)} \end{pmatrix} = \begin{pmatrix} h_1^{(k)} \\ h_2^{(k)} \\ h_3^{(k)} \end{pmatrix} + \begin{pmatrix} \bar{y}_1^{(k)} \\ \bar{y}_2^{(k)} \\ \bar{y}_3^{(k)} \end{pmatrix}, \tag{5.77}$$

where,

$$\begin{pmatrix} h_1^{(k)} \\ h_2^{(k)} \\ h_3^{(k)} \end{pmatrix} = A_k^{-1} \begin{pmatrix} (1 - \alpha_{k-1}) f_{\eta+1} \\ f_{\eta+2} \\ \alpha_k f_{\eta+3} \end{pmatrix}, \begin{pmatrix} \bar{y}_1^{(k)} \\ \bar{y}_2^{(k)} \\ \bar{y}_3^{(k)} \end{pmatrix} = A_k^{-1} \begin{pmatrix} -y_{k-1} \\ 0 \\ y_k \end{pmatrix}, \tag{5.78}$$

the residual can be written as

$$r = g - My = \begin{pmatrix} x_1^{(2)} - x_3^{(1)} \\ x_1^{(3)} - x_3^{(2)} \\ \vdots \\ x_1^{(p)} - x_3^{(p-1)} \end{pmatrix} = \begin{pmatrix} h_1^{(2)} - h_3^{(1)} \\ h_1^{(3)} - h_3^{(2)} \\ \vdots \\ h_1^{(p)} - h_3^{(p-1)} \end{pmatrix} + \begin{pmatrix} \bar{y}_1^{(2)} - \bar{y}_3^{(1)} \\ \bar{y}_1^{(3)} - \bar{y}_3^{(2)} \\ \vdots \\ \bar{y}_1^{(p)} - \bar{y}_3^{(p-1)} \end{pmatrix},$$

$$\tag{5.79}$$

where the second equality in (5.79) follows from the combination of (5.46), (5.48), (5.50) and (5.52). Let the initial guess be $y_{\text{init}} = 0$, then we have

$$
r_{\text{init}} = g = \begin{pmatrix} h_1^{(2)} - h_3^{(1)} \\ h_1^{(3)} - h_3^{(2)} \\ \vdots \\ h_1^{(p)} - h_3^{(p-1)} \end{pmatrix}.
\tag{5.80}
$$

Thus, to compute the initial residual we must solve the p independent linear systems (5.45) and subtract the bottom part of the solution vector of partition ζ, h_3^{ζ}, from the top part of the solution vector of partition $\zeta + 1$, $h_1^{(\zeta+1)}$, for $\zeta = 1, \ldots, p - 1$.

Finally, to compute matrix-vector products of the form $q = Mp$, we use (5.79) and (5.80), to obtain

$$
My = g - r = r_{\text{init}} - r = \begin{pmatrix} \bar{y}_3^{(1)} - \bar{y}_1^{(2)} \\ \bar{y}_3^{(2)} - \bar{y}_1^{(3)} \\ \vdots \\ \bar{y}_3^{(p-1)} - \bar{y}_1^{(p)} \end{pmatrix}.
\tag{5.81}
$$

Hence, we can compute the matrix-vector products Mp for any vector p in a fashion similar to computing the initial residual using (5.81) and (5.78). The modified Krylov subspace methods (e.g. CG, or BiCGstab) used to solve (5.51) are the standard ones except that the initial residual and the matrix-vector products are computed using (5.80) and (5.81), respectively. We call these solvers Domain-Decomposition-CG (DDCG) and Domain-Decomposition-BiCGstab (DDBiCGstab). They are schemes in which the solutions of the smaller independent linear systems in (5.45) are obtained via a direct solver, while the adjustment vector y is obtained using an iterative method, CG for SPD and BiCGStab for nonsymmetric linear systems. The outline of the DDCG algorithm is shown in Algorithm 5.2. The usual steps of CG are omitted, but the two modified steps are shown in detail. The outline of the DDBiCGstab scheme is similar and hence ommitted.

The balance system is preconditioned using a block-diagonal matrix of the form,

$$
\tilde{M} = \begin{pmatrix} \tilde{B}_{33}^{(1)} + \tilde{B}_{11}^{(2)} & & \\ & \ddots & \\ & & \tilde{B}_{33}^{(p-1)} + \tilde{B}_{11}^{(p)} \end{pmatrix},
\tag{5.82}
$$

where $\tilde{B}_{33}^{(\zeta)} = A_{33}^{(\zeta)^{-1}} \approx B_{33}^{(\zeta)}$ and $\tilde{B}_{11}^{(\zeta+1)} = A_{11}^{(\zeta+1)^{-1}} \approx B_{11}^{(\zeta+1)}$. Here, we are taking advantage of the fact that the elements of the inverse of the banded balance system decay as we move away from the diagonal, e.g., see [20]. Also such decay becomes more pronounced as the degree of diagonal dominance of the banded balance system becomes more pronounced. Using the Woodbury formula [22], we can write

---

Algorithm 5.2 Domain Decomposition Conjugate Gradient (DDCG)

Input: Banded SPD matrix A and right-hand side f
Output: Solution x of $Ax = f$
1: Tear the coefficient matrix A into partitions A_k for $k = 1, \ldots, p$.
2: Distribute the partitions across p processors.
3: Perform the Cholesky factorization of partition A_k on processor k.
4: Distribute the vector $y_{\text{init}}^\top = (y_{\text{init}_1}^\top, \ldots, y_{\text{init}_{p-1}}^\top)$ across $p - 1$ processors, so that processor ζ
   contains y_{init_ζ} for $\zeta = 1, \ldots, p - 1$ and the last processor is idle. All vectors in the modified
   iterative method are distributed similarly.
   //Perform the modified Conjugate Gradient:
5: Compute the initial residual $r_{\text{init}} = g$ using (5.80), in other words, on processor ζ, we compute
   $g_\zeta = \bar{h}_1^{(\zeta+1)} - \bar{h}_3^{(\zeta)}$, where $\bar{h}_3^{(\zeta)}$ and $\bar{h}_1^{(\zeta)}$ are computed on processor ζ by solving the first system
   in (5.78) directly.
6: **do** $i = 1, \ldots,$ until convergence
7:    Standard CG steps
8:    Compute the matrix-vector multiplication $q = Mp$ using (5.81), in other words, on processor
      ζ we compute $q_\zeta = \bar{y}_3^{(\zeta)} - \bar{y}_1^{(\zeta+1)}$, where $\bar{y}_3^{(\zeta)}$ and $\bar{y}_1^{(\zeta)}$ are computed on processor ζ by
      solving the second system in (5.78) directly.
9:    Standard CG steps
10: **end**
11: Solve the smaller independent linear systems in (5.45) directly in parallel.

$$(\tilde{B}_{33}^{(\zeta)} + \tilde{B}_{11}^{(\zeta+1)})^{-1} = (A_{33}^{(\zeta)^{-1}} + A_{11}^{(\zeta+1)^{-1}})^{-1} = A_{33}^{(\zeta)}(A_{33}^{(\zeta)} + A_{11}^{(\zeta+1)})^{-1}A_{11}^{(\zeta+1)},$$
$$= A_{33}^{(\zeta)} - A_{33}^{(\zeta)}(A_{33}^{(\zeta)} + A_{11}^{(\zeta+1)})^{-1}A_{33}^{(\zeta)},$$

$$(5.83)$$

The last equality is preferred as it avoids extra internode communication, assuming that the original overlapping block $A_{33}^{(\zeta)} + A_{11}^{(\zeta+1)}$ is stored separately on node ζ. Consequently, to precondition (5.51) we only need to perform matrix-vector products involving $A_{33}^{(\zeta)}$, and solve small linear systems with coefficient matrices $(A_{33}^{(\zeta)} + A_{11}^{(\zeta+1)})$.

5.5 Tridiagonal Systems

So far, we discussed algorithms for general banded systems; this section is devoted to tridiagonal systems because their simple structure makes possible the use of special purpose algorithms. Moreover, many applications and algorithms contain a tridiagonal system solver as a kernel, used directly or indirectly.

The parallel solution of linear systems of equations, $Ax = f$, with coefficient matrix A that is point (rather than block) tridiagonal,

$$A = \begin{pmatrix} \alpha_{1,1} & \alpha_{1,2} & & & \\ \alpha_{2,1} & \alpha_{2,2} & \alpha_{2,3} & & \\ & \ddots & \ddots & \ddots & \\ & & \ddots & \ddots & \alpha_{n-1,n} \\ & & & \alpha_{n,n-1} & \alpha_{n,n} \end{pmatrix} \tag{5.84}$$

and abbreviated as $[\alpha_{i,i-1}, \alpha_{i,i}, \alpha_{i,i+1}]$ when the dimension is known from the context, has been the subject of many studies. Even though the methods can be extended to handle multiple right-hand sides, here we focus on the case of only one, that we denote by f. Because of their importance in applications, tridiagonal solvers have been developed for practically every type of parallel computer system to date. The classic monographs [27–29] discuss the topic extensively. The activity is continuing, with refinements to algorithms and implementations in libraries for parallel computer systems such as SCALAPACK, and with implementations on different computer models and novel architectures; see e.g. [30–32].

In some cases, the matrix is not quite tridiagonal but it differs from one only by a low rank modification. It is then possible to express the solution (e.g. by means of the Sherman-Morrison-Woodbury formula [22]) in a manner that involves the solution of tridiagonal systems, possibly with multiple right-hand sides. The latter is a special case of a class of problems that requires the solution of multiple systems, all with the same or with different tridiagonal matrices. This provides the algorithm designer with more opportunities for parallelism. For very large matrices, it might be preferable to use a parallel algorithm for a single or just a few right-hand sides so as to be able to maintain all required coefficients in fast memory; see for instance remarks in [33, 34]. Sometimes the matrices have special properties such as Toeplitz structure, diagonal dominance, or symmetric positive definiteness. Algorithms that take advantage of these properties are more effective and sometimes safer in terms of their roundoff error behavior. We also note that because the cost of standard Gaussian elimination for tridiagonal systems is already linear, we do not expect impressive speedups. In particular, observe that in a general tridiagonal system of order n, each solution element, ξ_i, depends on the values of all inputs, that is $4n - 2$ elements. This is also evident by the fact that A^{-1} is dense (though data sparse, which could be helpful in designing inexact solvers). Therefore, with unbounded parallelism, we need $O(\log n)$ steps to compute each ξ_i. Without restricting the number of processors, the best possible speedup of a tridiagonal solver is expected to be $O(\frac{n}{\log n})$; see also [35] for this fan-in based argument.

The best known and earliest parallel algorithms for tridiagonal systems are *recursive doubling, cyclic reduction* and *parallel cyclic reduction*. We first review these methods as well as some variants. We then discuss hybrid and divide-and-conquer strategies that are more flexible as they offer hierarchical parallelism and ready adaptation for limited numbers of processors.

5.5.1 Solving by Marching

We first describe a very fast parallel but potentially unstable algorithm described in [12, Algorithm III]. This was inspired by the *shooting methods* [36–38] and by the *marching methods* described in [39, 40] and discussed also in Sect. 6.4.7 of Chap. 6 It is worth noting that the term "marching" has been used since the early days of numerical methods; cf. [41]. In fact, a definition can be found in the same book by Richardson [42, Chap. I, p. 2] that also contains the vision of the "human parallel computer" that we discussed in the Preface.

In the sequel we will assume that the matrix is irreducible, thus all elements in the super- and subdiagonal are nonzero; see also Definition 9.1 in Chap. 9. Otherwise the matrix can be brought into block-upper triangular (block diagonal, if symmetric) form with tridiagonal submatrices along the diagonal, possibly after row and column permutations and solved using block back substitution, where the major computation that has the leading cost at each step is the solution of a smaller tridiagonal system. In practice, we assume that there exists a preprocessing stage which detects such elements and if such are discovered, the system is reduced to smaller ones.

The key observation is the following: If we know the last element, ξ_n, of the solution $x = (\xi_i)_{1:n}$, then the remaining elements can be computed from the recurrence

$$\xi_{n-k-1} = \frac{1}{\alpha_{n-k,n-k-1}} (\phi_{n-k} - \alpha_{n-k,n-k}\xi_{n-k} - \alpha_{n-k,n-k+1}\xi_{n-k+1}). \quad (5.85)$$

for $k = 0, \ldots, n-2$, assuming that $\alpha_{n,n+1} = 0$. Note that because of the assumption of irreducibility, the denominator above is always nonzero. To compute ξ_n, consider the reordered system in which the first equation is ordered last. Then, assuming that $n \geq 3$ the system becomes

$$\begin{pmatrix} \tilde{R} & b \\ a^\top & 0 \end{pmatrix} \begin{pmatrix} \hat{x} \\ \xi_n \end{pmatrix} = \begin{pmatrix} g \\ \phi_1 \end{pmatrix}$$

where for simplicity we denote $\tilde{R} = A_{2:n,1:n-1}$ that is non-singular and banded upper triangular with bandwidth 3, $b = A_{2:n,n}$, $a^\top = A_{1,1:n-1}$, $\hat{x} = x_{1:n-1}$, and $g = f_{2:n}$. Applying block LU factorization

$$\begin{pmatrix} \tilde{R} & b \\ a^\top & 0 \end{pmatrix} = \begin{pmatrix} I & 0 \\ a^\top \tilde{R}^{-1} & 1 \end{pmatrix} \begin{pmatrix} \tilde{R} & b \\ 0 & -a^\top \tilde{R}^{-1}b \end{pmatrix}$$

it follows that

$$\xi_n = -(a^\top \tilde{R}^{-1}b)^{-1}(\phi_1 - a^\top \tilde{R}^{-1}g),$$

and

$$\hat{x} = \tilde{R}^{-1}g - \xi_n \tilde{R}^{-1}b.$$

Observe that because A is tridiagonal,

$$a^\top = (\alpha_{1,1}, \alpha_{1,2}, 0, \ldots, 0), b = (0, \ldots, 0, \alpha_{n-1,n}, \alpha_{n,n})^\top.$$

Taking into account that certain terms in the formulas for ξ_n and $\hat{x}$ are repeated, the leading cost is due to the solution of a linear system with coefficient matrix $\tilde{R}$ and two right-hand sides ($f_{2:n}$, $A_{2:n,n}$). From these results, ξ_1 and $x_{2:n}$ can be obtained with few parallel operations. If the algorithm of Theorem 3.2 is extended to solve non-unit triangular systems with two right-hand sides, the overall parallel cost is $\mathbf{T}_p = 3\log n + O(1)$ operations on $p = 4n + O(1)$ processors. The total number of operations is $\mathbf{O}_p = 11n + O(1)$, which is only slightly larger than Gaussian elimination, and the efficiency $\mathbf{E}_p = \frac{11}{12\log n}$.

Remark 5.2 The above methodology can also be applied to more general banded matrices. In particular, any matrix of bandwidth $(2m+1)$ (it is assumed that the upper and lower half-bandwidths are equal) of order n can be transformed by reordering the rows, into

$$\begin{pmatrix} R & B \\ C & 0 \end{pmatrix} \tag{5.86}$$

where R is of order $n - m$ and upper triangular with bandwidth $2m+1$, and the corner zero matrix is of order m. This property has been used in the design of other banded solvers, see e.g. [43]. As in Sect. "Notation", let $J = e_1 e_2^\top + \cdots + e_{n-1}e_n^\top$, and set $S = J + e_n e_1^\top$ be the "circular shift matrix". Then $S^m A$ is as (5.86). Indeed, this is straightforward to show by probing its elements, e.g. $e_i^\top S^m A e_j = 0$ if $j < i < n - m$ or if $i + 2m < j < n - m$. Therefore, instead of solving $Ax = b$, we examine the equivalent system $S^m Ax = S^m b$ and then use block LU factorization. We then solve by exploiting the banded upper triangular structure of R and utilizing the parallel algorithms described in Chap. 3.

The marching algorithm has the smallest parallel cost among the tridiagonal solvers described in this section. On the other hand, it has been noted in the literature that marching methods are prone to considerable error growth that render them unstable unless special precautions are taken. An analysis of this issue when solving block-tridiagonal systems that arise from elliptic problems was conducted in [40, 44]; cf. Sect. 6.4.7 in Chap. 6. We propose an explanation of the source of instability that is applicable to the tridiagonal case. Specifically, the main kernel of the algorithm is the solution of two banded triangular systems with the same coefficient matrix. The speed of the parallel marching algorithm is due to the fact that these systems are solved using a parallel algorithm, such as those described in Chap. 3. As is well known, serial substitution algorithms for solving triangular systems are extremely stable and the actual forward error is frequently much smaller than what a normwise or componentwise analysis would predict; cf. [45] and our discussion in Sect. 3.2.3. Here, however, we are interested in the use of parallel algorithms for solving the triangular systems. Unfortunately, normwise and componentwise analyses show that their forward error bounds, in the general case, depend on the cube of the condition number of the triangular matrix; cf. [46]. As was shown in [47], the 2-norm condition number, $\kappa_2(A_n)$, of order-n triangular matrices A_n whose nonzero entries are independent and normally distributed with mean 0 and variance 1 grows exponentially, in particular $\sqrt[n]{\kappa_2(A_n)} \to 2$ almost surely. Therefore, the condition is much worse than that for random dense matrices, where it grows linearly. There is therefore the possibility of exponentially fast increasing condition number compounded by the error's cubic dependence on it. Even though we suspect that banded random matrices are better behaved, our experimental results indicate that the increase in condition number is still rapid for triangular matrices such as $\tilde{R}$ that have bandwidth 3.

It is also worth noting that even in the serial application of marching, in which case the triangular systems are solved by stable back substitution, we cannot assume any special conditions for the triangular matrix except that it is banded. In fact, in experiments that we conducted with random matrices, the actual forward error also increases rapidly as the dimension of the system grows. Therefore, when the problem is large, the parallel marching algorithm is very likely to suffer from severe loss of accuracy.

5.5.2 Cyclic Reduction and Parallel Cyclic Reduction

Cyclic reduction (CR) relies on the combination of groups of three equations, each consisting of an even indexed one, say $2i$, together with its immediate neighbors, indexed $(2i \pm 1)$, and involve five unknowns. The equations are combined in order to eliminate odd-indexed unknowns and produce one equation with three unknowns per group. Thus, a reduced system with approximately half the unknowns is generated. Assuming that the system size is $n = 2^k - 1$, the process is repeated for k steps, until a single equation involving one unknown remains. After this is solved, $2, 2^2, \ldots, 2^{k-1}$ unknowns are computed and the system is fully solved. Consider, for instance, three adjacent equations from (5.84).

$$
\begin{pmatrix}
\alpha_{i-1,i-2} & \alpha_{i-1,i-1} & \alpha_{i-1,i} & & \\
 & \alpha_{i,i-1} & \alpha_{i,i} & \alpha_{i,i+1} & \\
 & & \alpha_{i+1,i} & \alpha_{i+1,i+1} & \alpha_{i+1,i+2}
\end{pmatrix}
\begin{pmatrix}
\xi_{i-2} \\
\xi_{i-1} \\
\xi_i \\
\xi_{i+1} \\
\xi_{i+2}
\end{pmatrix}
=
\begin{pmatrix}
\phi_{i-1} \\
\phi_i \\
\phi_{i+1}
\end{pmatrix}
$$

under the assumption that unknowns indexed n or above and 0 or below are zero. If both sides are multiplied by the row vector

$$\left(-\frac{\alpha_{i,i-1}}{\alpha_{i-1,i-1}}, \; 1, \; -\frac{\alpha_{i,i+1}}{\alpha_{i+1,i+1}}\right)$$

the result is the following equation:

$$-\frac{\alpha_{i,i-1}}{\alpha_{i-1,i-1}}\alpha_{i-1,i-2}\xi_{i-2} + (\alpha_{i,i} - \frac{\alpha_{i,i-1}}{\alpha_{i-1,i-1}}\alpha_{i-1,i} - \frac{\alpha_{i,i+1}}{\alpha_{i+1,i+1}}\alpha_{i+1,i})\xi_i$$
$$-\frac{\alpha_{i,i+1}}{\alpha_{i+1,i+1}}\alpha_{i+1,i+2}\xi_{i+2} = -\frac{\alpha_{i,i-1}}{\alpha_{i-1,i-1}}\phi_{i-1} + \phi_i - \frac{\alpha_{i,i+1}}{\alpha_{i+1,i+1}}\phi_{i+1}$$

This involves only the unknowns $\xi_{i-2}, \xi_i, \xi_{i+2}$. Note that 12 floating-point operations suffice to implement this transformation.

To simplify the description, unless mentioned otherwise we assume that CR is applied on a system of size $n = 2^k - 1$ for some k and that all the steps of the algorithm can be implemented without encountering division by 0. These transformations can be applied independently for $i = 2, 2 \times 2, \ldots, 2 \times (2^{k-1} - 1)$ to obtain a tridiagonal system that involves only the (even numbered) unknowns $\xi_2, \xi_4, \ldots, \xi_{2^{k-1}-2}$ and is of size $2^{k-1} - 1$ which is almost half the size of the previous one. If one were to compute these unknowns by solving this smaller tridiagonal system then the remaining ones can be recovered using substitution. Cyclic reduction proceeds recursively by applying the same transformation to the smaller tridiagonal system until a single scalar equation remains and the middle unknown, $\xi_{2^{k-1}}$ is readily obtained. From then on, using substitutions, the remaining unknowns are computed.

The seminal paper [48] introduced odd-even reduction for block-tridiagonal systems with the special structure resulting from discretizing Poisson's equation; we discuss this in detail in Sect. 6.4. That paper also presented an algorithm named "recursive cyclic reduction" for the multiple "point" tridiagonal systems that arise when solving Poisson's equation in 2 (or more) dimensions.

An key observation is that CR can be interpreted as Gaussian elimination with diagonal pivoting applied on a system that is obtained from the original one after renumbering the unknowns and equations so that those that are odd multiples of 2^0 are ordered first, then followed by the odd multiples of 2^1, the odd multiples of 2^2, etc.; cf. [49]. This equivalence is useful because it reveals that in order for CR to be numerically reliable, it must be applied to matrices for which Gaussian elimination without pivoting is applicable. See also [50–52].

The effect of the above reordering on the binary representation of the indices of the unknown and right-hand side vectors, is an *unshuffle* permutation of the *bit-permute-complement (BPC)* class; cf. [53, 54]. If Q denotes the matrix that implements the unshuffle, then

$$Q^\top A Q = \begin{pmatrix} D_o & B \\ C^\top & D_e \end{pmatrix},$$

where the diagonal blocks are diagonal matrices, specifically

$$D_o = \mathrm{diag}(\alpha_{1,1}, \alpha_{3,3}, \ldots, \alpha_{2^k-1,2^k-1}), \quad D_e = \mathrm{diag}(\alpha_{2,2}, \alpha_{4,4}, \ldots, \alpha_{2^k-2,2^k-2})$$

and the off-diagonal blocks are the rectangular matrices

$$B = \begin{pmatrix} \alpha_{1,2} & & & \\ \alpha_{3,2} & \ddots & & \\ & \ddots & \alpha_{2^k-2,2^k-2} & \\ & & \alpha_{2^k-1,2^k-2} \end{pmatrix}, \quad C^{\top} = \begin{pmatrix} \alpha_{2,1} & \alpha_{2,3} & & & \\ & \alpha_{4,3} & \alpha_{4,5} & & \\ & & \ddots & & \ddots \\ & & & \alpha_{2^k-2,2^k-3} & \alpha_{2^k-2,2^k-1} \end{pmatrix}.$$

Note that B and C are of size $2^{k-1} \times (2^{k-1} - 1)$. Then we can write

$$Q^{\top} A Q = \begin{pmatrix} I_{2^{k-1}} & 0 \\ C^{\top} D_o^{-1} & I_{2^{k-1}-1} \end{pmatrix} \begin{pmatrix} D_o & B \\ 0 & D_e - C^{\top} D_o^{-1} B \end{pmatrix}.$$

Therefore the system can be solved by computing the subvectors containing the odd and even numbered unknowns $x^{(o)}, x^{(e)}$

$$(D_e - C^{\top} D_o^{-1} B) x^{(e)} = b^{(e)} - C^{\top} D_o^{-1} f^{(0)}$$
$$D_o x^{(o)} = f^{(o)} - B x^{(e)},$$

where $f^{(o)}, f^{(e)}$ are the subvectors containing the odd and even numbered elements of the right-hand side. The crucial observation is that the Schur complement $D_e - C^{\top} D_o^{-1} B$ is of almost half the size, $2^{k-1} - 1$, and tridiagonal, because the term $C^{\top} D_o^{-1} B$ is tridiagonal and D_e diagonal. The tridiagonal structure of $C^{\top} D_o^{-1} B$ is due to the fact that $C^{\top}$ and B are upper and lower bidiagonal respectively (albeit not square). It also follows that CR is equivalent to Gaussian elimination with diagonal pivoting on a reordered matrix. Specifically, the unknowns are eliminated in the nested dissection order (cf. [55]) that is obtained by the repeated application of the above scheme: First the odd-numbered unknowns (indexed by $2k - 1$), followed by those indexed $2(2k - 1)$, followed by those indexed $2^2(2k - 1)$, and so on. The cost to form the Schur complement tridiagonal matrix and corresponding right-hand side is 12 operations per equation and so the cost for the first step is $12 \times (2^{k-1} - 1)$ operations. Applying the same method recursively on the tridiagonal system with the Schur complement matrix for the odd unknowns one obtains after $k - 1$ steps 1 scalar equation for ξ_{2k-1} that is the unknown in middle position. From then on, the remaining unknowns can be recovered using repeated applications of the second equation above. The cost is 5 operations per computed unknown, independent for each. On $p = O(n)$ processors, the cost is approximately $\mathbf{T}_p = 17 \log n$ parallel operations and $\mathbf{O}_p = 17n - 12 \log n$ in total, that is about twice the number of operations needed by Gaussian elimination with diagonal pivoting. CR is listed as Algorithm 5.3. The CR algorithm is a component in many other parallel tridiagonal

---

Algorithm 5.3 CR: cyclic reduction tridiagonal solver

Input: $A = [\alpha_{i,i-1}, \alpha_{i,i}, \alpha_{i,i+1}] \in \mathbb{R}^{n \times n}$ and $f \in \mathbb{R}^n$ where $n = 2^k - 1$.
Output: Solution x of $Ax = f$.
    //It is assumed that $\alpha_{i,i-1}^{(0)} = \alpha_{i,i-1}$, $\alpha_{i,i}^{(0)} = \alpha_{i,i}$, $\alpha_{i,i+1}^{(0)} = \alpha_{i,i+1}$ for $i = 1 : n$ //Reduction
    stage
1: **do** $l = 1 : k - 1$
2:   **doall** $i = 2^l : 2^l : n - 2^l$
3:     $\rho_i = -\alpha_{i,i-2^{l-1}}^{(l-1)}/\alpha_{i-2^{(l-1)}}^{(l-1)}, \ \tau_i = -\alpha_{i,i+2^{l-1}}^{(l-1)}/\alpha_{i+2^{(l-1)}}^{(l-1)}$
4:     $\alpha_{i,i-2^l}^{(l)} = \rho_i \alpha_{i,i-2^{l-1}}^{(l-1)}$
5:     $\alpha_{i,i+2^l}^{(l)} = \tau_i \alpha_{i,i+2^{l-1}}^{(l-1)}$
6:     $\alpha_{i,i}^{(l)} = \alpha_{i,i}^{(l-1)} + \rho_i \alpha_{i,i-2^{l-1}}^{(l-1)} + \tau_i \alpha_{i,i+2^{l-1}}^{(l-1)}$
7:     $\phi_i^{(l)} = \phi_i^{(l-1)} + \rho_i \phi_{i-2^{l-1}}^{(l-1)} + \tau_i \phi_{i+2^{l-1}}^{(l-1)}$
8:   **end**
9: **end**
    //Back substitution stage
10: **do** $l = k : -1 : 1$
11:   **doall** $i = 2^{l-1} : 2^{l-1} : n - 2^{l-1}$
12:     $\xi_i = (\phi_i^{(l-1)} - \alpha_{i,i-2^{l-1}}\xi_{i-2^{l-1}} - \alpha_{i,i+2^{l-1}}\xi_{i+2^{l-1}})/\phi_i^{(l-1)}$
13:   **end**
14: **end**

---

system solvers, e.g. [56–58]. A block extension of CR culminated in an important class of RES algorithms described in Sect. 6.4.5. The numerical stability of CR is analyzed in [59]. The complexity and stability results are summarized in the following proposition.

Proposition 5.1 *The* CR *method for solving* $Ax = f$ *when* $A \in \mathbb{R}^{n \times n}$ *is tridiagonal on* $O(n)$ *processors requires no more than* $\mathbf{T}_p = 17 \log n + O(1)$ *steps for a total of* $\mathbf{O}_p = 17n + O(\log n)$ *operations. Moreover, if* A *is diagonally dominant by rows or columns, then the computed solution satisfies*

$$(A + \delta A)\tilde{x} = f, \quad \|\delta A\|_\infty \leq 10(\log n)\|A\|_\infty \mathbf{u}$$

and the relative forward error satisfies the bound

$$\frac{\|\tilde{x} - x\|_\infty}{\|\tilde{x}\|_\infty} \leq 10(\log n)\kappa_\infty(A)\mathbf{u}$$

where $\kappa(A) = \|A\|_\infty \|A^{-1}\|_\infty$.

Note that even though the upper bound on the backward error depends on $\log n$, its effect was not seen in the experiments conducted in [59].

Since CR is equivalent to Guassian elimination with diagonal pivoting on a recorded matrix, it can be used for tridiagonal matrices for which the latter method is applicable These include symmetric positive definite matrices, M-matrices, totally nonnegative matrices, and matrices that can be written as $D_1 A D_2$ where

$|D_1| = |D_2| = I$ and A is of the aforementioned types. In those cases, it can be proved that Gaussian elimination with diagonal pivoting to solve $Ax = f$ succeeds and that the computed solution satisfies $(A + \delta A)\hat{x} = f$ where $|\delta A| \leq 4\mathbf{u}|A|$ ignoring second order terms. For diagonally dominant matrices (by rows or columns) the bound is multiplied by a small constant. See the relevant section in the treatise [45, Sect. 9.6] and the relevant paper [60] for a more detailed discussion.

One cause of inefficiency in this algorithm are memory bank conflicts that can arise in the course of the reduction and back substitution stages. We do not address this issue here but point to detailed discussions in [34, 61]. Another inefficiency, is that as the reduction proceeds as well as early in the back substitution stage, the number of equations that can be processed independently is smaller than the number of processors. At the last step of the reduction, for example, there is only one equation. It has been observed that this inefficiency emerges as an important bottleneck after bank conflicts are removed; see e.g. the discussion in [61] regarding CR performance on GPUs.

To address this issue, another algorithm, termed PARACR, has been designed (cf. [27]) to maintain the parallelism throughout at the expense of some extra arithmetic redundancy. Three equations are combined at a time, as in CR, however this combination is applied to subsequent equations, say $i - 1, i, i + 1$, irrespective of the parity of i, rather than only even ones. The eliminations that take place double the distance of the coupling between unknowns: we assume this time, for simplicity, that there are $n = 2^k$ equations, the distance is 1 at the first step (every unknown ξ_i is involved in an equation with its immediate neighbors, say $\xi_{i\pm1}$), to 2 (every unknown is involved in an equation with its neighbors at distance 2, say $\xi_{i\pm2}$), to 2^2, etc. keeping note that when the distance is such to make $i \pm d$ be smaller than 1 or larger than n, the variables are 0. Therefore, in $k = \log n$ steps there will be n equations, each connecting only one of the variables to the right-hand side.

As we explain next, each iteration of the algorithm can be elegantly described in terms of operations with the matrices D, L, U that arise in the splitting $A^{(j)} = D^{(j)} - L^{(j)} - U^{(j)}$ where $D^{(j)}$ is diagonal and $L^{(j)}, U^{(j)}$ are strictly lower and strictly upper triangular respectively and of a very special form. For this reason, we will refer to our formulation as *matrix splitting-based* PARACR.

Definition 5.1 A lower (resp. upper) triangular matrix is called t-upper (resp. lower) diagonal if its only non-zero elements are at the tth diagonal below (resp. above) the main diagonal. A matrix that is the sum of a diagonal matrix, a t-upper diagonal and a t-lower diagonal matrix is called t-tridiagonal.

Clearly, if $t \geq n$, then the t-tridiagonal matrix will be diagonal and the t-diagonal matrices will be zero. Note that t-tridiagonal matrices appear in the study of special Fibonacci numbers (cf. [62]) and are a special case of "triadic matrices", defined in [63] to be matrices for which the number of non-zero off-diagonal elements in each column is bounded by 2.

Lemma 5.1 *The product of two t-lower (resp. upper) diagonal matrices of equal size is 2t-lower (resp. upper) diagonal. Also if $2t > n$ then the product is the zero* *matrix. If L is t-lower diagonal and U is t-upper diagonal, then LU and UL are both diagonal. Moreover the first t elements of the diagonal of LU are 0 and so are the last t diagonal elements of UL.*

Proof The nonzero structure of a 1-upper diagonal matrix is the same (without loss of generality, we ignore the effect of zero values along the superdiagonal) with that of $J^\top$, where as usual $J^\top = e_1 e_2^\top + e_2 e_3^\top + \cdots + e_{n-1} e_n^\top$ is the 1-upper diagonal matrix with all nonzero elements equal to 1. A similar result holds for 1-lower diagonal matrices, except that we use J. Observe that $(J^\top)^2 = e_1 e_3^\top + \cdots + e_{n-2} e_n^\top$ which is 2-upper diagonal and in general

$$(J^\top)^t = e_1 e_{t+1}^\top + \cdots e_{n-t} e_n^\top.$$

which is t-upper diagonal and has the same nonzero structure as any t-upper diagonal U. If we multiply two t-upper diagonal matrices, the result has the same nonzero structure as $(J^\top)^{2t}$ and thus will be $2t$-upper diagonal. A similar argument holds for the t-lower diagonal case, which proves the first part of the lemma. Note next that

$$(J^\top)^t J^t = (e_1 e_{t+1}^\top + \cdots e_{n-t} e_n^\top)(e_1 e_{t+1}^\top + \cdots e_{n-t} e_n^\top)^\top$$
$$= e_1 e_1^\top + \cdots + e_{n-t} e_{n-t}^\top$$

which is a diagonal matrix with zeros in the last t elements of its diagonal. In the same way, we can show that $J^t (J^\top)^t$ is diagonal with zeros in the first t elements.

Corollary 5.3 *Let $A = D - L - U$ be a t-tridiagonal matrix for some nonnegative integer t such that $t < n$, L (resp. U) is t-lower (resp. upper) diagonal and all diagonal elements of D are nonzero. Then $(D + L + U)D^{-1}A$ is 2t-tridiagonal. If $2t > n$ then the result of all the above products is the zero matrix.*

Proof After some algebraic simplifications we obtain

$$(I + LD^{-1} + UD^{-1})A = D - (LD^{-1}U + UD^{-1}L) - (LD^{-1}L + UD^{-1}U). \quad (5.87)$$

Multiplications with D^{-1} have no effect on the nonzero structure of the results. Using the previous lemma and the nonzero structure of U and L, it follows that $UD^{-1}U$ is $2t$-upper diagonal, $LD^{-1}L$ is $2t$-lower diagonal. Also $UD^{-1}L$ is diagonal with its last t diagonal elements zero and $LD^{-1}U$ is diagonal with its first t diagonal elements zero. Therefore, in formula (5.87), the second right-hand side term (in parentheses) is diagonal and the third is the sum of a $2t$-lower diagonal and a $2t$-upper diagonal matrix, proving the claim.

Let now $n = 2^k$ and consider the sequence of transformations

$$(A^{(j+1)}, f^{(j+1)}) = (I + L^{(j)}(D^{(j)})^{-1} + U^{(j)}(D^{(j)})^{-1})(A^{(j)}, f^{(j)}) \quad (5.88)$$

for $j = 1, \ldots, k-1$, where $(A^{(1)}, f^{(1)}) = (A, f)$. Observe the structure of $A^{(j)}$. The initial $A^{(1)}$ is tridiagonal and so from Corollary 5.3, matrix $A^{(2)}$ is 2-tridiagonal, $A^{(3)}$

is 2^2-tridiagonal, and so on; finally $A^{(k)}$ is diagonal. Therefore (A, f) is transformed to $(A^{(k)}, f^{(k)})$ with $A^{(k)}$ diagonal from which the solution can be computed in 1 vector division. Figure 5.9 shows the nonzero structure of $A^{(1)}$ up to $A^{(4)}$ for $A \in \mathbb{R}^{16 \times 16}$. Also note that at each step. the algorithm only needs the elements of $A^{(j)}$ which can overwrite A.

We based the construction of PARACR in terms of the splitting $A = D - L - U$, that is central in classical iterative methods. In Algorithm 5.4 we describe the algorithm in terms of vector operations. At each stage, the algorithm computes the three diagonals of the new 2^j-tridiagonal matrix from the current ones and updates the right-hand side. Observe that there appear to be at most 12 vector operations per iteration. These affect vectors of size $n - 2^j$ at steps $j = 1 : k - 1$. Finally, there is one extra operation of length n. So the total number of operations is approximately $12n \log n$. All elementwise multiplications in lines 5–8 can be computed concurrently if sufficiently many arithmetic units are available. In this way, there are only 8 parallel operations in each iteration of PARACR.

Proposition 5.2 *The matrix splitting-based* PARACR *algorithm can be implemented in* $\mathbf{T}_p = 8 \log n + O(1)$ *parallel operations on* $p = O(n)$ *processors.*

Algorithm 5.4 PARACR: matrix splitting-based PARACR (using transformation 5.88). Operators $\odot$, $\oslash$ denote elementwise multiplication and division respectively.

Input: $A = [\lambda_i, \delta_i, \upsilon_{i+1}] \in \mathbb{R}^{n \times n}$ and $f \in \mathbb{R}^n$ where $n = 2^k$.
Output: Solution x of $Ax = f$.
1: $l = (\lambda_2, \ldots, \lambda_n)^\mathsf{T}, d = (\delta_1, \ldots, \delta_n)^\mathsf{T}, c = (\upsilon_2, \ldots, \upsilon_n)^\mathsf{T}.$
2: $p = -l; q = -u;$
3: **do** $j = 1 : k$
4:     $\sigma = 2^{j-1};$
5:     $p = l \oslash d_{1:n-\sigma}; q = u \oslash d_{1+\sigma:n}$
6:     $f = f + \begin{pmatrix} 0_\rho \\ p \odot f_{1:n-\sigma} \end{pmatrix} + \begin{pmatrix} q \odot f_{\sigma+1:n} \\ 0_\sigma \end{pmatrix}$
7:     $d = d - \begin{pmatrix} 0_\sigma \\ p \odot u \end{pmatrix} - \begin{pmatrix} q \odot l \\ 0_\sigma \end{pmatrix}$
8:     $l = p_{\sigma+1:n-\sigma} \odot l_{1:n-2\sigma}; u = q_{1:n-2\sigma} \odot u_{1+\sigma:n-\sigma}$
9: **end**
10: $x = f \oslash d$

It is frequently the case that the dominance of the diagonal terms becomes more pronounced as CR and PARACR progress. One can thus consider truncated forms of the above algorithms to compute approximate solutions. This idea was explored in [51] where it was called *incomplete cyclic reduction*, in the context of cyclic reduction for block-tridiagonal matrices. The idea is to stop the reduction stage before the $\log n$ steps, and instead of solving the reduced system exactly, obtain an approximation followed by the necessary back substitution steps. It is worth noting that terminating the reduction phase early alleviates or avoids completely the loss of parallelism of CR.

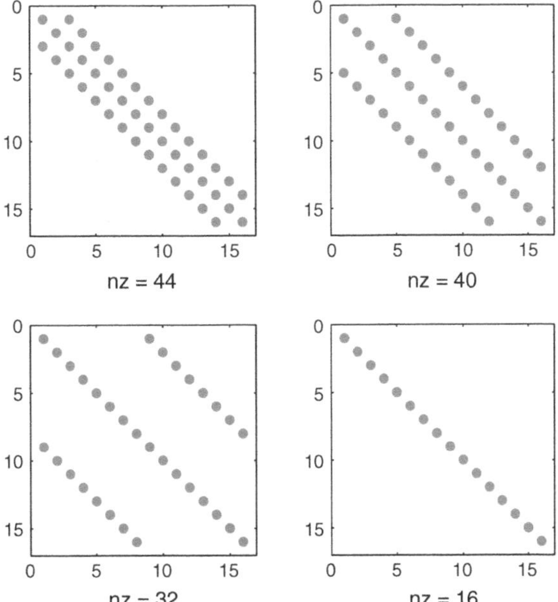

Fig. 5.9 Nonzero structure of $A^{(j)}$ as PARACR progresses from $j = 1$ (*upper left*) to $j = 4$ (*lower right*)

Incomplete point and block cyclic reduction were studied in detail in [64, 65]; it was shown that if the matrix is diagonally dominant by rows then if the *row dominance factor* (revealing the degree of diagonal dominance) of A, defined by

$$\mathrm{rdf}(A) := \max_i \left\{ \frac{1}{|\alpha_{i,i}|} \sum_{j \neq i} |\alpha_{i,j}| \right\}$$

is sufficiently smaller than 1, then incomplete cyclic reduction becomes a viable alternative to the SCALAPACK algorithms for tridiagonal and banded systems.

We next show the feasibility of an *incomplete matrix splitting-based* PARACR algorithm for matrices that are strictly diagonally dominant by rows based on the previously established formulation for PARACR and a result from [64]. We illustrate this idea by considering the first step of PARACR. Recall that

$$A^{(2)} = (D - LD^{-1}U - UD^{-1}L) - (LD^{-1}L + UD^{-1}U).$$

is 2-tridiagonal (and thus triadic). The first term in parentheses provides the diagonal elements and the second term the off-diagonal ones. Next note that $\|A\|_\infty = \||A|\|_\infty$ and the terms $LD^{-1}L$ and $UD^{-1}U$ have their nonzeros at different position. Therefore, the row dominance factor of $A^{(2)}$ is equal to

$$\mathrm{rdf}(A^{(2)}) = \|(D - LD^{-1}U - UD^{-1}L)^{-1}(LD^{-1}L + UD^{-1}U)\|_\infty.$$

The element in diagonal position i of $D - LD^{-1}U - UD^{-1}L$ is

$$\alpha_{i,i} - \frac{\alpha_{i,i-1}}{\alpha_{i-1,i-1}}\alpha_{i-1,i} - \frac{\alpha_{i,i+1}}{\alpha_{i+1,i+1}}\alpha_{i+1,i}.$$

Also the sum of the magnitudes of the off-diagonal elements at row i is

$$|\alpha_{i-1,i-2}\frac{\alpha_{i,i-1}}{\alpha_{i-1,i-1}}| + |\alpha_{i+1,i+2}\frac{\alpha_{i,i+1}}{\alpha_{i+1,i+1}}|.$$

We follow the convention that elements whose indices are 0 or $n+1$ are 0.

Then the row dominance factor of $A^{(2)}$ is

$$\mathrm{rdf}(A^{(2)}) = \max_i \left\{ \frac{|\alpha_{i,i-1}\frac{\alpha_{i-1,i-2}}{\alpha_{i-1,i-1}}| + |\alpha_{i,i+1}\frac{\alpha_{i+1,i+2}}{\alpha_{i+1,i+1}}|}{|\alpha_{i,i} - \frac{\alpha_{i,i-1}}{\alpha_{i-1,i-1}}\alpha_{i-1,i} - \frac{\alpha_{i,i+1}}{\alpha_{i+1,i+1}}\alpha_{i+1,i}|} \right\}$$

$$= \max_i \left\{ \frac{|\frac{\alpha_{i,i-1}}{\alpha_{i,i}}\frac{\alpha_{i-1,i-2}}{\alpha_{i-1,i-1}}| + |\frac{\alpha_{i,i+1}}{\alpha_{i,i}}\frac{\alpha_{i+1,i+2}}{\alpha_{i+1,i+1}}|}{|1 - \frac{\alpha_{i,i-1}}{\alpha_{i,i}}\frac{\alpha_{i-1,i}}{\alpha_{i-1,i-1}} - \frac{\alpha_{i,i+1}}{\alpha_{i,i}}\frac{\alpha_{i+1,i}}{\alpha_{i+1,i+1}}|} \right\}$$

$$\leq \max_i \left\{ \frac{|\frac{\alpha_{i,i-1}}{\alpha_{i,i}}\frac{\alpha_{i-1,i-2}}{\alpha_{i-1,i-1}}| + |\frac{\alpha_{i,i+1}}{\alpha_{i,i}}\frac{\alpha_{i+1,i+2}}{\alpha_{i+1,i+1}}|}{|1 - |\frac{\alpha_{i,i-1}}{\alpha_{i,i}}\frac{\alpha_{i-1,i}}{\alpha_{i-1,i-1}}| - |\frac{\alpha_{i,i+1}}{\alpha_{i,i}}\frac{\alpha_{i+1,i}}{\alpha_{i+1,i+1}}||} \right\}$$

For row i set

$$\hat{\psi}_i = |\frac{\alpha_{i,i-1}}{\alpha_{i,i}}|, \; \psi_i = |\frac{\alpha_{i,i+1}}{\alpha_{i,i}}|, \; \zeta_i = |\frac{\alpha_{i-1,i-2}}{\alpha_{i-1,i-1}}|, \; \eta_i = |\frac{\alpha_{i+1,i+2}}{\alpha_{i+1,i+1}}|,$$

$$\hat{\theta}_i = |\frac{\alpha_{i-1,i}}{\alpha_{i-1,i-1}}|, \; \theta_i = |\frac{\alpha_{i+1,i}}{\alpha_{i+1,i+1}}|.$$

and let $\varepsilon = \mathrm{rdf}(A)$. Then,

$$\hat{\psi}_i + \psi_i \leq \varepsilon, \; \zeta_i + \hat{\theta}_i \leq \varepsilon, \; \eta_i + \theta_i \leq \varepsilon \tag{5.89}$$

and because of strict row diagonal dominance, $\hat{\psi}_i\hat{\theta}_i + \psi_i\theta_i < 1$. Therefore if we compute for $i = 1, \ldots, n$ the maximum values of the function

$$g_i(\hat{\psi}, \psi) = \frac{\hat{\psi}\zeta_i + \psi\eta_i}{1 - \hat{\psi}\hat{\theta}_i - \psi\theta_i}$$

over $(\psi, \hat{\psi})$ assuming that conditions such as (5.89) hold, then the row dominance factor of $A^{(2)}$ is less than or equal to the maximum of these values. It was shown in

[64] that if conditions such as (5.89) hold then

$$\frac{\hat{\psi}\zeta_i + \psi\eta_i}{1 - \hat{\psi}\hat{\theta}_i - \psi\theta_i} \le \varepsilon^2. \tag{5.90}$$

Therefore, $\mathrm{rdf}(A^{(2)}) \le \varepsilon^2$. The following proposition can be used to establish a termination criterion for incomplete matrix splitting-based PARACR. We omit the details of the proof.

Proposition 5.3 *Let A be tridiagonal and strictly diagonally dominant by rows with dominance factor ε. Then*

$$\mathrm{rdf}(A^{(j+1)}) \le \varepsilon^{2^j}, \ j = 1, \ldots, \log n - 2.$$

We also observe that it is possible to use the matrix splitting framework in order to describe CR. We sketch the basic idea, assuming this time that $n = 2^k - 1$. At each step $j = 1, \ldots, k - 1$, of the reduction phase the following transformation is implemented:

$$(A^{(j+1)}, f^{(j+1)}) = (I + L_e^{(j)}(D^{(j)})^{-1} + U_e^{(j)}(D^{(j)})^{-1})(A^{(j)}, f^{(j)}), \tag{5.91}$$

where initially $(A^{(1)}, f^{(1)}) = (A, f)$. We assume that the steps of the algorithm can be brought to completion without division by zero. Let $A^{(j)} = D^{(j)} - L^{(j)} - U^{(j)}$ denote the splitting of $A^{(j)}$ into its diagonal and strictly lower and upper triangular parts. From $L^{(j)}$ and $U^{(j)}$ we extract the strictly triangular matrices $L_e^{(j)}$ and $U_e^{(j)}$ following the rule that they contain the values of $L^{(j)}$ and $U^{(j)}$ at locations $(i2^j, i2^j - 2^{j-1})$ and $(i2^j - 2^{j-1}, i2^j)$ respectively for $i = 1, \ldots, 2^{k-j} - 1$ and zero everywhere else. Then at the end of the reduction phase, row 2^{k-1} (at the middle) of matrix $A^{(k-1)}$ will only have its diagonal (middle) element nonzero and the unknown $\xi_{2^{k-1}}$ can be computed in one division. This is the first step of the back substitution phase which consists of k steps. At step $j = 1, \ldots, k$, there is a vector division of length 2^{j-1} to compute the unknowns indexed by $(1, 3, \ldots, 2^j - 1) \cdot 2^{k-j}$ and 2^j _AXPY, BLAS1, operations on vectors of length $2^{k-j} - 1$ for the updates. The panels in Fig. 5.10 illustrate the matrix structures that result after k steps of cyclic reduction ($k = 3$ in the left and $k = 4$ in the right panel).

The previous procedure can be extended to block-tridiagonal systems. An analysis similar to ours for the case of block-tridiagonal systems and block cyclic reduction was described in [66]; cf. Sect. 6.4.

5.5.3 LDU Factorization by Recurrence Linearization

Assume that all leading principal submatrices of A are nonsingular. Then there exists a factorization $A = LDU$ where $D = \mathrm{diag}(\delta_1, \ldots, \delta_n)$ is diagonal and

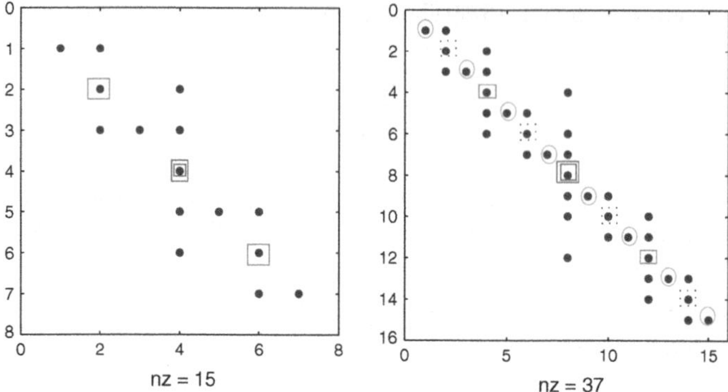

Fig. 5.10 Nonzero structure of $A^{(3)} \in \mathbb{R}^{7 \times 7}$ (*left*) and $A^{(4)} \in \mathbb{R}^{15 \times 15}$ (*right*). In both *panels*, the unknown corresponding to the middle equation is obtained using the middle value of each matrix enclosed by the *rectangle* with double border. The next set of computed unknowns (2 of them) correspond to the diagonal elements enclosed by the simple *rectangle*, the next set of computed unknowns (2^2 of them) correspond to the diagonal elements enclosed by the *dotted rectangles*. For $A^{(4)}$, the final set of 2^3 unknowns correspond to the *encircled elements*

$$
L = \begin{pmatrix} 1 & & & \\ \lambda_2 & & & \\ & \ddots & \ddots & \\ & & \lambda_n & 1 \end{pmatrix}, \quad U = \begin{pmatrix} 1 & \upsilon_2 & & \\ & \ddots & \ddots & \\ & & & \upsilon_n \\ & & & 1 \end{pmatrix}.
$$

We can write the matrix A as a sum of the rank-1 matrices formed by the columns of L and the rows of U multiplied by the corresponding diagonal element of D.

$$
A = \delta_1 L_{:,1} U_{1,:} + \cdots + \delta_n L_{:,n} U_{n,:}
$$

Observing the equalities along the diagonal, the following recurrence of degree 1 holds:

$$
\delta_i = \alpha_{i,i} - \frac{\alpha_{i-1,i}\alpha_{i,i-1}}{\delta_{i-1}}, \quad i = 2:n. \tag{5.92}
$$

This recurrence is linearized as described in Sect. 3.4. Specifically, using new variables τ_i and setting $\delta_i = \tau_i / \tau_{i-1}$ with $\tau_0 = 1, \tau_1 = \alpha_{1,1}$ it follows that

$$
\tau_i = \alpha_{i,i}\tau_{i-1} - \alpha_{i-1,i}\alpha_{i,i-1}\tau_{i-2}, \quad \tau_0 = 1, \tau_1 = \alpha_{1,1}. \tag{5.93}
$$

The values $\tau_1, \ldots, \tau_n$ satisfy the system

$$
\begin{pmatrix}
1 & & & & \\
-\alpha_{1,1} & 1 & & & \\
\alpha_{1,2}\alpha_{2,1} & -\alpha_{2,2} & 1 & & \\
& \ddots & & \ddots & \\
& & \alpha_{n-1,n}\alpha_{n,n-1} & -\alpha_{n,n} & 1
\end{pmatrix}
\begin{pmatrix}
\tau_0 \\ \tau_1 \\ \tau_2 \\ \vdots \\ \tau_n
\end{pmatrix}
=
\begin{pmatrix}
1 \\ 0 \\ 0 \\ \vdots \\ 0
\end{pmatrix}
\tag{5.94}
$$

The right-hand side is the unit vector, thus the solution is the first column of the inverse of the coefficient matrix, that is lower triangular with bandwidth 3. This specific system, as was shown in [18, Lemma 2], can be solved in $2\log n + O(1)$ steps using at most $2n$ processors. It is worth noting that this is faster than the approximate cost of $3\log n$ steps predicted by Theorem 3.2 and this is due to the unit vector in the right-hand side. Once the τ_i's are available, the elements of D, L and U are computable in $O(1)$ parallel steps using n processors. Specifically

$$
\delta_i = \tau_i / \tau_{i-1}, \quad \lambda_{i+1} = \frac{\alpha_{i+1,i}}{d_i}, \quad \upsilon_i = \frac{\alpha_{i,i+1}}{d_i}.
$$

In theory, therefore, if the factorization exists, we can compute the factors L, D, U in $2\log n + O(1)$ steps using at most $2n$ processors. To solve a linear system, it only remains to perform the backward and forward substitutions and one parallel division. Using the same theorem, each of the bidiagonal systems can be solved in $2\log n$ steps on $n-1$ processors. We thus obtain the following result for the parallel computation of the LDU factorization and its application to solving tridiagonal systems.

Proposition 5.4 [67] *If a tridiagonal matrix $A \in \mathbb{R}^{n \times n}$ is non-singular and so are all its leading principal submatrices then all the elements of its LDU factorization can be computed on $p = 2n$ processors using $\mathbf{T}_p = 2\log n + O(1)$ parallel steps. For such an A, the linear system $Ax = f$ can thus be solved using $6\log n + O(1)$ steps.*

Algorithm 5.5 shows the steps for the LDU factorization. In practice, the above algorithm must be used with care. In particular, there is the possibility of overflow or underflow because $\tau_i = \delta_i \delta_{i-1} \cdots \delta_1$. For example, when $\delta_i = i$ making $\tau_i = i!$ [67].

It is worth noting that the τ_i's can also be computed by rewriting relations (5.93) as

$$
\begin{pmatrix} \tau_i & \tau_{i-1} \end{pmatrix} = \begin{pmatrix} \tau_{i-1} & \tau_{i-2} \end{pmatrix}
\begin{pmatrix} \alpha_{i,i}^2 & 1 \\ -\alpha_{i-1,i}\alpha_{i,i-1} & 0 \end{pmatrix}, \quad \begin{pmatrix} \tau_1 & \tau_0 \end{pmatrix} = \begin{pmatrix} \alpha_{1,1} & 1 \end{pmatrix}
$$

This is a vector recurrence for the row vector $t^{(i)} = (\tau_i, \tau_{i-1})$ that we can write as

$$
t^{(i)} = t^{(i-1)} S_i,
$$

Algorithm 5.5 TRID_LDU: LDU factorization of tridiagonal matrix.

Input: $A = [\alpha_{i,i-1}, \alpha_{i,i}, \alpha_{i,i+1}] \in \mathbb{R}^{n \times n}$ irreducible and such that all leading principal submatrices are nonsingular.

Output: The elements $\delta_{1:n}, \lambda_{2:n}$ and $\upsilon_{2:n}$ of the diagonal D, lower bidiagonal L and upper bidiagonal U.

    //Construct triangular matrix $C = [\gamma_{i,j}] \in \mathbb{R}^{(n+1)\times(n+1)}$

1: $C = e_1^{(n+1)}(e_1^{(n+1)})^\top$, $\gamma_{n+1,n-1} = \alpha_{n-1,n}\alpha_{n,n-1}$

2: **doall** $i = 1 : n$

3:     $\gamma_{i+1,i+1} = 1, \gamma_{i+1,i} = -\alpha_{i,i}$

4: **end**

    //Solve using parallel algorithm for banded lower triangular systems

5: Solve $Ct = e_1^{(n+1)}$ where $C \in \mathbb{R}^{(n+1)\times(n+1)}$

    //Compute the unknown elements of L, D, U

6: **doall** $i = 1 : n$

7:     $\delta_i = \frac{t_{i+1}}{t_i}$

8: **end**

9: **doall** $i = 1 : n - 1$

10:     $\lambda_{i+1} = \frac{\alpha_{i+1,i}}{\delta_i}, \upsilon_{i+1} = \frac{\alpha_{i,i+1}}{\delta_i}$

11: **end**

---

where S_i denotes the 2×2 multiplier. Therefore, the unknowns can be recovered by computing the terms

$$t^{(2)} = t^{(1)}S_1, t^{(3)} = t^{(1)}S_1S_2, \ldots, t^{(n)} = t^{(1)}S_1 \cdots S_n$$

and selecting the first element from each $t^{(i)}$,

$$\begin{pmatrix} \tau_1 \\ \vdots \\ \tau_n \end{pmatrix} = \begin{pmatrix} t^{(1)} \\ \vdots \\ t^{(n)} \end{pmatrix} e_1.$$

This can be done using a matrix (product) parallel prefix algorithm on the set $t^{(1)}, S_1, \ldots, S_n$. Noticing that $t^{(i)}$ and $t^{(i+1)}$ both contain the value τ_i, it is sufficient to compute approximately half the terms, say $t^{(1)}, t^{(3)}, \ldots$. For convenience assume that n is odd. Then this can be accomplished by first computing the products $t^{(1)}S_1$ and $S_{2i}S_{2i+1}$ for $i = 1, \ldots, (n-1)/2$ and then applying parallel prefix matrix product on these elements. For example, when $n = 7$, the computation can be accomplished in 3 parallel steps, shown in Table 5.1.

Table 5.1 Steps of a parallel prefix matrix product algorithm to compute the recurrence (5.93)

Step 1	$t^{(1)}S_1$	$S_{2:3}$	$S_{4:5}$	$S_{6:7}$
Step 2		$t^{(1)}S_{1:3}$	$S_{2:5}$	$S_{4:7}$
Step 3			$t^{(1)}S_{1:5}$	$t^{(1)}S_{1:7}$

For $j > i$, the term $S_{i:j}$ denotes the product $S_i S_{i+1} \cdots S_j$

Regarding stability in finite precision, to achieve the stated complexity, the algorithm uses either parallel prefix matrix products or a lower banded triangular solver (that also uses parallel prefix). An analysis of the latter process in [18] and some improvements in [68], show that the bound for the 2-norm of the absolute forward error contains a factor σ^{n+1}, where $\sigma = \max_i \|S_i\|_2$. This suggests that the absolute forward error can be very large in norm. Focusing on the parallel prefix approach, it is concluded that the bound might be pessimistic, but deriving a tighter one, for the general case, would be hard. Therefore, the algorithm must be used with precaution, possibly warning the user when the error becomes large.

5.5.4 Recursive Doubling

Let a tridiagonal matrix be irreducible; then the following recurrence holds:

$$\xi_{i+1} = -\frac{\alpha_{i,i}}{\alpha i, i+1}\xi_i - \frac{\alpha_{i,i-1}}{\alpha i, i+1}\xi_{i-1} + \frac{\phi_i}{\alpha_{i,i+1}}$$

If we set $\hat{x}_i = \begin{pmatrix} \xi_i \\ \xi_{i-1} \\ 1 \end{pmatrix}$ then we can write the matrix recurrence

$$\hat{x}_{i+1} = M_i \hat{x}_i, \text{ where } M_i = \begin{pmatrix} \rho_i & \sigma_i & \tau_i \\ 1 & 0 & 0 \\ 0 & 0 & 1 \end{pmatrix}, \; \rho_i = -\frac{\alpha_{i,i}}{\alpha_{i,i+1}}, \sigma_i = -\frac{\alpha_{i,i-1}}{\alpha_{i,i+1}}, \tau_i = \frac{\phi_i}{\alpha_{i,i+1}}.$$

The initial value is $\hat{x}_1 = (\xi_1, 0, 1)^\top$. Observe that ξ_1 is yet unknown. If it were available, then the matrix recurrence can be used to compute all elements of x from $\hat{x}_2, \ldots, \hat{x}_n$:

$$\hat{x}_2 = M_1 \hat{x}_1, \; \hat{x}_3 = M_2 \hat{x}_2 = M_2 M_1 \hat{x}_1, \; \ldots, \hat{x}_n = M_{n-1} \cdots M_1 \hat{x}_1.$$

We can thus write

$$\hat{x}_{n+1} = P_n \hat{x}_1, \text{ where } P_n = M_n \cdots M_1.$$

From the structure of M_j we can write

$$P_n = \begin{pmatrix} \pi_{1,1} & \pi_{1,2} & \pi_{1,3} \\ \pi_{2,1} & \pi_{2,2} & \pi_{3,3} \\ 0 & 0 & 1 \end{pmatrix}.$$

To find ξ_1 we observe that $\xi_0 = \xi_{n+1} = 0$ and so $0 = \pi_{1,1}\xi_1 + \pi_{1,3}$ from which it follows that $\xi_1 = -\pi_{1,3}/\pi_{1,1}$ once the elements of P_n are computed. In [69], a tridiagonal solver is proposed based on this formulation. We list this as Algorithm 5.6. The algorithm uses matrix parallel prefix on the matrix sequence $M_1, \ldots, M_n$ in order to compute P_n and then $\hat{x}_1, \ldots, \hat{x}_n$ to recover x, and takes a total of $15n + O(1)$ arithmetic operations on one processor. However, the matrix needs to be diagonally dominant for the method to proceed without numerical problems; some relevant error analysis was presented in [70].

5.5.5 Solving by Givens Rotations

Givens rotations can be used to implement the QR factorization in place of Householder reflections in order to bring a matrix into upper triangular form [22, 71]. Givens rotations become more attractive for matrices with special structure, e.g. upper Hessenberg and tridiagonal. Givens rotations have been widely used in parallel algorithms for eigenvalues; see for example [18, 72] and our discussion and references in Chaps. 7 and 8. Parallel algorithms for linear systems based on Givens rotations were proposed early on in [12, 18].

Algorithm 5.6 RD_PREF: tridiagonal system solver using RD and matrix parallel prefix.

Input: Irreducible $A = [\alpha_{i,i-1}, \alpha_{i,i}, \alpha_{i,i+1}] \in \mathbb{R}^{n \times n}$, and right-hand side $f \in \mathbb{R}^n$.
Output: Solution of $Ax = f$.
1: **doall** $i = 1, \ldots, n$
2:     $\rho_i = -\frac{\alpha_{i,i}}{\alpha_{i,i+1}}, \sigma_i = -\frac{\alpha_{i,i-1}}{\alpha_{i,i+1}}, \tau_i = \frac{\phi_i}{\alpha_{i,i+1}}.$
3: **end**
4: Compute the products $P_2 = M_2 M_1, \ldots, P_n = M_n \cdots M_1$ using a parallel prefix matrix product
   algorithm, where $M_i = \begin{pmatrix} \rho_i & \sigma_i & \tau_i \\ 1 & 0 & 0 \\ 0 & 0 & 1 \end{pmatrix}.$
5: Compute $\xi_1 = -(P_n)_{1,3}/(P_n)_{1,1}$
6: **doall** $i = 2, \ldots, n$
7:     $\hat{x}_i = P_i \hat{x}_1$ where $\hat{x}_1 = (\xi_1, 0, 1)^{\top}.$
8: **end**
9: Gather the elements of x from $\{\hat{x}_1, \ldots, \hat{x}_n\}$

We first consider the Givens based parallel algorithm from [12, Algorithm I]). As will become evident, the algorithm shares many features with the parallel LDU factorization (Algorithm 5.5). Then in Sect. 5.5.5 we revisit the Givens rotation solver based on Spike partitioning from [12, Algorithm II].

We will assume that the matrix is irreducible and will denote the elementary 2×2 Givens rotation submatrix that will be used to eliminate the element in position $(i + 1, i)$ of the tridiagonal matrix by

$$G_{i+1,i} = \begin{pmatrix} c_i & s_i \\ -s_i & c_i. \end{pmatrix} \tag{5.95}$$

To perform the elimination of subdiagonal element in position $(i + 1, i)$ we use the block-diagonal matrix

$$
G^{(i)}_{i+1,i} = \begin{pmatrix} I_{i-1} & & \\ & G_{i+1,i} & \\ & & I_{n-i-1} \end{pmatrix}.
$$

Let $A^{(0)} = A$ and $A^{(i)} = G^{(i)}_{i+1,i} \cdots G^{(1)}_{2,1} A^{(0)}$ be the matrix after $i = 1, \ldots, n - 1$ rotation steps so that the subdiagonal elements in positions $(2, 1), \ldots, (i, i - 1)$ of $A^{(i)}$ are 0 and $A^{(n-1)} = R$ is upper triangular. Let

$$
A^{(i-1)} = \begin{pmatrix}
\lambda_1 & \mu_1 & \nu_1 & & & & & \\
 & \lambda_2 & \mu_2 & \nu_2 & & & & \\
 & & \ddots & \ddots & \ddots & & & \\
 & & & \lambda_{i-1} & \mu_{i-1} & \nu_{i-1} & & \\
 & & & & \pi_i & c_{i-1}\alpha_{i,i+1} & & \\
 & & & & \alpha_{i+1,i} & \alpha_{i+1,i+1} & \alpha_{i+1,i+2} & \\
 & & & & & \ddots & \ddots & \ddots
\end{pmatrix}
\tag{5.96}
$$

be the result after the subdiagonal entries in rows $2, \ldots, i$ have been annihilated. Observe that R is banded, with bandwidth 3. If $i = n - 1$, the process terminates and the element in position (n, n) is π_n; otherwise, if $i < n - 1$ then rows i and $i + 1$ need to be multiplied by a rotation matrix to zero out the subdiagonal element in row $i + 1$.

Next, rows i and $i + 1$ of $A^{(i-1)}$ need to be brought into their final form. The following relations hold for $i = 1, \ldots, n - 1$ and initial values $c_0 = 1, s_0 = 0$:

$$
c_i = \frac{\pi_i}{\lambda_i}, \quad s_i = \frac{\alpha_{i+1,i}}{\lambda_i}
\tag{5.97}
$$

$$
\pi_i = c_{i-1}\alpha_{i,i} - s_{i-1}c_{i-2}\alpha_{i-1,i}
\tag{5.98}
$$

$$
\mu_i = s_i\alpha_{i+1,i+1} + c_ic_{i-1}\alpha_{i,i+1}
\tag{5.99}
$$

$$
\nu_i = s_i\alpha_{i+1,i+2}
\tag{5.100}
$$

From the above we obtain

$$
\begin{aligned}
c_i\lambda_i &= c_{i-1}\alpha_{i,i} - s_{i-1}c_{i-2}\alpha_{i-1,i} \\
&= c_{i-1}\alpha_{i,i} - \frac{\alpha_{i,i-1}}{\lambda_{i-1}}c_{i-2}\alpha_{i-1,i}
\end{aligned}
$$

that we write as

$$
\frac{c_i}{c_{i-1}}\lambda_i = \alpha_{i,i} - \frac{\alpha_{i,i-1}\alpha_{i-1,i}}{\frac{c_{i-1}}{c_{i-2}}\lambda_{i-1}}.
$$

Observe the similarity with the nonlinear recurrence (5.92) we encountered in the steps leading to the parallel LDU factorization Algorithm 5.5. Following the same approach, we apply the change of variables

$$\frac{\tau_i}{\tau_{i-1}} = \frac{c_i}{c_{i-1}} \lambda_i \tag{5.101}$$

in order to obtain the linear recurrence

$$\tau_i = \alpha_{i,i} \tau_{i-1} - \alpha_{i,i-1}\alpha_{i-1,i}\tau_{i-2}, \quad \tau_0 = 1, \tau_1 = \alpha_{1,1}, \tag{5.102}$$

and thus the system

$$\begin{pmatrix} 1 & & & & \\ -\alpha_{1,1} & 1 & & & \\ \alpha_{1,2}\alpha_{2,1} & -\alpha_{2,2} & 1 & & \\ & & \ddots & & \\ Gant & & \alpha_{n-2,n-1}\alpha_{n-1,n-2} & -\alpha_{n-1,n-1} & 1 \end{pmatrix} \begin{pmatrix} \tau_0 \\ \tau_1 \\ \tau_2 \\ \vdots \\ \tau_{n-1} \end{pmatrix} = \begin{pmatrix} 1 \\ 0 \\ 0 \\ \vdots \\ 0 \end{pmatrix} \tag{5.103}$$

The coefficient matrix can be generated in one parallel multiplication. As noted in our earlier discussion for Algorithm 5.5, forming and solving the system takes $2 \log n + O(1)$ steps using $2n - 4$ processors (cf. Algorithm 3.3). Another way is to observe that the recurrence (5.102) can also be expressed as follows

$$\begin{pmatrix} \tau_i & \tau_{i-1} \end{pmatrix} = \begin{pmatrix} \tau_{i-1} & \tau_{i-2} \end{pmatrix} \begin{pmatrix} \alpha_{i,i} & 1 \\ -\alpha_{i,i-1}\alpha_{i-1,i} & 0 \end{pmatrix}, \tag{5.104}$$

where $\begin{pmatrix} \tau_1 & \tau_0 \end{pmatrix} = \begin{pmatrix} \alpha_{1,1} & 1 \end{pmatrix}$ \tag{5.105}

We already discussed the application of parallel prefix matrix multiplication to compute the terms of this recurrence in the context of the LDU algorithm; cf. Sect. 5.5.3. From relation (5.101) and the initial values for $c_0 = 1, \tau_0 = 1$, it follows that

$$\tau_i = \frac{c_i}{c_{i-1}} \lambda_i \tau_{i-1} = \lambda_i \cdots \lambda_1 \frac{c_i}{c_{i-1}} \cdots \frac{c_1}{c_0} \tau_0$$
$$= \lambda_i \cdots \lambda_1 c_i.$$

Setting $\theta_i = (\prod_{j=1}^{i} \lambda_j)^2$ it follows from trigonometric identities that

$$\theta_i = \alpha_{i+1,i}^2 \theta_{i-1} + \tau_i^2, \quad \text{for } i = 1, \ldots, n-1 \tag{5.106}$$

with initial values $\theta_0 = \tau_0 = 1$. This amounts to a linear recurrence for θ_i that can be expressed as the unit lower bidiagonal system

$$
\begin{pmatrix}
1 & & & & \\
-\alpha_{2,1}^2 & 1 & & & \\
& -\alpha_{3,2}^2 & 1 & & \\
& & \ddots & \ddots & \\
& & & -\alpha_{n,n-1}^2 & 1
\end{pmatrix}
\begin{pmatrix}
\theta_0 \\ \theta_1 \\ \theta_2 \\ \vdots \\ \theta_{n-1}
\end{pmatrix}
=
\begin{pmatrix}
\tau_0^2 \\ \tau_1^2 \\ \tau_2^2 \\ \vdots \\ \tau_{n-1}^2
\end{pmatrix}
\tag{5.107}
$$

On $2(n-1)$ processors, one step is needed to form the coefficient matrix and right-hand side. This lower bidiagonal system can be solved in $2\log n$ steps on $n-1$ processors, (cf. [18, Lemma 1] and Theorem 3.2).

As before, we can also write the recurrence (5.106) as

$$
(\theta_i \ \ 1) = (\theta_{i-1} \ \ 1)
\begin{pmatrix}
\alpha_{i+1,i}^2 & 0 \\
\tau_i^2 & 1
\end{pmatrix}, \quad \theta_0 = 1.
\tag{5.108}
$$

It is easy to see that all θ_i's can be obtained using a parallel algorithm for the prefix matrix product. Moreover, the matrices are all nonnegative, and so the componentwise relative errors are expected to be small; cf. [68]. This further strengthens the finding in [12, Lemma 3] that the normwise relative forward error bound in computing the θ_i's only grows linearly with n.

Using the values $\{\theta_i, \tau_i\}$ the following values can be computed in 3 parallel steps for $i = 0, \ldots, n-1$:

$$
c_i^2 = \frac{\tau_i^2}{\theta_i}, \quad \lambda_i^2 = \frac{\theta_i}{\theta_{i-1}}, \quad s_i^2 = \frac{\alpha_{i+1,i}^2}{\lambda_i^2}.
\tag{5.109}
$$

From these, the values μ_i, ν_i along the superdiagonals of $R = A^{(n-1)}$ can be obtained in 3 parallel steps on $2n$ processors.

The solution of a system using QR factorization also requires the multiplication of the right-hand side vector by the matrix Q that is the product of the $n-1$ associated Givens rotations. By inspection, it can be easily seen that this product is lower Hessenberg; cf. [73]. An observation with practical implications made in [12] that is not difficult is that Q can be expressed as $Q = WLY + SJ^\top$ [12], where L is the lower triangular matrix with all nonzero elements equal to 1, $J = (e_2, \ldots, e_n, 0)$ and $W = \mathrm{diag}(\omega_1, \ldots, \omega_n)$, $Y = \mathrm{diag}(\eta_1, \ldots, \eta_n)$ and $S = \mathrm{diag}(s_1, \ldots, s_{n-1}, 0)$, where the $\{c_i, s_i\}$ are as before and

$$
\omega_i = c_i \rho_{i-1}, \eta_i = \frac{c_{i-1}}{\rho_{i-1}}, \quad i = 1, \ldots, n,
\tag{5.110}
$$

and $\rho_i = (-1)^i \prod_{j=1}^{i} s_j$, $i = 1, \ldots, n-1$ with $\rho_0 = c_0 = c_n = 1$.

The elements ρ_i are obtained from a parallel prefix product in $\log n$ steps using $\frac{n-2}{2}$ processors. It follows that multiplication by Q can be decomposed into five easy steps involving arithmetic, 3 being multiplications of a vector with the diagonal matrices W, Y and S, and a computation of all partial sums of a vector (the effect of L). The latter is a prefix sum and can be accomplished in $\log n$ parallel steps using n processors; see for instance [74].

Algorithm 5.7, called PARGIV, incorporates all these steps to solve the tridiagonal system using the parallel generation and application of Givens rotations.

Algorithm 5.7 PARGIV: tridiagonal system solver using Givens rotations.

Input: Irreducible $A = [\alpha_{i,i-1}, \alpha_{i,i}, \alpha_{i,i+1}] \in \mathbb{R}^{n \times n}$, and right-hand side $f \in \mathbb{R}^n$.
Output: Solution of $Ax = f$.
    //Stage I:
1: Compute $(\tau_1 \ldots, \tau_{n-1})$ by solving the banded lower triangular system (5.103) or using parallel prefix matrix product for (5.104).
    //Stage II:
2: Compute $(\theta_1, \ldots, \theta_{n-1})$ by solving the banded lower triangular system (5.107) or using parallel prefix matrix product for (5.108).
    //Stage III:
3: Compute $\{c_i, \lambda_i\}$ for $i = 1 : n$ followed by $\{s_i\}$ from Eqs. 5.109.
    //Stage IV:
4: Compute W and Y from (5.110).
5: Compute $\hat{f} = WLYf + SJf$
6: Solve $Rx = \hat{f}$ where $R = A^{(n-1)}$ is the banded upper triangular matrix in (5.96).

The leading cost of PARGIV is $9 \log n$ parallel operations: There are $2 \log n$ in each of stages I, II because of the special triangular systems that need to be solved, $2 \log n$ in Stage IV (line 4) because of the special structure of the factors W, L, Y, S, J and finally, another $3 \log n$ in stage IV (line 6) also because of the banded trangular system. Stage III contributes only a constant to the cost.

From the preceding discussion the following theorem holds:

Theorem 5.6 ([12, Theorem 3.1, Lemma 3.1]) *Let A be a nonsingular irreducible tridiagonal matrix of order n. Algorithm* PARGIV *for solving $Ax = f$ based on the orthogonal factorization $QA = R$ constructed by means of Givens rotations takes* $\mathbf{T}_p = 9 \log n + O(1)$ *parallel steps on $3n$ processors. The resulting speedup is* $\mathbf{S}_p = O(\frac{n}{\log n})$ *and the efficiency* $\mathbf{E}_p = O(\frac{1}{\log n})$. *Specifically, the speedup over the algorithm implemented on a single processor is approximately* $\frac{8n}{3 \log n}$ *and over Gaussian elimination with partial pivoting approximately* $\frac{4n}{3 \log n}$. *Moreover, if $B \in \mathbb{R}^{n \times k}$, then the systems $AX = B$ can be solved in the same number of parallel operations using approximately $(2 + k)n$ processors.*

Regarding the numerical properties of the algorithm, we have already seen that the result of stage I can have large absolute forward error. Since the bound grows exponentially with n if the largest 2-norm of the multipliers S_i is larger than 1, care is required in using the method in the general case. Another difficulty is that the computation of the diagonal elements of W and Y in Stage IV can lead to underflow and overflow. Both of the above become less severe when n is small. We next describe a partitioning approach for circumventing these problems.

5.5.6 Partitioning and Hybrids

Using the tridiagonal solvers we encountered so far as basic components, we can construct more complex methods either by combination into *hybrids* that attempt to combine the advantages of each component method (see e.g. [30]), and/or by applying *partitioning* in order to create another layer of parallelism. In addition, in the spirit of the discussion for Spike in Sect. 5.2.2, we could create *polyalgorithms* that adapt according to the characteristics of the problem and the underlying architecture.

Many algorithms for tridiagonal systems described in the literature adopt the partitioning approach. Both Spike (cf. Sect. 5.1) and the algorithms used in SCALAPACK [2, 3, 75] for tridiagonal and more general banded systems are of this type. In a well implemented partitioning scheme it appears relatively easy to obtain an algorithm that returns almost linear speedup for a moderate number of processors, say p, by choosing the number of partitions equal to the number of processors. The sequential complexity of tridiagonal solvers is linear in the size of the problem, therefore, since the reduced system is of size $O(p)$ and $p \ll n$, then the parallel cost of the partitioning method would be $O(n/p)$, resulting in linear speedup.

We distinguish three major characteristics of partitioning methods: (i) the partitioning scheme for the matrix, (ii) the method used to solve the subsystems, and (iii) the method used to solve the reduced system. The actual partitioning of the matrix, for example, leads to submatrices and a reduced system that can be anything from well conditioned to almost singular. In some cases, the original matrix is SPD or diagonally dominant, and so the submatrices along the diagonal inherit this property and thus partitioning can be guided solely by issues such as load balancing and parallel efficiency. In the general case, however, there is little known beforehand about the submatrices. One possibility would be to attempt to detect singularity of the submatrices corresponding to a specific partitioning, and repartition if this is the case. Note that it is possible to detect in linear time whether a given tridiagonal matrix can become singular due to small perturbations of its coefficients [76, 77]. We next show a parallel algorithm based on Spike that detects exact singularity in any subsystem and implements a low-cost correction procedure to compute the solution of the original (nonsingular) tridiagonal system. From the partitionings proposed in the literature (see for instance [5, 12, 15, 56, 57, 78, 79]) we consider the Spike partitioning that we already encountered for banded matrices earlier in this chapter and write the tridiagonal system $Ax = f$ as follows:

$$
\begin{pmatrix}
A_{1,1} & A_{1,2} & & & \\
A_{2,1} & A_{2,2} & A_{2,3} & & \\
& \ddots & \ddots & \ddots & \\
& & A_{p,p-1} & A_{p,p}
\end{pmatrix}
\begin{pmatrix}
x_1 \\ x_2 \\ \vdots \\ x_p
\end{pmatrix}
=
\begin{pmatrix}
f_1 \\ f_2 \\ \vdots \\ f_p
\end{pmatrix},
\tag{5.111}
$$

with all the diagonal submatrices being square. Furthermore, to render the exposition simpler we assume that p divides n and that each $A_{i,i}$ is of size, $m = n/p$. Under these assumptions, the $A_{i,i}$'s are tridiagonal, the subdiagonal blocks $A_{2,1}, \ldots, A_{p,p-1}$ are multiples of $e_1 e_m^\top$ and each superdiagonal block $A_{1,2}, \ldots, A_{p-1,p}$ is a multiple of $e_m e_1^\top$.

The algorithm we describe here uses the Spike methodology based on a Givens-QR tridiagonal solver applied to each subsystem. In fact, the first version of this method was proposed in [12] as a way to restrict the size of systems on which PARGIV (Algorithm 5.7) is applied and thus prevent the forward error from growing too large. This partitioning for stabilization was inspired by work on marching methods; cf. [12, p. 87] and Sect. 6.4.7 of Chap. 6. Today, of course, partitioning is considered primarily as a method for adding an extra level of parallelism and greater opportunities for exploiting the underlying architecture. Our Spike algorithm, however, does not have to use PARGIV to solve each subsystem. It can be built, for example, using a sequential method for each subsystem.

Key to the method are the following two results that restrict the rank deficiency of tridiagonal systems.

Proposition 5.5 *Let the nonsingular tridiagonal matrix A be of order n, where $n = pm$ and let it be partitioned into a block-tridiagonal matrix of p blocks of order m each in the diagonal with rank-1 off-diagonal blocks. Then each of the tridiagonal submatrices along the diagonal and in block positions 2 up to $p - 1$ have rank at least $m - 2$ and the first and last submatrices have rank at least $m - 1$. Moreover, if A is irreducible, then the rank of each diagonal submatrix is at least $m - 1$.*

The proof is omitted; cf. related results in [12, 80]. Therefore, the rank of each diagonal block of an irreducible tridiagonal matrix is at least $n - 1$. With this property in mind, in infinite precision at least, we can correct rank deficiencies by only adding a rank-1 matrix. Another result will also be useful.

Proposition 5.6 ([12, Lemma 3.2]) *Let the order n irreducible tridiagonal matrix A be singular. Then its rank is exactly $n - 1$ and the last row of the triangular factor R in any QR decomposition of A will be zero, that is $\pi_n = 0$. Moreover, if this is computed via the sequence of Givens rotations (5.95), the last element satisfies $c_{n-1} \neq 0$.*

The method, we call SP_GIVENS, proceeds as follows. Given the partitioned system (5.111), each of the submatrices $A_{i,i}$ along the diagonal is reduced to upper triangular form using Givens transformations. This can be done, for example, by means of Algorithm 5.7 (PARGIV).

In the course of this reduction, we obtain (in implicit form) the orthogonal matrices $Q_1, \ldots, Q_p$ such that $Q_i A_{i,i} = R_i$, where R_i is upper triangular. Applying the same transformations on the right-hand side, the original system becomes

$$
\begin{pmatrix}
R_1 & B_2 & & \\
C_2 & R_2 & B_3 & \\
& \ddots & \ddots & \ddots \\
& & C_p & R_p
\end{pmatrix}
\begin{pmatrix}
x_1 \\ x_2 \\ \vdots \\ x_p
\end{pmatrix}
=
\begin{pmatrix}
\tilde{f}_1 \\ \tilde{f}_2 \\ \vdots \\ \tilde{f}_p
\end{pmatrix}
\tag{5.112}
$$

where

$$
C_i = Q_i A_{i,i-1}, \quad B_{i+1} = Q_i A_{i,i+1}, \quad \tilde{f}_i = Q_i f_i.
$$

We denote the transformed system by $\tilde{A}x = \tilde{b}$.

Recall from the presentation of PARGIV that the orthogonal matrices Q_i can be written in the form

$$
Q_i = W_i L Y_i + S_i J^{\top}
$$

where the elements of the diagonal matrices W_i, Y_i, S_i, see (5.109) and (5.110) correspond to the values obtained in the course of the upper triangularization of $A_{i,i}$ by Givens rotations. Recalling that $A_{i,i-1} = \alpha_{(i-1)m+1,(i-1)m} e_1 e_m^{\top}$ and $A_{i,i+1} = \alpha_{im,im+1} e_m e_1^{\top}$, then C_i and B_{i+1} are given by:

$$
\begin{aligned}
C_i = Q_i A_{i,i-1} &= \alpha_{(i-1)m+1,(i-1)m}(W_i L Y_i e_1 + S_i J^{\top} e_1)e_m^{\top} \\
&= \alpha_{(i-1)m+1,(i-1)m} w^{(i)} e_m^{\top}, \quad \text{where } w^{(i)} = \text{diag}(W_i),
\end{aligned}
$$

(note that $J^{\top} e_1 = 0$ and $e_1^{\top} Y_i e_1 = \eta_1^{(i)} = 1$ from (5.110)) and

$$
\begin{aligned}
B_{i+1} = Q_i A_{i,i+1} &= \alpha_{im,im+1}(W_i L Y_i e_m + S_i J^{\top} e_m)e_1^{\top} \\
&= \alpha_{im,im+1}(\eta_m^{(i)} \omega_m^{(i)} e_m + s_{m-1}^{(i)} e_{m-1})e_1^{\top}.
\end{aligned}
$$

From relations (5.110), and the fact that $c_m^{(i)} = 1$, it holds that

$$
B_{i+1} = \alpha_{im,im+1}(c_{m-1}^{(i)} e_m + s_{m-1}^{(i)} e_{m-1})e_1^{\top}.
\tag{5.113}
$$

Note that the only nonzero terms of each $C_i = Q_i A_{i,i-1}$ are in the last column whereas the only nonzero elements of $B_{i+1} = Q_i A_{i,i+1}$ are the last two elements of the first column.

Consider now any block row of the above partition. If all diagonal blocks R_i are invertible, then we proceed exactly as with the Spike algorithm. Specifically, for each block row, consisting of (C_i, R_i, B_{i+1}) and the right-hand side $\tilde{f}_i$, the following $(3p - 2)$ independent triangular systems can be solved simultaneously:

$$
\begin{aligned}
&R_1^{-1}(s_{m-1}^{(1)} e_{m-1} + \eta_m^{(1)} \omega_m^{(1)}), \quad R_1^{-1}\tilde{f}_1, \\
&R_i^{-1} w^{(i)}, \qquad\qquad R_i^{-1}(s_{m-1}^{(i)} e_{m-1} + \eta_m^{(i)} \omega_m^{(i)}), \quad R_i^{-1}\tilde{f}_i, \quad (i = 2, \ldots, p - 1), \\
&\qquad\qquad\qquad R_p^{-1} w^{(p)}, \qquad\qquad\qquad R_p^{-1}\tilde{f}_p
\end{aligned}
\tag{5.114}
$$

Note that we need to solve upper triangular systems with 2 or 3 right-hand sides to generate the spikes as well as update the right-hand side systems.

We next consider the case when one or more of the tridiagonal blocks $A_{i,i}$ is singular. Then it is possible to modify $\tilde{A}$ so that the triangular matrices above are made invertible. In particular, since the tridiagonal matrix A is invertible, so wiil be the coefficient matrix $\tilde{A}$ in (5.112). When one or more of the of the submatrices $A_{i,i}$ is singular, then so will be the corresponding submatrices R_i and, according to Proposition 5.6, this would manifest itself with a zero value appearing at the lower right corner of the corresponding R_i's. To handle this situation, the algorithm applies multiple *boostings* to shift away from zero the values at the corners of the triangular blocks R_i along the diagonal.) As we will see, we do not have to do this for the last block R_p. Moreover, all these boostings are independent and can be applied in parallel. This step for blocks other than the last one is represented as a multiplication of $\tilde{A}$ with a matrix, P_{boost}, of the form

$$
P_{\text{boost}} = I_n + \sum_{i=1}^{p-1} \zeta_i e_{im+1} e_{im}^\top,
$$

$$
\text{where } \zeta_i = \begin{cases} 1 \text{ if } |(R_i)_{m,m}| < \text{threshold} \\ 0 \text{ otherwise.} \end{cases}
$$

In particular, consider any $k = 1, \ldots, p - 1$, then

$$
\begin{aligned}
e_{km}^\top \tilde{A} P_{\text{boost}} e_{km} &= (R_k)_{m,m} + \zeta_k e_{km}^\top \tilde{A} e_{km+1} \\
&= (R_k)_{m,m} + \zeta_k \alpha_{(k-1)m,(k-1)m+1} c_{m-1}^{(k)}.
\end{aligned}
\tag{5.115}
$$

Observe that the right-hand side is nonzero since in case $(R_k)_{m,m} = 0$, the term immediately to its right (that is in position $(km, km + 1)$) is nonzero. In this way, matrix $\tilde{A} P_{\text{boost}}$ has all diagonal blocks nonsingular except possibly the last one. Call these (modified in case of singularity, unmodified otherwise) diagonal blocks $\tilde{R}_i$ and set the bock diagonal matrix $\tilde{R} = \text{diag}[\tilde{R}_1, \ldots, \tilde{R}_{p-1}, \tilde{R}_p]$. If $A_{p,p}$ is found to be singular, then $\tilde{R}_p = \text{diag}[\hat{R}_p, 1]$, where $\hat{R}_p$ is the leading principal submatrix of R_p of order $m - 1$ which will be nonsingular in case R_p was found to be singular. Thus, as constructed, $\tilde{R}$ is nonsingular.

It is not hard to verify that

$$P_{\text{boost}}^{-1} = I_n - \sum_{i=1}^{p-1} \zeta_i e_{im+1} e_{im}^{\top}.$$

Taking the above into account, the original system is equivalent to

$$\tilde{A} P_{\text{boost}} P_{\text{boost}}^{-1} x = \tilde{f}.$$

where the matrix $\tilde{A} P_{\text{boost}}$ is block-tridiagonal with all its p diagonal blocks upper triangular and invertible with the possible exception of the last one, which might contain a zero in the last diagonal element. It is also worth noting that the above amounts to the Spike DS factorization of a modified matrix, in particular

$$A P_{\text{boost}} = \tilde{D} \tilde{S} \tag{5.116}$$

where $\tilde{D} = \text{diag}[\tilde{Q}_1, \ldots, \tilde{Q}_p]^{\top} \tilde{R}$ and $\tilde{S}$ is the Spike matrix.

The reduced system of order $2(p-1)$ or $2p-1$ is then obtained by reordering the matrix and numbering first the unknowns $m, m+1, 2m, 2m+1, \ldots, (p-1)m$, $(p-1)m+1$. The system is of order $2p-1$ if R_p is found to be singular in which case the last component of the solution is ordered last in the reduced system. The resulting system has the form

$$\begin{pmatrix} T & 0 \\ S & I \end{pmatrix} \begin{pmatrix} \hat{x}_1 \\ \hat{x}_2 \end{pmatrix} = \begin{pmatrix} \hat{f}_1 \\ \hat{f}_2 \end{pmatrix}$$

where T is tridiagonal and nonsingular. The reduced system $T\hat{x}_1 = \hat{f}_1$ is solved first. Note that even when its order is $2p-1$ and the last diagonal element of T is zero, the reduced system is invertible.

Finally, once $\hat{x}_1$ is computed, the remaining unknowns $\hat{x}_2$ are easily computed in parallel.

We next consider the solution of the reduced system. This is essentially tridiagonal and thus a hybrid scheme can be used: That is, apply the same methodology and use partitioning and DS factorization producing a new reduced system, until the system is small enough to switch to a parallel method not based on partitioning, e.g. PARGIV, and later to a serial method.

The idea of boosting was originally described in [81] as a way to reduce row interchanges and preserve sparsity in the course of Gaussian elimination. Boosting was used in sparse solvers including Spike; cf. [16, 82]. The mechanism consists of adding a suitable value whenever a small diagonal element appears in the pivot position so as to avoid instability. This is equivalent to a rank-1 modification of the matrix. For instance, if before the first step of LU on a matrix A, element $\alpha_{1,1}$ were found to be small enough, then A would be replaced by $A + \gamma_1 e_1 e_1^{\top}$ with γ chosen so as to make $\alpha_{1,1} + \gamma$ large enough to be an acceptable pivot. If no other boosting were required in the course of the LU decomposition, the factors computed would satisfy $LU = A + \gamma_1 e_1 e_1^\top$. From these factors it is possible to recover the solution $A^{-1} f$ using the Sherman-Morrison-Woodbury formula.

In our case, boosting is applied on diagonal blocks R_i, of $\tilde{A}$, that contain a zero (in practice, a value below some tolerance threshold) at their corner position. The boosting is based on the spike columns of $\tilde{A}$, thus avoiding the need for solving auxiliary systems in the Sherman-Morrison-Woodbury formula. Therefore, boosting can be expressed as the product

$$\tilde{A}(I + \sum_{\{i \mid |(R_i)_{m,m}| \approx 0\}} \gamma_i e_{im+1} e_{im}^\top)$$

which amounts to right preconditioning with a low rank modification of the identity and this as well as the inverse transformation can be applied without solving auxiliary systems.

We note that the method is appropriate when the partitioning results in blocks along the diagonal whose singularities are revealed by the Givens QR factorization. In this case, the blocks can be rendered nonsingular by rank-1 modifications. It will not be effective, however, if some blocks are almost, but not exactly singular, and QR is not rank revealing without extra precautions, such as pivoting. See also [83] for more details regarding this algorithm and its implementation on GPUs.

Another partition-based method that is also applicable to general banded matrices and does not necessitate that the diagonal submatrices are invertible is proposed in [80]. The algorithm uses row and possibly column pivoting when factoring the diagonal blocks so as to prevent loss of stability. The linear system is partitioned as in [5], that is differently from the Spike.

Factorization based on block diagonal pivoting without interchanges

If the tridiagonal matrix is symmetric and all subsystems $A_{i,i}$ are nonsingular, then it is possible to avoid interchanges by computing an $LBL^\top$ factorization, where L is unit lower triangular and B is block-diagonal, with 1×1 and 2×2 diagonal blocks. This method avoids the loss of symmetry caused by partial pivoting when the matrix is indefinite; cf. [84, 85] as well as [45]. This strategy can be extended for nonsymmetric systems computing an $LBM^\top$ factorization with M also unit lower triangular; cf. [86, 87]. These methods become attractive in the context of high performance systems where interchanges can be detrimental to performance. In [33] diagonal pivoting of this type was combined with Spike as well as "on the fly" recursive Spike (cf. Sect. 5.2.2 of this chapter) to solve large tridiagonal systems on GPUs, clusters of GPUs and systems consisting of CPUs and GPUs. We next briefly describe the aforementioned $LBM^\top$ factorization.

The first step of the algorithm illustrates the basic idea. The crucial observation is that if a tridiagonal matrix, say A, is nonsingular, then it must hold that either $\alpha_{1,1}$ or the determinant of its 2×2 leading principal submatrix, that is $\alpha_{1,1}\alpha_{2,2} - \alpha_{1,2}\alpha_{2,1}$ must be nonzero. Therefore, if we partition the matrix as

$$A = \begin{pmatrix} A_{1,1} & A_{1,2} \\ A_{2,1} & A_{2,2} \end{pmatrix},$$

where $A_{1,1}$ is $d \times d$, that is $A_{1,1} = \alpha_{1,1}$ or whereas in the latter it is

$$\begin{pmatrix} \alpha_{1,1} & \alpha_{1,2} \\ \alpha_{2,1} & \alpha_{2,2} \end{pmatrix}.$$

Then the following factorization will be valid for either $d = 1$ or $d = 2$:

$$A = \begin{pmatrix} I_d & 0 \\ A_{2,1} A_{1,1}^{-1} & I_{n-d} \end{pmatrix} \begin{pmatrix} A_{1,1} & 0 \\ 0 & A_{1,1} - A_{2,1} A_{1,1}^{-1} A_{1,2} \end{pmatrix} \begin{pmatrix} I_d & A_{1,1}^{-1} A_{1,2} \\ 0 & I_{n-d} \end{pmatrix}. \quad (5.117)$$

Moreover, the corresponding Schur complement, $S_d = A_{1,1} - A_{2,1} A_{1,1}^{-1} A_{1,2}$ is tridiagonal and nonsingular and so the same strategy can be recursively applied until the final factorization is computed. In finite precision, instead of testing if any of these values are exactly 0, a different approach is used. In this, the root $\mu = (\sqrt{5}-1)/2 \approx 0.62$ of the quadratic equation $\mu^2 + \mu - 1 = 0$ plays an important role. Two pivoting strategies proposed in the literature are as follows. If the largest element, say τ, of A is known, then select $d = 1$ if

$$|\alpha_{1,1}|\tau \geq \mu|\alpha_{2,1}\alpha_{1,2}|. \quad (5.118)$$

Otherwise, $d = 2$. In both cases, the factorization is updated as in (5.117). Another strategy that is based on information from only the adjacent equations, is to set

$$\tau = \max\{|\alpha_{2,2}|, |\alpha_{1,2}|, |\alpha_{2,1}|, |\alpha_{2,3}|, |\alpha_{3,2}|\}.$$

and apply the same strategy, (5.118). From the error analysis conducted in [86], it is claimed that the method is backward stable.

5.5.7 Using Determinants and Other Special Forms

It is one of the basic tenets of numerical linear algebra that determinants and Cramer's rule are not to be used for solving general linear systems. When the matrix is tridiagonal, however, determinental equalities can be used to show that the inverse has a special form. This can be used to design methods for computing elements of the inverse, and for solving linear systems. These methods are particularly useful when we want to compute one or few selected elements of the inverse or the solution, something that is not possible with standard direct methods. In fact, there are several applications in science and engineering that require such selective computations, e.g. see some of the examples in [88].

It has long been known, for example, that the inverse of symmetric irreducible tridiagonal matrices, has special structure; reference [89] provides several remarks on the history of this topic, and the relevant discovery made in [90]. An algorithm for inverting such matrices was presented in [91]. Inspired by that work, a parallel algorithm for computing selected elements of the solution or the inverse that is applicable for general tridiagonal matrices was introduced in [67]. A similar algorithm for solving tridiagonal linear systems was presented in [92]. We briefly outline the basic idea of these methods.

For the purpose of this discussion we recall the notation $A = [\alpha_{i,i-1}, \alpha_{i,i}, \alpha_{i,i+1}]$. As before we assume that it is nonsingular and irreducible. Let

$$A^{-1} = [\tau_{i,j}] \qquad i, j = 1, 2, \ldots, n,$$

then

$$\tau_{i,j} = (-1)^{i+j} \det(A(j, i))/\det(A)$$

where $A(j, i)$ is that matrix of order $(n - 1)$ obtained from A by deleting the ith column and jth row. We next use the notation A_{i-1} for the leading principal tridiagonal submatrix of A and $\tilde{A}_{j+1}$ for the trailing tridiagonal submatrix of order $n - j$. Assuming that we pick $i \leq j$, we can write

$$A(j, i) = \begin{pmatrix} A_{i-1} & & \\ Z_1 & S_{j-i} & \\ & Z_2 & \tilde{A}_{j+1} \end{pmatrix}$$

where A_{i-1} and $\tilde{A}_{j+1}$ are tridiagonal of order $i-1$ and $n-j$ respectively, S_{j-i} is of order $j-i$ and is lower triangular with diagonal elements $\alpha_{i,i+1}, \alpha_{i+1,i+2}, \ldots, \alpha_{j-1,j}$. Moreover, Z_1 is of size $(j - i) \times (i - 1)$ with first row $\alpha_{i,i-1}(e_{i-1}^{(i-1)})^\top$ and Z_2 of order $(n - j) \times (j - i)$ with first row $\alpha_{j+1,j}(e_{j-i}^{(j-i)})^\top$ Hence, for $i \leq j$,

$$\tau_{i,j} = (-1)^{i+j} \frac{\det(A_{i-1})\det(\tilde{A}_{j+1}) \prod_{k=i}^{j-1} \alpha_{k,k+1}}{\det(A)}.$$

Rearranging the terms in the above expression we obtain

$$\tau_{i,j} = \left(\frac{(-1)^i \det(A_{i-1})}{\prod_{k=2}^{i} \alpha_{k,k-1}} \right) \left(\frac{(-1)^j \det(\tilde{A}_{j+1})}{\prod_{k=j}^{n-1} \alpha_{k,k+1}} \right) \left(\frac{\prod_{k=i}^{n-1} \alpha_{k,k+1} \prod_{k=2}^{i} \alpha_{k,k-1}}{\det(A)} \right)$$

or $\tau_{i,j} = v_i v_j \omega_i$. Similarly, $\tau_{i,j} = v_j v_i \omega_i$ for $i \geq j$. Therefore

$$\tau_{i,j} = \begin{cases} v_i v_j \omega_i & \text{if } i \leq j \\ v_j v_i \omega_i & \text{if } i \geq j \end{cases}$$

This shows that the lower (resp. upper) triangular part of the inverse of an irreducible tridiagonal matrix is the lower (resp. upper) triangular part of a rank-1 matrix. The respective rank-1 matrices are generated by the vectors $u = (\upsilon_1, \ldots, \upsilon_n)^\top$ $v = (v_1, \ldots, v_n)^\top$ and $w = (\omega_1, \ldots, \omega_n)^\top$. Moreover, to compute one element of the inverse we only need a single element from each of u, v and w. Obviously, if the vectors u, v and w are known, then we can compute any of the elements of the inverse and with little extra work, any element of the solution. Fortunately, these vectors can be computed independently from the solution of three lower triangular systems of very small bandwidth and very special right-hand sides, using, for example, the algorithm of Theorem 3.2. Let us see how to do this.

Let $d_k = \det(A_k)$. Observing that in block form

$$A_k = \begin{pmatrix} A_{k-1} & \alpha_{k-1,k}e_{k-1}^{(k-1)} \\ \alpha_{k,k-1}(e_{k-1}^{(k-1)})^\top & \alpha_{k,k} \end{pmatrix},$$

expanding along the last row, we obtain then

$$d_k = \alpha_{k,k}d_{k-1} - \alpha_{k,k-1}\alpha_{k-1,k}d_{k-2} \quad k = 2, 3, \ldots, n,$$

where $d_0 = 1$ and $d_1 = \alpha_{1,1}$. Now, υ_i can be written as

$$\upsilon_i = (-1)^i \left(\frac{\alpha_{i-1,i-1}}{\alpha_{i,i-1}} \frac{d_{i-2}}{\prod_{k=2}^{i-1}\alpha_{k,k-1}} - \frac{\alpha_{i-2,i-1}}{\alpha_{i,i-1}} \frac{d_{i-3}}{\prod_{k=2}^{i-2}\alpha_{k-1,k}} \right).$$

Thus,

$$\upsilon_i = -\alpha_{i-1,i-1}\upsilon_{i-1} - \alpha_{i-2,i-1}\upsilon_{i-2} \quad i = 2, 3, \ldots, n$$

in which

$$\hat{\alpha}_i = \alpha_{i,i}/\alpha_{i+1,i}$$
$$\hat{\beta}_i = \alpha_{i,i+1}/\alpha_{i+2,i+1}$$
$$\upsilon_0 = 0, \quad \upsilon_1 = -1.$$

This is equivalent to the unit lower triangular system

$$L_1 u = e_1^{(n)} \tag{5.119}$$

where

$$L_1 = \begin{pmatrix} 1 & & & & \\ \hat{\alpha}_1 & 1 & & & \\ \hat{\beta}_1 & \hat{\alpha}_2 & 1 & & \\ & \ddots & \ddots & \ddots & \\ & & \hat{\beta}_{n-2} & \hat{\alpha}_{n-1} & 1 \end{pmatrix}.$$

We next follow a similar approach considering instead the trailing tridiagonal sub-matrix

$$\tilde{A}_k = \begin{pmatrix} \alpha_{k,k} & \alpha_{k,k+1}(e_1^{(n-k)})^\top \\ \alpha_{k+1,k}e_1^{(n-k)} & \tilde{A}_{k+1} \end{pmatrix}.$$

Expanding along the first row we obtain

$$\hat{d}_k = \alpha_{k,k}\hat{d}_{k+1} - \alpha_{k,k+1}\alpha_{k+1,k}\hat{d}_{k+2} \quad k = n-1,\ldots,3,2$$

in which $\hat{d}_i = \det(\tilde{A}_i)$, $d_{n+1} = 1$ and $d_n = a_n$. Now, v_j is given by

$$v_j = -\hat{\gamma}_{j+1}v_{j+1} - \hat{\delta}_{j+2}v_{j+2} \quad j = n-1,\ldots,1,0$$

where,

$$\begin{aligned} \hat{\gamma}_j &= \alpha_{j,j}/\alpha_{j-1,j}, \\ \hat{\delta}_j &= \alpha_{j,j-1}/\alpha_{j-2,j-1}, \quad \text{and} \\ v_{n+1} &= 0, \quad v_n = (-1)^n. \end{aligned}$$

Again, this may be expressed as the triangular system of order $(n+1)$,

$$L_2 v = (-1)^n e_1^{(n+1)} \tag{5.120}$$

where

$$L_2 = \begin{pmatrix} 1 & & & & \\ \hat{\gamma}_n & 1 & & & \\ \hat{\delta}_n & \hat{\gamma}_{n-1} & 1 & & \\ & \ddots & \ddots & \ddots & \\ & & \hat{\delta}_2 & \hat{\gamma}_1 & 1 \end{pmatrix}.$$

Finally, from

$$\omega_i = \prod_{k=i}^{n-1} \alpha_{k,k+1} \prod_{k=2}^{i} \alpha_{k,k-1}/\det(T)$$

we have

$$\omega_{k+1} = \theta_{k+1}\omega_k \quad k = 1,2,\ldots,n-1$$

where,

$$\begin{aligned} \theta_i &= \alpha_{i,i-1}/\alpha_{i,i+1} \\ \omega_1 &= (\alpha_{1,2}\alpha_{2,3}\cdots\alpha_{k,k+1})/\det(T) = (1/v_0). \end{aligned}$$

In other words, w is the solution of the triangular system

$$L_3 w = (1/v_0)e_1^{(n)} \tag{5.121}$$

where,

$$L_3 = \begin{pmatrix} 1 & & & \\ -\theta_2 & 1 & & \\ & \ddots & \ddots & \\ & & -\theta_n & 1 \end{pmatrix}.$$

Observe that the systems

$$\begin{aligned} L_1 u &= e_1^{(n)}, \\ L_2 v &= (-1)^n e_1^{(n+1)}, \quad \text{and} \\ L_3 \tilde{w} &= e_1^{(n)} \end{aligned}$$

where $\tilde{w} = v_0 w$, can be solved independently for u, v, and $\tilde{w}$ using the parallel algorithms of this chapter from which we obtain $w = \tilde{w}/v_0$. One needs to be careful, however, because as in other cases where parallel triangular solvers that require the least number of parallel arithmetic operations are used, there is the danger of underflow, overflow or inaccuracies in the computed results; cf. the relevant discussions in Sect. 3.2.3.

Remarks

The formulas for the inverse of a symmetric tridiagonal matrix with nonzero elements [91] were also extended independently to the nonsymmetric tridiagonal case in [93, Theorem 2]. Subsequently, there have been many investigations concerning the form of the inverse of tridiagonal and more general banded matrices. For extensive discussion of these topics see [94] as well as [89] which is a seminal monograph in the area of structured matrices.

References

1. Arbenz, P., Hegland, M.: On the stable parallel solution of general narrow banded linear systems. High Perform. Algorithms Struct. Matrix Probl. 47–73 (1998)
2. Arbenz, P., Cleary, A., Dongarra, J., Hegland, M.: A comparison of parallel solvers for general narrow banded linear systems. Parallel Distrib. Comput. Pract. 2(4), 385–400 (1999)
3. Blackford, L., Choi, J., Cleary, A., D'Azevedo, E., Demmel, J., Dhillon, I., Dongarra, J., Hammarling, S., Henry, G., Petitet, A., Stanley, K., Walker, D., Whaley, R.: ScaLAPACK User's Guide. SIAM, Philadelphia (1997). URL http://www.netlib.org/scalapack
4. Conroy, J.: Parallel algorithms for the solution of narrow banded systems. Appl. Numer. Math. 5, 409–421 (1989)
5. Dongarra, J., Johnsson, L.: Solving banded systems on a parallel processor. Parallel Comput. 5(1–2), 219–246 (1987)

6. George, A.: Numerical experiments using dissection methods to solve n by n grid problems. SIAM J. Numer. Anal. **14**, 161–179 (1977)
7. Golub, G., Sameh, A., Sarin, V.: A parallel balance scheme for banded linear systems. Numer. Linear Algebra Appl. **8**, 297–316 (2001)
8. Johnsson, S.: Solving narrow banded systems on ensemble architectures. ACM Trans. Math. Softw. **11**, 271–288 (1985)
9. Meier, U.: A parallel partition method for solving banded systems of linear equations. Parallel Comput. **2**, 33–43 (1985)
10. Tang, W.: Generalized Schwarz splittings. SIAM J. Sci. Stat. Comput. **13**, 573–595 (1992)
11. Wright, S.: Parallel algorithms for banded linear systems. SIAM J. Sci. Stat. Comput. **12**, 824–842 (1991)
12. Sameh, A., Kuck, D.: On stable parallel linear system solvers. J. Assoc. Comput. Mach. **25**(1), 81–91 (1978)
13. Dongarra, J.J., Sameh, A.: On some parallel banded system solvers. Technical Report ANL/MCS-TM-27, Mathematics Computer Science Division at Argonne National Laboratory (1984)
14. Gallivan, K., Gallopoulos, E., Sameh, A.: CEDAR—an experiment in parallel computing. Comput. Math. Appl. **1**(1), 77–98 (1994)
15. Lawrie, D.H., Sameh, A.: The computation and communication complexity of a parallel banded system solver. ACM TOMS **10**(2), 185–195 (1984)
16. Polizzi, E., Sameh, A.: A parallel hybrid banded system solver: the SPIKE algorithm. Parallel Comput. **32**, 177–194 (2006)
17. Polizzi, E., Sameh, A.: SPIKE: a parallel environment for solving banded linear systems. Compon. Fluids **36**, 113–120 (2007)
18. Sameh, A., Kuck, D.: A parallel QR algorithm for symmetric tridiagonal matrices. IEEE Trans. Comput. **26**(2), 147–153 (1977)
19. Anderson, E., Bai, Z., Bischof, C., Blackford, S., Demmel, J., Dongarra, J., Du Croz, J., Greenbaum, A., Hammarling, S., McKenney, A., Sorensen, D.: LAPACK Users' Guide, 3rd edn. Society for Industrial and Applied Mathematics, Philadelphia (1999)
20. Demko, S., Moss, W., Smith, P.: Decay rates for inverses of band matrices. Math. Comput. **43**(168), 491–499 (1984)
21. Björck, Å.: Numerical Methods for Least Squares Problems. SIAM, Philadelphia (1996)
22. Golub, G., Van Loan, C.: Matrix Computations, 4th edn. Johns Hopkins. University Press, Baltimore (2013)
23. Davis, T.: Algorithm 915, SuiteSparseQR: multifrontal multithreaded rank-revealing sparse QR factorization. ACM Trans. Math. Softw. **38**(1), 8:1–8:22 (2011). doi:10.1145/2049662. 2049670, URL http://doi.acm.org/10.1145/2049662.2049670
24. Lou, G.: Parallel methods for solving linear systems via overlapping decompositions. Ph.D. thesis, University of Illinois at Urbana-Champaign (1989)
25. Naumov, M., Sameh, A.: A tearing-based hybrid parallel banded linear system solver. J. Comput. Appl. Math. **226**, 306–318 (2009)
26. Benzi, M., Golub, G., Liesen, J.: Numerical solution of saddle-point problems. Acta Numer. 1–137 (2005)
27. Hockney, R., Jesshope, C.: Parallel Computers. Adam Hilger (1983)
28. Ortega, J.M.: Introduction to Parallel and Vector Solution of Linear Systems. Plenum Press, New York (1988)
29. Golub, G., Ortega, J.: Scientific Computing: An Introduction with Parallel Computing. Academic Press Inc., San Diego (1993)
30. Davidson, A., Zhang, Y., Owens, J.: An auto-tuned method for solving large tridiagonal systems on the GPU. In: Proceedings of IEEE IPDPS, pp. 956–965 (2011)
31. Lopez, J., Zapata, E.: Unified architecture for divide and conquer based tridiagonal system solvers. IEEE Trans. Comput. **43**(12), 1413–1425 (1994). doi:10.1109/12.338101
32. Santos, E.: Optimal and efficient parallel tridiagonal solvers using direct methods. J. Supercomput. **30**(2), 97–115 (2004). doi:10.1023/B:SUPE.0000040615.60545.c6, URL http://dx. doi.org/10.1023/B:SUPE.0000040615.60545.c6

33. Chang, L.W., Stratton, J., Kim, H., Hwu, W.M.: A scalable, numerically stable, high-performance tridiagonal solver using GPUs. In: Proceedings International Conference High Performance Computing, Networking Storage and Analysis, SC'12, pp. 27:1–27:11. IEEE Computer Society Press, Los Alamitos (2012). URL http://dl.acm.org/citation.cfm?id=2388996.2389033

34. Goeddeke, D., Strzodka, R.: Cyclic reduction tridiagonal solvers on GPUs applied to mixed-precision multigrid. IEEE Trans. Parallel Distrib. Syst. **22**(1), 22–32 (2011)

35. Codenotti, B., Leoncini, M.: Parallel Complexity of Linear System Solution. World Scientific, Singapore (1991)

36. Ascher, U., Mattheij, R., Russell, R.: Numerical Solution of Boundary Value Problems for Ordinary Differential Equations. Classics in Applied Mathematics. SIAM, Philadelphia (1995)

37. Isaacson, E., Keller, H.B.: Analysis of Numerical Methods. Wiley, New York (1966)

38. Keller, H.B.: Numerical Methods for Two-Point Boundary-Value Problems. Dover Publications, New York (1992)

39. Bank, R.E.: Marching algorithms and block Gaussian elimination. In: Bunch, J.R., Rose, D. (eds.) Sparse Matrix Computations, pp. 293–307. Academic Press, New York (1976)

40. Bank, R.E., Rose, D.: Marching algorithms for elliptic boundary value problems. I: the constant coefficient case. SIAM J. Numer. Anal. **14**(5), 792–829 (1977)

41. Roache, P.: Elliptic Marching Methods and Domain Decomposition. CRC Press Inc., Boca Raton (1995)

42. Richardson, L.F.: Weather Prediction by Numerical Process. Cambridge University Press. Reprinted by Dover Publications, 1965 (1922)

43. Arbenz, P., Hegland, M.: The stable parallel solution of narrow banded linear systems. In: Heath, M., et al. (eds.) Proceedings of Eighth SIAM Conference Parallel Processing and Scientific Computing SIAM, Philadelphia (1997)

44. Bank, R.E., Rose, D.: Marching algorithms for elliptic boundary value problems. II: the variable coefficient case. SIAM J. Numer. Anal. **14**(5), 950–969 (1977)

45. Higham, N.: Accuracy and Stability of Numerical Algorithms, 2nd edn. SIAM, Philadelphia (2002)

46. Higham, N.: Stability of parallel triangular system solvers. SIAM J. Sci. Comput. **16**(2), 400–413 (1995)

47. Viswanath, D., Trefethen, L.: Condition numbers of random triangular matrices. SIAM J. Matrix Anal. Appl. **19**(2), 564–581 (1998)

48. Hockney, R.: A fast direct solution of Poisson's equation using Fourier analysis. J. Assoc. Comput. Mach. **12**, 95–113 (1965)

49. Gander, W., Golub, G.H.: Cyclic reduction: history and applications. In: Luk, F., Plemmons, R. (eds.) Proceedings of the Workshop on Scientific Computing, pp. 73–85. Springer, New York (1997). URL http://people.inf.ethz.ch/gander/papers/cyclic.pdf

50. Amodio, P., Brugnano, L.: Parallel factorizations and parallel solvers for tridiagonal linear systems. Linear Algebra Appl. **172**, 347–364 (1992). doi:10.1016/0024-3795(92)90034-8, URL http://www.sciencedirect.com/science/article/pii/0024379592900348

51. Heller, D.: Some aspects of the cyclic reduction algorithm for block tridiagonal linear systems. SIAM J. Numer. Anal. **13**(4), 484–496 (1976)

52. Lambiotte Jr, J., Voigt, R.: The solution of tridiagonal linear systems on the CDC STAR 100 computer. ACM Trans. Math. Softw. **1**(4), 308–329 (1975). doi:10.1145/355656.355658, URL http://doi.acm.org/10.1145/355656.355658

53. Nassimi, D., Sahni, S.: An optimal routing algorithm for mesh-connected parallel computers. J. Assoc. Comput. Mach. **27**(1), 6–29 (1980)

54. Nassimi, D., Sahni, S.: Parallel permutation and sorting algorithms and a new generalized connection network. J. Assoc. Comput. Mach. **29**(3), 642–667 (1982)

55. George, A.: Nested dissection of a regular finite element mesh. SIAM J. Numer. Anal. **10**(2), 345–363 (1973). URL http://www.jstor.org/stable/2156361

56. Amodio, P., Brugnano, L., Politi, T.: Parallel factorization for tridiagonal matrices. SIAM J. Numer. Anal. **30**(3), 813–823 (1993)

57. Johnsson, S.: Solving tridiagonal systems on ensemble architectures. SIAM J. Sci. Stat. Comput. **8**, 354–392 (1987)
58. Zhang, Y., Cohen, J., Owens, J.: Fast tridiagonal solvers on the GPU. ACM SIGPLAN Not. **45**(5), 127–136 (2010)
59. Amodio, P., Mazzia, F.: Backward error analysis of cyclic reduction for the solution of tridiagonal systems. Math. Comput. **62**(206), 601–617 (1994)
60. Higham, N.: Bounding the error in Gaussian elimination for tridiagonal systems. SIAM J. Matrix Anal. Appl. **11**(4), 521–530 (1990)
61. Zhang, Y., Owens, J.: A quantitative performance analysis model for GPU architectures. In: Proceedings of the 17th IEEE International Symposium on High-Performance Computer Architecture (HPCA 17) (2011)
62. El-Mikkawy, M., Sogabe, T.: A new family of k-Fibonacci numbers. Appl. Math. Comput. **215**(12), 4456–4461 (2010). URL http://www.sciencedirect.com/science/article/pii/S009630031000007X
63. Fang, H.R., O'Leary, D.: Stable factorizations of symmetric tridiagonal and triadic matrices. SIAM J. Math. Anal. Appl. **28**(2), 576–595 (2006)
64. Mikkelsen, C., Kågström, B.: Parallel solution of narrow banded diagonally dominant linear systems. In: Jónasson, L. (ed.) PARA 2010. LNCS, vol. 7134, pp. 280–290. Springer (2012). doi:10.1007/978-3-642-28145-7_28, URL http://dx.doi.org/10.1007/978-3-642-28145-7_28
65. Mikkelsen, C., Kågström, B.: Approximate incomplete cyclic reduction for systems which are tridiagonal and strictly diagonally dominant by rows. In: Manninen, P., Öster, P. (eds.) PARA 2012. LNCS, vol. 7782, pp. 250–264. Springer (2013). doi:10.1007/978-3-642-36803-5_18, URL http://dx.doi.org/10.1007/978-3-642-36803-5_18
66. Bini, D., Meini, B.: The cyclic reduction algorithm: from Poisson equation to stochastic processes and beyond. Numer. Algorithms **51**(1), 23–60 (2008). doi:10.1007/s11075-008-9253-0, URL http://www.springerlink.com/index/10.1007/s11075-008-9253-0; http://www.springerlink.com/content/m40t072h273w8841/fulltext.pdf
67. Sameh, A.: Numerical parallel algorithms—a survey. In: Kuck, D., Lawrie, D., Sameh, A. (eds.) High Speed Computer and Algorithm Optimization, pp. 207–228. Academic Press, Sans Diego (1977)
68. Mathias, R.: The instability of parallel prefix matrix multiplication. SIAM J. Sci. Comput. **16**(4) (1995), to appear
69. Eğecioğlu, O., Koç, C., Laub, A.: A recursive doubling algorithm for solution of tridiagonal systems on hypercube multiprocessors. J. Comput. Appl. Math. **27**, 95–108 (1989)
70. Dubois, P., Rodrigue, G.: An analysis of the recursive doubling algorithm. In: Kuck, D., Lawrie, D., Sameh, A. (eds.) High Speed Computer and Algorithm Organization, pp. 299–305. Academic Press, San Diego (1977)
71. Hammarling, S.: A survey of numerical aspects of plane rotations. Report Maths. 1, Middlesex Polytechnic (1977). URL http://eprints.ma.man.ac.uk/1122/. Available as Manchester Institute for Mathematical Sciences MIMS EPrint 2008.69
72. Bar-On, I., Codenotti, B.: A fast and stable parallel QR algorithm for symmetric tridiagonal matrices. Linear Algebra Appl. **220**, 63–95 (1995). doi:10.1016/0024-3795(93)00360-C, URL http://www.sciencedirect.com/science/article/pii/002437959300360C
73. Gill, P.E., Golub, G., Murray, W., Saunders, M.: Methods for modifying matrix factorizations. Math. Comput. **28**, 505–535 (1974)
74. Lakshmivarahan, S., Dhall, S.: Parallelism in the Prefix Problem. Oxford University Press, New York (1994)
75. Cleary, A., Dongarra, J.: Implementation in ScaLAPACK of divide and conquer algorithms for banded and tridiagonal linear systems. Technical Report UT-CS-97-358, University of Tennessee Computer Science Technical Report (1997)
76. Bar-On, I., Codenotti, B., Leoncini, M.: Checking robust nonsingularity of tridiagonal matrices in linear time. BIT Numer. Math. **36**(2), 206–220 (1996). doi:10.1007/BF01731979, URL http://dx.doi.org/10.1007/BF01731979

77. Bar-On, I.: Checking non-singularity of tridiagonal matrices. Electron. J. Linear Algebra **6**, 11–19 (1999). URL http://math.technion.ac.il/iic/ela
78. Bondeli, S.: Divide and conquer: a parallel algorithm for the solution of a tridiagonal system of equations. Parallel Comput. **17**, 419–434 (1991)
79. Wang, H.: A parallel method for tridiagonal equations. ACM Trans. Math. Softw. **7**, 170–183 (1981)
80. Wright, S.: Parallel algorithms for banded linear systems. SIAM J. Sci. Stat. Comput. **12**(4), 824–842 (1991)
81. Stewart, G.: Modifying pivot elements in Gaussian elimination. Math. Comput. **28**(126), 537–542 (1974)
82. Li, X., Demmel, J.: SuperLU-DIST: A scalable distributed-memory sparse direct solver for unsymmetric linear systems. ACM TOMS **29**(2), 110–140 (2003). URL http://doi.acm.org/10.1145/779359.779361
83. Venetis, I.E., Kouris, A., Sobczyk, A., Gallopoulos, E., Sameh, A.: A direct tridiagonal solver based on Givens rotations for GPU-based architectures. Technical Report HPCLAB-SCG-06/11-14, CEID, University of Patras (2014)
84. Bunch, J.: Partial pivoting strategies for symmetric matrices. SIAM J. Numer. Anal. **11**(3), 521–528 (1974)
85. Bunch, J., Kaufman, K.: Some stable methods for calculating inertia and solving symmetric linear systems. Math. Comput. **31**, 162–179 (1977)
86. Erway, J., Marcia, R.: A backward stability analysis of diagonal pivoting methods for solving unsymmetric tridiagonal systems without interchanges. Numer. Linear Algebra Appl. **18**, 41–54 (2011). doi:10.1002/nla.674, URL http://dx.doi.org/10.1002/nla.674
87. Erway, J.B., Marcia, R.F., Tyson, J.: Generalized diagonal pivoting methods for tridiagonal systems without interchanges. IAENG Int. J. Appl. Math. **4**(40), 269–275 (2010)
88. Golub, G.H., Meurant, G.: Matrices, Moments and Quadrature with Applications. Princeton University Press, Princeton (2009)
89. Vandebril, R., Van Barel, M., Mastronardi, N.: Matrix Computations and Semiseparable Matrices. Volume I: Linear Systems. Johns Hopkins University Press (2008)
90. Gantmacher, F., Krein, M.: Sur les matrices oscillatoires et complèments non négatives. Composition Mathematica **4**, 445–476 (1937)
91. Bukhberger, B., Emelyneko, G.: Methods of inverting tridiagonal matrices. USSR Comput. Math. Math. Phys. **13**, 10–20 (1973)
92. Swarztrauber, P.N.: A parallel algorithm for solving general tridiagonal equations. Math. Comput. **33**, 185–199 (1979)
93. Yamamoto, T., Ikebe, Y.: Inversion of band matrices. Linear Algebra Appl. **24**, 105–111 (1979). doi:10.1016/0024-3795(79)90151-4, URL http://www.sciencedirect.com/science/article/pii/0024379579901514
94. Strang, G., Nguyen, T.: The interplay of ranks of submatrices. SIAM Rev. **46**(4), 637–646 (2004). URL http://www.jstor.org/stable/20453569

Chapter 6
Special Linear Systems

One key idea when attempting to build algorithms for large scale matrix problems is to detect if the matrix has special properties, possibly due to the characteristics of the application, that could be taken into account in order to design faster solution methods. This possibility was highlighted early on by Turing himself, when he noted in his report for the Automatic Computing Engine (ACE) that even though with the storage capacities available at that time it would be hard to store and handle systems larger than 50×50,

> ... the majority of problems have very degenerate matrices and we do not need to store anything like as much (...) the coefficients in these equations are very systematic and mostly zero. [1].

The special systems discussed in this chapter encompass those that Turing characterized as "degenerate" in that they can be represented and stored much more economically than general matrices as their entries are systematic (in ways that will be made precise later), and frequently, most are zero. Because the matrices can be represented with fewer parameters, they are also termed *structured* [2] or *data sparse*.

In this chapter, we are concerned with the solution of linear systems with methods that are designed to exploit the matrix structure. In particular, we show the opportunities for parallel processing when solving linear systems with Vandermonde matrices, banded Toeplitz matrices, a class of matrices that are called SAS-decomposable, and special matrices that arise when solving elliptic partial differential equations which are amenable to the application of fast direct methods, commonly referred as rapid elliptic solvers (RES).

Observe that to some degree, getting high speedup and efficiency out of parallel algorithms for matrices with special structure is more challenging than for general ones since the gains are measured vis-a-vis serial solvers of reduced complexity.

It is also worth noting that in some cases, the matrix structure is not known a priori or is hidden and it becomes necessary to convert the matrix into a special representation permitting the construction of fast algorithms; see for example [3–5]. This can be a delicate task because arithmetic and data representation are in finite precision.

© Springer Science+Business Media Dordrecht 2016

E. Gallopoulos et al., *Parallelism in Matrix Computations*,

Scientific Computation, DOI 10.1007/978-94-017-7188-7_6

Another type of structure that is present in the Vandermonde and Toeplitz matrices is that they have small *displacement rank*, a property introduced in [6] to characterize matrices for which it is possible to construct low complexity algorithms; cf. [7]. What this means is that if A is the matrix under consideration, there exist lower triangular matrices P, Q such that the rank of either $A - PAQ^\top$ or $PA - AQ^\top$ (called displacements) is small. A similar notion, of block displacement, exists for block matrices. For this reason, such matrices are also characterized as "low displacement".

Finally note that even if a matrix is only approximately but not exactly structured, that is it can be expressed as $A = S + E$, where S structured and E is nonzero but small, in some sense, (e.g. has small rank or small norm), this can be valuable because then the corresponding structured matrix S could be an effective preconditioner in an iterative scheme.

A detailed treatment of structured matrices (in particular *structured rank* matrices, that is matrices for which any submatrix that lies entirely below or above the main diagonal has rank that is bounded above by some fixed value smaller than its size) can be found in [8]. See also [9, 10] regarding data sparse matrices.

6.1 Vandermonde Solvers

We recall that Vandermonde matrices are determined from one vector, say $x = (\xi_1, \ldots, \xi_n)^\top$, as

$$
V_m(x) = \begin{pmatrix}
1 & 1 & \cdots & 1 \\
\xi_1 & \xi_2 & \cdots & \xi_n \\
\vdots & \vdots & & \vdots \\
\xi_1^{m-1} & \xi_2^{m-1} & \cdots & \xi_n^{m-1}
\end{pmatrix},
$$

where m indicates the number of rows and the number of columns is the size of x. When the underlying vector or the row dimension are implied by the context, the symbols are omitted.

If $V(x)$ is a square Vandermonde matrix of order n and $Q = \operatorname{diag}(\xi_1, \ldots, \xi_n)$, then $\operatorname{rank}(V(x) - JV(x)Q) = 1$. It follows that Vandermonde matrices have small displacement rank (equal to 1). We are interested in the following problems for any given nonsingular Vandermonde matrix V and vector of compatible size b.

1. Compute the inverse V^{-1}.
2. Solve the primal Vandermonde system $Va = b$.
3. Solve the dual Vandermonde system $V^\top a = b$.

The inversion of Vandermonde matrices (as described in [11]) and the solution of Vandermonde systems (using algorithms in [12] or via inversion and multiplication as proposed in [13]) can be accomplished with fast and practical algorithms that require only $O(n^2)$ arithmetic operations rather than the $O(n^3)$ predicted by (structure-oblivious) Gaussian elimination. See also [14, 15] and historical remarks therein. Key to many algorithms is a fundamental result from the theory of polynomial interpolation, namely that given $n + 1$ interpolation (node, value)-pairs, $\{(\xi_k, \beta_k)\}_{k=0:n}$, where the ξ_k are all distinct, there exists a unique polynomial of degree at most n, say p_n, that satisfies $p_n(\xi_k) = \beta_k$ for $k = 0, \ldots, n$. Writing the polynomial in power form, $p_n(\xi) = \sum_{j=0}^{n} \alpha_j \xi^j$, the vector of coefficients $a = (\alpha_0, \ldots, \alpha_n)^{\top}$ is the solution of problem (3). We also recall the most common representations for the interpolating polynomial (see for example [16, 17]):

Lagrange form:

$$p_n(\xi) = \sum_{k=0}^{n} \beta_k l_k(\xi), \tag{6.1}$$

where l_k are the Lagrange basis polynomials, defined by

$$l_k(\xi) = \prod_{\substack{j=0 \\ k \neq j}}^{n} \frac{(\xi - \xi_j)}{(\xi_k - \xi_j)}. \tag{6.2}$$

Newton form:

$$p_n(\xi) = \gamma_0 + \gamma_1(\xi - \xi_0) + \cdots + \gamma_n(\xi - \xi_0) \cdots (\xi - \xi_{n-1}), \tag{6.3}$$

where γ_j are the divided differences coefficients.

A word of warning: Vandermonde matrices are notoriously ill-conditioned so manipulating them in floating-point can be extremely error prone unless special conditions hold. It has been shown in [18] that the condition with respect to the ∞-norm when all nodes are positive grows as $O(2^n)$ for Vandermonde matrices of order n. The best known, well-conditioned Vandermonde matrices, are those defined on the n roots of unity. To distinguish this case we replace x by $w = (1, \omega, \ldots, \omega^n)^{\top}$ where $\omega = \exp(\iota 2\pi \frac{1}{n+1})$. The matrix $V(w)$ of order $n + 1$, is unitary (up to scaling) and thus is perfectly conditioned, symmetric, and moreover corresponds to the discrete Fourier transform. We will need this matrix in the sequel so we show it here, in simplified notation:

$$V_{n+1} = \begin{pmatrix} 1 & 1 & 1 & \cdots & 1 \\ 1 & \omega & \omega^2 & \cdots & \omega^n \\ 1 & \omega^2 & \omega^4 & \cdots & \omega^{2n} \\ \vdots & \vdots & & & \vdots \\ 1 & \omega^n & \omega^{2n} & \cdots & \omega^{n^2} \end{pmatrix}. \tag{6.4}$$

The action of V either in multiplication by a vector or in solving a linear system on a uniprocessor can be computed using $O((n+1)\log(n+1))$ arithmetic operations using the fast Fourier transform (FFT); see for example [19] for a comprehensive treatise on the subject. We call $\frac{1}{\sqrt{n+1}}V(w)^\top$ a *unitary Fourier matrix*. Good conditioning can also be obtained for other sets of interpolation nodes on the unit circle, [20, 21]. Some improvements are also possible for sequences of real nodes, though the growth in the condition number remains exponential; cf. [22].

The algorithms in this section are suitable for Vandermonde matrices that are not extremely ill-conditioned and that cannot be manipulated with the FFT. On the other hand they must be large enough to justify the use of parallelism.

Finally, it is worth mentioning that the solution of Vandermonde systems of order in the thousands using very high-precision on a certain parallel architecture has been described in [23]. It would be of interest to investigate whether the parallel algorithms we present here can be used, with extended arithmetic, to solve large general Vandermonde systems.

Product to Power Form Conversion

All algorithms for Vandermonde matrices that we describe require, at some stage, transforming a polynomial from product to power form. Given (real or complex) values $\{\xi_1,\ldots,\xi_n\}$ and a nonzero γ then if

$$p(\xi) = \gamma \prod_{j=1}^{n}(\xi - \xi_j) \tag{6.5}$$

is the product form representation of the polynomial, we seek fast and practical algorithms for computing the transformation

$$\mathcal{F} : (\xi_1,\ldots,\xi_n,\gamma) \rightarrow (\alpha_0,\ldots,\alpha_n),$$

where $\sum_{i=0}^{n}\alpha_i\xi^i$ is the power form representation of $p(\xi)$. The importance of making available the power form representation of the interpolation polynomials was noted early on in [11]. By implication, it is important to provide fast and practical transformations between representations. The standard serial algorithm, implemented for example by MATLAB's `poly` function, takes approximately n^2 arithmetic operations. The algorithms presented in this section for converting from product to power form (6.1) and (6.2) are based on the following lemma.

Proposition 6.1 *Let* $u_j = e_2^{(n+1)} - \rho_j e_1^{(n+1)}$ *for* $j = 1,\ldots,n$ *be the vectors containing the coefficients for term* $x - \rho_j$ *of* $p(x) = \prod_{j=1}^{n}(x - \rho_j)$, *padded with zeros. Denote by* DFT_k *(resp.* DFT_k^{-1}*) the discrete (resp. inverse discrete) Fourier transform of length* k *and let* $a = (\alpha_0,\ldots,\alpha_n)^\top$ *be the vector of coefficients of the power form of* $p_n(x)$*. Then*

$$a = \mathrm{DFT}_{n+1}^{-1}\left(\mathrm{DFT}_{n+1}(u_1) \odot \cdots \odot \mathrm{DFT}_{n+1}(u_n)\right).$$

The proposition can be proved from classical results for polynomial multiplication using convolution (e.g. see [24] and [25, Problem 2.4]).

In the following observe that if V is as given in (6.4), then

$$\mathrm{DFT}_{n+1}(u) = Vu, \quad \text{and} \quad \mathrm{DFT}_{n+1}^{-1}(u) = \frac{1}{n+1} V^*u, \tag{6.6}$$

Algorithm 6.1 (PR2PW) uses the previous proposition to compute the coefficients of the power form from the roots of the polynomial.

Algorithm 6.1 PR2PW: Conversion of polynomial from product form to power form.

Input: $r = (\rho_1, \ldots, \rho_n)^\top$ //product form $\prod_{j=1}^n (x - \rho_j)$
Output: coefficients $a = (\alpha_0, \ldots, \alpha_n)$ //power form $\sum_{i=0}^n \alpha_i \xi^j$
1: $U = -e_1^{(n+1)} r^\top + e_2^{(n+1)} (e^{(n)})^\top$
2: **doall** $j = 1 : n$
3:    $\hat{U}_{:,j} = \mathrm{DFT}_{n+1}(U_{:,j})$
4: **end**
5: **doall** $i = 1 : n + 1$
6:    $\hat{\alpha}_i = \mathtt{prod}\ (\hat{U}_{i,:})$
7: **end**
8: $a = \mathrm{DFT}_{n+1}^{-1}(\hat{a})$ //$\hat{a} = (\hat{\alpha}_1, \ldots, \hat{\alpha}_{n+1})^\top$

We next comment on Algorithm 6.1 (PR2PW). Both DFT and $\mathtt{prod}$ can be implemented using parallel algorithms. If $p = O(n^2)$ processors are available, the cost is $O(\log n)$. On a distributed memory system with $p \leq n$ processors, using a one-dimensional block-column distribution for U, then in the first loop (lines 2–4) only local computations in each processor are performed. In the second loop (lines 5–7) there are $(n + 1)/p$ independent products of vectors of length n performed sequentially on each processor at a total cost of $(n - 1)(n + 1)/p$. The result is a vector of length $n + 1$ distributed across the p processors. Vector $\hat{a}$ would then be distributed across the processors, so the final step consists of a single transform of length $n + 1$ that can be performed in parallel over the p processors at a cost of $O(\frac{n+1}{p} \log(n + 1))$ parallel operations. So the parallel cost for Algorithm 6.1 (PR2PW) is

$$\mathbf{T}_p = \frac{n}{p}\tau_1 + \frac{(n + 1)}{p}(n - 1) + \tau_p, \tag{6.7}$$

where τ_1, τ_p are the times for an DFT_{n+1} on 1 and p processors respectively.

Using $p = n$ processors, the dominant parallel cost is $O(n \log n)$ and is caused by the first term, that is the computation of the DFTs via the FFT (lines 2–4).

We next show that the previous algorithm can be modified to reduce the cost of the dominant, first step. Observe that the coefficient vectors u_j in Proposition 6.1 are sparse and in particular, linear combinations of $e_1^{(n+1)}$ and $e_2^{(n+1)}$ because the polynomials being multiplied are all linear and monic. From (6.6) it follows that the DFT of each u_j is then

$$V(w)u_j = V(w)e_2 - \rho_j V(w)e_1$$
$$= w - \rho_j e = (1 - \rho_j, \omega - \rho_j, \ldots, \omega^n - \rho_j)^\top. \tag{6.8}$$

Therefore each DFT can be computed by means a DAXPY-type operation without using the FFT. Moreover, the computations can be implemented in a parallel architecture in a straightforward manner. Assuming that the powers of ω are readily available, the arithmetic cost of this step on $p \leq n$ processors is $n(n+1)/p$ parallel operations. If there are $p = O(n^2)$ processors, the cost is 1 parallel operation. Algorithm 6.2 (POWFORM) implements this idea and also incorporates the multiplication with a scalar coefficient γ when the polynomial is not monic. Note that the major difference from Algorithm PR2PW is the replacement of the first major loop with the doubly nested loop that essentially computes the size $(n+1) \times n$ matrix product

$$\begin{pmatrix} 1 & 1 \\ 1 & \omega \\ \vdots & \vdots \\ 1 & \omega^n \end{pmatrix} \begin{pmatrix} -\rho_1 & -\rho_2 & \ldots & -\rho_n \\ 1 & 1 & \ldots & 1 \end{pmatrix}.$$

Since all terms are independent, this can be computed in $2n(n+1)/p$ parallel operations. It follows that the overall cost to convert from product to power form on p processors is approximately

$$\mathbf{T}_p = 2\frac{n(n+1)}{p} + \tau_p, \tag{6.9}$$

where τ_p is as in (6.7). When $p = n$, the dominant cost of POWFORM is $O(n)$ instead of $O(n \log n)$ for PR2PW. On n^2 processors, the parallel cost of both algorithms is $O(\log n)$. We note that the replacement of the FFT-based multiplication with $V(w)$ with matrix-multiplication implemented as linear combination of the columns was inspired by early work in [26] on rapid elliptic solvers with sparse right-hand sides.

6.1.1 Vandermonde Matrix Inversion

We make use of the following result (see for example [15]):

Proposition 6.2 *Consider the order $n + 1$ Vandermonde matrix $V(x)$, where $x = (\xi_0, \ldots, \xi_n)^\top$. Then if its inverse is written in row form,*

Algorithm 6.2 POWFORM: Conversion from full product (roots and leading coefficient) (6.5) to power form (coefficients) using the explicit formula (6.8) for the transforms.

Function: $r = \text{POWFORM}(r, \gamma)$
Input: vector $r = (\rho_1, \dots, \rho_n)^\top$ and scalar γ
Output: power form coefficients $a = (\alpha_0, \dots, \alpha_n)^\top$
1: $\omega = \exp(-\iota 2\pi/(n+1))$
2: **doall** $i = 1 : n + 1$
3:    **doall** $j = 1 : n$
4:      $\hat{v}_{i,j} = \omega^{i-1} - \rho_j$
5:    **end**
6: **end**
7: **doall** $i = 1 : n + 1$
8:    $\hat{\alpha}_{i-1} = \texttt{prod}\ (\hat{U}_{i,:})$
9: **end**
10: $a = \text{DFT}_{n+1}^{-1}(\hat{a})$
11: **if** $\gamma \neq 1$ **then**
12:    $a = \gamma r$
13: **end if**

$$V^{-1} = \begin{pmatrix} \hat{v}_1^\top \\ \vdots \\ \hat{v}_{n+1}^\top \end{pmatrix}$$

then each row $\hat{v}_i^\top$, $i = 1, \dots, n + 1$, is the vector of coefficients of the power form representation of the corresponding Lagrange basis polynomial $l_{i-1}(\xi)$.

Algorithm 6.3 for computing the inverse of V, by rows, is based on the previous proposition and Algorithm 6.2 (POWFORM).

Most operations of IVAND occur in the last loop, where $n + 1$ conversions to power form are computed. Using PR2PW on $O(n^3)$ processors, these can be done in $O(\log n)$ operations; the remaining computations can be accomplished in equal or fewer steps so the total parallel cost of IVAND is $O(\log n)$.

Algorithm 6.3 IVAND: computing the Vandermonde inverse.

Input: $x = (\xi_0, \dots, \xi_n)^\top$ with pairwise distinct values   // Vandermonde matrix is $V(x)$
Output: $\hat{v}_1, \dots, \hat{v}_{n+1}$ $//V^{-1} = (\hat{v}_1, \dots, \hat{v}_{n+1})^\top$
1: $U = x(e^{(n+1)})^\top - e^{(n+1)}x^\top + I$
2: **doall** $i = 1 : n + 1$
3:    $\pi_i = \texttt{prod}\ (U_{i,:})$ $//\pi$ are power form coefficients of $\prod_{j\neq i}(\xi_i - \xi_j)$
4: **end**
5: **doall** $i = 1 : n + 1$
6:    $\hat{x}^{(i)} = (x_{1:i-1}; x_{i+1:n})$
7:    $\hat{v}^{(i)} = \text{POWFORM}(\hat{x}^{(i)}, 1/\pi_i)$
8: **end**

On $(n + 1)^2$ processors, the cost is dominated by $O(n)$ parallel operations for the last loop according to the preceding analysis of PR2PW for the case of n processors. Note that dedicating all processors for each invocation of PR2PW incurs higher cost, namely $O(n \log n)$. The remaining steps of IVAND incur smaller cost, namely $O(\log n)$ for the second loop and $O(1)$ for the first. So on $(n + 1)^2$ processors, the parallel cost of IVAND is $O(n)$ operations. It is also easy to see that on $p \le n$ processors, the cost becomes $O(n^3/p)$.

6.1.2 Solving Vandermonde Systems and Parallel Prefix

The solution of primal or dual Vandermonde systems can be computed by first obtaining V^{-1} using Algorithm 6.3 (IVAND) and then using a dense BLAS2 (matrix vector multiplication) to obtain $V^{-1}b$ or $V^{-\top}b$.

We next describe an alternative approach for dual Vandermonde systems, $(V(x))^\top a = b$, that avoids the computation of the inverse. The algorithm consists of the following major stages.

1. From the values in (x, b) compute the divided difference coefficients $g = (\gamma_0, \ldots, \gamma_n)^\top$ for the Newton form representation (6.3).
2. Using g and x construct the coefficients of the power form representation for each term $\gamma_j \prod_{i=0}^{j-1}(\xi - \xi_i)$ of (6.3).
3. Combine these terms to produce the solution a as the vector of coefficients of the power form representation for the Newton polynomial (6.3).

Step 1: Computing the divided differences

Table-based Approach

In this method, divided difference coefficients $g = (\gamma_0, \ldots, \gamma_n)^\top$ emerge as elements of a table whose first two columns are the vectors x and b. The elements of the table are built recursively, column by column while the elements in each column can be computed independently. One version of this approach, NEVILLE, is shown as Algorithm 6.4. A matrix formulation of the generation of divided differences can be found in [14]. NEVILLE lends itself to parallel processing using e.g. the *systolic array model;* cf. [27, 28].

Algorithm 6.4 NEVILLE: computing the divided differences by the Neville method.

Input: $x = (\xi_0, \ldots, \xi_n)^\top$, $b = (\beta_0, \ldots, \beta_n)^\top$ where the ξ_j are pairwise distinct.
Output: Divided difference coefficients $c = (\gamma_0, \ldots, \gamma_n)^\top$
1: $c = b$ //initialization
2: **do** $k = 0 : n - 1$
3:    **doall** $i = k + 1 : n$
4:       $\gamma_i = (\gamma_i - \gamma_{i-1})/(\xi_i - \xi_{i-k-1})$
5:    **end**
6: **end**

Sums of Fractions Approach

We next describe a method of logarithmic complexity for constructing divided differences. The algorithm uses the sum of fractions representation for divided differences. In particular, consider the order $n + 1$ matrix

$$U = xe^\top - ex^\top + I, \tag{6.10}$$

where $x = (\xi_0, \ldots, \xi_n)^\top$. Its elements are

$$\upsilon_{i,j} = \begin{cases} \xi_{i-1} - \xi_{j-1} & \text{if } i \neq j \\ 1 & \text{otherwise} \end{cases}, \quad i, j = 1, \ldots, n + 1$$

Observe that the matrix is shifted skew-symmetric and that it is also used in Algorithm 6.3 (line 1). The divided differences can be expresssed as a linear combination of the values β_j with coefficients computed from U. Specifically, for $l = 0, \ldots, n$,

$$\gamma_l = \frac{1}{\upsilon_{1,1} \cdots \upsilon_{1,l+1}} \beta_0 + \frac{1}{\upsilon_{2,1} \cdots \upsilon_{2,l+1}} \beta_1 + \cdots + \frac{1}{\upsilon_{l+1,1} \cdots \upsilon_{l+1,l+1}} \beta_l. \tag{6.11}$$

The coefficients of the terms $\beta_0, \ldots, \beta_l$ for γ_l are the (inverses) of the products of the elements in rows 1 to $l+1$ of the order $l+1$ leading principal submatrix, $U_{1:l+1,1:l+1}$, of U. The key to the parallel algorithm is the following observation [29, 30]:

> For fixed j, the coefficients of β_j in the divided difference coefficients $\gamma_0, \gamma_1, \ldots, \gamma_n$ are the (inverses) of the $n + 1$ prefix products $\{\upsilon_{j,1}, \upsilon_{j,1}\upsilon_{j,2}, \ldots, \upsilon_{j,1} \cdots \upsilon_{j,n+1}\}$.

The need to compute prefixes arises frequently in Discrete and Computational Mathematics and is recognized as a kernel and as such has been studied extensively in the literature. For completeness, we provide a short description of the problem and parallel algorithms in Sect. 6.1.3.

We now return to computing the divided differences using (6.11). We list the steps as Algorithm 6.5 (DD_PPREFIX). First U is calculated. Noting that the matrix is shifted skew-symmetric, this needs 1 parallel subtraction on $n(n + 1)/2$. Applying $n + 1$ independent instances of parallel prefix (Algorithm 6.8) and 1 parallel division, all of the inverse coefficients of the divided differences are computed in $\log n$ steps using $(n + 1)n$ processors. Finally, the application of n independent instances of a logarithmic depth tree addition algorithm yields the sought values in $\log(n + 1)$ arithmetic steps on $(n + 1)n/2$ processors. Therefore, the following result holds.

Proposition 6.3 ([30]) *The divided difference coefficients of the Newton interpolating polynomial of degree n can be computed from the $n + 1$ value pairs $\{(\xi_k, \beta_k), k = 0, \ldots, n\}$ in $2\log(n+1) + 2$ parallel arithmetic steps using $(n+1)n$ processors.*

Algorithm 6.5 DD_PPREFIX: computing divided difference coefficients by sums of rationals and parallel prefix.

Input: $x = (\xi_0, \ldots, \xi_n)^\top$ and $b = (\beta_0, \ldots, \beta_n)^\top$ where the ξ_j's are pairwise distinct
Output: divided difference coefficients $\{\gamma_j\}_{j=0}^n$
1: $U = xe^\top - ex^\top + I; \hat{U} = 1_{n+1,n+1} \oslash U;$
2: **doall** $j = 1 : n + 1$
3:     $P_{j,:} = \text{PREFIX\_OPT}(\hat{U}_{j,:});$ //It is assumed that $n = 2^k$ and that PREFIX_OPT (cf. Sect. 6.1.3) is used to compute prefix products
4: **end**
5: **doall** $l = 1 : n + 1$
6:     $\gamma_{l-1} = b_{1:l}^\top P_{1:l,l}$
7: **end**

Steps 2–3: Constructing power form coefficients from the Newton form

The solution a of the dual Vandermonde system $V^\top(x)a = b$ contains the coeffients of the power form representation of the Newton interpolation polynomial in (6.3). Based on the previous discussion, this will be accomplished by independent invocations of Algorithm 6.2 (POWFORM). Each of these returns the set of coefficients for the power form of an addend, $\gamma_l \prod_{j=0}^{l-1}(\xi - \xi_j)$, in (6.3). Finally, each of these intermediate vectors is summed to return the corresponding element of a. When $O(n^2)$ processors are available, the parallel cost of this algorithm is $\mathbf{T}_p = O(\log n)$. These steps are listed in Algorithm 6.6 (DD2PW).

Algorithm 6.6 DD2PW: converting divided differences to power form coefficients

---

Input: vector of nodes $x = (\xi_0, \ldots, \xi_{n-1})$ and divided difference coefficients $g = (\gamma_0, \ldots, \gamma_n)$.
Output: power form coefficients $a = (\alpha_0, \ldots, \alpha_n)$.
1: $R = 0_{n+1,n+1}; R_{1,1} = \gamma_0;$
2: **doall** $j = 1 : n$
3:     $R_{1:j+1,j+1} = \text{POWFORM}((\xi_0, \ldots, \xi_{j-1}), \gamma_j)$
4: **end**
5: **doall** $j = 1 : n + 1$
6:     $a = \text{sum }(R_{j,:}, 2)$
7: **end**

6.1.3 A Brief Excursion into Parallel Prefix

Definition 6.1 Let $\star$ be an associative binary operation on a set $\mathscr{S}$. The *prefix computation* problem is as follows: Given an ordered n-tuple $(\alpha_1, \alpha_2, \ldots, \alpha_n)$ of elements of $\mathscr{S}$, compute all n prefixes $\alpha_1 \star \cdots \star \alpha_i$ for $i = 1, 2, \ldots, n$.

The computations of interest in this book are prefix sums on scalars and prefix products on scalars and matrices. Algorithm 6.7 accomplishes the prefix computation; to simplify the presentation, it is assumed that $n = 2^k$. The cost of the algorithm is $\mathbf{T}_p = O(\log n)$ parallel "$\star$" operations.

Algorithm 6.7 PREFIX: prefix computation

Input: $a = (\alpha_1, \ldots, \alpha_n)$ where $n = 2^k$ and an associative operation $\star$.
Output: $p = (\pi_1, \ldots, \pi_n)$ where $\pi_i = \alpha_1 \star \cdots \star \alpha_i$.
1: $p = a$;
2: **do** $j = 1 : \log n$
3:    $h = 2^{j-1}$
4:    **doall** $i = 1 : n - h$
5:      $\hat{\pi}_{i+h} = p_{i+h} \star p_i$
6:    **end**
7:    **doall** $i = 1 : n - h$
8:      $\pi_{i+h} = \hat{\pi}_{i+h}$
9:    **end**
10: **end**

Algorithm 6.7 (PREFIX) is straightforward but the total number of operations is of the order $n \log n$, so there is $O(\log n)$ redundancy over the simple serial computation.

A parallel prefix algorithm of $O(1)$ redundancy can be built by a small modification of Algorithm 6.7 at the cost of a few additional steps that do not alter the logarithmic complexity. We list this as Algorithm 6.8 (PREFIX_OPT). The cost is $\mathbf{T}_p = 2 \log n - 1$ parallel steps and $\mathbf{O}_p = 2n - \log n - 2$ operations. The computations are organized in two phases, the first performing k steps of a reduction using a fan-in approach to compute final and intermediate prefixes at even indexed elements $2, 4, \ldots, 2^k$ and the second performing another $k - 1$ steps to compute the remaining prefixes. See also [31–33] for more details on this and other implementations.

Algorithm 6.8 PREFIX_OPT: prefix computation by odd-even reduction.

Input: Vector $a = (\alpha_1, \ldots, \alpha_n)$ where n is a power of 2.   //$\star$ is an associative operation
Output: Vector p of prefixes $\pi_i = \alpha_1 \star \cdots \star \alpha_i$.
1: $p = a$
   //forward phase
2: **do** $j = 1 : \log n$
3:    **doall** $i = 2^j : 2^j : n$
4:      $\hat{\pi}_i = \pi_i \star \pi_{i-2^{j-1}}$
5:    **end**
6:    **doall** $i = 2^j : 2^j : n$
7:      $\pi_i = \hat{\pi}_i$
8:    **end**
9: **end**
   //backward phase
10: **do** $j = \log n - 1 : -1 : 1$
11:    **doall** $i = 3 \cdot 2^{j-1} : 2^j : n - 2^{j-1}$
12:      $\hat{\pi}_i = \pi_i \star \pi_{i-2^{j-1}}$
13:    **end**
14:    **doall** $i = 3 \cdot 2^{j-1} : 2^j : n - 2^{j-1}$
15:      $\pi_i = \hat{\pi}_i$
16:    **end**
17: **end**

The pioneering array oriented programming language APL included the *scan* instruction for computing prefixes [34]. In [35] many benefits of making prefix available as a primitive instruction are outlined. The monograph [32] surveys early work on parallel prefix. Implementations have been described for several parallel architectures, e.g. see [36–40], and via using the Intel Threading Building Blocks C++ template library (TBB) in [41].

6.2 Banded Toeplitz Linear Systems Solvers

6.2.1 Introduction

Problems in mathematical physics and statistics give rise to Toeplitz, block-Toeplitz, and the related Hankel and block-Hankel systems of linear equations. Examples include problems involving convolutions, integral equations with difference kernels, least squares approximations using polynomials as basis functions, and stationary time series. Toeplitz matrices are in some sense the prototypical structured matrices. In fact, in certain situations, banded Toeplitz matrices could be used as effective preconditioners especially if fast methods for solving systems involving such preconditioners are available.

We already discussed algorithms for linear systems with triangular Toeplitz and banded triangular Toeplitz matrices in Sect. 3.2.4 of Chap. 3. Here, we present algorithms for systems with more general banded Toeplitz matrices. Several uniprocessors algorithms have been developed for the inversion of dense Toeplitz and Hankel matrices of order n, all requiring $O(n^2)$ arithmetic operations, e.g. see [42–50]. Extending these algorithms to block-Toeplitz and block-Hankel matrices of order np with square blocks of order p shows that the inverse can be obtained in $O(p^3n^2)$ operations; see [51, 52]. One of the early surveys dealing with inverses of Toeplitz operators is given in [53]. Additional uniprocessor algorithms have been developed for solving dense Toeplitz systems or computing the inverse of dense Toeplitz matrices, e.g. see [54–58] in $O(n \log^2 n)$ arithmetic operations. These *superfast* schemes must be used with care because sometimes they exhibit stability problems.

In what follows, we consider solving banded Toeplitz linear systems of equations of bandwidth $(2m + 1)$, where the system size n is much larger than m. These systems arise in the numerical solution of certain initial-value problems, or boundary-value problems, for example. Uniprocessor banded solvers require $O(n \log m)$ arithmetic operations, e.g. see [7]. We describe three algorithms for solving banded Toeplitz systems, originally presented in [59], that are suitable for parallel implementation. Essentially, in terms of complexity, we show that a symmetric positive-definite banded Toeplitz system may be solved in $O(m \log n)$ parallel arithmetic operations given enough processors. Considering that a banded Toeplitz matrix of bandwidth $2m + 1$ can be described by at most $(2m + 1)$ floating-point numbers, a storage scheme may be adopted so as to avoid excessive internode communications and maximize conflict-free access within each multicore node.

The banded Toeplitz algorithms to be presented in what follows rely on several facts concerning Toeplitz matrices and the closely related circulant matrices. We review some of these facts and present some of the basic parallel kernels that are used in Sect. 6.2.2.

Definition 6.2 Let $A = [\alpha_{ij}] \in \mathbb{R}^{n \times n}$. Then A is *Toeplitz* if $\alpha_{ij} = \alpha_{i-j}$, $i, j = 1, 2, \ldots, n$. In other words, the entries of A along each diagonal are the same.

It follows from the definition that Toeplitz matrices are low displacement matrices. In particular it holds that

$$A - JAJ^\top = (A - (e_1^\top A e_1)I)e_1 e_1^\top + e_1 e_1^\top A,$$

where J is the lower bidiagonal matrix defined in section and the rank of the term on the right-hand side of the equality above is at most 2.

Definition 6.3 Let $A, E_n \in \mathbb{R}^{n \times n}$, where $E_n = [e_n, e_{n-1}, \ldots, e_1]$, in which e_i is the ith column of the identity matrix I_n. A is then said to be *persymmetric* if $E_n A E_n = A^\top$; in other words, if A is symmetric about the cross-diagonal.

Definition 6.4 ([60]) Let $A \in \mathbb{R}^{2n \times 2n}$. Hence, A is called *centrosymmetric* if it can be written as

$$A = \begin{pmatrix} B & C \\ C^\top & E_n B E_n \end{pmatrix}$$

where $B, C \in \mathbb{R}^{n \times n}$, $B = B^\top$, and $C^\top = E_n C E_n$. Equivalently, $E_{2n} A E_{2n} = A$.

Lemma 6.1 ([50]) *Let $A \in \mathbb{R}^{n \times n}$ be a nonsingular Toeplitz matrix. Then, both A and A^{-1} are persymmetric; i.e., $E_n A E_n = A^\top$, and $E_n A^{-1} E_n = A^{-\top}$.*

The proof follows directly from the fact that A is nonsingular, and $E_n^2 = I$.

Lemma 6.2 ([61]) *Let $A \in \mathbb{R}^{2n \times 2n}$ be a symmetric Toeplitz matrix of the form*

$$A = \begin{pmatrix} B & C \\ C^\top & B \end{pmatrix}.$$

Thus, A is also centrosymmetric. Furthermore, if

$$P_{2n} = \frac{1}{\sqrt{2}} \begin{pmatrix} I_n & E_n \\ I_n & -E_n \end{pmatrix}$$

we have

$$P_{2n} A P_{2n}^\top = \begin{pmatrix} B + CE_n & 0 \\ 0 & B - CE_n \end{pmatrix}.$$

The proof is a direct consequence of Definition 6.4 and Lemma 6.1.

Theorem 6.1 ([53, 62, 63]) *Let $A \in \mathbb{R}^{n \times n}$ be a Toeplitz matrix with all its leading principal submatrices being nonsingular. Let also, $Au = \alpha e_1$ and $Av = \beta e_n$, where*

$$u = (1, \mu_1, \mu_2, \ldots, \mu_{n-1})^{\top},$$
$$v = (\nu_{n-1}, \nu_{n-2}, \ldots, \nu_1, 1)^{\top}.$$

Since A^{-1} is persymmetric, then $\alpha = \beta$, and A^{-1} can be written as

$$\alpha A^{-1} = UV - \tilde{V}\tilde{U} \tag{6.12}$$

in which the right-hand side consists of the triangular Toeplitz matrices

$$U = \begin{pmatrix} 1 & & & & \\ \mu_1 & 1 & & & \\ \vdots & & \ddots & & \\ \mu_{n-2} & & & \ddots & \\ \mu_{n-1} & \mu_{n-2} & \cdots & \mu_1 & 1 \end{pmatrix} \qquad V = \begin{pmatrix} 1 & \nu_1 & \cdots & \nu_{n-2} & \nu_{n-1} \\ & \ddots & & & \nu_{n-2} \\ & & \ddots & & \vdots \\ & & & \ddots & \nu_1 \\ & & & & 1 \end{pmatrix}$$

$$\tilde{V} = \begin{pmatrix} 0 & & & & \\ \nu_{n-1} & 0 & & & \\ \vdots & & \ddots & & \\ \nu_2 & & & \ddots & \\ \nu_1 & \nu_2 & \cdots & \nu_{n-1} & 0 \end{pmatrix} \qquad \tilde{U} = \begin{pmatrix} 0 & \mu_{n-1} & \cdots & \mu_2 & \mu_1 \\ & \ddots & & & \mu_2 \\ & & \ddots & & \vdots \\ & & & \ddots & \mu_{n-1} \\ & & & & 0 \end{pmatrix}$$

Equation (6.12) was first given in [62]. It shows that A^{-1} is completely determined by its first and last columns, and can be used to compute any of its elements. Cancellation, however, is assured if one attempts to compute an element of A^{-1} below the cross-diagonal. Since A^{-1} is persymmetric, this situation is avoided by computing the corresponding element above the cross-diagonal. Moreover, if A is symmetric, its inverse is both symmetric and persymmetric, and $u = E_n v$ completely determines A^{-1} via

$$\alpha A^{-1} = UU^{\top} - \tilde{U}^{\top}\tilde{U}. \tag{6.13}$$

Finally, if A is a lower triangular Toeplitz matrix, then $A^{-1} = \alpha^{-1}U$, see Sect. 3.2.3.

Definition 6.5 Let $A \in \mathbb{R}^{n \times n}$ be a banded symmetric Toeplitz matrix with elements $\alpha_{ij} = \alpha_{|i-j|} = \alpha_k, k = 0, 1, \ldots, m.$ where $m < n, \alpha_m \neq 0$ and $\alpha_k = 0$ for $k > m$. Then the complex rational polynomial

$$\phi(\xi) = \alpha_m \xi^{-m} + \cdots + \alpha_1 \xi^{-1} + \alpha_0 + \alpha_1 \xi + \cdots + \alpha_m \xi^m \tag{6.14}$$

is called the *symbol* of A [64].

Lemma 6.3 *Let A be as in Definition 6.5. Then A is positive definite for all $n > m$ if and only if:*

(i) $\phi(e^{i\theta}) = \alpha_0 + 2\sum_{k=1}^{m} \alpha_k \cos k\theta \geqq 0$ *for all θ, and*

(ii) $\phi(e^{i\theta})$ *is not identically zero.*

   Note that condition (ii) is readily satisfied since $\alpha_m \neq 0$.

Theorem 6.2 *([65]) Let A in Definition 6.5 be positive semi-definite for all $n > m$. Then the symbol $\phi(\xi)$ can be factored as*

$$\phi(\xi) = \psi(\xi)\psi(1/\xi) \tag{6.15}$$

where

$$\psi(\xi) = \beta_0 + \beta_1 \xi + \cdots + \beta_m \xi^m \tag{6.16}$$

is a real polynomial with $\beta_m \neq 0$, $\beta_0 > 0$, and with no roots strictly inside the unit circle.

This result is from [66], see also [67, 68]. The factor $\psi(\xi)$ is called the *Hurwitz factor*. Such a factorization arises mainly in the time series analysis and the realization of dynamic systems. The importance of the Hurwitz factor will be apparent from the following theorem.

Theorem 6.3 *Let $\phi(e^{i\theta}) > 0$, for all θ, i.e. A is positive definite and the symbol (6.14) has no roots on the unit circle. Consider the Cholesky factorization*

$$A = LL^\top \tag{6.17}$$

where,

$$L = \begin{pmatrix}
\lambda_0^{(1)} & & & & & & & \\
\lambda_1^{(2)} & \lambda_0^{(2)} & & & & & & \\
\lambda_2^{(3)} & \lambda_1^{(3)} & \lambda_0^{(3)} & & & & & \\
\vdots & \vdots & \vdots & & & & & \\
\lambda_m^{(m+1)} & \lambda_{m-1}^{(m+1)} & \lambda_{m-2}^{(m+1)} & \cdots & \lambda_0^{(m+1)} & & & \\
 & \lambda_m^{(m-2)} & \lambda_{m-1}^{(m+2)} & \cdots & \lambda_1^{(m+2)} & \lambda_0^{(m+2)} & & \\
 & & \ddots & \ddots & & \ddots & \ddots & \\
\mathbf{0} & & & \lambda_m^{(k)} & \lambda_{m-1}^{(k)} & & \lambda_1^{(k)} & \lambda_0^{(k)}
\end{pmatrix} \tag{6.18}$$

Then,

$$\lim_{k \to \infty} \lambda_j^{(k)} = \beta_j, \qquad 0 \leqq j \leqq m,$$

where β_j is given by (6.16). In fact, if τ is the root of $\psi(\xi)$ closest to the unit circle (note that $|\tau| > 1$) with multiplicity p, then we have

$$\lambda_j^{(k)} = \beta_j + O[(k-j)^{2(p-1)}/|\tau|^{2(k-j)}]. \tag{6.19}$$

This theorem shows that the rows of the Cholesky factor L converge, linearly, with an asymptotic convergence factor that depends on the magnitude of the root of $\psi(\xi)$ closest to the unit circle. The larger this magnitude, the faster the convergence.

Next, we consider circulant matrices that play an important role in one of the algorithms in Sect. 6.2.2. In particular, we consider these circulants associated with banded symmetric or nonsymmetric Toeplitz matrices. Let $A = [\alpha_{-m}, \ldots, \alpha_{-1}, \alpha_0, \alpha_1, \ldots, \alpha_m] \in \mathbb{R}^{n \times n}$ denote a banded Toeplitz matrix of bandwidth $(2m+1)$ where $2m+1 \leq n$, and $\alpha_m, \alpha_{-m} \neq 0$. Writing this matrix as

$$A = \begin{pmatrix} B & C & & & & \\ D & B & C & & & \\ & \ddots & \ddots & \ddots & & \\ & & D & B & C \\ & & \cdots & D & B \end{pmatrix}$$

$$\tag{6.20}$$

where $B \in \mathbb{R}^{m \times m}$ is the Toeplitz matrix

$$B = \begin{pmatrix} \alpha_0 & \alpha_1 & \cdots & \alpha_{m-1} \\ \alpha_{-1} & \alpha_0 & & \vdots \\ \vdots & & \ddots & \alpha_1 \\ \alpha_{-m+1} & & \alpha_{-1} & \alpha_0 \end{pmatrix}$$

and $C, D \in \mathbb{R}^{m \times m}$ are the triangular Toeplitz matrices

$$C = \begin{pmatrix} \alpha_m & & & \\ \alpha_{m-1} & \alpha_m & & \\ \vdots & & \ddots & \\ \alpha_1 & \cdots & \alpha_{m-1} & \alpha_m \end{pmatrix} \quad \text{and} \quad D = \begin{pmatrix} \alpha_{-m} & \alpha_{-m+1} & \cdots & & \alpha_{-1} \\ & \ddots & & & \alpha_{-2} \\ & & \ddots & & \vdots \\ & & & \ddots & \alpha_{-m+1} \\ & & & & \alpha_{-m} \end{pmatrix}$$

The circulant matrix $\tilde{A}$ corresponding to A is therefore given by

$$\tilde{A} = \begin{pmatrix} B & C & & & & D \\ D & B & C & & & \\ & \ddots & \ddots & \ddots & & \\ & & & D & B & C \\ C & & & \cdots & D & B \end{pmatrix}$$

$$(6.21)$$

which may also be written as the matrix polynomial

$$\tilde{A} = \sum_{j=0}^{m} \alpha_j K^j + \sum_{j=1}^{m} \alpha_{-j} K^{n-j},$$

where $K = J_n^\top + e_n e_1^\top$, that is

$$K = \begin{pmatrix} 0 & 1 & & & \\ & 0 & 1 & & \\ & & \ddots & \ddots & \\ 0 & & & \ddots & 1 \\ 1 & 0 & & & 0 \end{pmatrix}$$

$$(6.22)$$

It is of interest to note that $K^\top = K^{-1} = K^{n-1}$, and $K^n = I_n$.

In the next lemma we make use of the unitary Fourier matrix introduced in Sect. 6.1.

Lemma 6.4 *Let W be the unitary Fourier matrix of order n*

$$W = \frac{1}{\sqrt{n}} \begin{pmatrix} 1 & 1 & 1 & \cdots & 1 \\ 1 & \omega & \omega^2 & \cdots & \omega^{(n-1)} \\ 1 & \omega^2 & \omega^4 & \cdots & \omega^{2(n-1)} \\ \vdots & \vdots & \vdots & & \vdots \\ 1 & \omega^{(n-1)} & \omega^{2(n-1)} & \cdots & \omega^{(n-1)^2} \end{pmatrix}$$

$$(6.23)$$

in which $\omega = e^{i(2\pi/n)}$ is an nth root of unity. Then

$$\begin{aligned} W^* K W &= \Omega \\ &= \mathrm{diag}(1, \omega, \omega^2, \ldots, \omega^{n-1}), \end{aligned}$$

$$(6.24)$$

and

$$W^* \tilde{A} W = \Gamma$$
$$= \sum_{j=0}^{m} \alpha_j \Omega^j + \sum_{j=1}^{m} \alpha_{-j} \Omega^{-j} \tag{6.25}$$

where $\Gamma = \mathrm{diag}(\gamma_0, \gamma_1, \ldots, \gamma_{n-1})$.

In other words, the kth eigenvalue of $\tilde{A}$, $k = 0, 1, \ldots, n - 1$, is given by

$$\gamma_k = \tilde{\phi}(\omega^k) \tag{6.26}$$

where

$$\tilde{\phi}(\xi) = \sum_{j=-m}^{m} \alpha_j \xi^j$$

is the symbol of the nonsymmetric Toeplitz matrix (6.20), or

$$\gamma_k = \sqrt{n} e_{k+1}^{\top} W a, \tag{6.27}$$

in which $a^{\top} = e_1^{\top} \tilde{A}$.

In what follows, we make use of the following well-known result.

Theorem 6.4 ([69]) *Let $W \in \mathbb{C}^{m \times n}$ be as in Lemma 6.4, $y \in \mathbb{C}^n$ and n be a power of 2. Then the product Wy may be obtained in $\mathbf{T}_p = 3 \log n$ parallel arithmetic operations using $p = 2n$ processors.*

6.2.2 Computational Schemes

In this section we present three algorithms for solving the banded Toeplitz linear system

$$Ax = f \tag{6.28}$$

where A is given by (6.20) and $n \gg m$. These have been presented first in [59]. We state clearly the conditions under which each algorithm is applicable, and give complexity upper bounds on the number of parallel arithmetic operations and processors required.

The first algorithm, listed as Algorithm 6.9, requires the least number of parallel arithmetic operations of all three, $6 \log n + O(1)$. It is applicable only when the corresponding circulant matrix is nonsingular. The second algorithm, numbered Algorithm 6.10 may be used if A, in (6.28), is positive definite or if all its principal minors are non-zero. It solves (6.28) in $O(m \log n)$ parallel arithmetic operations. The third algorithm, Algorithm 6.11, uses a modification of the second algorithm to compute the Hurwitz factorization of the symbol $\phi(\xi)$ of a positive definite Toeplitz matrix. The Hurwitz factor, in turn, is then used by an algorithm proposed in [65] to solve (6.28). This last algorithm is applicable only if none of the roots of $\phi(\xi)$ lie on the unit circle. In fact, the root of the factor $\psi(\xi)$ nearest to the unit circle should be far enough away to assure early convergence of the modified Algorithm 6.10. The third algorithm requires $O(\log m \log n)$ parallel arithmetic operations provided that the Hurwitz factor has already been computed. It also requires the least storage of all three algorithms. In the next section we discuss another algorithm that is useful for block-Toeplitz systems that result from the discretization of certain elliptic partial differential equations.

A Banded Toeplitz Solver for Nonsingular Associated Circulant Matrices: Algorithm 6.9

First, we express the linear system (6.28), in which the banded Toeplitz matrix A is given by (6.20), as

$$(\tilde{A} - S)x = f \tag{6.29}$$

where $\tilde{A}$ is the circulant matrix (6.21) associated with A, and

$$S = U \begin{pmatrix} 0 & D \\ C & 0 \end{pmatrix} U^\top,$$

in which

$$U^\top = \begin{pmatrix} I_m & 0 & \cdots & 0 & 0 \\ 0 & 0 & \cdots & 0 & I_m \end{pmatrix}.$$

Assuming $\tilde{A}$ is nonsingular, the Sherman-Morrison-Woodbury formula [70] yields the solution

$$x = \tilde{A}^{-1} f - \tilde{A}^{-1} U G^{-1} U^\top \tilde{A}^{-1} f, \tag{6.30}$$

where

$$G = U^\top \tilde{A}^{-1} U - \begin{pmatrix} 0 & D \\ C & 0 \end{pmatrix}^{-1}.$$

The number of parallel arithmetic operations required by this algorithm, Algorithm 6.9, which is clearly dominated by the first stage, is $6 \log n + O(m \log m)$ employing no more than $4n$ processors. The corresponding sequential algorithm consumes $O(n \log n)$ arithmetic operations and hence realize $O(n)$ speedup. Note that at no time do we need more than $2n + O(m^2)$ storage locations.

The notion of inverting a Toeplitz matrix via correction of the corresponding circulant-inverse is well known and has been used many times in the past; cf. [65, 71].

Such an algorithm, when applicable, is very attractive on parallel architectures that allow highly efficient FFT's. See for example [72] and references therein. As we have seen, solving (6.28) on a parallel architecture can be so inexpensive that one is tempted to improve the solution via one step of iterative refinement. Since v_1, which determines $\tilde{A}^{-1}$ completely, is already available, each step of iterative refinement

---

Algorithm 6.9 A banded Toeplitz solver for nonsymmetric systems with nonsingular associated circulant matrices

---

Input: Banded nonsymmetric Toeplitz matrix A as in (6.20) and the right-hand side f.

Output: Solution of the linear system $Ax = f$

//Stage 1   //Consider the circulant matrix $\tilde{A}$ associated with A as given by Eq. (6.21). First, determine whether $\tilde{A}$ is nonsingular and, if so, determine $\tilde{A}^{-1}$ and $y = \tilde{A}^{-1} f$. Since the inverse of a circulant matrix is also circulant, $\tilde{A}^{-1}$ is completely determined by solving $\tilde{A} v_1 = e_1$. This is accomplished via (6.25), i.e., $y = W \Gamma^{-1} W^* f$ and $v_1 = W \Gamma^{-1} W^* e_1$. This computation is organized as follows:

1: Simultaneously form $\sqrt{n} W a$ and $W^* f$ //see (6.27) and Theorem 6.4 (FFT). This is an inexpensive test for the nonsingularity of $\tilde{A}$. If none of the elements of $\sqrt{n} W a$ (eigenvalues of $\tilde{A}$) vanish, we proceed to step (2).

2: Simultaneously, obtain $\Gamma^{-1}(W^* e_1)$ and $\Gamma^{-1}(W^* f)$.   //Note that $\sqrt{n} W^* e_1 = (1, 1, \ldots, 1)^{\mathsf{T}}$.

3: Simultaneously obtain $v_1 = W(\Gamma^{-1} W^* e_1)$ and $y = W(\Gamma^{-1} W^* f)$ via the FFT.   //see Theorem 6.4

//Stage 2   //solve the linear system

$$Gz = U^{\mathsf{T}} y = \begin{pmatrix} y_1 \\ y_v \end{pmatrix} \tag{6.31}$$

where y_1, $y_v \in \mathbb{R}^m$ contain the first and last m elements of y, respectively.

4: form the matrix $G \in \mathbb{R}^{2m \times 2m}$.   //From Stage 1, v_1 completely determines $\tilde{A}^{-1}$, in particular we have

$$U^{\mathsf{T}} \tilde{A}^{-1} U = \begin{pmatrix} F & M \\ N & F \end{pmatrix},$$

where F, M, and N are Toeplitz matrices each of order m.

5: Compute $G = \begin{pmatrix} F & M - C^{-1} \\ N - D^{-1} & F \end{pmatrix}$.   //Recall that C^{-1} and D^{-1} are completely determined by their first columns; see also Theorem 6.1

6: compute the solution z of (6.31)

//Stage 3

7: compute $u = \tilde{A}^{-1} U z$

//Stage 4

8: compute $x = y - u$

---

is equally inexpensive. The desirability of iterative refinement for this algorithm is discussed later.

A Banded Toeplitz Solver for Symmetric Positive Definite Systems: Algorithm 6.10

Here, we consider symmetric positive definite systems which implies that for the banded Toeplitz matrix A in (6.20), we have $D = C^{\mathsf{T}}$, and B is symmetric positive definite. For the sake of ease of illustration, we assume that $n = 2^q p$, where p and q are integers with $p = O(\log n) \geq m$. On a uniprocessor, the Cholesky factorization is the preferred algorithm for solving the linear system (6.28). Unless the corresponding circulant matrix is also positive definite, however, the rows of L (the Cholesky factors of A) will not converge, see Theorem 6.3. This means that one would have to store

$O(mn)$ elements, which may not be acceptable for large n and relatively large m. Even if the corresponding circulant matrix is positive definite, Theorem 6.3 indicates that convergence can indeed be slow if the magnitude of that root of the Hurwitz factor $\psi(\xi)$ (see Theorem 6.2) nearest to the unit circle is only slightly greater than 1. If s is that row of L at which convergence takes place, the parallel Cholesky factorization and the subsequent forward and backward sweeps needed for solving (6.28) are more efficient the smaller s is compared to n.

In the following, we present an alternative parallel algorithm that solves the same positive definite system in $O(m \log n)$ parallel arithmetic operations with $O(n)$ processors, which requires no more than $2n + O(m^2)$ temporary storage locations. For $n = 2^q p$, the algorithm consists of $(q + 1)$ stages which we outline as follows.

Stage 0.

Let the pth leading principal submatrix of A be denoted by A_0, and the right-hand side of (6.28) be partitioned as

$$f^\top = (f_1^{(0)^\top}, f_2^{(0)^\top}, \ldots, f_\eta^{(0)^\top}),$$

where $f_i^{(0)} \in R^p$ and $\eta = n/p = 2^q$. In this initial stage, we simultaneously solve the $(\eta + 1)$ linear systems with the same coefficient matrix:

$$A_0(z_0, y_1^{(0)}, \ldots, y_\eta^{(0)}) = (e_1^{(p)}, f_1^{(0)}, \ldots, f_p^{(0)}) \tag{6.32}$$

where $e_1^{(p)}$ is the first column of the identity I_p. From Theorem 4.1 and the discussion of the column sweep algorithm (CSWEEP) in Sect. 3.2.1 it follows that the above systems can be solved in $9(p-1)$ parallel arithmetic operations using $m\eta$ processors.

Stage $(j = 1, 2, \ldots, q)$.

Let

$$A_j = \left(\begin{array}{c|c} A_{j-1} & \begin{array}{cc} & 0 \\ & C \end{array} \\ \hline \begin{array}{cc} & C^\top \\ 0 & \end{array} & A_{j-1} \end{array} \right) \tag{6.33}$$

be the leading $2r \times 2r$ principal submatrix of A, where $r = 2^{j-1} p$. Also, let f be partitioned as

$$f^\top = (f_1^{(j)^\top}, f_2^{(j)^\top}, \ldots, f_\nu^{(j)^\top})$$

where $f_i^{(j)} \in \mathbb{R}^{2r}$, and $v = n/2r$. Here, we simultaneously solve the $(v+1)$ linear systems

$$A_j z_j = e_1^{(2r)}, \tag{6.34}$$

$$A_j y_i^{(j)} = f_i^{(j)}, \qquad i = 1, 2, \ldots, v, \tag{6.35}$$

where stage $(j-1)$ has already yielded $z_{j-1} = A_{j-1}^{-1} e_1^{(r)}$ and $y_i^{(j-1)} = A_{j-1}^{-1} f_i^{(j-1)}$, $i = 1, 2, \ldots, 2v$. Next, we consider solving the ith linear system in (6.35). Observing that

$$f_i^{(j)} = \begin{pmatrix} f_{2i-1}^{(j-1)} \\ f_{2i}^{(j-1)} \end{pmatrix},$$

and premultiplying both sides of (6.35) by the positive definite matrix

$$D_j = \mathrm{diag}(A_{j-1}^{-1}, A_{j-1}^{-1})$$

we obtain the linear system

$$\left(\begin{array}{c|c} I_r & G_j \;\; 0 \\ \hline 0 \;\; H_j & I_r \end{array} \right) y_i^{(j)} = \begin{pmatrix} y_{2i-1}^{(j-1)} \\ y_{2i}^{(j-1)} \end{pmatrix} \tag{6.36}$$

where

$$G_j = A_{j-1}^{-1} \begin{pmatrix} 0 \\ C \end{pmatrix} \quad \text{and} \quad H_j = A_{j-1}^{-1} \begin{pmatrix} C^\top \\ 0 \end{pmatrix}.$$

From Lemma 6.1, both A_{j-1}^{-1} and C are persymmetric, i.e., $E_r A_{j-1}^{-1} E_r = A_{j-1}^{-1}$ and $E_m C E_m = C^\top$, thus $H_j = E_r G_j E_m$. Using the Gohberg-Semencul formula (6.13), see Theorem 6.1, H_j may be expressed as follows. Let

$$\begin{aligned} u &= \alpha z_{j-1} \\ &= (1, \mu_1, \mu_2, \ldots, \mu_{r-1})^\top, \end{aligned}$$

and

$$\tilde{u} = J_r E_r u.$$

Hence, H_j is given by

$$\alpha H_j = (Y Y_1^\top - \tilde{Y} \tilde{Y}_1^\top) C^\top, \tag{6.37}$$

in which the $m \times r$ matrices Y and $\tilde{Y}$ are given by

$$\begin{aligned} Y &= (u, J_r u, \ldots, J_r^{m-1} u), \\ \tilde{Y} &= (\tilde{u}, J_r \tilde{u}, \ldots, J_r^{m-1} \tilde{u}), \end{aligned}$$

$$Y_1 = (I_m, 0)Y = \begin{pmatrix} 1 & & & & \\ \mu_1 & 1 & & & \\ \mu_2 & \mu_1 & 1 & & \\ \vdots & & \ddots & \ddots & \\ \mu_{m-1} & \cdots & \mu_2 & \mu_1 & 1 \end{pmatrix}$$

and

$$\tilde{Y}_1 = (I_m, 0)\tilde{Y} = \begin{pmatrix} 0 & & & & \\ \mu_{r-1} & 0 & & & \\ \mu_{r-2} & \mu_{r-1} & 0 & & \\ \vdots & & \ddots & \ddots & \\ \mu_{r-m+1} & \cdots & \mu_{r-2} & \mu_{r-1} & 0 \end{pmatrix}$$

The coefficient matrix in (6.36), $D_j^{-1} A_j$, may be written as

$$\begin{pmatrix} I_{r-m} & N_1^{(j)} & 0 \\ 0 & N_2^{(j)} & 0 \\ 0 & N_3^{(j)} & I_{r-m} \end{pmatrix},$$

where the central block $N_2^{(j)}$ is clearly nonsingular since A_j is invertible. Note also that the eigenvalues of $N_2^{(j)} \in \mathbb{R}^{2m \times 2m}$ are the same as those eigenvalues of $(D_j^{-1} A_j)$ that are different from 1. Since $D_j^{-1} A_j$ is similar to the positive definite matrix $D_j^{-1/2} A_j D_j^{-1/2}$, $N_2^{(j)}$ is not only nonsingular, but also has all its eigenvalues positive. Hence, the solution of (6.36) is trivially obtained if we first solve the middle $2m$ equations,

$$N_2^{(j)} h = g, \tag{6.38}$$

or

$$\begin{pmatrix} I_m & E_m M_j E_m \\ M_j & I_m \end{pmatrix} \begin{pmatrix} h_{1,i} \\ h_{2,i} \end{pmatrix} = \begin{pmatrix} g_{1,i} \\ g_{2,i} \end{pmatrix}, \tag{6.39}$$

where

$$M_j = (I_m, 0)H_j = \alpha^{-1}(Y_1 Y_1^\top - \tilde{Y}_1 \tilde{Y}_1^\top)C^\top$$

and $g_{k,i}, h_{k,i}, k = 1, 2$, are the corresponding partitions of $f_i^{(j)}$ and $y_i^{(j)}$, respectively. Observing that $N_2^{(j)}$ is centrosymmetric (see Definition 6.4), then from Lemma 6.2 we reduce (6.39) to the two independent linear systems

$$(I_m + E_m M_j)(h_{1,i} + E_m h_{2,i}) = (g_{1,i} + E_m g_{2,i}),$$
$$(I_m - E_m M_j)(h_{1,i} - E_m h_{2,i}) = (g_{1,i} - E_m g_{2,i}). \tag{6.40}$$

Since $P_{2m} N_2^{(j)} P_{2m}^{\top}$ has all its leading principal minors positive, these two systems may be simultaneously solved using Gaussian elimination without partial pivoting. Once h_1 and $E_m h_2$ are obtained, $y_i^{(j)}$ is readily available.

$$y_i^{(j)} = \begin{pmatrix} y_{2i-1}^{(j)} - E_r H_j E_m h_{2,i} \\ y_{2i}^{(j-1)} - H_j h_{1,i} \end{pmatrix}. \tag{6.41}$$

A Banded Toeplitz Solver for Symmetric Positive Definite Systems with Positive Definite Associated Circulant Matrices: Algorithm 6.11

The last banded Toeplitz solver we discuss in this section is applicable only to those symmetric positive definite systems that have positive definite associated circulant matrices. In other words, we assume that both A in (6.20) and $\tilde{A}$ in (6.21) are symmetric positive definite. Hence, from Lemma 6.4 and Theorem 6.3, row $\hat{\imath}$ of the Cholesky factor of A will converge to the coefficients of $\psi(\xi)$, where $\hat{\imath} < n$. Consider the Cholesky factorization (6.17), $A = LL^{\top}$, where

$$L = \begin{pmatrix} S_1 & & & & & \\ V_1 & S_2 & & & & \\ & \ddots & \ddots & & & \\ & & V_{i-1} & S_i & & \\ & & & V & S & \\ & & & & V & S \\ & & & & & \ddots & \ddots \end{pmatrix} \tag{6.42}$$

in which $\hat{\imath} = mi + 1$, S_j and $V_j \in \mathbb{R}^{m \times m}$ are lower and upper triangular, respectively, and S, V are Toeplitz and given by

$$
S = \begin{pmatrix}
\beta_0 & & & & \\
\beta_1 & \ddots & & & \\
\vdots & & \ddots & & \\
\vdots & & & \ddots & \\
\beta_{m-1} & \cdots & \cdots & \beta_1 & \beta_0
\end{pmatrix}
\quad \text{and} \quad
V = \begin{pmatrix}
\beta_m & \beta_{m-1} & \cdots & \cdots & \beta_1 \\
& \ddots & & & \beta_2 \\
& & \ddots & & \vdots \\
& & & \ddots & \beta_{m-1} \\
& & & & \beta_m
\end{pmatrix}
$$

Equating both sides of (6.17), we get

$$
B = SS^\top + VV^\top,
$$

and

$$
C = SV^\top,
$$

where B, and C are given in (6.20), note that in this case $D = C^\top$. Hence, A can be expressed as

$$
A = R^\top R + \begin{pmatrix} I_m \\ 0 \end{pmatrix} VV^\top (I_m, 0) \tag{6.43}
$$

where $R^\top \in \mathbb{R}^{n \times n}$ is the Toeplitz lower triangular matrix

$$
R^\top = \begin{pmatrix}
S & & & & \\
V & S & & & \\
& \ddots & \ddots & & \\
& & V & S & \\
& & \cdots & V & S
\end{pmatrix}
$$

$$\tag{6.44}$$

Assuming that an approximate Hurwitz factor, $\tilde{\psi}(\xi) = \sum_{j=0}^{m} \tilde{\beta}_j \xi^j$, is obtained from the Cholesky factorization of A, i.e.,

$$\tilde{R}^{\top} = \begin{pmatrix} \tilde{S} & & & & \\ \tilde{V} & \tilde{S} & & & \\ & \ddots & \ddots & & \\ & & \tilde{V} & \tilde{S} & \\ & & \cdots & \tilde{V} & \tilde{S} \end{pmatrix}$$

is available, the solution of the linear system (6.28) may be computed via the Sherman-Morrison-Woodbury formula [14], e.g. see [65],

$$x = F^{-1}f - F^{-1}\begin{pmatrix} \tilde{V} \\ 0 \end{pmatrix} Q^{-1}[\tilde{V}^{\top}, 0]F^{-1}f \qquad (6.45)$$

where

$$F = \tilde{R}^{\top}\tilde{R},$$

and

$$Q = I_m + (\tilde{V}^{\top}, 0)F^{-1}\begin{pmatrix} \tilde{V} \\ 0 \end{pmatrix}.$$

Computing x in (6.45) may then be organized as follows:

(a) Solve the linear systems $Fv = f$, via the forward and backward sweeps $\tilde{R}^{\top}w = f$ and $\tilde{R}v = w$. Note that algorithm BTS (Algorithm 3.5) which can be used for these forward and backward sweeps corresponds to Theorem 3.4. Further, by computing $\tilde{R}^{-1}e_1$, we completely determine $\tilde{R}^{-1}$.
(b) Let $\tilde{L}_1 = \tilde{R}^{-\top}\begin{pmatrix} I_m \\ 0 \end{pmatrix}$ and $v_1 = (I_m, 0)v$, and simultaneously form $T = \tilde{L}_1\tilde{V}$ and $u = \tilde{V}^{\top}v_1$. Thus, we can easily compute $Q = I_m + T^{\top}T$.
(c) Solve the linear system $Qa = u$ and obtain the column vector $b = Ta$. See Theorem 4.1 and Algorithms 3.1 and 3.2 in Sect. 3.2.1.
(d) Finally, solve the Toeplitz triangular system $\tilde{R}c = b$, and obtain the solution $x = v - c$.

From the relevant basic lemmas and theorems we used in this section, we see that, given the approximation to the Hurwitz factor $\tilde{R}$, obtaining the solution x in (6.45) requires $(10 + 6\log m)\log n + O(m)$ parallel arithmetic operations using no more than mn processors, and only $n + O(m^2)$ storage locations, the lowest among all three algorithms in this section.

Since the lowest number of parallel arithmetic operations for obtaining $\tilde{S}$ and $\tilde{V}$ via the Cholesky factorization of A using no less than m^2 processors is $O(\hat{\imath})$, the cost of computing the Hurwitz factorization can indeed be very high unless $\hat{\imath} \ll n$. This will happen only if the Hurwitz factor $\psi(\xi)$ has its nearest root to the unit circle of magnitude well above 1.

Next, we present an alternative to the Cholesky factorization. A simple modification of the banded Toeplitz positive definite systems solver (Algorithm 6.10), yields an efficient method for computing the Hurwitz factor. Let $\tilde{r} = 2^k p \geq m(i+1) = \hat{i} + m - 1$. Then, from (6.17), (6.36)–(6.42) we see that

$$E_m M_k E_m = S^{-\top} S^{-1} C. \tag{6.46}$$

In other words, the matrices M_j in (6.39) converge to a matrix M (say), and the elements $\tilde{\beta}_j, 0 \leq j \leq m-1$, of $\tilde{S}$ are obtained by computing the Cholesky factorization of

$$E_m C^\top M^{-1} E_m = \left((0, I_m) A_{j-1}^{-1} \binom{0}{I_m} \right)^{-1}.$$

Now, the only remaining element of $\tilde{V}$, namely $\tilde{\beta}_m$ is easily obtained as $\alpha_m / \tilde{\beta}_0$, which can be verified from the relation $C = SV^\top$.

Observing that we need only to compute the matrices M_j, Algorithm 6.10 may be modified so that in any stage j, we solve only the linear systems $A_j z_j = e_1^{(2r)}$, where $j = 0, 1, 2, \ldots$, and $2r = 2^{j+1} p$. Since we assume that convergence takes place in the kth stage, where $\tilde{r} = 2^k p \leq n/2$, the number of parallel arithmetic operations required for computing the Hurwitz factor is approximately $6mk \leq 6m \log(n/2p)$ using $2m\tilde{r} \leq mn$ processors.

Using positive definite Toeplitz systems with bandwidths 5, i.e. $m = 2$. Lemma 6.3 states that all sufficiently large pentadiagonal Toeplitz matrices of the form $[1, \sigma, \delta, \sigma, 1]$ are positive definite provided that the symbol function $\phi(e^{i\theta}) = \delta + 2\sigma \cos\theta + 2\cos 2\theta$ has no negative values. For positive δ, this condition is satisfied when (σ, δ) corresponds to a point on or above the lower curve in Fig. 6.1. The matrix is diagonally dominant when the point lies above the upper curve. For example, if we choose test matrices for which $\delta = 6$ and $0 \leq \sigma \leq 4$; every point on the dashed line in Fig. 6.1 represents one of these matrices. In [59], numerical experiments were conducted to

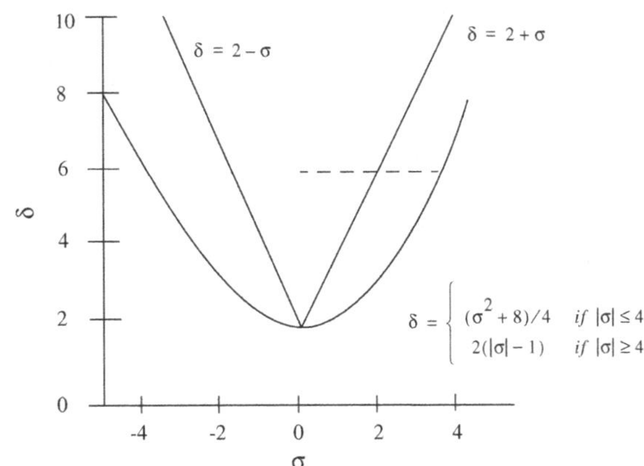

Fig. 6.1 Regions in the (σ, δ)-plane of positive definiteness and diagonal dominance of pentadiagonal Toeplitz matrices $(1, \sigma, \delta, \sigma, 1)$

Table 6.1 Parallel arithmetic operations, number of processors, and overall operation counts for Algorithms 6.9, 2 and 3

	Parallel arith. ops	Number of processors	Overall ops.	Storage
Algorithm 6.9	$6\log n$	$2n$	$10n\log n$	$2n$
[a]Algorithm 6.10	$18\log n$	$4n$	$4n\log n$	$2n$
[b]Algorithm 6.11	$16\log n$	$2n$	$12n\log n$	n

[a]The 2-by-2 linear systems (6.40) are solved via Cramer's rule
[b]Algorithm 6.11 does not include the computation of the Hurwitz factors

compare the relative errors in the computed solution achieved by the above three algorithm and other solvers. Algorithm 6.9 with only one step of iterative refinement semed to yield the lowest relative error.

Table 6.1, summarizes the number of parallel arithmetic operations, and the number of required processors for each of the three pentadiagonal Toeplitz solvers (Algorithms 6.9, 6.10, and 6.11). In addition, Table 6.1 lists the overall number of arithmetic operations required by each solver if implemented on a uniprocessor, together with the required storage. Here, the pentadiagonal test matrices are of order n, with the various entries showing only the leading term.

It is clear, however, that implementation details of each algorithm on a given parallel architecture will determine the cost of internode communications, and the cost of memory references within each multicore node. It is such cost, rather than the cost of arithmetic operations, that will determine the most scalable parallel banded Toeplitz solver on a given architecture.

6.3 Symmetric and Antisymmetric Decomposition (SAS)

The symmetric-and-antisymmetric (SAS) decomposition method is aimed at certain classes of coefficient matrices. Let A and $B \in \mathbb{R}^{n\times n}$, satisfy the relations $A = PAP$ and $B = -PBP$ where P is a signed symmetric permutation matrix, i.e. a permutation matrix in which its nonzero elements can be either 1 or -1, with $P^2 = I$. We call such a matrix P a *reflection* matrix, and the matrices A and B are referred to as reflexive and antireflexive, respectively, e.g. see [73–75]. The above matrices A and B include centrosymmetric matrices C as special cases since a centrosymmetric matrix C satisfies the relation $CE = EC$ where the permutation matrix E is given by Definition 6.3, i.e. the only nonzero elements of E are ones on the cross diagonal with $E^2 = I$. In this section, we address some fundamental properties of such special matrices A and B starting with the following basic definitions:

- *Symmetric and antisymmetric vectors.* Let P be a reflection matrix of order n. A vector $x \in \mathbb{R}^n$ is called symmetric with respect to P if $x = Px$. Likewise, we say a vector $z \in \mathbb{R}^n$ is antisymmetric if $z = -Pz$.

- *Reflexive and antireflexive matrices.* A matrix $A \in \mathbb{R}^{n \times n}$ is said to be reflexive (or antireflexive) with respect to a reflection matrix of P of order n if $A = PAP$ (or $A = -PAP$).
- A matrix $A \in \mathbb{R}^{n \times n}$ is said to possess the SAS (or anti-SAS) property with respect to a reflection matrix P if A is reflexive (or antireflexive) with respect to P.

Theorem 6.5 *Given a reflection matrix P of order n, any vector $b \in \mathbb{R}^n$ can be decomposed into two parts, u and v, such that*

$$u + v = b \tag{6.47}$$

where

$$u = Pu \quad and \quad v = -Pv. \tag{6.48}$$

The proof is readily established by taking $u = \frac{1}{2}(b + Pb)$ and $v = \frac{1}{2}(b - Pb)$.

Corollary 6.1 *Given a reflection matrix P of order n, any matrix $A \in \mathbb{R}^{n \times n}$ can be decomposed into two parts, U and V, such that*

$$U + V = A \tag{6.49}$$

where

$$U = PUP \quad and \quad V = -PVP. \tag{6.50}$$

Proof Similar to Theorem 6.5, the proof is easily established if one takes $U = \frac{1}{2}(A + PAP)$ and $V = \frac{1}{2}(A - PAP)$.

Theorem 6.6 *Given a linear system $Ax = f$, $A \in \mathbb{R}^{n \times n}$, and $f, x \in \mathbb{R}^n$, with A nonsingular and reflexive with respect to some reflection matrix P, then A^{-1} is also reflexive with respect to P and x is symmetric (or antisymmetric) with respect to P if and only if f is symmetric (or antisymmetric) with respect to P.*

Proof Since P is a reflection matrix, i.e. $P^{-1} = P$, we have

$$A^{-1} = (PAP)^{-1} = PA^{-1}P. \tag{6.51}$$

Therefore, A^{-1} is reflexive with respect to P. Now, if x is symmetric with respect to P, i.e. $x = Px$, we have

$$f = Ax = PAPPx = PAx = Pf, \tag{6.52}$$

Further, since A is reflexive with respect to P, and f is symmetric with respect to P, then from $f = Pf$, $Ax = f$, and (6.51) we obtain

$$x = A^{-1}f = PA^{-1}Pf = PA^{-1}f = Px. \tag{6.53}$$

completing the proof.

Corollary 6.2 *Given a linear system $Ax = f$, $A \in \mathbb{R}^{n \times n}$, and $f, x \in \mathbb{R}^n$, with A nonsingular and antireflexive with respect to some reflection matrix P, then A^{-1} is also antireflexive with respect to P and x is antisymmetric (or symmetric) with respect to P if and only if f is symmetric (or antisymmetric) with respect to P.*

Theorem 6.7 *Given two matrices A and B where $A, B \in \mathbb{R}^{n \times n}$, the following relations hold:*

1. *if both A and B are reflexive with respect to P, then*

$$(\alpha A)(\beta B) = P(\alpha A)(\beta B)P \quad \text{and} \quad (\alpha A + \beta B) = P(\alpha A + \beta B)P;$$

2. *if both A and B are antireflexive with respect to P, then*

$$(\alpha A)(\beta B) = P(\alpha A)(\beta B)P \quad \text{and} \quad (\alpha A + \beta B) = -P(\alpha A + \beta B)P; \quad \text{and}$$

3. *if A is reflexive and B is antireflexive, or vise versa, with respect to P, then*

$$(\alpha A)(\beta B) = -P(\alpha A)(\beta B)P$$

where α and β are scalars.

Three of the most common forms of P are given by the Kronecker products,

$$P = E_r \otimes \pm E_s, \quad P = E_r \otimes \pm I_s, \quad P = I_r \otimes \pm E_s, \tag{6.54}$$

where E_r is as defined earlier, I_r is the identity of order r, and $r \times s = n$.

6.3.1 Reflexive Matrices as Preconditioners

Reflexive matrices, or low rank perturbations of reflexive matrices, arise often in several areas in computational engineering such as structural mechanics, see for example [75]. As an illustration, we consider solving the linear system $Az = f$ where the stiffness matrix $A \in \mathbb{R}^{2n \times 2n}$ is reflexive, i.e., $A = PAP$, and is given by,

$$A = \begin{pmatrix} A_{11} & A_{12} \\ A_{21} & A_{22} \end{pmatrix}, \tag{6.55}$$

in which the reflection matrix P is of the form,

$$\begin{pmatrix} 0 & P_1^\top \\ P_1 & 0 \end{pmatrix} \tag{6.56}$$

with P_1 being some signed permutation matrix of order n, for instance. Now, consider the orthogonal matrix

$$X = \frac{1}{\sqrt{2}} \begin{pmatrix} I & -P_1^\top \\ P_1 & I \end{pmatrix}. \tag{6.57}$$

Instead of solving $Az = f$ directly, the SAS decomposition method leads to the linear system,

$$\tilde{A}\tilde{x} = \tilde{f} \tag{6.58}$$

where $\tilde{A} = X^\top A X$, $\tilde{x} = X^\top x$, and $\tilde{f} = X^\top f$. It can be easily verified that $\tilde{A}$ is of the form,

$$\tilde{A} = \begin{pmatrix} A_{11} + A_{12}P_1 & 0 \\ 0 & A_{22} - A_{21}P_1^\top \end{pmatrix}. \tag{6.59}$$

From (6.56), it is clear that the linear system has been decomposed into two independent subsystems that can be solved simultaneously. This decoupling is a direct consequence of the assumption that the matrix A is reflexive with respect to P.

In many cases, both of the submatrices $A_{11} + A_{12}P_1$ and $A_{22} - A_{21}P_1^\top$ still possess the SAS property, with respect to some other reflection matrix. For example, in three-dimensional linear isotropic or orthotropic elasticity problems that are symmetrically discretized using rectangular hexahedral elements, the decomposition can be further carried out to yield eight independent subsystems each of order $n/4$, and not possessing the SAS property; e.g. see [73, 75]. Now, the smaller decoupled subsystems in (6.58) can each be solved by either direct or preconditioned iterative methods offering a second level of parallelism.

While for a large number of orthotropic elasticity problems, the resulting stiffness matrices can be shown to possess the special SAS property, some problems yield stiffness matrices A that are low-rank perturbations of matrices that possess the SAS property. As an example consider a three-dimensional isotropic elastic long bar with asymmetry arising from the boundary conditions. Here, the bar is fixed at its left end, $\xi = 0$, and supported by two linear springs at its free end, $\xi = L$, as shown in Fig. 6.2. The spring elastic constants K_1 and K_2 are different. The dimensionless constants and material properties are given as: length L, width b, height c, Young's Modulus E, and the Poisson's Ratio v. The loading applied to

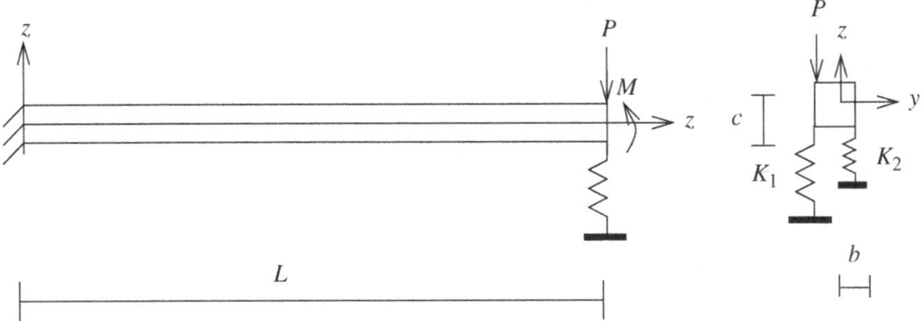

Fig. 6.2 Prismatic bar with one end fixed and the other elastically supported

this bar is a uniform simple bending moment M across the cross section at its right end, and a concentrated force P at $\left(L, \frac{-b}{2}, \frac{c}{2}\right)$. For finite element discretization, we use the basic 8-node rectangular hexahedral elements to generate an $N_1 \times N_2 \times N_3$ grid in which all discretized elements are of identical size. Here, N_d ($d = 1, 2,$ or 3) denotes the number of grid spacings in each direction d. The resulting system stiffness matrix is not globally SAS-decomposable due to the presence of the two springs of unequal strength. This problem would become SAS-decomposable, however, into four independent subproblems (the domain is symmetrical only about two axes) if the springs were absent. In other words, the stiffness matrix is SAS-decomposable into only two independent subproblems if the two springs had identical stiffness. Based on these observations, we consider splitting the stiffness matrix A into two parts, U and W, in which U is reflexive and can be decomposed into four disjoint submatrices via the SAS technique, and W is a rank-2 perturbation that contains only the stiffness contributions from the two springs. Using the reflexive matrix U as a preconditioner for the Conjugate Gradient method for solving a system of the form $Ax = f$, convergence is realized in no more than three iterations to obtain an approximate solution with a very small relative residual.

6.3.2 Eigenvalue Problems

If the matrix A, in the eigenvalue problem $Ax = \lambda x$, possesses the SAS property, it can be shown (via similarity transformations) that the proposed decomposition approach can be used for solving the problem much more efficiently. For instance, if P is of the form

$$P = E_2 \otimes E_{n/2} \tag{6.60}$$

with $A = PAP$ partitioned as in (6.55), then using the orthogonal matrix,

$$X = \frac{1}{\sqrt{2}} \begin{pmatrix} I & -E_{n/2} \\ E_{n/2} & I \end{pmatrix}. \tag{6.61}$$

we see that $X^{\top}AX$ is a block-diagonal matrix consisting of the two blocks: $A_1 = A_{11} + A_{12}E_{n/2}$, and $A_2 = A_{22} - A_{21}E_{n/2}$.

Thus first, all the eigenvalues of the original matrix A can be obtained from those of the decomposed submatrices A_1 and A_2, which are of lower order than A. Further, the amount of work for computing those eigenpairs is greatly reduced even for sequential computations. Second, the extraction of the eigenvalues of the different submatrices can be performed independently, which implies that high parallelism can be achieved. Third, the eigenvalues of each submatrix are in general better separated, which implies faster convergence for schemes such as the QR algorithm (e.g., see [76]).

To see how much effort can be saved, we consider the QR iterations for a real-valued full matrix (or order N). In using the QR iterations for obtaining the eigenpairs of a matrix, we first reduce this matrix to the upper Hessenberg form (or a tridiagonal matrix for symmetric problems). On a uniprocessor, the reduction step takes about cN^3 flops [14] for some constant c. If the matrix A satisfying $PAP = A$ can be decomposed into four submatrices each of order $N/4$; then the amount of floating point operations required in the reduction step is reduced to $cN^3/16$. In addition, because of the fully independent nature of the four subproblems, we can further reduce the computing time on parallel architectures.

Depending on the form of the signed reflection matrix P, several similarity transformations can be derived for this special class of matrices $A = PAP$. Next, we present another computationally useful similarity transformation.

Theorem 6.8 ([75]) *Let $A \in \mathbb{R}^{n \times n}$ be partitioned as $(A_{i,j})$, $i,j = 1,2,$ and 3 with A_{11} and A_{33} of order r and A_{22} of order s, where $2r + s = n$. If $A = PAP$ where P is of the form*

$$P = \begin{pmatrix} 0 & 0 & P_1^\top \\ 0 & I_s & 0 \\ P_1 & 0 & 0 \end{pmatrix}$$

in which P_1 is some signed permutation matrix of order r, then there exists an orthogonal matrix,

$$X = \frac{1}{\sqrt{2}} \begin{pmatrix} I & 0 & -P_1^\top \\ 0 & \sqrt{2}I_s & 0 \\ P_1 & 0 & I \end{pmatrix}. \tag{6.62}$$

that yields $X^\top AX$ as a block-diagonal matrix of the form,

$$\begin{pmatrix} A_{11} + A_{12}P_1 & \sqrt{2}A_{12} & 0 \\ \sqrt{2}A_{21} & A_{22} & 0 \\ 0 & 0 & A_{33} - A_{31}P_1^\top \end{pmatrix}. \tag{6.63}$$

It should be noted that if A is symmetric, then both diagonal blocks in (6.63) are also symmetric. The same argument holds for the two diagonal blocks in (6.59).

Note that the application of the above decomposition method can be extended to the generalized eigenvalue problem $Ax = \lambda Bx$ if B also satisfies the SAS property, namely $B = PBP$.

6.4 Rapid Elliptic Solvers

In this section we consider the parallelism in the solution of linear systems with matrices that result from the discretization of certain elliptic partial differential equations. As it turns out, there are times when the equations, boundary conditions and discretization are such that the structure of the coefficient matrix for the linear system allow one to develop very fast direct solution methods, collectively known as Rapid Elliptic Solvers (RES for short) or Fast Poisson Solvers. The terms are due to the fact that their computational complexity on a uniprocessor is only $O(mn \log mn)$ or less for systems of mn unknowns compared to the $O((mn)^3)$ complexity of Gaussian elimination for dense systems.

Here we are concerned with direct methods, in the sense that in the absence of roundoff the solvers return an exact solution. When well implemented, RES are faster than other direct and iterative methods [77, 78]. The downside is their limited applicability: RES can only be used directly for special elliptic PDEs (meaning the equation, domain of definition) under suitable discretization; moreover, their performance in general depends on the boundary conditions and problem size. On the other hand, there are many cases when RES cannot be used directly but can be helpful as preconditioners. Interest in the design and implementation of parallel algorithms for RES started in the early 1970s; see Sect. 6.4.9 for some historical notes. Theoretically, parallel RES can solve the linear systems under consideration in $O(\log mn)$ parallel operations on $O(mn)$ processors instead of the fastest but impractical algorithm [79] for general linear systems that requires $O(\log^2 mn)$ parallel operations on $O((mn)^4)$ processors. In the sequel we describe and evaluate the properties of some interesting algorithms from this class.

6.4.1 Preliminaries

The focus of this chapter is on parallel RES for the following model problem:

$$
\begin{pmatrix}
T & -I & & & \\
-I & T & -I & & \\
 & \ddots & \ddots & \ddots & \\
 & & -I & T & -I \\
 & & & -I & T
\end{pmatrix}
\begin{pmatrix}
u_1 \\
u_2 \\
\vdots \\
\vdots \\
u_n
\end{pmatrix}
=
\begin{pmatrix}
f_1 \\
f_2 \\
\vdots \\
\vdots \\
f_n
\end{pmatrix},
\tag{6.64}
$$

where $T = [-1, 4, -1]_m$ and the unknowns and right-hand side u, f are conformally partitioned into subvectors $u_i = (v_{i,1}, \ldots, v_{i,m})^\top \in \mathbb{R}^m$ and $f_i = (\phi_{i,1}, \ldots, \phi_{i,m})^\top$. See Sect. 9.1 of Chap. 9 for some details regarding the derivation of this system but also books such as [80, 81]. We sometimes refer to the linear system as *discrete Poisson system* and to the block-tridiagonal matrix A in (6.64) as *Poisson matrix*. Note that this is a 2-level Toeplitz matrix, that is a block-Toeplitz matrix with Toeplitz blocks [82]. The 2-level Toeplitz structure is a consequence of constant coefficients in the differential equation and the Dirichlet boundary conditions. It is worth noting that the discrete Poisson system (6.64) can be reordered and rewritten as

$$[-I_n, \tilde{T}, -I_n]\tilde{u} = \tilde{f}.$$

This time the Poisson matrix has m blocks of order n each. The two systems are equivalent; for that reason, in the algorithms that follow, the user can select the formulation that minimizes the cost. For example, if the cost is modeled by $\mathbf{T}_p = \gamma_1 \log m \log n + \gamma_2 \log^2 n$ for positive γ_1, γ_2, then it is preferable to choose $n \le m$.

The aforementioned problem is a special version of the block-tridiagonal system

$$Au = f, \quad \text{where} \quad A = [W, T, W]_n, \tag{6.65}$$

where $T, W \in \mathbb{R}^{m \times m}$ are symmetric and commute under multiplication so that they are simultaneously diagonalizable by the same orthogonal similarity transformation. Many of the parallel methods we discuss here also apply with minor modifications to the solution of system (6.65).

It is also of interest to note that A is of low *block displacement rank*. In particular if we use our established notation for matrices I_m and J_n then

$$A - (J_n^\top \otimes I_m)^\top A (J_n^\top \otimes I_m) = XY^\top + YX^\top, \tag{6.66}$$

where

$$X = \left(\frac{1}{2}T, W, 0, \ldots, 0\right)^\top, Y = (I_m, 0, \ldots, \ldots, 0)^\top.$$

Therefore the rank of the result in (6.66) is at most $2m$. This can also be a starting point for constructing iterative solvers using tools from the theory of low displacement rank matrices; see for example [57, 82].

6.4.2 Mathematical and Algorithmic Infrastructure

We define the Chebyshev polynomials that are useful here and in later chapters. See also refs. [83–86].

Definition 6.6 The degree-k Chebyshev polynomial of the 1st kind is defined as:

$$\mathscr{T}_k(\xi) = \begin{cases} \cos(k \arccos \xi) & \text{when } |\xi| \le 1 \\ \cosh(k \operatorname{arccosh} \xi) & \text{when } |\xi| > 1. \end{cases}$$

The following recurrence holds:

$$\mathscr{T}_{k+1}(\xi) = 2\xi \, \mathscr{T}_k(\xi) - \mathscr{T}_{k-1}(\xi)$$
$$\text{where} \quad \mathscr{T}_0(\xi) = 1, \mathscr{T}_1(\xi) = \xi.$$

The degree-k modified Chebyshev polynomial of the 2nd kind is defined as:

$$\hat{\mathscr{U}}_k(\xi) = \begin{cases} \frac{\sin((k+1)\theta)}{\sin(\theta)} \text{ where } \cos(\theta) = \frac{\xi}{2} & \text{when } 0 \leq \xi < 2 \\ k+1 & \text{when } \xi = 2 \\ \frac{\sinh((k+1)\psi)}{\sinh(\psi)} \text{ where } \cosh(\psi) = \frac{\xi}{2} & \text{when } \xi > 2. \end{cases}$$

The following recurrence holds:

$$\hat{\mathscr{U}}_{k+1}(\xi) = \xi \hat{\mathscr{U}}_k(\xi) - \hat{\mathscr{U}}_{k-1}(\xi)$$
$$\text{where} \quad \hat{\mathscr{U}}_0(\xi) = 1, \hat{\mathscr{U}}_1(\xi) = \xi.$$

Because of the special structure of matrices A and T, several important quantities, like their inverse, eigenvalues and eigenvectors, can be expressed analytically; cf. [83, 87–89].

Proposition 6.4 *Let $T = [-1, \alpha, -1]_n$ and $\alpha \geq 2$. For $j \geq i$*

$$(T^{-1})_{i,j} = \begin{cases} \frac{\sinh(i\xi)\sinh((n-j+1)\xi)}{\sinh(\xi)\sinh((n+1)\xi)} \text{ where } \cosh(\xi) = \frac{\alpha}{2} & \text{when } \alpha > 2 \\ i\left(\frac{n-j+1}{n+1}\right) & \text{when } \alpha = 2. \end{cases}$$

The eigenvalues and eigenvectors of T are given by

$$\lambda_j = \alpha - 2\cos\left(\frac{j\pi}{m+1}\right), \tag{6.67}$$

$$q_j = \sqrt{\frac{2}{m+1}}\left(\sin\left(\frac{j\pi}{m+1}\right), \ldots, \sin\left(\frac{jm\pi}{m+1}\right)\right)^{\top}, \quad j = 1, \ldots, m.$$

A similar formula, involving Chebyshev polynomials can be written for the inverse of the Poisson matrix.

Proposition 6.5 *For any nonsingular $T \in \mathbb{R}^{m \times m}$, the matrix $A = [-I, T, -I]_n$ is nonsingular if and only if $\hat{\mathscr{U}}_n(T)$ is nonsingular. Then, A^{-1} can be written as a block matrix that has, as block (i, j) the order m submatrix*

$$(A^{-1})_{i,j} = \begin{cases} \hat{\mathscr{U}}_n^{-1}(T)\hat{\mathscr{U}}_{i-1}(T)\hat{\mathscr{U}}_{n-j}(T), \ j \geq i, \\ \hat{\mathscr{U}}_n^{-1}(T)\hat{\mathscr{U}}_{j-1}(T)\hat{\mathscr{U}}_{n-i}(T), \ i \geq j. \end{cases} \tag{6.68}$$

The eigenvalues and eigenvectors of the block-tridiagonal matrix A in Proposition 6.5 can be obtained using the properties of Kronecker products and sums; cf. [14, 90].

Multiplication of the matrix of eigenvectors $Q = (q_1, \ldots, q_m)$ of T with a vector $y = (\psi_1, \ldots, \psi_m)^{\top}$ amounts to computing the elements

$$\sqrt{\frac{2}{m+1}} \sum_{j=1}^{m} \psi_j \sin\left(\frac{ij\pi}{m+1}\right), \quad i = 1, \ldots, m.$$

Therefore this multiplication is essentially equivalent to the *discrete sine transform* (DST) of y. The DST and its inverse can be computed in $O(m \log m)$ operations using FFT-type methods. For this, in counting complexities in the remainder of this section we do not distinguish between the forward and inverse transforms.

The following operations are frequently needed in the context of RES:

($1.i$, $1.ii$) Given $Y = (y_1, \ldots, y_s) \in \mathbb{R}^{m \times s}$, solve $TX = Y$, that is solve a linear system with banded coefficient matrix for (i) $s = 1$ or (ii) $s > 1$ right-hand sides;
($1.iii$) Given $y \in \mathbb{R}^m$ and scalars $\mu_1, \ldots, \mu_d$ solve the d linear systems $(T - \mu_j I)x_j = y$;
($1.iv$) Given $Y \in \mathbb{R}^{m \times d}$ and scalars μ_j as before solve the d linear systems $(T - \mu_j I)x_j = y_j$;
($1.v$) Given Y and scalars μ_j as before, solve the sd linear systems $(T - \mu_j I)x_j^{(k)} = y_k$, $j = 1, \ldots, d$ and $k = 1, \ldots, s$;
(2) Given $Y \in \mathbb{R}^{m \times n}$ compute the discrete Fourier-type transform (DST, DCT, DFT) of all columns or of all rows.

For an extensive discussion of the fast Fourier and other discrete transforms used in RES see the monograph [19] and the survey [72] and references therein for efficient FFT algorithms.

6.4.3 *Matrix Decomposition*

Matrix decomposition (MD) refers to a large class of methods for solving the model problem we introduced earlier as well as more general problems, including (6.65); cf. [91]. As was stated in [92] "It seldom happens that the application of L processors would yield an L-fold increase in efficiency relative to a single processor, but that is the case with the MD algorithm." The first detailed study of parallel MD for the Poisson equation was presented in [93]. The Poisson matrix can be written as a *Kronecker sum* of Toeplitz tridiagonal matrices. Specifically,

$$A = I_n \otimes \tilde{T}_m + \tilde{T}_n \otimes I_m, \tag{6.69}$$

where $\tilde{T}_k = [-1, 2, -1]_k$ for $k = m, n$. To describe MD, we make use of this representation. We also use the VEC operator: Acting on a matrix $Y = (y_1, \ldots, y_n)$, it returns the vector that is formed by stacking the columns of X, that is

$$\text{VEC}(y_1, \ldots, y_n) = \begin{pmatrix} y_1 \\ \vdots \\ y_n \end{pmatrix}.$$

The UNVEC_n operation does the reverse: For a vector of mn elements, it selects its n contiguous subvectors of length m and returns the matrix with these as columns. Finally, the permutation matrix $\Pi_{m,n} \in \mathbb{R}^{mn \times mn}$ is defined as the unique matrix, sometimes called vec-permutation, such that $\text{VEC}(A) = \Pi_{m,n} \text{VEC}(A^\top)$, where $A \in \mathbb{R}^{m \times n}$.

MD algorithms consist of 3 major stages. The first amounts to transforming A and f, then solving a set of independent systems, followed by back transforming to compute the final solution. These 3 stages are characteristic of MD algorithms; cf. [91]. Moreover, directly or after transformations, they consist of several independent subproblems that enable straightforward implementation on parallel architectures.

Denote by Q_x the matrix of eigenvectors of $\tilde{T}_m$. In the first stage, both sides of the discrete Poisson system are multiplied by the block-diagonal orthogonal matrix $I_n \otimes Q_x^\top$. We now write

$$(I_n \otimes Q_x^\top)(I_n \otimes \tilde{T}_m + \tilde{T}_n \otimes I_m)(I_n \otimes Q_x)(I_n \otimes Q_x^\top)u = (I_n \otimes Q_x^\top)f$$

$$(I_n \otimes \tilde{\Lambda}_m + \tilde{T}_n \otimes I_m)(I_n \otimes Q_x^\top)u = (I_n \otimes Q_x^\top)f. \quad (6.70)$$

The computation of $(I_n \otimes Q_x^\top)f$ amounts to n independent multiplications with $Q_x^\top$, one for each subvector of f.

Using the vec-permutation matrix $\Pi_{m,n}$, (6.70) is transformed to

$$\underbrace{(\tilde{\Lambda}_m \otimes I_n + I_m \otimes \tilde{T}_n)}_{B}(\Pi_{m,n}(I_n \otimes Q_x^\top)u) = \Pi_{m,n}(I_n \otimes Q_x^\top)f. \quad (6.71)$$

Matrix B is block-diagonal with diagonal blocks of the form $\tilde{T}_n + \lambda_i^{(m)}I$ where $\lambda_1^{(m)}, \ldots, \lambda_m^{(m)}$ are the eigenvalues of $\tilde{T}_m$. Recall also from Proposition 6.4 that the eigenvalues are computable from closed formulas. Therefore, the transformed system is equivalent to m independent subsystems of order n, each of which has the same structure as $\tilde{T}_n$. These are solved in the second stage of MD. The third stage of MD consists of the multiplication of the result of the previous stage with $I_n \otimes Q_x$. From (6.71), it follows that

$$u = (I_n \otimes Q_x)\Pi_{m,n}^\top B^{-1}\Pi_{m,n}(I_n \otimes Q_x^\top)f. \quad (6.72)$$

where $\Pi_{m,n}$ is as before.

The steps that we described can be applied to more general systems with coefficient matrices of the form $T_1 \otimes W_2 + W_1 \otimes T_2$ where the order n matrices T_1, W_1 commute and W_2, T_2 are order m; cf. [91]. Observe also that if we only need few of the u_is, it is sufficient to first compute $\Pi_{m,n}^\top B^{-1} \Pi_{m,n} (I_n \otimes Q_x^\top) f$ and then apply Q_x to the appropriate subvectors.

Each of the 3 stages of the MD algorithm consists of independent operations with matrices of order m and n rather than mn. In the case of the discrete Poisson system, the multiplications with Q_x can be implemented with fast, Fourier-type transformations implementing the DST. We list the resulting method as Algorithm 6.12 and refer to it as MD-FOURIER. Its cost on a uniprocessor is $\mathbf{T}_1 = O(mn \log m)$.

Consider the cost of each step on $p = mn$ processors. Stage (I) consists of n independent applications of a length m DST at cost $O(\log m)$. Stage (II) amounts to the solution of m tridiagonal systems of order n. Using CR or PARACR (see Sect. 5.5 of Chap. 5), these are solved in $O(\log n)$ steps. Stage (III) has a cost identical to stage (I). The total cost of MD-FOURIER on mn processors is $\mathbf{T}_p = O(\log m + \log n)$. If the number of processors is $p < mn$, the cost for stages (I, III) becomes $O(\frac{n}{p} m \log m)$. In stage (II), if the tridiagonal systems are distributed among the processors, and a sequential algorithm is applied to solve each system, then the cost is $O(\frac{m}{p} n)$. Asymptotically, the dominant cost is determined by the first and last stages with a total cost of $\mathbf{T}_p = O(\frac{mn}{p} \log n)$.

Algorithm 6.12 MD-FOURIER: matrix decomposition method for the discrete Poisson system.

Input: Block tridiagonal matrix $A = [-I_m, T, I_m]_n$, where $T = [-1, 4, -1]_n$ and the right-hand side $f = (f_1; \dots; f_n)$

    //Stage I: apply fast DST on each subvector $f_1, \dots, f_n$

1: **doall** $j = 1 : n$

2:    $\hat{f}_j = Q_x^\top f_j$

3:    set $\hat{F} = (\hat{f}_1, \dots, \hat{f}_n)$

4: **end**

    //Stage II: implemented with suitable solver (Toeplitz, tridiagonal, multiple shifts, multiple right-hand sides)

5: **doall** $i = 1 : m$

6:    compute $(\tilde{T}_n + \lambda_i^{(m)} I)^{-1} \hat{F}_{i,:}^\top$ and store the result in ith row of a temporary matrix $\hat{U}$; $\lambda_1^{(m)}, \dots, \lambda_m^{(m)}$ are the eigenvalues of $\tilde{T}_m$

7: **end**

    //Stage III: apply fast DST on each column of $\hat{U}$

8: **doall** $j = 1 : n$

9:    $u_j = Q_x \hat{U}_{:,j}$

10: **end**

6.4.4 Complete Fourier Transform

When both $\tilde{T}_m$ and $\tilde{T}_n$ in (6.69) are diagonalizable with Fourier-type transforms, it becomes possible to solve Eq. (6.64) using another approach, called the complete

Fourier transform method (CFT) that we list as Algorithm 6.13 [94]. Here, instead of (6.72) we write the solution of (6.64) as

$$\mathscr{U} = (Q_y \otimes Q_x)(I_n \otimes \tilde{\Lambda}_m + \tilde{\Lambda}_n \otimes I_m)^{-1}(Q_y^\top \otimes Q_x^\top)f,$$

where matrix $(I_n \otimes \tilde{\Lambda}_m + \tilde{\Lambda}_n \otimes I_m)$ is diagonal. Multiplication by $Q_y^\top \otimes Q_x^\top$ amounts to performing n independent DSTs of length m and m independent DSTs of length n; similarly for $Q_y \otimes Q_x$. The middle stage is an element-by-element division by the diagonal of $I_n \otimes \tilde{\Lambda}_m + \tilde{\Lambda}_n \otimes I_m$ that contains the eigenvalues of A.

CFT is rich in operations that can be performed in parallel. As in MD, we distinguish three stages, the first and last of which are a combination of two steps, one consisting of n independent DSTs of length m each, the other consisting of m independent DSTs. The middle step consists of mn independent divisions. The total cost on mn processors is $\mathbf{T}_p = O(\log m + \log n)$.

Algorithm 6.13 CFT: complete Fourier transform method for the discrete Poisson system.

Input: Block tridiagonal matrix $A = [-I_m, T, I_m]_n$, where $T = [-1, 4, -1]_n$ and the right-hand side f
      //The right-hand side arranged as the $m \times n$ matrix $F = (f_1, \ldots, f_n)$
Output: Solution $u = (u_1; \ldots; u_n)$
      //Stage Ia: DST on columns of F
1: **doall** $j = 1 : n$
2:    $\hat{f}_j = Q_x^\top f_j$
3:    set $\hat{F} = (\hat{f}_1, \ldots, \hat{f}_n)$
4: **end**   //Stage Ib: DST on rows of $\hat{F}$
5: **doall** $i = 1 : m$
6:    $\tilde{F}_{i,:} = Q_y^\top \hat{F}_{i,:}$
7: **end**   //Stage II: elementwise division of $\tilde{F}$ by eigenvalues of A
8: **doall** $i = 1 : m$
9:    **doall** $j = 1 : n$
10:      $\tilde{F}_{i,j} = \hat{F}_{i,j}/(\lambda_i^{(m)} + \lambda_j^{(n)})$
11:    **end**
12: **end**   //Stage IIIa: DST on rows of $\tilde{F}$
13: **doall** $i = 1 : m$
14:    $\hat{F}_{i,:} = Q_y \tilde{F}_{i,:}$
15: **end**   //Stage IIIb: DST on columns of $\hat{F}$
16: **doall** $j = 1 : n$
17:    $u_j = Q_x \hat{F}_{:,j}$
18: **end**

6.4.5 Block Cyclic Reduction

BCR for the discrete Poisson system (6.65) is a method that generalizes the CR
algorithm (cf. 5.5) for point tridiagonal systems (cf. Chap. 5) while taking advantage
of its 2-level Toeplitz structure; cf. [95]. BCR is more general than Fourier-MD in the
sense that it does not require knowledge of the eigenstructure of T nor does it deploy
Fourier-type transforms. We outline it next for the case that the number of blocks
is $n = 2^k - 1$; any other value can be accommodated following the modifications
proposed in [96].

For steps $r = 1, \ldots, k - 1$, adjacent blocks of equations are combined in groups
of 3 to eliminate 2 blocks of unknowns; in the first step, for instance, unknowns from
even numbered blocks are eliminated and a reduced system with block-tridiagonal
coefficient matrix $A^{(1)} = [-I, T^2 - 2I, -I]_{2^{k-1}-1}$, containing approximately only
half of the blocks remains. The right-hand side is transformed accordingly. Setting
$T^{(0)} = T$, $f^{(0)} = f$, and $T^{(r)} = (T^{(r-1)})^2 - 2I$, the reduced system at the rth step
is

$$[-I, T^{(r)}, -I]_{2^{k-r}-1} u^{(r)} = f^{(r)},$$

where $f^{(r)} = \mathrm{vec}[f^{(r)}_{2^r.1}, \ldots, f^{(r)}_{2^r.(2^{k-r}-1)}]$ and

$$f^{(r)}_{j2^r} = f^{(r-1)}_{j2^r - 2^{r-1}} + f^{(r-1)}_{j2^r + 2^{r-1}} + T^{(r-1)} f^{(r-1)}_{j2^r}. \tag{6.73}$$

The key observation here is that the matrix $T^{(r)}$ can be written as

$$T^{(r)} = 2\mathscr{T}_{2^r}\left(\frac{T}{2}\right). \tag{6.74}$$

From the closed form expressions of the eigenvalues of T and the roots of the Chebyshev polynomials of the 1st kind $\mathscr{T}_{2^r}$, and relation (6.74), it is straightforward to show
that the roots of the matrix polynomial $T^{(r)}$ are

$$\rho_i^{(r)} = 2\cos\left(\frac{(2j-1)}{2^{r+1}}\pi\right). \tag{6.75}$$

Therefore, the polynomial in product form is

$$T^{(r)} = (T - \rho_{2^r}^{(r)} I) \cdots (T - \rho_1^{(r)} I). \tag{6.76}$$

The roots (cf. (6.75)) are distinct therefore the inverse can also be expressed in terms
of the partial fraction representation of the rational function $1/2\mathscr{T}_{2^r}(\frac{\xi}{2})$:

$$(T^{(r)})^{-1} = \sum_{i=1}^{2^r} \gamma_i^{(r)} (T - \rho_i^{(r)} I)^{-1}. \tag{6.77}$$

From the analytic expression for $T^{(r)}$ in (6.74) and standard formulas, the partial fraction coefficients are equal to

$$\gamma_i^{(r)} = (-1)^{i+1} \frac{1}{2^r} \sin\left(\frac{(2i-1)\pi}{2^{r+1}}\right), \quad i = 1, \ldots, 2^r.$$

The partial fraction approach for solving linear systems with rational matrix coefficients is discussed in detail in Sect. 12.1 of Chap. 12. Here we list as Algorithm 6.14 (SIMPLESOLVE_PF), one version that can be applied to solving systems of the form $\left(\prod_{j=1}^{d}(T - \rho_j I)\right) x = b$ for mutually distinct ρ_j's. This will be applied to solve any systems with coefficient matrix such as (6.76).

Algorithm 6.14 SIMPLESOLVE_PF: solving $\left(\prod_{j=1}^{d}(T - \rho_j I)\right) x = b$ for mutually distinct values ρ_j from partial fraction expansions.

Input: $T \in \mathbb{R}^{m \times m}, b \in \mathbb{R}^m$ and distinct values $\{\rho_1, \ldots, \rho_d\}$ none of them equal to an eigenvalue of T.

Output: Solution $x = \left(\prod_{j=1}^{d}(T - \rho_j I)\right)^{-1} b$.

1: **doall** $j = 1 : d$
2:    compute coefficient $\gamma_j = \frac{1}{p'(\tau_j)}$, where $p(\zeta) = \prod_{j=1}^{d}(\zeta - \rho_j)$
3:    solve $(T - \rho_j I)x_j = b$
4: **end**
5: set $c = (\gamma_1, \ldots, \gamma_d)^\top$, $X = (x_1, \ldots, x_d)$
6: compute and return $x = Xc$

Next implementations of block cyclic reduction are presented that deploy the product (6.76) and partial fraction representations (6.77) for operations with $T^{(r)}$.

After $r < k - 1$ block cyclic reduction steps the system has the form:

$$[-I, T^{(r)}, -I]u^{(r)} = f^{(r)}, \quad \text{where } f^{(r)} = \begin{pmatrix} f_{2^r}^{(r)} \\ f_{2 \cdot 2^r}^{(r)} \\ \vdots \\ f_{(2^{k-r}-1) \cdot 2^r}^{(r)} \end{pmatrix}. \tag{6.78}$$

Like the Poisson matrix, $T^{(r)}$ is block-tridiagonal, it has the same eigenvectors as T and its eigenvalues are available analytically from the polynomial form of $T^{(r)}$ and the eigenvalues of T. If we stop after step $r = \hat{r}$ and compute $u^{(\hat{r})}$, then a back substitution stage consists of solving consecutive block-diagonal systems to compute the missing subvectors, as shown next for $r = \hat{r}, \hat{r} - 1, \ldots, 1$.

$$\text{diag}[T^{(r-1)}] \begin{pmatrix} u_{1\cdot 2^{r-1}} \\ u_{3\cdot 2^{r-1}} \\ \vdots \\ u_{(2^{\hat{r}-r+1}-1)\cdot 2^{r-1}} \end{pmatrix} = \begin{pmatrix} f^{(r-1)}_{1\cdot 2^{r-1}} + u_{1\cdot 2^{r}} \\ f^{(r-1)}_{3\cdot 2^{r-1}} + u_{3\cdot 2^{r}-2^{r-1}} + u_{3\cdot 2^{r}+2^{r-1}} \\ \vdots \\ f^{(r-1)}_{(2^{\hat{r}-r+1}-1)\cdot 2^{r-1}} + u_{(2^{\hat{r}-r+1}-1)\cdot 2^{r-1}-2^{r-1}} \end{pmatrix}.$$

If $\hat{r} = k - 1$ reduction steps are applied, a single tridiagonal system remains,

$$T^{(k-1)} u_{2^{k-1}} = f^{(k-1)}_{2^{k-1}}, \tag{6.79}$$

so back substitution can be used to recover all the subvectors.

As described, the algorithm is called Cyclic Odd-Even Reduction and Factorization (CORF) and listed as Algorithm 6.15 [95]. Consider its cost on a uniprocessor. Each reduction step $r = 1, \dots, k-1$, requires $2^{k-r} - 1$ independent matrix-vector multiplications, each with $T^{(r-1)}$ (in product form); cf. (6.73). To compute the "middle subvector" $u_{2^{k-1}}$ from (6.79), requires solving 2^{k} tridiagonal systems. Finally, each step $r = k, k-1, \dots, 1$, of back substitution, consists of solving 2^{k-r} independent tridiagonal systems with coefficient matrix $T^{(r-1)}$. The cost of the first and last stages is $O(nm \log n)$ while the cost of solving (6.79) is $O(mn)$, for an overall cost of $\mathbf{T}_1 = O(mn \log n)$. Note that the algorithm makes no use of fast transforms to achieve its low complexity. Instead, this depends on the factored form of the $T^{(r)}$ term. In both the reduction and back substitution stages, CORF involves the independent application in multiplications and system solves respectively, of matrices $T^{(r)}$; see lines 3 and 9 respectively. CORF, therefore, provides two levels of parallelism, one from the individual matrix operations, and another from their independent application. Parallelism at this latter level, however, varies considerably. In the reduction, the

---

Algorithm 6.15 CORF: Block cyclic reduction for the discrete Poisson system

---

Input: Block tridiagonal matrix $A = [-I_m, T, I_m]_n$, where $T = [-1, 4, -1]_n$ and the right-hand side f. It is assumed that $n = 2^k - 1$.

Output: Solution $u = (u_1; \dots; u_n)$

    //Stage I: Reduction

1: **do** $r = 1 : k - 1$

2:    **doall** $j = 1 : 2^{k-r} - 1$

3:       $f^{(r)}_{j2^r} = f^{(r-1)}_{j2^r-2^{r-1}} + f^{(r-1)}_{j2^r+2^{r-1}} + T^{(r-1)} f^{(r-1)}_{j2^r}$

       //multiply $T^{(r-1)} f^{(r-1)}_{2^r j}$ exploiting the product form (6.76)

4:    **end**

5: **end**

    //Stage II: Solution by back substitution

6: Solve $T^{(k-1)} u_{2^{k-1}} = f^{(k-1)}_{2^{k-1}}$

7: **do** $r = k - 1 : -1 : 1$

8:    **doall** $j = 1 : 2^{k-r}$

9:       solve $T^{(r-1)} u_{(2j-1)\cdot 2^{r-1}} = f^{(r-1)}_{(2j-1)\cdot 2^{r-1}} + u_{(2j-1)\cdot 2^{r-1}-2^{r-1}} + u_{(2j-1)\cdot 2^{r-1}+2^{r-1}}$

10:   **end**

11: **end**

---

number of independent matrix operations is halved at every step, ranging from $2^{k-1} - 1$ down to 1. In back substitution, the number is doubled at every step, from 1 up to 2^{k-1}. So in the first step of back substitution only one system with a coefficient matrix (that is a product of 2^{k-1} matrices) is solved. Therefore, in the later reduction and earlier back substitution steps, the opportunity for independent manipulations using the product form representation lowers dramatically and eventually disappears. In the back substitution stage (II), this problem can be resolved by replacing the product form (6.76) when solving with $T^{(r)}$ with the partial fraction representation (6.77) of its inverse. In particular, the solution is computed via Algorithm 6.14.

On mn processors, the first step of back substitution (line 6) can be implemented in $O(\log m)$ operations to solve each system in parallel and then $O(\log n)$ operations to sum the solutions by computing their linear combination using the appropriate partial fraction coefficients as weight vector. The next steps of back substitution, consist of the loop in lines 7–11. The first loop iteration involves the solution of two systems with coefficient matrix $T^{(k-2)}$ each. Using the partial fraction representation for $T^{(k-2)}$, the solution of each of system can be written as the sum of 2^{k-2} vectors that are solutions of tridiagonal systems. Using no more than mn processors, this step can be accomplished in $O(\log m)$ parallel steps to solve all the systems and $\log(n/2)$ for the summation. Continuing in this manner for all steps of back substitution, its cost is $O(\log n(\log n + \log m))$ parallel operations. On the other hand, during the reduction stage, the multiplication with $T^{(r)}$, involves 2^r consecutive multiplications with tridiagonal matrices of the form $T - \rho I$ and these multiplications cannot be readily implemented independently. In the last reduction step, for example, one must compute $T^{(k-2)}$ times a single vector. This takes $O(n)$ parallel operations, irrespective of the number of processors, since they are performed consecutively. Therefore, when $m \approx n$, the parallel cost lower bound of CORF is $O(n)$, causing a performance bottleneck.

An even more serious problem that hampers CORF is that the computation of (6.73) leads to catastrophic errors. To see that, first observe that the eigenvalues of $T^{(r)}$ are

$$2\mathcal{T}_{2^r}\left(\frac{\lambda_j}{2}\right), \, j = 1, \ldots m, \text{ where } \lambda_j = 4 - 2\cos\left(\frac{\pi j}{m+1}\right).$$

Therefore all eigenvalues of T lie in the interval $(2, 6)$ so the largest eigenvalue of $T^{(r)}$ will be of the order of $2\mathcal{T}_{2^r}(3 - \delta)$, for some small δ. This is very large even for moderate values of r since $\mathcal{T}_{2^r}$ is known to grow very large for arguments of magnitude greater than 1. Therefore, computing (6.73) will involve the combination of elements of greatly varying magnitude causing loss of information. Fortunately, both the computational bottleneck and the stability problem in the reduction phase can be overcome. In fact, as we show in the sequel, resolving the stability problem also enables enhanced parallelism.

BCR **Stabilization and Parallelism**

The loss of information in BCR is reduced and its numerical properties considerably improved if one applies a modification proposed by Buneman; we refer to [95] for a detailed dicussion of the scheme and a proof of its numerical properties. The idea is: instead of computing $f^{(r)}$ directly, to express it indirectly in terms of 2 vectors, $p^{(r)}, q^{(r)}$, updated so that at every step they satisfy

$$f_{j2^r}^{(r)} = f_{j2^r-2^{r-1}}^{(r-1)} + f_{j2^r+2^{r-1}}^{(r-1)} + T^{(r-1)} f_{j2^r}^{(r-1)}$$
$$= T^{(r-1)} p_{j2^r}^{(r-1)} + q_{j2^r}^{(r-1)}$$

In one version of this scheme, the recurrence (6.73) for f_{j2^r} is replaced by the recurrences

$$p_j^{(r)} = p_j^{(r-1)} - (T^{(r-1)})^{-1}(p_{j-2^{r-1}}^{(r-1)} + p_{j+2^{r-1}}^{(r-1)} - q_j^{(r-1)})$$
$$q_j^{(r)} = (q_{j-2^{r-1}}^{(r-1)} + q_{j+2^{r-1}}^{(r-1)} - 2p_j^{(r)}).$$

The steps of stabilized BCR are listed as Algorithm 6.16 (BCR). Note that because the sequential costs of multiplication and solution with tridiagonal matrices are both linear, the number of arithmetic operations in Buneman stabilized BCR is the same as that of CORF.

Algorithm 6.16 BCR: Block cyclic reduction with Buneman stabilization for the discrete Poisson system

---

Input: Block tridiagonal matrix $A = [-I_m, T, I_m]_n$, where $T = [-1, 4, -1]_n$ and the right-hand
    side f. It is assumed $n = 2^k - 1$.
Output: Solution $u = (u_1; \ldots; u_n)$
    //Initialization
1: $p_j^{(0)} = 0_{m,1}$ and $q_j^{(0)} = f_j$ $(j = 1 : n)$
    //Stage I: Reduction. Vectors with subscript 0 or 2^k are taken to be 0
2: **do** $r = 1 : k - 1$
3:     **doall** $j = 1 : 2^{k-r} - 1$
4:       $p_{j2^r}^{(r)} = p_{j2^r}^{(r-1)} - (T^{(r-1)})^{-1}(p_{j2^r-2^{r-1}}^{(r-1)} + p_{j2^r+2^{r-1}}^{(r-1)} - q_{j2^r}^{(r-1)})$
5:       $q_{j2^r}^{(r)} = (q_{j2^r-2^{r-1}}^{(r-1)} + q_{j2^r+2^{r-1}}^{(r-1)} - 2p_{j2^r}^{(r)})$
6:     **end**
7: **end**
    //Stage II: Solution by back substitution. It is assumed that $u_0 = u_{2^k} = 0$
8: Solve $T^{(k)} u_{2^{k-1}} = q_{2^{k-1}}^{(k)}$
9: **do** $r = k - 1 : -1 : 1$
10:     **doall** $j = 1 : 2^{k-r}$
11:       solve $T^{(r-1)} \hat{u}_{(2j-1)2^{r-1}} = q_{(2j-1)2^{r-1}}^{(k)} - (u_{(j-1)2^{r-1}} + u_{(j+1)2^{r-1}})$
12:       $u_{(2j-1)2^{r-1}} = \hat{u}_{(2j-1)2^{r-1}} + p_{(2j-1)2^{r-1}}$
13:     **end**
14: **end**

In terms of operations with $T^{(r)}$, the reduction phase (line 4) now consists only of applications of $(T^{(r)})^{-1}$ so parallelism is enabled by utilizing the partial fraction representation (6.77) as in the back substitution phase of CORF.

Therefore, solutions with coefficient matrix $T^{(r)}$ and $2^{k-r} - 1$ right-hand sides for $r = 1, \ldots, k - 1$ can be accomplished by solving 2^r independent tridiagonal systems for each right-hand side and then combining the partial solutions by multiplying $2^{k-r} - 1$ matrices, each of size $m \times 2^r$ with the vector of 2^r partial fraction coefficients. Therefore, Algorithm 6.16, can be efficiently implemented on parallel architectures using partial fractions to solve one or more independent linear systems with coefficient matrix $T^{(r)}$ in lines 4, 8 and 11. If we assume that the cost of solving a tridiagonal system of order m using m processors is $\tau(m)$, then if there are $P = mn$ processors, the parallel cost of BCR is approximately equal for the 2 stages: It is easy to see that there is a total of $2k\tau(m) + k^2 + O(k)$ operations. If we use PARACR to solve the tridiagonal systems, the cost becomes $16 \log n \log m + \log^2 n + O(\log n)$. This is somewhat more than the cost of parallel Fourier-MD and CFT, but BCR is applicable to a wider range of problems, as we noted earlier.

Historically, the invention of stabilized BCR by Buneman preceded the parallelization of BCR based on partial fractions [97, 98]. In light of the preceding discussion, we can view the Buneman scheme as a method for resolving the parallelization bottleneck in the reduction stage of CORF, that also handles the instability and an example where the introduction of multiple levels of parallelism stabilizes a numerical process. This is interesting, especially in view of discussions regarding the interplay between numerical stability and parallelism; cf. [99].

It was assumed so far for convenience that $n = 2^k - 1$ and that BCR was applied for the discrete Poisson problem that originated from a PDE with Dirichlet boundary conditions. For general values of n or other boundary conditions, the use of reduction can be shown to generate more general matrix rational functions numerator that has nonzero degree that is smaller than than of the demoninator. The systems with these matrices are then solved using the more general method listed as Algorithm 12.3 of Chap. 12. This not only enables parallelization but also eliminates the need for multiplications with the numerator polynomial; cf. the discussion in the Notes and References Sect. 6.4.9. In terms of the kernels of Sect. 6.4.2 one calls from category $(1.iii)$ (when $r = k - 1$) or $(1.v)$ for other values of r.

It is also worth noting that the use of partial fractions in these cases is numerically safe; cf. [100] as well as the discussion in Sect. 12.1.3 of Chap. 12 for more details on the numerical issues that arise when using partial fractions.

6.4.6 Fourier Analysis-Cyclic Reduction

There is another way to resolve the computational bottleneck of CORF and BCR and to stabilize the process. This is to monitor the reduction stage and terminate it before the available parallelism is greatly reduced and accuracy is compromised and switch to another method for the smaller system. Therefore the algorithm is made to adapt to the available computational resources yielding acceptable solutions.

The method presented next is in this spirit but combines early stopping with the MD-Fourier technique. The Fourier analysis-cyclic reduction method (FACR) is a hybrid method consisting of l block cyclic reduction steps as in CORF, followed by MD-Fourier for the smaller block-tridiagonal system and back substitution to compute the final solution. To account for limiting the number of reduction steps, l, the method is denoted by FACR(l). It is based on the fact that at any step of the reduction stage of BCR, the coefficient matrix is $A^{(r)} = [-I, T^{(r)}, -I]_{2^{k-r}-1}$ and since $T^{(r)} = 2\mathscr{T}_{2^r}(\frac{T}{2})$ has the same eigenvectors as T with eigenvalues $2\mathscr{T}_{2^r}(\frac{\lambda_i^{(m)}}{2})$, the reduced system can be solved using Fourier-MD. If reduction is applied without stabilization, l must be small. Several analyses of the computational complexity of the algorithm (see e.g. [101]) indicate that for the $l \approx \log\log m$, the sequential complexity is $O(mn\log(\log n))$. Therefore, properly designed FACR is faster than MD-Fourier and BCR. In practice, the best choice for l depends on the relative performance of the underlying kernels and other characteristics of the target computer platform. The parallel implementation of all steps can proceed using the techniques deployed for BCR and MD; FACR(l) can be viewed as an alternative to partial fractions to avoid the parallel implementation bottleneck that was observed after a few steps of reduction in BCR. For example, the number of systems solved in parallel in BCR the algorithm can be monitored in order to trigger a switch to MD-Fourier before they become so few as not make full use of available parallel resources.

6.4.7 Sparse Selection and Marching

In many applications using a matrix or matrix function as an operator acting on a sparse vector, one often desires to obtain only very few elements of the result (probing). In such a case, it is possible to realize great savings compared to when all elements of the results are required. We have already seen this in Sect. 6.1, where the DFT of vectors with only 2 nonzero elements was computed with BLAS1, _AXPY, operations between columns of a Vandermonde matrix instead of the FFT. We mentioned then that this was inspired by the work in [26] in the context of RES. The situation is especially favorable when the matrix is structured and there are closed formulas for the inverse as the case for the Poisson matrix, that can be computed, albeit at high serial cost. A case where there could be further significant reductions in cost is when only few elements of the solution are sought. The next lemma shows that the number of arithmetic operations to compute $c^\top A b$ when the matrix A is dense and the vector b, c are sparse depends only on the sparsity of the vectors.

Proposition 6.6 *Given an arbitrary matrix A and sparse vectors b, c then $c^\top A b$ can be computed with leading cost $2\mathrm{n_{nz}}(b)\mathrm{n_{nz}}(c)$ operations.*

Proof Let P_b and P_c be the permutations that order b and c so that their nonzero elements are listed first in $P_b b$ and $P_c c$ respectively. Then $c^\top A b = (P_c c)^\top P_c A P_b^\top P_b b$ where

$$c^\top A b = (P_c c)^\top P_c A P_b^\top P_b b = \begin{pmatrix} c_{nz}^\top & 0 \end{pmatrix} \begin{pmatrix} \hat{A} & 0 \\ 0 & 0_{\hat{\mu},\hat{v}} \end{pmatrix} \begin{pmatrix} b_{nz}^\top \\ 0 \end{pmatrix}$$

$$= c_{nz}^\top \hat{A} b_{nz},$$

where $\hat{A}$ is of dimension $n_{nz}(c) \times n_{nz}(b)$, $\hat{\mu} = n_{nz}(c)$, $\hat{v} = n_{nz}(b)$, and b_{nz} and c_{nz} are the subvectors of nonzero elements of b and c. This can be computed in

$$\min((2n_{nz}(b) - 1)n_{nz}(c), (2n_{nz}(c) - 1)n_{nz}(b))$$

operations, proving the lemma. This cost can be reduced even further if A is also sparse.

From the explicit formulas in Proposition 6.4 we obtain the following result.

Proposition 6.7 *Let $T = [-1, \alpha, -1]_m$ be a tridiagonal Toeplitz matrix and consider the computation of $\xi = c^\top T^{-1} b$ for sparse vectors b, c. Then ξ can be computed in $O(n_{nz}(b)n_{nz}(c))$ arithmetic operations.*

This holds if the participating elements of T^{-1} are already available or if they can be computed in $O(n_{nz}(b)n_{nz}(c))$ operations as well.

When we seek only k elements of the solution vector, then the cost becomes $O(k\, n_{nz}(b))$. From Proposition 6.5 it follows that these ideas can also be used to reduce costs when applying the inverse of A. To compute only u_n, we write

$$u_n = C^\top A^{-1} f, \quad \text{where} \quad C = (0, 0, \ldots, I_m)^\top \in \mathbb{R}^{(mn) \times m}$$

hence

$$u_n = \sum_{j=1}^{n} \hat{\mathcal{U}}_n^{-1}(T) \hat{\mathcal{U}}_{j-1}(T) f_j \tag{6.80}$$

since $\hat{\mathcal{U}}_0(T) = I$. It is worth noting that on uniprocessors or computing platforms with limited parallelism, instead of the previous explicit formula, it is more economical to compute u_n as the solution of the following system:

$$\hat{\mathcal{U}}_n(T) u_n = f_1 - [T, -I, 0, \ldots, 0] H^{-1} \bar{f}, \quad \text{where} \quad \bar{f} = [f_2^\top, \ldots, f_n^\top]^\top,$$

where

$$H = \begin{pmatrix} -I & T & -I & & \\ & -I & T & -I & \\ & & \ddots & \ddots & \\ & & & -I & T \\ & & & & -I \end{pmatrix}.$$

The term $H^{-1}\bar{f}$ can be computed first using a block recurrence, each step of which consists of a multiplication of T with a vector and some other simple vector operations, followed by the solution with $\hat{\mathcal{U}}_n(T)$ which can be computed by utilizing its product form or the partial fraction representation of its inverse. Either way, this requires solving n linear systems with coefficient matrices that are simple shifts of T, a case of kernel ($1.iv$).

If the right-hand side is also very sparse, so that $f_j = 0$ except for l values of the index j, then the sum for u_n consists of only l block terms. Further cost reductions are possible if the nonzero f_j's are also sparse and if we only seek few elements from u_n. One way to compute u_n based on these observations is to diagonalize T as in the Fourier-MD methods, applying the DFT on each of the non-zero f_j terms. Further savings are possible if these f_j's are also sparse, since DFTs could then be computed directly as BLAS, without recourse to FFT kernels.

After the transform, the coefficients in the sum (6.80) are diagonal matrices with entries $\hat{\mathcal{U}}_n^{-1}(\lambda_j^{(m)})\hat{\mathcal{U}}_{j-1}(\lambda_j^{(m)})$. The terms can be computed directly from Definition 6.6. Therefore, each term of the sum (6.80) is the result of the element-by-element multiplication of the aforementioned diagonal with the vector $\hat{f}_j$, which is the DFT of f_j. Thus

$$u_n = Q \sum_{j=1}^{n} \begin{pmatrix} \hat{\mathcal{U}}_n^{-1}(\lambda_1^{(m)})\hat{\mathcal{U}}_{j-1}(\lambda_1^{(m)})\hat{f}_{j,1} \\ \vdots \\ \hat{\mathcal{U}}_n^{-1}(\lambda_m^{(m)})\hat{\mathcal{U}}_{j-1}(\lambda_m^{(m)})\hat{f}_{j,m} \end{pmatrix}$$

Diagonalization decouples the computations and facilitates parallel implementation as in the case with the Fourier-MD approach. Specifically, with $O(mn)$ processors, vector u_n can be computed at the cost of $O(\log m)$ to perform the n independent length-m transforms of f_j, $j = 1, \ldots, n$, followed by few parallel operations to prepare the partial subvectors, $O(\log n)$ operations to add them, followed by $O(\log m)$ operations to do the back transform, for a total of $O(\log n + \log m)$ parallel arithmetic operations. This cost can be lowered even further when most of the f_j's are zero.

The use of u_n to obtain $u_{n-1} = -f_n - Tu_n$ and then the values of the previous subvectors using the simple block recurrence

$$u_{j-1} = -f_j + Tu_j - u_{j+1}, j = n-1, \ldots, 2, \tag{6.81}$$

is an example of what is frequently referred as marching (in fact here we march backwards). The overall procedure of computing u_n as described and then the remaining subvectors from the block recurrence is known to cost only $O(mn)$ operations on a uniprocessor. Unfortunately, the method is unstable. A much better approach, called *generalized marching*, was described in [83]. The idea is to partition the original (6.64) into subproblems that are sufficiently small in size (to prevent instability from manifesting itself). We do not describe this approach any further but note that the partitioning into subproblems is a form of algebraic domain decomposition that results in increased opportunities for parallelism, since each block recurrence can be evaluated independently and thus assigned to a different processor. Generalized marching is one more case where partitioning and the resulting hierarchical parallelism that it induces, introduces more parallelism while it serves to reduce the risk of instability. In fact, generalized marching was the inspiration behind the stabilizing transformations that also led from the Givens based parallel tridiagonal solver of [102] to the Spike algorithm; cf. our discussion in Sects. 5.2 and 5.5.

6.4.8 Poisson Inverse in Partial Fraction Representation

Proposition 6.5, in combination with partial fractions makes possible the design of a method for computing all or selected subvectors of the solution vector of the discrete Poisson system.

From Definition 6.6, the roots of the Chebyshev polynomial $\hat{\mathscr{U}}_n$ are

$$\rho_j = 2\cos\frac{j\pi}{n+1}, \quad j = 1, \ldots, n.$$

From Proposition 6.5, we can write the subvector u_i of the solution of the discrete Poisson system as

$$
\begin{aligned}
u_i &= \sum_{j=1}^{n} (A_{ij})^{-1} f_j \\
&= \sum_{j=1}^{i-1} \hat{\mathscr{U}}_n^{-1}(T)\hat{\mathscr{U}}_{j-1}(T)\hat{\mathscr{U}}_{n-i}(T) f_j + \sum_{j=i}^{n} \hat{\mathscr{U}}_n^{-1}(T)\hat{\mathscr{U}}_{i-1}(T)\hat{\mathscr{U}}_{n-j}(T) f_j.
\end{aligned}
$$

Because each block $(A^{-1})_{i,j}$ is a rational matrix function with denominator of degree n and numerator of smaller degree and, with the roots of the denominator being distinct, $(A_{ij})^{-1}$ can be expressed as a partial fraction sum of n terms.

$$u_i = \sum_{j=1}^{n} \sum_{k=1}^{n} \gamma_{j,k}^{(i)} (T - \rho_k I)^{-1} f_j \tag{6.82}$$

$$= \sum_{k=1}^{n} (T - \rho_k I)^{-1} \sum_{j=1}^{n} \gamma_{j,k}^{(i)} f_j \tag{6.83}$$

$$= \sum_{k=1}^{n} (T - \rho_k I)^{-1} \tilde{f}_k^{(i)}, \tag{6.84}$$

where $F = (f_1, \ldots, f_n) \in \mathbb{R}^{m \times n}$, and $\rho_1, \ldots, \rho_n$ and $\gamma_{j,1}^{(i)}, \ldots, \gamma_{j,n}^{(i)}$ are the roots of the denominator and partial fraction coefficients for the rational polynomial for $(A^{-1})_{i,j}$, respectively. Let $G^{(i)} = [\gamma_{j,k}^{(i)}]$ be the matrix that contains along each row the partial fraction coefficients $(\gamma_{j,1}^{(i)}, \ldots, \gamma_{j,n}^{(i)})$ used in the inner sum of (6.82). In column format $G^{(i)} = (g_1^{(i)}, \ldots, g_n^{(i)})$, where in (6.84), $\tilde{f}_k^{(i)} = F g_k^{(i)}$. Based on this formula, it is straightforward to compute the u_i's as is shown in Algorithm 6.17; cf. [103]. Algorithm EES offers abundant parallelism and all u_i's can be evaluated

---

Algorithm 6.17 EES: Explicit Elliptic Solver for the discrete Poisson system

---

Input: Block tridiagonal matrix $\mathscr{A} = [-I_m, T, I_m]_n$, where $T = [-1, 4, -1]_n$ and the right-hand side $f = (f_1; \ldots; f_n)$.
Output: Solution $u = (u_1; \ldots; u_n)$ //the method can be readily modified to produce only selected subvectors
1: Compute n roots ρ_k of $\hat{\mathscr{U}}_n(x)$ in (6.82)
2: Compute coefficients $\gamma_{j,k}^{(i)}$ in (6.83)
3: **doall** $i = 1 : n$
4:    **doall** $k = 1 : n$
5:       compute $\tilde{f}_k^{(i)} = \sum_{j=1}^{n} \gamma_{j,k}^{(i)} f_j$
6:       compute $\bar{u}_k^{(i)} = (T - \rho_k I)^{-1} \tilde{f}_k^{(i)}$
7:    **end**
8:    compute $u_i = \sum_{k=1}^{n} \bar{u}_k^{(i)}$
9: **end**

---

in $O(\log n + \log m)$. For this, however, $O(n^3 m)$ processors appear to be necessary, which is too high. An $O(n)$ reduction in the processor count is possible by first observing that for each i, the matrix $G^{(i)}$ of partial fraction coefficients is the sum of a Toeplitz matrix and a Hankel matrix and then using fast multiplications with these special matrices; cf. [103].

MD-Fourier Based on Partial Fraction Representation of the Inverse

We next show that EES can be reorganized in such a way that its steps are the same as those of MD- FOURIER, that is independent DSTs and tridiagonal solves, and thus few or all u_i's can be computed in $O(\log n + \log m)$ time using only $O(mn)$ processors. This is not surprising since MD- FOURIER is also based on an explicit formula for the Poisson matrix inverse (cf. (6.72)) but has some interest as it reveals a direct connection between MD- FOURIER and EES.

The elements and structure of $G^{(i)}$ are key to establishing the connection. For $i \geq j$ (the case $i < j$ can be treated similarly)

$$\gamma_{j,k}^{(i)} = \frac{\hat{\mathcal{U}}_{j-1}(\rho_k)\hat{\mathcal{U}}_{n-i}(\rho_k)}{\hat{\mathcal{U}}_n'(\rho_k)} = \frac{\sin(j\theta_k)\sin((n+1-i)\theta_k)}{\sin^2\theta_k\,\hat{\mathcal{U}}_n'(\rho_k)} \tag{6.85}$$

where $\hat{\mathcal{U}}_n'$ denotes the derivative of $\hat{\mathcal{U}}_n$. From standard trigonometric identities it follows that for $i \geq j$ the numerator of (6.85) is equal to

$$\sin(j\theta_k)\sin((n+1-i)\theta_k) = (-1)^{k+1}\sin\frac{jk\pi}{n+1}\sin\frac{ik\pi}{n+1}. \tag{6.86}$$

The same equality holds when $i < j$ since relation (6.86) is symmetric with respect to i and j. So from now on we use this numerator irrespective of the relative ordering of i and j. With some further algebraic manipulations it can be shown that

$$\hat{\mathcal{U}}_n'(x)\,|_{x=\rho_k} = (-1)^{k+1}\frac{(n+1)}{2\sin^2\theta_k}.$$

From (6.85) and (6.86) it follows that the elements of $G^{(i)}$ are

$$\gamma_{j,k}^{(i)} = \frac{2}{n+1}\sin\frac{jk\pi}{n+1}\sin\frac{ik\pi}{n+1}.$$

Hence, we can write

$$G^{(i)} = \frac{2}{n+1}\begin{pmatrix} \sin(\frac{\pi}{n+1}) & \cdots & \sin(\frac{n\pi}{n+1}) \\ \vdots & \ddots & \vdots \\ \sin(\frac{n\pi}{n+1}) & \cdots & \sin(\frac{n^2\pi}{n+1}) \end{pmatrix}\begin{pmatrix} \sin\frac{i\pi}{n+1} & & \\ & \ddots & \\ & & \sin\frac{in\pi}{n+1} \end{pmatrix}.$$

We can thus write the factorization

$$G^{(i)} = QD^{(i)}, \tag{6.87}$$

where matrix Q and the diagonal matrix $D^{(i)}$ are

$$Q = \sqrt{\frac{2}{n+1}}\left(\sin\frac{jk\pi}{n+1}\right)_{j,k}, \quad D^{(i)} = \sqrt{\frac{2}{n+1}}\,\mathrm{diag}\left(\sin\frac{i\pi}{n+1},\ldots,\sin\frac{in\pi}{n+1}\right).$$

Observe that multiplication of a vector with the symmetric matrix Q is a DST. From (6.83) it follows that

$$u_i = \sum_{k=1}^{n} (T - \rho_k I)^{-1} h_k^{(i)} \tag{6.88}$$

where $h_k^{(i)}$ is the kth column of $FG^{(i)}$. Setting $H^{(i)} = (h_1^{(i)}, \ldots, h_n^{(i)})$ and recalling that Q encodes a DST, it is preferable to compute by rows

$$(H^{(i)})^\top = (FQD^{(i)})^\top = D^{(i)}QF^\top.$$

This amounts to applying m independent DSTs, one for each row of F. These can be accomplished in $O(\log n)$ steps if $O(mn)$ processors are available. Following that and the multiplication with $D^{(i)}$, it remains to solve n independent tridiagonal systems, each with coefficient matrix $(T - \rho_k I)$ and right-hand side consisting of the kth column of $H^{(i)}$. With $O(nm)$ processors this can be done in $O(\log m)$ operations. Finally, u_i is obtained by adding the n partial results, in $O(\log n)$ steps using the same number of processors. Therefore, the overall parallel cost for computing any u_i is $O(\log n + \log m)$ operations using $O(mn)$ processors.

It is actually possible to compute all subvectors $u_1, \ldots, u_n$ without exceeding the $O(\log n + \log m)$ parallel cost while still using $O(mn)$ processors. Set $\hat{H} = FQ$ and denote the columns of this matrix by $\hat{h}_k$. These terms do not depend on the index i. Therefore, in the above steps, the multiplication with the diagonal matrix $D^{(i)}$ can be deferred until after the solution of the independent linear systems in (6.88). Specifically, we first compute the columns of the $m \times n$ matrix

$$\check{H} = [(T - \rho_1 I)^{-1} \hat{h}_1, \ldots, (T - \rho_n I)^{-1} \hat{h}_n].$$

We then multiply each of the columns of $\check{H}$ with the appropriate diagonal element of $D^{(i)}$ and then sum the partial results to obtain u_i. However, this is equivalent to multiplying $\check{H}$ with the column vector consisting of the diagonal elements of $D^{(i)}$. Let us call this vector $d^{(i)}$, then from (6.87) it follows that the matrix $(d^{(1)}, \ldots, d^{(n)})$ is equal to the DST matrix Q. Therefore, the computation of

$$\check{H}(d^{(1)}, \ldots, d^{(n)}) = \check{H}Q,$$

can be performed using independent DSTs on the m rows of $\check{H}$. Finally note that the coefficient matrices of the systems we need to solve are given by,

$$T - \rho_k I = T - 2\cos\left(\frac{k\pi}{n+1}\right) I = \left[-1, 2 + \left(2 - 2\cos\left(\frac{k\pi}{n+1}\right)\right), -1\right]_m$$
$$= \tilde{T}_m + \lambda_k^{(n)} I.$$

It follows that the major steps of EES can be implemented as follows. First apply m independent DSTs, one for each row of F, and assemble the results in matrix $\hat{H}$. Then solve n independent tridiagonal linear systems, each corresponding to the application of $(T - \rho_k I)^{-1}$ on the kth column of $\hat{H}$. If the results are stored in $\check{H}$, finally compute the m independent DSTs, one for each row of $\check{H}$. The cost is cost $\mathbf{T}_p = O(\log m + \log n)$ using $O(mn)$ processors. We summarize as follows:

Proposition 6.8 *The direct application of Proposition 6.5 to solve (6.64) is essentially equivalent to the Fourier-MD method, in particular with a reordered version in which the DSTs are first applied across the rows (rather than the columns) of the matrix of right-hand sides and the independent tridiagonal systems are $\tilde{T}_m + \lambda_k^{(m)} I_m$ instead of $\tilde{T}_n + \lambda_k^{(m)} I_n$.*

6.4.9 Notes

Some important milestones in the development of RES are (*a*) reference [104], where Fourier analysis and marching were proposed to solve Poisson's equation; (*b*) reference [87] appears to be the first to provide the explicit formula for the Poisson matrix inverse, and also [88–90], where this formula, spectral decomposition, and a first version of MD and CFT were described (called the semi-rational and rational solutions respectively in [88]); (*c*) reference [105], on Kronecker (tensor) product solvers.

The invention of the FFT together with reference [106] (describing FACR(1) and the solution of tridiagonal systems with cyclic reduction) mark the beginning of the modern era of RES. These and reference [95], with its detailed analysis of the major RES were extremely influential in the development of the field; see also [107, 108] and the survey of [80] on the numerical solution of elliptic problems including the topic of RES.

The first discussions of parallel RES date from the early 1970s, in particular references [92, 93] while the first implementation on the Illiac IV was reported in [109]. There exist descriptions of implementations of RES for most important high performance computing platforms, including vector processors, vector multiprocessors, shared memory symmetric multiprocessors, distributed memory multiprocessors, SIMD and MIMD processor arrays, clusters of heterogeneous processors and in Grid environments. References for specific systems can be found for the Cray-1 [110, 111]; the ICL DAP [112]; Alliant FX/8 [97, 113]; Caltech and Intel hypercubes [114–118]; Thinking Machines CM-2 [119]; Denelcor HEP [120, 121]; the University of Illinois Cedar machine [122, 123]; Cray X-MP [110]; Cray Y-MP [124]; Cray T3E [125, 126]; Grid environments [127]; Intel multicore processors and processor clusters [128]; GPUs [129, 130]. Regarding the latter, it is worth noting the extensive studies conducted at Yale on an early GPU, the FPS-164 [131]. Proposals for special purpose hardware can be found in [132]. RES were included in the PELLPACK parallel problem-solving environment for elliptic PDEs [133].

The inverses of the special Toeplitz tridiagonal and 2-level Toeplitz matrices that we encountered in this chapter follow directly from usual expansion formula for determinants and the adjoint formula for the matrix inverse; see [134].

MD-Fourier algorithms have been designed and implemented on vector and parallel architectures and are frequently used as the baseline in evaluating new Poisson solvers; cf. [92, 93, 109, 111, 112, 115–117, 119, 124, 125, 131, 135, 136].

It is interesting to note that CFT was discussed in [104] long before the discovery of the FFT. Studies of parallel CFT were conducted in [94, 137], and it was found to have lower computational and communication complexity than MD and BCR for systems with $O(mn)$ processors connected in a hypercube. Specific implementations were described in reference [131] for the FPS-164 attached array processor and in [116] for Intel hypercubes.

BCR was first analyzed in the influential early paper [95]. Many formulations and uses of the algorithm were studied by the authors of [138]. They interpret BCR, applied on a block-Toeplitz tridiagonal system such as the discrete Poisson system, by means of a set of matrix recurrences that resemble the ones we derived in Sect. 5.5.2 for simple tridiagonal systems. The approach, however, is different in that they consider Schur complements rather than the structural interpretation provided by the t-tridiagonal matrices (cf. Definition 5.1). They also consider a functional form of these recurrences that provides an elegant, alternative interpretation of the splitting in Corollary 5.2, if that were to be used with CR to solve block-Toeplitz systems.

It took more than 10 years for the ingenious "parallelism via partial fractions" idea[1] to be applied to speed-up the application of rational functions of matrices on vectors, in which case the gains can be really substantial. The application of partial fractions to parallelize BCR was described in [98] and independently in [97]. References [94, 113, 123] also discuss the application of partial fraction based parallel BCR and evaluate its performance on a variety of vector and parallel architectures. Parallel versions of MD and BCR were implemented in CRAYFISHPAK, a package that contained most of the functionality of FISHPAK. Results from using the package on a Cray Y-MP were presented in [110].

FACR was first proposed in [106]. The issue of how to choose l for parameterized vector and parallel architectures has been studied at length in [137] and for a shared memory multiprocessor in [120] where it was proposed to monitor the number of systems that can be solved in parallel.

A general conclusion is that the number of steps should be small (so that BCR can be applied without stabilization) especially for high levels of parallelism. Its optimal value for a specific configuration can be determined empirically. FACR(l) can be viewed as an alternative to partial fractions to avoid the bottlenecks in parallel implementation that are observed after a few steps of reduction in BCR. For example, one can monitor the number of systems that can be solved in parallel in BCR and before that number of independent computations becomes too small and no longer acceptable, switch to MD. This approach was analyzed in [120] on a shared-memory multiprocessor (Denelcor HEP). FACR is frequently found to be faster than Fourier-MD and CFT; see e.g. [131] for results on an attached array processor (FPS-164). See also [111, 112] for designs of FACR on vector and parallel architectures (Cray-1, Cyber-205, ICL DAP).

---

[1]Due to H.T. Kung.

Exploiting sparsity and solution probing was observed early on for RES in reference [26] as we already noted in Sect. 6.1. These ideas were developed further in [126, 139–141]. A RES for an SGI with 8 processors and a 16-processo Beowulf cluster based on [139, 140] was presented in [142].

References

1. Turing, A.: Proposed electronic calculator. www.emula3.com/docs/Turing_Report_on_ACE. pdf (1946)
2. Dewilde, P.: Minimal complexity realization of structured matrices. In: Kailath, T., Sayed, A. (eds.) Fast Reliable Algorithms for Matrices with Structure, Chapter 10, pp. 277–295. SIAM (1999)
3. Chandrasekaran, S., Dewilde, P., Gu, M., Somasunderam, N.: On the numerical rank of the off-diagonal blocks of Schur complements of discretized elliptic PDEs. SIAM J. Matrix Anal. Appl. **31**(5), 2261–2290 (2010)
4. Lin, L., Lu, J., Ying, L.: Fast construction of hierarchical matrix representation from matrix-vector multiplication. J. Comput. Phys. **230**(10), 4071–4087 (2011). doi:10.1016/j.jcp.2011. 02.033. http://dx.doi.org/10.1016/j.jcp.2011.02.033
5. Martinsson, P.: A fast randomized algorithm for computing a hierarchically semiseparable representation of a matrix. SIAM J. Matrix Anal. Appl. **32**(4), 1251–1274 (2011)
6. Kailath, T., Kung, S.-Y., Morf, M.: Displacement ranks of matrices and linear equations. J. Math. Anal. Appl. **68**(2), 395–407 (1979)
7. Kailath, T., Sayed, A.: Displacement structure: theory and applications. SIAM Rev. **37**(3), 297–386 (1995)
8. Vandebril, R., Van Barel, M., Mastronardi, N.: Matrix Computations and Semiseparable Matrices. Volume I: Linear Systems. Johns Hopkins University Press, Baltimore (2008)
9. Bebendorf, M.: Hierarchical Matrices: A Means to Efficiently Solve Elliptic Boundary Value Problems. Lecture Notes in Computational Science and Engineering (LNCSE), vol. 63. Springer, Berlin (2008). ISBN 978-3-540-77146-3
10. Hackbusch, W., Borm, S.: Data-sparse approximation by adaptive H2-matrices. Computing **69**(1), 1–35 (2002)
11. Traub, J.: Associated polynomials and uniform methods for the solution of linear problems. SIAM Rev. **8**(3), 277–301 (1966)
12. Björck, A., Pereyra, V.: Solution of Vandermonde systems of equations. Math. Comput. **24**, 893–903 (1971)
13. Gohberg, I., Olshevsky, V.: The fast generalized Parker-Traub algorithm for inversion of Vandermonde and related matrices. J. Complex. **13**(2), 208–234 (1997)
14. Golub, G., Van Loan, C.: Matrix Computations, 4th edn. Johns Hopkins, Baltimore (2013)
15. Higham, N.: Accuracy and Stability of Numerical Algorithms, 2nd edn. SIAM, Philadelphia (2002)
16. Davis, P.J.: Interpolation and Approximation. Dover, New York (1975)
17. Gautschi, W.: Numerical Analysis: An Introduction. Birkhauser, Boston (1997)
18. Gautschi, W., Inglese, G.: Lower bounds for the condition number of Vandermonde matrices. Numer. Math. **52**, 241–250 (1988)
19. Van Loan, C.: Computational Frameworks for the Fast Fourier Transform. SIAM, Philadelphia (1992)
20. Córdova, A., Gautschi, W., Ruscheweyh, S.: Vandermonde matrices on the circle: spectral properties and conditioning. Numer. Math. **57**, 577–591 (1990)
21. Berman, L., Feuer, A.: On perfect conditioning of Vandermonde matrices on the unit circle. Electron. J. Linear Algebra **16**, 157–161 (2007)

22. Gautschi, W.: Optimally scaled and optimally conditioned Vandermonde and Vandermonde-like matrices. BIT Numer. Math. **51**, 103–125 (2011)
23. Gunnels, J., Lee, J., Margulies, S.: Efficient high-precision matrix algebra on parallel architectures for nonlinear combinatorial optimization. Math. Program. Comput. **2**(2), 103–124 (2010)
24. Aho, A., Hopcroft, J.E., Ullman, J.D.: The Design and Analysis of Computer Algorithms. Addison-Wesley, Reading (1974)
25. Pan, V.: Complexity of computations with matrices and polynomials. SIAM Rev. **34**(2), 255–262 (1992)
26. Banegas, A.: Fast Poisson solvers for problems with sparsity. Math. Comput. **32**(142), 441–446 (1978). http://www.jstor.org/stable/2006156
27. Cappello, P., Gallopoulos, E., Koç, Ç.: Systolic computation of interpolating polynomials. Computing **45**, 95–118 (1990)
28. Koç, Ç., Cappello, P., Gallopoulos, E.: Decomposing polynomial interpolation for systolic arrays. Int. J. Comput. Math. **38**, 219–239 (1991)
29. Koç, Ç.: Parallel algorithms for interpolation and approximation. Ph.D. thesis, Department of Electrical and Computer Engineering, University of California, Santa Barbara, June 1988
30. Eğecioğlu, Ö., Gallopoulos, E., Koç, Ç.: A parallel method for fast and practical high-order Newton interpolation. BIT **30**, 268–288 (1990)
31. Breshaers, C.: The Art of Concurrency—A Thread Monkey's Guide to Writing Parallel Applications. O'Reilly, Cambridge (2009)
32. Lakshmivarahan, S., Dhall, S.: Parallelism in the Prefix Problem. Oxford University Press, New York (1994)
33. Harris, M., Sengupta, S., Owens, J.: Parallel prefix sum (scan) with CUDA. GPU Gems **3**(39), 851–876 (2007)
34. Falkoff, A., Iverson, K.: The evolution of APL. SIGPLAN Not. **13**(8), 47–57 (1978). doi:10.1145/960118.808372. http://doi.acm.org/10.1145/960118.808372
35. Blelloch, G.E.: Scans as primitive operations. IEEE Trans. Comput. **38**(11), 1526–1538 (1989)
36. Chatterjee, S., Blelloch, G., Zagha, M.: Scan primitives for vector computers. In: Proceedings of the 1990 ACM/IEEE Conference on Supercomputing, pp. 666–675. IEEE Computer Society Press, Los Alamitos (1990). http://dl.acm.org/citation.cfm?id=110382.110597
37. Hillis, W., Steele Jr, G.: Data parallel algorithms. Commun. ACM **29**(12), 1170–1183 (1986). doi:10.1145/7902.7903. http://doi.acm.org/10.1145/7902.7903
38. Dotsenko, Y., Govindaraju, N., Sloan, P.P., Boyd, C., Manferdelli, J.: Fast scan algorithms on graphics processors. In: Proceedings of the 22nd International Conference on Supercomputing ICS'08, pp. 205–213. ACM, New York (2008). doi:10.1145/1375527.1375559. http://doi.acm.org/10.1145/1375527.1375559
39. Sengupta, S., Harris, M., Zhang, Y., Owens, J.: Scan primitives for GPU computing. Graphics Hardware 2007, pp. 97–106. ACM, New York (2007)
40. Sengupta, S., Harris, M., Garland, M., Owens, J.: Efficient parallel scan algorithms for many-core GPUs. In: Kurzak, J., Bader, D., Dongarra, J. (eds.) Scientific Computing with Multicore and Accelerators, pp. 413–442. CRC Press, Boca Raton (2010). doi:10.1201/b10376-29
41. Intel Corporation: Intel(R) Threading Building Blocks Reference Manual, revision 1.6 edn. (2007). Document number 315415-001US
42. Bareiss, E.: Numerical solutions of linear equations with Toeplitz and vector Toeplitz matrices. Numer. Math. **13**, 404–424 (1969)
43. Gallivan, K.A., Thirumalai, S., Van Dooren, P., Varmaut, V.: High performance algorithms for Toeplitz and block Toeplitz matrices. Linear Algebra Appl. **241–243**, 343–388 (1996)
44. Justice, J.: The Szegö recurrence relation and inverses of positive definite Toeplitz matrices. SIAM J. Math. Anal. **5**, 503–508 (1974)
45. Trench, W.: An algorithm for the inversion of finite Toeplitz matrices. J. Soc. Ind. Appl. Math. **12**, 515–522 (1964)
46. Trench, W.: An algorithm for the inversion of finite Hankel matrices. J. Soc. Ind. Appl. Math. **13**, 1102–1107 (1965)

47. Phillips, J.: The triangular decomposition of Hankel matrices. Math. Comput. **25**, 599–602 (1971)
48. Rissanen, J.: Solving of linear equations with Hankel and Toeplitz matrices. Numer. Math. **22**, 361–366 (1974)
49. Xi, Y., Xia, J., Cauley, S., Balakrishnan, V.: Superfast and stable structured solvers for Toeplitz least squares via randomized sampling. SIAM J. Matrix Anal. Appl. **35**(1), 44–72 (2014)
50. Zohar, S.: Toeplitz matrix inversion: The algorithm of W. Trench. J. Assoc. Comput. Mach. **16**, 592–701 (1969)
51. Watson, G.: An algorithm for the inversion of block matrices of Toeplitz form. J. Assoc. Comput. Mach. **20**, 409–415 (1973)
52. Rissanen, J.: Algorithms for triangular decomposition of block Hankel and Toeplitz matrices with applications to factoring positive matrix polynomials. Math. Comput. **27**, 147–154 (1973)
53. Kailath, T., Vieira, A., Morf, M.: Inverses of Toeplitz operators, innovations, and orthogonal polynomials. SIAM Rev. **20**, 106–119 (1978)
54. Gustavson, F., Yun, D.: Fast computation of Padé approximants and Toeplitz systems of equations via the extended Euclidean algorithm. Technical report 7551, IBM T.J. Watson Research Center, New York (1979)
55. Brent, R., Gustavson, F., Yun, D.: Fast solution of Toeplitz systems of equations and computation of Padé approximants. J. Algorithms **1**, 259–295 (1980)
56. Morf, M.: Doubling algorithms for Toeplitz and related equations. In: Proceedings of the IEEE International Conference on Acoustics, Speech and Signal Processing, pp. 954–959 (1980)
57. Chandrasekaran, S., Gu, M., Sun, X., Xia, J., Zhu, J.: A superfast algorithm for Toeplitz systems of linear equations. SIAM J. Matrix Anal. Appl. **29**(4), 1247–1266 (2007). doi:10.1137/040617200. http://dx.doi.org/10.1137/040617200
58. Xia, J., Xi, Y., Gu, M.: A superfast structured solver for Toeplitz linear systems via randomized sampling. SIAM J. Matrix Anal. Appl. **33**(3), 837–858 (2012)
59. Grcar, J., Sameh, A.: On certain parallel Toeplitz linear system solvers. SIAM J. Sci. Stat. Comput. **2**(2), 238–256 (1981)
60. Aitken, A.: Determinants and Matrices. Oliver Boyd, London (1939)
61. Cantoni, A., Butler, P.: Eigenvalues and eigenvectors of symmetric centrosymmetric matrices. Numer. Linear Algebra Appl. **13**, 275–288 (1976)
62. Gohberg, I., Semencul, A.: On the inversion of finite Toeplitz matrices and their continuous analogues. Mat. Issled **2**, 201–233 (1972)
63. Gohberg, I., Feldman, I.: Convolution equations and projection methods for their solution. Translations of Mathematical Monographs, vol. 41. AMS, Providence (1974)
64. Gohberg, I., Levin, S.: Asymptotic properties of Toeplitz matrix factorization. Mat. Issled **1**, 519–538 (1978)
65. Fischer, D., Golub, G., Hald, O., Leiva, C., Widlund, O.: On Fourier-Toeplitz methods for separable elliptic problems. Math. Comput. **28**(126), 349–368 (1974)
66. Riesz, F., Sz-Nagy, B.: Functional Analysis. Frederick Ungar, New York (1956). (Translated from second French edition by L. Boron)
67. Szegö, G.: Orthogonal Polynomials. Technical Report, AMS, Rhode Island (1959). (Revised edition AMS Colloquium Publication)
68. Grenander, U., Szegö, G.: Toeplitz Forms and their Applications. University of California Press, California (1958)
69. Pease, M.: The adaptation of the fast Fourier transform for parallel processing. J. Assoc. Comput. Mach. **15**(2), 252–264 (1968)
70. Householder, A.S.: The Theory of Matrices in Numerical Analysis. Dover Publications, New York (1964)
71. Morf, M., Kailath, T.: Recent results in least-squares estimation theory. Ann. Econ. Soc. Meas. **6**, 261–274 (1977)
72. Franchetti, F., Püschel, M.: Fast Fourier transform. In: Padua, D. (ed.) Encyclopedia of Parallel Computing. Springer, New York (2011)

73. Chen, H.C.: The SAS domain decomposition method. Ph.D. thesis, University of Illinois at Urbana-Champaign (1988)
74. Chen, H.C., Sameh, A.: Numerical linear algebra algorithms on the Cedar system. In: Noor, A. (ed.) Parallel Computations and Their Impact on Mechanics. Applied Mechanics Division, vol. 86, pp. 101–125. American Society of Mechanical Engineers, New York (1987)
75. Chen, H.C., Sameh, A.: A matrix decomposition method for orthotropic elasticity problems. SIAM J. Matrix Anal. Appl. **10**(1), 39–64 (1989)
76. Wilkinson, J.H.: The Algebraic Eigenvalue Problem. Oxford University Press, Oxford (1965)
77. Botta, E.: How fast the Laplace equation was solved in 1995. Appl. Numer. Math. **24**(4), 439–455 (1997). doi:10.1016/S01689274(97)00041X. http://dx.doi.org/10.1016/S0168-9274(97)00041-X
78. Knightley, J.R., Thompson, C.P.: On the performance of some rapid elliptic solvers on a vector processor. SIAM J. Sci. Stat. Comput. **8**(5), 701–715 (1987)
79. Csansky, L.: Fast parallel matrix inversion algorithms. SIAM J. Comput. **5**, 618–623 (1977)
80. Birkhoff, G., Lynch, R.: Numerical Solution of Elliptic Problems. SIAM, Philadelphia (1984)
81. Iserles, A.: Introduction to Numerical Methods for Differential Equations. Cambridge University Press, Cambridge (1996)
82. Olshevsky, V., Oseledets, I., Tyrtyshnikov, E.: Superfast inversion of two-level Toeplitz matrices using Newton iteration and tensor-displacement structure. Recent Advances in Matrix and Operator Theory. Birkhäuser Verlag, Basel (2007)
83. Bank, R.E., Rose, D.: Marching algorithms for elliptic boundary value problems. I: the constant coefficient case. SIAM J. Numer. Anal. **14**(5), 792–829 (1977)
84. Lanczos, C.: Tables of the Chebyshev Polynomials $S_n(x)$ and $C_n(x)$. Applied Mathematics Series, vol. 9. National Bureau of Standards, New York (1952)
85. Rivlin, T.: The Chebyshev Polynomials. Wiley-Interscience, New York (1974)
86. Abramowitz, M., Stegun, I.: Handbook of Mathematical Functions. Dover, New York (1965)
87. Karlqvist, O.: Numerical solution of elliptic difference equations by matrix methods. Tellus **4**(4), 374–384 (1952). doi:10.1111/j.2153-3490.1952.tb01025.x. http://dx.doi.org/10.1111/j.2153-3490.1952.tb01025.x
88. Bickley, W.G., McNamee, J.: Matrix and other direct methods for the solution of systems of linear difference equations. Philos. Trans. R. Soc. A: Math. Phys. Eng. Sci. **252**(1005), 69–131 (1960). doi:10.1098/rsta.1960.0001. http://rsta.royalsocietypublishing.org/cgi/doi/10.1098/rsta.1960.0001
89. Egerváry, E.: On rank-diminishing operations and their application to the solution of linear equations. Zeitschrift fuer angew. Math. und Phys. **11**, 376–386 (1960)
90. Egerváry, E.: On hypermatrices whose blocks are computable in pair and their application in lattice dynamics. Acta Sci. Math. Szeged **15**, 211–222 (1953/1954)
91. Bialecki, B., Fairweather, G., Karageorghis, A.: Matrix decomposition algorithms for elliptic boundary value problems: a survey. Numer. Algorithms (2010). doi:10.1007/s11075-010-9384-y. http://www.springerlink.com/index/10.1007/s11075-010-9384-y
92. Buzbee, B.: A fast Poisson solver amenable to parallel computation. IEEE Trans. Comput. **C-22**(8), 793–796 (1973)
93. Sameh, A., Chen, S.C., Kuck, D.: Parallel Poisson and biharmonic solvers. Computing **17**, 219–230 (1976)
94. Swarztrauber, P.N., Sweet, R.A.: Vector and parallel methods for the direct solution of Poisson's equation. J. Comput. Appl. Math. **27**, 241–263 (1989)
95. Buzbee, B., Golub, G., Nielson, C.: On direct methods for solving Poisson's equation. SIAM J. Numer. Anal. **7**(4), 627–656 (1970)
96. Sweet, R.A.: A cyclic reduction algorithm for solving block tridiagonal systems of arbitrary dimension. SIAM J. Numer. Anal. **14**(4), 707–720 (1977)
97. Gallopoulos, E., Saad, Y.: Parallel block cyclic reduction algorithm for the fast solution of elliptic equations. Parallel Comput. **10**(2), 143–160 (1989)
98. Sweet, R.A.: A parallel and vector cyclic reduction algorithm. SIAM J. Sci. Stat. Comput. **9**(4), 761–765 (1988)

99. Demmel, J.: Trading off parallelism and numerical stability. In: Moonen, M.S., Golub, G.H., Moor, B.L.D. (eds.) Linear Algebra for Large Scale and Real-Time Applications. NATO ASI Series E, vol. 232, pp. 49–68. Kluwer Academic Publishers, Dordrecht (1993)

100. Calvetti, D., Gallopoulos, E., Reichel, L.: Incomplete partial fractions for parallel evaluation of rational matrix functions. J. Comput. Appl. Math. **59**, 349–380 (1995)

101. Temperton, C.: On the FACR(l) algorithm for the discrete Poisson equation. J. Comput. Phys. **34**, 314–329 (1980)

102. Sameh, A., Kuck, D.: On stable parallel linear system solvers. J. Assoc. Comput. Mach. **25**(1), 81–91 (1978)

103. Gallopoulos, E., Saad, Y.: Some fast elliptic solvers for parallel architectures and their complexities. Int. J. High Speed Comput. **1**(1), 113–141 (1989)

104. Hyman, M.: Non-iterative numerical solution of boundary-value problems. Appl. Sci. Res. B **2**, 325–351 (1951–1952)

105. Lynch, R., Rice, J., Thomas, D.: Tensor product analysis of partial differential equations. Bull. Am. Math. Soc. **70**, 378–384 (1964)

106. Hockney, R.: A fast direct solution of Poisson's equation using Fourier analysis. J. Assoc. Comput. Mach. **12**, 95–113 (1965)

107. Haigh, T.: Bill Buzbee, Oral History Interview (2005). http://history.siam.org/buzbee.htm

108. Cooley, J.: The re-discovery of the fast Fourier transform algorithm. Mikrochim. Acta **III**, 33–45 (1987)

109. Ericksen, J.: Iterative and direct methods for solving Poisson's equation and their adaptability to Illiac IV. Technical report UIUCDCS-R-72-574, Department of Computer Science, University of Illinois at Urbana-Champaign (1972)

110. Sweet, R.: Vectorization and parallelization of FISHPAK. In: Dongarra, J., Kennedy, K., Messina, P., Sorensen, D., Voigt, R. (eds.) Proceedings of the Fifth SIAM Conference on Parallel Processing for Scientific Computing, pp. 637–642. SIAM, Philadelphia (1992)

111. Temperton, C.: Fast Fourier transforms and Poisson solvers on Cray-1. In: Hockney, R., Jesshope, C. (eds.) Infotech State of the Art Report: Supercomputers, vol. 2, pp. 359–379. Infotech Int. Ltd., Maidenhead (1979)

112. Hockney, R.W.: Characterizing computers and optimizing the FACR(l) Poisson solver on parallel unicomputers. IEEE Trans. Comput. **C-32**(10), 933–941 (1983)

113. Jwo, J.S., Lakshmivarahan, S., Dhall, S.K., Lewis, J.M.: Comparison of performance of three parallel versions of the block cyclic reduction algorithm for solving linear elliptic partial differential equations. Comput. Math. Appl. **24**(5–6), 83–101 (1992)

114. Chan, T., Resasco, D.: Hypercube implementation of domain-decomposed fast Poisson solvers. In: Heath, M. (ed.) Proceedings of the 2nd Conference on Hypercube Multiprocessors, pp. 738–746. SIAM (1987)

115. Resasco, D.: Domain decomposition algorithms for elliptic partial differential equations. Ph.D. thesis, Yale University (1990). http://www.cs.yale.edu/publications/techreports/tr776.pdf. YALEU/DCS/RR-776

116. Cote, S.: Solving partial differential equations on a MIMD hypercube: fast Poisson solvers and the alternating direction method. Technical report UIUCDCS-R-91-1694, University of Illinois at Urbana-Champaign (1991)

117. McBryan, O., Van De Velde, E.: Hypercube algorithms and implementations. SIAM J. Sci. Stat. Comput. **8**(2), s227–s287 (1987)

118. Sweet, R., Briggs, W., Oliveira, S., Porsche, J., Turnbull, T.: FFTs and three-dimensional Poisson solvers for hypercubes. Parallel Comput. **17**, 121–131 (1991)

119. McBryan, O.: Connection machine application performance. Technical report CH-CS-434-89, Department of Computer Science, University of Colorado, Boulder (1989)

120. Briggs, W.L., Turnbull, T.: Fast Poisson solvers for MIMD computers. Parallel Comput. **6**, 265–274 (1988)

121. McBryan, O., Van de Velde, E.: Elliptic equation algorithms on parallel computers. Commun. Appl. Numer. Math. **2**, 311–318 (1986)

122. Gallivan, K.A., Heath, M.T., Ng, E., Ortega, J.M., Peyton, B.W., Plemmons, R.J., Romine, C.H., Sameh, A., Voigt, R.G.: Parallel Algorithms for Matrix Computations. SIAM, Philadelphia (1990)
123. Gallopoulos, E., Sameh, A.: Solving elliptic equations on the Cedar multiprocessor. In: Wright, M.H. (ed.) Aspects of Computation on Asynchronous Parallel Processors, pp. 1–12. Elsevier Science Publishers B.V. (North-Holland), Amsterdam (1989)
124. Chan, T.F., Fatoohi, R.: Multitasking domain decomposition fast Poisson solvers on the Cray Y-MP. In: Proceedings of the Fourth SIAM Conference on Parallel Processing for Scientific Computing. SIAM (1989) (to appear)
125. Giraud, L.: Parallel distributed FFT-based solvers for 3-D Poisson problems in meso-scale atmospheric simulations. Int. J. High Perform. Comput. Appl. **15**(1), 36–46 (2001). doi:10.1177/109434200101500104. http://hpc.sagepub.com/cgi/content/abstract/15/1/36
126. Rossi, T., Toivanen, J.: A parallel fast direct solver for block tridiagonal systems with separable matrices of arbitrary dimension. SIAM J. Sci. Stat. Comput. **20**(5), 1778–1796 (1999)
127. Tromeur-Dervout, D., Toivanen, J., Garbey, M., Hess, M., Resch, M., Barberou, N., Rossi, T.: Efficient metacomputing of elliptic linear and non-linear problems. J. Parallel Distrib. Comput. **63**(5), 564–577 (2003). doi:10.1016/S0743-7315(03)00003-0
128. Intel Cluster Poisson Solver Library—Intel Software Network. http://software.intel.com/en-us/articles/intel-cluster-poisson-solver-library/
129. Rossinelli, D., Bergdorf, M., Cottet, G.H., Koumoutsakos, P.: GPU accelerated simulations of bluff body flows using vortex particle methods. J. Comput. Phys. **229**(9), 3316–3333 (2010)
130. Wu, J., JaJa, J., Balaras, E.: An optimized FFT-based direct Poisson solver on CUDA GPUs. IEEE Trans. Parallel Distrib. Comput. **25**(3), 550–559 (2014). doi:10.1109/TPDS.2013.53
131. O'Donnell, S.T., Geiger, P., Schultz, M.H.: Solving the Poisson equation on the FPS-164. Technical report, Yale University, Department of Computer Science (1983)
132. Vajteršic, M.: Algorithms for Elliptic Problems: Efficient Sequential and Parallel Solvers. Kluwer Academic Publishers, Dordrecht (1993)
133. Houstis, E.N., Rice, J.R., Weerawarana, S., Catlin, A.C., Papachiou, P., Wang, K.Y., Gaitatzes, M.: PELLPACK: a problem-solving environment for PDE-based applications on multicomputer platforms. ACM Trans. Math. Softw. (TOMS) **24**(1) (1998). http://portal.acm.org/citation.cfm?id=285864
134. Meurant, G.: A review on the inverse of symmetric tridiagonal and block tridiagonal matrices. SIAM J. Matrix Anal. Appl. **13**(3), 707–728 (1992)
135. Hoffmann, G.R., Swarztrauber, P., Sweet, R.: Aspects of using multiprocessors for meteorological modelling. In: Hoffmann, G.R., Snelling, D. (eds.) Multiprocessing in Meteorological Models, pp. 125–196. Springer, New York (1988)
136. Johnsson, S.: The FFT and fast Poisson solvers on parallel architectures. Technical Report 583, Yale University, Department of Computer Science (1987)
137. Hockney, R., Jesshope, C.: Parallel Computers. Adam Hilger, Bristol (1983)
138. Bini, D., Meini, B.: The cyclic reduction algorithm: from Poisson equation to stochastic processes and beyond. Numerical Algorithms **51**(1), 23–60 (2008). doi:10.1007/s11075-008-9253-0. http://www.springerlink.com/content/m40t072h273w8841/fulltext.pdf
139. Kuznetsov, Y.A., Matsokin, A.M.: On partial solution of systems of linear algebraic equations. Sov. J. Numer. Anal. Math. Model. **4**(6), 453–467 (1989)
140. Vassilevski, P.: An optimal stabilization of the marching algorithm. Comptes Rendus Acad. Bulg. Sci. **41**, 29–32 (1988)
141. Rossi, T., Toivanen, J.: A nonstandard cyclic reduction method, its variants and stability. SIAM J. Matrix Anal. Appl. **20**(3), 628–645 (1999)
142. Bencheva, G.: Parallel performance comparison of three direct separable elliptic solvers. In: Lirkov, I., Margenov, S., Wasniewski, J., Yalamov, P. (eds.) Large-Scale Scientific Computing. Lecture Notes in Computer Science, vol. 2907, pp. 421–428. Springer, Berlin (2004). http://dx.doi.org/10.1007/978-3-540-24588-9_48

Chapter 7
Orthogonal Factorization and Linear Least Squares Problems

Orthogonal factorization (or QR factorization) of a dense matrix is an essential tool in several matrix computations. In this chapter we consider algorithms for QR as well as its application to the solution of linear least squares problems. The QR factorization is also a major step in some methods for computing eigenvalues; cf. Chap. 8. Algorithms for both the QR factorization and the solution of linear least squares problems are major components in data analysis and areas such as computational statistics, data mining and machine learning; cf. [1–3].

7.1 Definitions

We begin by introducing the two problems we address in this chapter.

Definition 7.1 (*QR factorization*) For any matrix $A \in \mathbb{R}^{m \times n}$, there exists a pair of matrices $Q \in \mathbb{R}^{m \times m}$ and $R \in \mathbb{R}^{m \times n}$ such that:

$$A = QR, \tag{7.1}$$

where Q is orthogonal and R is either upper triangular when $m \geq n$ (i.e. R is of the form $R = \begin{pmatrix} R_1 \\ 0 \end{pmatrix}$, or upper trapezoidal when $m < n$).

If A is of maximal column rank n, and if we require that the nonzero diagonal elements of R_1 to be positive, then the QR-factorization is unique. Further, R_1 is the transpose of the Cholesky factor of the matrix $A^\top A$. Computing the factorization (7.1) consists of pre-multiplying A by a finite number of elementary orthogonal transformations $Q_1^\top, \ldots, Q_q^\top$, such that

$$Q_q^\top \cdots Q_2^\top Q_1^\top A = R, \tag{7.2}$$

© Springer Science+Business Media Dordrecht 2016
E. Gallopoulos et al., *Parallelism in Matrix Computations*,
Scientific Computation, DOI 10.1007/978-94-017-7188-7_7

where $Q = Q_1 Q_2 \cdots Q_q$ satisfies (7.2). When $m > n$, partitioning the columns of Q as $Q = [U, V]$ with $U \in \mathbb{R}^{m \times n}$ consisting of n orthonormal columns, the factorization can be expressed as

$$A = UR_1, \tag{7.3}$$

This is referred to as *thin* factorization, whereas the one obtained in (7.2) is referred to as *thick* factorization.

Definition 7.2 (*Linear least squares problem*)

$$\text{Obtain } x \in \mathbb{R}^n, \text{ so as to minimize the 2-norm of } (b - Ax), \tag{7.4}$$

where $A \in \mathbb{R}^{m \times n}$ and $b \in \mathbb{R}^m$.

The vector x is a solution if and only if the residual $r = b - Ax$ is orthogonal to the range of A: $A^\top r = 0$. When A is of maximal column-rank n, the solution is unique. When $\text{rank}(A) < n$, the solution of choice is that one with the smallest 2-norm. In this case $x = A^+ b$, where $A^+ \in \mathbb{R}^{n \times m}$ is the Moore-Penrose generalized inverse of A.

Thus, solving (7.4) consists of first computing the QR factorization (7.1) of A. Further, since

$$\|b - Ax\| = \|Q^\top (b - Ax)\| = \|Q^\top b - Rx\| = \|c - Rx\|, \tag{7.5}$$

the solution x is chosen so as to minimize $\|c - Rx\|$. When A is of full column rank, the solution is unique. By partitioning the right hand side $c = \begin{pmatrix} c_1 \\ c_2 \end{pmatrix}$ conformally with $Q = [U, V]$, the least squares solution x is obtained by solving the upper triangular system $R_1 x = c_1$. When A is rank deficient a strategy of column pivoting is necessary for realizing a rank-revealing orthogonal factorization.

In this chapter, we first review the existing parallel algorithms for computing the QR-factorization of a full rank matrix, and then we discuss algorithms for the parallel solution of rank deficient linear least squares problems. We also consider a method for solving systems with tall-narrow matrices and its application to the adjustment of geodetic networks.

Both the QR factorization and the linear least squares problem have been discussed extensively in the literature, e.g. see [4, 5] and references therein.

7.2 QR Factorization via Givens Rotations

For the sake of illustrating this parallel factorization scheme, we consider only square matrices $A \in \mathbb{R}^{n \times n}$ even though the procedure can be easily extended to rectangular matrices $A \in \mathbb{R}^{m \times n}$, with $m > n$. In this procedure, each orthogonal matrix $Q_j^\top$ which appears in (7.2) is built as the product of several plane rotations of the form

$$R_{ij} = \begin{pmatrix} I_{i-1} & & & & \\ & c_i^{(j)} & & s_i^{(j)} & \\ & & I_{j-i-2} & & \\ & -s_i^{(j)} & & c_i^{(j)} & \\ & & & & I_{j-i-1} I_{n-j} \end{pmatrix}, \tag{7.6}$$

$$\underset{i^{th}col.}{\uparrow} \qquad \underset{j^{th}col.}{\uparrow}$$

where $c_i^{(j)^2} + s_i^{(j)^2} = 1$.

For example, one of the simplest organization for successively eliminating entries below the main diagonal column by column starting with the first column, i.e. $j = 1, 2, \ldots, n-1$, i.e. $Q^\top$ is the product of $n(n-1)/2$ plane rotations.

Let us denote by $R_{i,i+1}^{(j)}$ that plane rotation which uses rows i and $(i+1)$ of A to annihilate the off-diagonal element in position $(i+1, j)$. Such a rotation is given by:

$$
\begin{aligned}
&\text{if} \quad |\alpha_{i+1,j}| > |\alpha_{i,j}|, \\
&\qquad \tau = -\frac{\alpha_{i,j}}{\alpha_{i+1,j}}, \; s_i^{(j)} = \frac{1}{\sqrt{1+\tau^2}}, \; c_i^{(j)} = s_i^{(j)} \tau, \\
&\text{else} \\
&\qquad \tau = -\frac{\alpha_{i+1,j}}{\alpha_{i,j}}, \; c_i^{(j)} = \frac{1}{\sqrt{1+\tau^2}}, \; s_i^{(j)} = c_i^{(j)} \tau. \\
&\text{end}
\end{aligned}
\tag{7.7}
$$

Thus, the annihilation of the entries of column j is obtained by the sequence of rotations $Q_j^\top = R_{j,j+1}^{(j)} \cdots R_{n-2,n-1}^{(j)} R_{n-1,n}^{(j)}$. Hence, the orthogonal matrix $Q = Q_1 \cdots Q_{n-2} Q_{n-1}$ is such that $Q^\top A$ is upper-triangular. This sequential algorithm requires $n(4n^2 - n - 3)/2$ arithmetic operations, and $(n-1)/2$ square roots.

Next, we construct a multiprocessor algorithm that achieves a speedup of $O(n^2)$ using $O(n^2)$ processors.

Theorem 7.1 ([6]) *Let A be a nonsingular matrix of even order n, then we can obtain the factorization (7.2) in $T_p = 10n - 15$ parallel arithmetic operations and $(2n - 3)$ square roots using $p = n(3n - 2)/2$ processors. This results in a speedup of $O(n^2)$, efficiency $O(1)$ and no redundant arithmetic operations.*

Proof Let

$$U_{i,i+1}^{(j)} = \begin{pmatrix} c_i^{(j)} & s_i^{(j)} \\ -s_i^{(j)} & c_i^{(j)} \end{pmatrix}$$

be that plane rotation that acts on rows i and $i+1$ for annihilating the element in position $(i+1, j)$. Starting with $A_1 = A$, we construct the sequence $A_{k+1} = Q_k^\top A_k$, $k = 1, \ldots, 2n - 3$, where $Q_k^\top$ is the direct sum of the independent plane rotations $U_{i,i+1}^{(j)}$ where the indices i, j are as given in Algorithm 7.1. Thus, each orthogonal

Fig. 7.1 Ordering for the Givens rotations: all the entries of label k are annihilated by Q_k

```
                          *
      15   *
      14 16   *
      13 15 17   *
      12 14 16 18   *
      11 13 15 17 19   *
      10 12 14 16 18 20   *
       9 11 13 15 17 19 21   *
       8 10 12 14 16 18 20 22   *
       7  9 11 13 15 17 19 21 23   *
       6  8 10 12 14 16 18 20 22 24   *
       5  7  9 11 13 15 17 19 21 23 25   *
       4  6  8 10 12 14 16 18 20 22 24 26   *
       3  5  7  9 11 13 15 17 19 21 23 25 27   *
       2  4  6  8 10 12 14 16 18 20 22 24 26 28   *
       1  3  5  7  9 11 13 15 17 19 21 23 25 27 29   *
```

matrix $Q_k^\top$ annihilates simultaneously several off-diagonal elements without destroying zeros introduced in a previous step. Specifically, $Q_k^\top$ annihilates the entries $(i_j + 1, j)$ where $i_j = 2j + n - k - 2$, for $j = 1, 2, \ldots, n - 1$. Note that elements annihilated by each $Q_k^\top$ relate to one another by the knight's move on a Chess board, see Fig. 7.1 for $n = 16$, and Algorithm 7.1.

Hence, $A_{2n-2} \equiv Q_{2n-3}^\top \ldots Q_2^\top Q_1^\top A$ is upper triangular. Let

$$U = \begin{pmatrix} c & s \\ -s & c \end{pmatrix}$$

rotate the two row vectors $u^\top$ and $v^\top \in \mathbb{R}^q$ so as to annihilate v_1, the first component of v. Then U can be determined in 3 parallel steps and one square root using two processors. The resulting vectors $\hat{u}$ and $\hat{v}$ have the elements

$$\begin{aligned}
\hat{u}_1 &= (u_1^2 + v_1^2)^{1/2} \\
\hat{u}_i &= c u_i + s v_i, \qquad 2 \le i \le q, \\
\hat{v}_i &= -s u_i + c v_i, \qquad 2 \le i \le q.
\end{aligned} \qquad (7.8)$$

Each pair $\hat{u}_i$ and $\hat{v}_i$ in (7.8) is evaluated in 2 parallel arithmetic operations using 4 processors. Note that the cost of computing $\hat{u}_1$ has already been accounted for in determining U. Consequently, the total cost of each transformation $A_{k+1} = Q_k A_k$ is 5 parallel arithmetic operations and one square root. Thus we can triangularize A in $5(2n-3)$ parallel arithmetic operations and $(2n-3)$ square roots. The maximum number of processors is needed for $k = n - 1$, and is given by

$$p = 4 \sum_{j=1}^{n/2} n - j = n(3n - 2)/2.$$

Ignoring square roots in both the sequential and multiprocessor algorithms, we obtain a speedup of roughly $\frac{n^2}{5}$, and a corresponding efficiency of roughly $\frac{2}{15}$. If we compare this parallel Givens reduction with the sequential Gaussian elimination scheme, we obtain a speedup and efficiency that are one third of the above.

Algorithm 7.1 QR by Givens rotations

Input: $A \in \mathbb{R}^{n \times n}$ (for even n).
Output: A is upper triangular.
1: **do** $k = 1 : n - 1$,
2:    **doall** $j = 1 : \lceil k/2 \rceil$,
3:      $i = 2j + n - k - 2$;
4:        Apply on the left side of A the rotation $R_{i,i+1}^{(j)}$ to annihilate the entry $(i + 1, j)$;
5:    **end**
6: **end**
7: **do** $k = n : 2n - 3$,
8:    **doall** $j = k - n + 2 : \lceil k/2 \rceil$,
9:      $i = 2j + n - k - 2$;
10:        Apply on the left side of A the rotation $R_{i,i+1}^{(j)}$ to annihilate the entry $(i + 1, j)$;
11:    **end**
12: **end**

Other orderings for parallel Givens reduction are given in [7–9], with the ordering presented here judged as being asymptotically optimal [10].

If n is not even, we simply consider the orthogonal factorization of $\mathrm{diag}(A, 1)$. Also, using square root-free Givens' rotations, e.g., see [11] or [12], we can obtain the positive diagonal matrix D and the unit upper triangular matrix R in the factorization $QA = D^{1/2}R$ in $O(n)$ parallel arithmetic operations (and no square roots) employing $O(n^2)$ processors.

Observe that Givens rotations are also useful for determining QR factorization of matrices with structures such as Hessenberg matrices or more special patterns (see e.g. [13]).

7.3 QR Factorization via Householder Reductions

Given a vector $a \in \mathbb{R}^s$, an elementary reflector $H = H(v) = I_s - \beta v v^\top$ can be determined such that $Ha = \pm \|a\| e_1$. Here, $\beta = 2/v^\top v$, $v \in \mathbb{R}^s$, and e_1 is the first column of the identity I_s. Note that H is orthogonal and symmetric, i.e. H is involutory. Application of H to a given vector $x \in \mathbb{R}^s$, i.e. $H(v)x = x - \beta(v^\top x)v$ can be realized on a uniprocessor using two BLAS1 routines [14], namely _DOT and _AXPY, involving only $4s$ arithmetic operations.

The orthogonal factorization of A is obtained by successively applying such orthogonal transformations to A: $A_1 = A$ and for $k = 1, \ldots, \min(n, m - 1)$, $A_{k+1} = P_k A_k$ where

$$P_k = \begin{pmatrix} I_{k-1} & 0 \\ 0 & H(v_k) \end{pmatrix}, \tag{7.9}$$

with v_k chosen such that all but the first entry of the column vector $H(v_k)A_k(k : m, k)$ (in Matlab notations) are zero.

At step k, the rest of $A_{k+1} = P_k A_k$ is obtained using the multiplication $H(v_k)A_k(k : m, k : n)$. It can be implemented by two BLAS2 routines [15], namely the matrix-vector multiplication DGEMV and the rank-one update DGER. The procedure stores the matrix A_{k+1} in place of A_k.

On a uniprocessor, the total procedure involves $2n^2(m - n/3) + O(mn)$ arithmetic operations for obtaining R. When necessary for subsequent calculations, the sequence of the vectors $(v_k)_{1,n}$, can be stored for instance in the lower part of the transformed matrix A_k. When an orthogonal basis $\tilde{Q} = [q_1, \ldots, q_n] \in \mathbb{R}^{m \times n}$ of the range of A is needed, the basis is obtained by pre-multiplying the matrix $\begin{pmatrix} I_n \\ 0 \end{pmatrix}$ successively by $P_n, P_{n-1}, \ldots, P_1$. On a uniprocessor, this procedure involves $2n^2(m - n/3) + O(mn)$ additional arithmetic operations instead of $4(m^2 n m n^2 + n^3/3) + O(mn)$ operations when the whole matrix $Q = [q_1, \cdots, q_m] \in \mathbb{R}^{m \times m}$ must be assembled.

Fine Grain Parallelism

The general pipelined implementation has already been presented in Sect. 2.3. Additional details are given in [16]. Here, however, we explore the finest grain tasks that can be used in this orthogonal factorization.

Similar to Sect. 7.2, and in order to present an account similar to that in [17], we assume that A is a square matrix of order n.

The classical Householder reduction produces the sequence $A_{k+1} = P_k A_k$, $k = 1, 2, \ldots, n - 1$, so that A_n is upper triangular, with A_{k+1} given by,

$$A_{k+1} = \begin{pmatrix} R_{k-1} & b_{k-1} & B_{k-1} \\ 0 & \rho_{kk} e_1 & H_k C_{k-1} \end{pmatrix}.$$

where R_{k-1} is upper triangular of order $k-1$. The elementary reflector $H_k = H_k(v_k)$, and the scalar ρ_{kk} can be obtained in $(3 + \log(n - k + 1))$ parallel arithmetic operations and one square root, using $(n - k + 1)$ processors. In addition, $H_k C_{k-1}$ can be computed in $(4 + \log(n - k + 1))$ parallel arithmetic operations using $(n - k + 1)^2$ processors. Thus, the total cost of obtaining the orthogonal factorization $Q^\top A = R$, where $Q = P_1 P_2 \cdots P_{n-1}$ is given by,

$$\sum_{r=1}^{n-1}(7 + 2\log(n - r + 1)) = 2n \log n + O(n),$$

parallel arithmetic operations and $(n - 1)$ square roots, using no more than n^2 processors. Since the sequential algorithm requires $T_1 = O(n^3)$ arithmetic operations, we obtain a speedup of $S_p = O(n^2/\log n)$ and an efficiency E_p proportional to $(1/\log n)$ using $p = O(n^2)$ processors. Such a speedup is not as good as that realized by parallel Givens' reduction.

Block Reflectors

Block versions of this Householder reduction have been introduced in [18] (the WY form), and in [19] (the $GG^\top$ form). The generation and application of the WY form are implemented in the routines of LAPACK [20]. They involve BLAS3 primitives [21]. More specifically, the WY form consists of considering a narrow window of few columns that is reduced to the upper triangular form using elementary reflectors. If s is the window width, the s elementary reflectors are accumulated first in the form $P_{k+s} \cdots P_{k+1} = I + WY^\top$ where $W, Y \in \mathbb{R}^{m \times s}$. This expression allows the use of BLAS3 in updating the remaining part of A.

On a limited number of processors, the block version of the algorithm is the preferred parallel scheme; it is implemented in SCALAPACK [22] where the matrix A is distributed on a two-dimensional grid of processes according to the block cyclic scheme. The block size is often chosen large enough to allow BLAS3 routines to achieve maximum performance on each involved uniprocessor. Note, however, that while increasing the block size improves the granularity of the computation, it may negatively affect concurrency. An optimal tradeoff therefore must be found depending on the architecture of the computing platform.

7.4 Gram-Schmidt Orthogonalization

The goal is to compute an orthonormal basis $Q = [q_1, \ldots, q_n]$ of the subspace spanned by the columns of the matrix $A = [a_1, \ldots, a_n]$, in such a way that, for $1 \leq k \leq n$, the columns of $Q_k = [q_1, \ldots, q_k]$ is a basis of the subspace spanned by the first k columns of A. The Gram-Schmidt schemes consists of applying successively the transformations $P_k, k = 1, \ldots, n - 1$, where P_k is the orthogonal projector onto the orthogonal complement of the subspace spanned by $\{a_1, \ldots, a_k\}$. $R = Q^\top A$ can be built step-by-step during the process.

Two different versions of the algorithm are obtained by expressing P_k in distinct ways:

CGS: In the classical Gram-Schmidt (Algorithm 7.2): $P_k = I - Q_k Q_k^\top$, where $Q_k = [q_1, \ldots, q_k]$. Unfortunately, it was proven in [23] to be numerically unreliable except when it is applied two times which makes it numerically equivalent to the following, modified Gram-Schmidt (MGS) procedure.

MGS: In the modified Gram-Schmidt (Algorithm 7.3): $P_k = (I - q_k q_k^\top) \cdots (I - q_1 q_1^\top)$. Numerically, the columns of Q are orthogonal up to a tolerance determined by the machine precision multiplied by the condition number of A [23]. By applying the algorithm a second time (i.e. with a complete reorthogonalization), the columns of Q become orthogonal up to the tolerance determined by the machine precision parameter similar to the Householder or Givens reductions.

Algorithm 7.2 CGS: classical Gram-Schmidt

Input: $A = [a_1, \cdots, a_n] \in \mathbb{R}^{m \times n}$.
Output: $Q = [q_1, \cdots, q_n] \in \mathbb{R}^{m \times n}$, orthonormal basis of $\mathscr{R}(A)$.
1: $q_1 = a_1 / \|a_1\|$;
2: **do** $k = 1 : n - 1$,
3:     $r = Q_k^\top a_{k+1}$
4:     $w = a_{k+1} - Q_k r$;
5:     $q_{k+1} = w / \|w\|$;
6: **end**

Algorithm 7.3 MGS: modified Gram-Schmidt

Input: $A = [a_1, \cdots, a_n] \in \mathbb{R}^{m \times n}$.
Output: $Q = [q_1, \cdots, q_n] \in \mathbb{R}^{m \times n}$, orthonormal basis of $\mathscr{R}(A)$.
1: $q_1 = a_1 / \|a_1\|$;
2: **do** $k = 1 : n - 1$,
3:     $w = a_{k+1}$;
4:     **do** $j = 1 : k$,
5:       $\alpha_{j,k+1} = q_j^\top w$;
6:       $w = w - \alpha_{j,k+1} q_j$;
7:     **end**
8:     $q_{k+1} = w / \|w\|$;
9: **end**

The basic procedures involve $2mn^2 + O(mn)$ arithmetic operations on a uniprocessor. CGS is based on BLAS2 procedures but it must be applied two times while MGS is based on BLAS1 routines. As mentioned in Sect. 2.3, by inverting the two loops of MGS, the procedure can proceed with a BLAS2 routine. This is only possible when A is available explicitly (for instance, this is not possible with the Arnoldi process as defined in Algorithm 9.3).

The parallel implementations of Gram-Schmidt orthogonalization has been discussed in Sect. 2.3. For distributed memory architectures, however, a block partitioning must be considered, similar to that of SCALAPACK for MGS. In order to proceed with BLAS3 routines, a block version BGS is obtained by considering blocks of vectors instead of single vectors as in MGS , and by replacing the normalizing step by an application of MGS on the individual blocks. Here, the matrix A is partitioned as $A = (A_1, \ldots, A_\ell)$, where we assume that ℓ divides n $(n = \ell q)$. The basic primitives used in that algorithm belong to BLAS3 thus insuring high performance on most architectures. Note that it is necessary to apply the normalizing step twice to reach the numerical accuracy of MGS, e.g. see [24]. The resulting method B2GS is given in Algorithm 7.4. The number of arithmetic operations required in BGS is the same as

---

Algorithm 7.4 B2GS: block Gram-Schmidt

Input: $A = [A_1, \cdots , A_\ell] \in \mathbb{R}^{m \times n}$.
Output: $Q = [Q_1, \cdots , Q_\ell] \in \mathbb{R}^{m \times n}$, orthonormal basis of $\mathscr{R}(A)$.
1: $W = \mathrm{MGS}(A_1)$;
2: $Q_1 = \mathrm{MGS}(W)$;
3: **do** $k = 1 : \ell - 1$,
4:     $W = A_{k+1}$;
5:     **do** $i = 1 : k$,
6:         $R_{i,k+1} = Q_i^\top W$;
7:         $W = W - Q_i R_{i,k+1}$;
8:     **end**
9:     $W = \mathrm{MGS}(W)$;
10:     $Q_{k+1} = \mathrm{MGS}(W)$;
11: **end**

---

that of MGS $(2mn^2)$ while that of B2GS involves $2m\ell q^2 = 2mnq$ additional arithmetic operations to reorthogonalize the ℓ blocks of size $m \times q$. Thus, the number of arithmetic operations required by B2GS is $(1 + \frac{1}{\ell})$ times that of MGS. Consequently, for B2GS to be competitive, we need ℓ to be large or the number of columns in each block to be small. Having blocks with a small number of columns, however, will not allow us to capitalize on the higher performance afforded by BLAS3. Using blocks of 32 or 64 columns is often a reasonable compromise on many architectures.

7.5 Normal Equations Versus Orthogonal Reductions

For linear least squares problems (7.4) in which $m \gg n$, normal equations are very often used in several applications including statistical data analysis. In matrix computation literature, however, one is warned of possible loss of information in forming $A^\top A$ explicitly, and solving the linear system,

$$A^\top A x = A^\top b. \tag{7.10}$$

whose condition number is the square of that of A.

If it is known a priori that A is properly scaled and extremely well conditioned, one may take advantage of the high parallel scalability of the matrix multiplication $A^\top(A, b)$, and solving a small linear system of order n using any of the orthogonal factorization schemes discussed so far. Note that the total number of arithmetic operations required for solving the linear least squares problem using the normal equations approach is $2n^2 m + O(n^3)$, while that required using Householder's reduction is $2n^2 m + O(mn)$. Hence, from the sequential complexity point of view and considering the high parallel scalability of the multiplication $A^\top(A, b)$, the normal equations approach has an advantage of realizing more parallel scalability whenever $m \gg n^2$.

7.6 Hybrid Algorithms When $m \gg n$

As an alternative to the normal equations approach for solving the linear least squares problem (7.4) when $m \gg n$, we offer the following orthogonal factorization scheme for tall-narrow matrices of maximal column rank. Such matrices arise in a variety of applications. In this book, we consider two examples: (i) the adjustment of geodetic networks (see Sect. 7.7), and (ii) orthogonalization of a nonorthogonal Krylov basis (see Sect. 9.3.2). In both cases, m could be larger than 10^6 while n is as small as several hundreds.

The algorithm for the orthogonal factorization of a tall-narrow matrix A on a distributed memory architecture was first introduced in [25]. For the sake of illustration, consider the case of using two multicore nodes. Partitioning A as follows: $A^\top = [A_1^\top, A_2^\top]$, In the first stage of the algorithm, the orthogonal factorization of A_1 and A_2 is accomplished simultaneously to result in the upper triangular factors T_1 and T_2, respectively (Fig. 7.2a). Thus, the first most time consuming stage

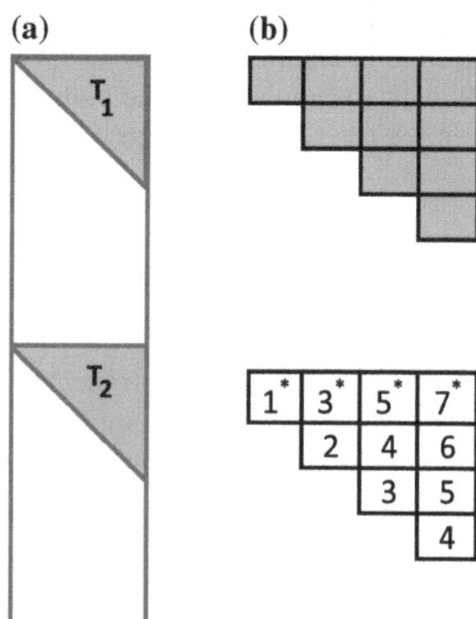

Fig. 7.2 Parallel annihilation when $m \gg n$: a two-step procedure

is accomplished without any communication overhead. In Fig. 7.2b, T_1 and T_2 are depicted as upper triangular matrices of size 4 in order to simplify describing the second stage – how Givens rotations can be used to annihilate T_2 and update T_1.

Clearly, the entry $(1, 1)$ of the first row of T_2 can be annihilated by rotating the first rows of T_1 and T_2 incurring communication overhead. Once the $(1, 1)$ entry of T_2 is annihilated, it is possible to annihilate all entries on the first diagonal of T_2 by rotating the appropriate rows of T_2 without the need for communication with the first node containing T_1. This is followed by rotating row 2 of T_1 with row 1 of T_2 to annihilate the entry $(1, 2)$ in T_2 incurring communication overhead. Consequently, local rotations in the second node can annihilate all elements on the second diagonal of T_2, and so on. Overlapping the communication-free steps with those requiring communication, we can accomplish the annihilation of T_2 much faster. The ordering of rotations in such a strategy is shown in Fig. 7.2b. Here, as soon as entry $(2, 2)$ of T_2 is annihilated, the entry $(1, 2)$ of T_2 can be annihilated (via internode communications) simultaneously with the communication-free annihilation of entry $(3, 3)$ of T_2, and so on. Rotations indicated with stars are the only ones which require internode communication. By following this strategy, the method RODDEC in [26] performs the orthogonal factorization of a matrix on a ring of processors. It is clear that for a ring of p processors, the method is weakly scalable since the first stage is void of any communication, while in the second stage we have only nearest neighbor communications (i.e. no global communication). Variants of this approach have also been recently considered in [27].

7.7 Orthogonal Factorization of Block Angular Matrices

Here we consider the orthogonal factorization of structured matrices of the form,

$$
A = \begin{pmatrix} B_1 & & & & C_1 \\ & B_2 & & & C_2 \\ & & \ddots & & \vdots \\ & & & B_p & C_p \end{pmatrix}
\tag{7.11}
$$

where A, as well as each block B_i, for $i = 1, \ldots, p$, is of full column rank. The block angular form implies that the QR factorization of A yields an upper triangular matrix with the same block structure; this can be seen from the Cholesky factorization of the block arrowhead matrix $A^\top A$. Therefore, similar to the first stage of the algorithm described above in Sect. 7.6, for $i = 1, \ldots, p$, multicore node i first performs the orthogonal factorization of the block (B_i, C_i). The transformed block is now of the form $\begin{pmatrix} R_i & E_{i1} \\ 0 & E_{i2} \end{pmatrix}$. As a result, after the first orthogonal transformation, and an appropriate permutation, we have a matrix of the form:

$$A_1 = \begin{pmatrix} R_1 & & & & E_{11} \\ & R_2 & & & E_{21} \\ & & \ddots & & \vdots \\ & & & R_p & E_{p1} \\ & & & & E_{12} \\ & & & & E_{22} \\ & & & & \vdots \\ & & & & E_{p2} \end{pmatrix}. \tag{7.12}$$

Thus, to complete the orthogonal factorization process, we need to obtain the QR-factorization of the tall and narrow matrix consisting of the sub-blocks E_{i2}, $i = 1, \ldots, p$. using our orthogonal factorization scheme described in (7.6). Other approaches for the orthogonal factorization of block angular matrices are considered in [5].

Adjustment of Geodetic Networks

As an illustration of the application of our tall and narrow orthogonal factorization scheme, we consider the problem of adjustment of geodetic networks which has enjoyed a revival with the advent of the Global Positioning System (GPS), e.g. see [28, 29]. The geodetic least squares adjustment problem is to compute accurately the coordinates of points (or stations) on the surface of the Earth from a collection of measured distances and angles between these points. The computations are based upon a geodetic position network which is a mathematical model consisting of several mesh points or geodetic stations, with unknown positions over a reference surface or in 3-dimensional space. These observations then lead to a system of overdetermined nonlinear equations involving, for example, trigonometric identities and distance formulas relating the unknown coordinates. Each equation typically involves only a small number of unknowns. Thus, resulting in sparse overdetermined system of nonlinear equations,

$$F(x) = q \tag{7.13}$$

where x is the vector containing the unknown coordinates, and q represents the observation vector. Using the Gauss-Newton method for solving the nonlinear system (7.13), we will need to solve a sequence of linear least squares problems of the form (7.4), where A denotes the Jacobian of F at the current vector of unknown coordinates, initially x^0, and $r^0 = q - F(x^0)$. The least squares solution vector y is the adjustment vector, i.e., $x^1 = x^0 + y$ is the improved approximation of the coordinate vector x, e.g. see [30, 31].

The geodetic adjustment problem just described has the computationally convenient feature that the geodetic network domain can be readily decomposed into smaller subproblems. This decomposition is based upon the Helmert blocking of the network as described in [32]. Here, the observation matrix A is assembled into the block angular form (7.11) by the way in which the data is collected by regions.

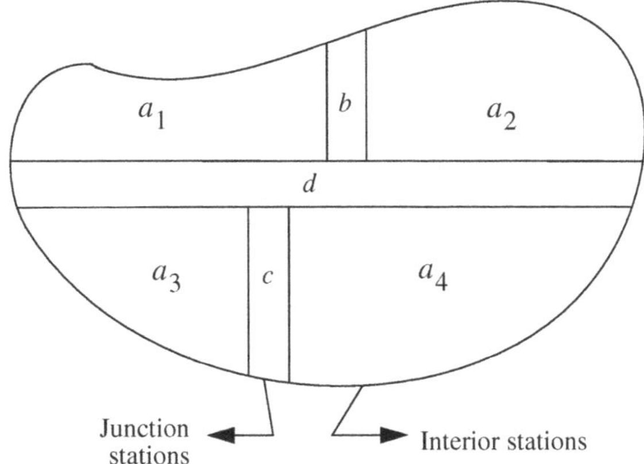

Fig. 7.3 Geographical partition of a geodetic network

If the geodetic position network on a geographical region is partitioned as shown in Fig. 7.3, with two levels of nested bisection, the corresponding observation matrix A has the particular block-angular form shown in (7.14),

$$
A = \begin{pmatrix}
A_1 & & & B_1 & D_1 \\
& A_2 & & B_2 & D_2 \\
& & A_3 & C_1 & D_3 \\
& & A_4 & C_2 & D_4
\end{pmatrix}
\tag{7.14}
$$

where for the sake of illustration we assume that $A_i \in \mathbb{R}^{m_i \times n_i}$, $D_i \in \mathbb{R}^{m_i \times n_d}$, for $i = 1, \ldots, 4$, and $B_j \in \mathbb{R}^{m_j \times n_b}$, $C_j \in \mathbb{R}^{m_{j+2} \times n_c}$, for $j = 1, 2$ are all with more rows than columns and of maximal column rank.

In order to solve this linear least squares problem on a 4-cluster system, in which each cluster consisting of several multicore nodes, via orthogonal factorization of A, we proceed as follows:

Stage 1: Let cluster i (or rather its global memory) contain the observations corresponding to region a_i and its share of regions (b or c) and d. Now, each cluster i proceeds with the orthogonal factorization of A_i and updating its portion of (B_i or C_{i-2}) and D_i. Note that in this stage, each cluster operates in total independence of the other three clusters. The orthogonal factorization on each cluster can be performed by one of the algorithms described in Sects. 7.3 or 7.6.

The result of this reduction is a matrix $A^{(1)}$ with the following structure

$$
A^{(1)} = \left(\begin{array}{ccccc}
R_1 & & & * & * \\
0 & & & * & * \\ \hline
& R_2 & & * & * \\
& 0 & & * & * \\ \hline
& & R_3 & * & * \\
& & 0 & * & * \\ \hline
& & R_4 & * & * \\
& & 0 & * & *
\end{array}\right)
\begin{array}{l}
\text{cluster 1} \\[4pt] \hline
\text{cluster 2} \\[4pt] \hline
\text{cluster 3} \\[4pt] \hline
\text{cluster 4}
\end{array}
\tag{7.15}
$$

where $R_i \in \mathbb{R}^{n_i \times n_i}$ are upper triangular and nonsingular for $i = 1, \ldots, 4$. The stars represent the resulting blocks after the orthogonal transformations are applied to the blocks B_*, C_* and D_*.

Stage 2: Each cluster continues, independently, the orthogonal factorization procedure: clusters 1 and 2 factor their portions of the transformed blocks B and clusters 3 and 4 factor their portions of the transformed blocks C. The resulting matrix, $A^{(2a)}$ is of the form,

$$
A^{(2a)} =
\left(
\begin{array}{ccc}
R_1 & * & * \\
0 & T_1 & * \\
0 & 0 & * \\
\hline
R_2 & * & * \\
0 & T_2 & * \\
0 & 0 & * \\
\hline
R_3 & * & * \\
0 & T_3 & * \\
0 & 0 & * \\
\hline
R_4 & * & * \\
0 & T_4 & * \\
0 & 0 & *
\end{array}
\right)
\begin{array}{l}
\text{cluster 1} \\ \\ \\
\text{cluster 2} \\ \\ \\
\text{cluster 3} \\ \\ \\
\text{cluster 4}
\end{array}
\tag{7.16}
$$

where the matrices $T_i \in \mathbb{R}^{n_b \times n_b}$ and $T_{i+2} \in \mathbb{R}^{n_c \times n_c}$ are upper triangular, $i = 1, 2$. Now clusters 1, and 2, and clusters 3 and 4 need to cooperate (i.e. communicate) so as to annihilate the upper triangular matrices T_1 and T_4, respectively, to obtain the matrix $A^{(2b)}$ which is of the following structure,

$$
A^{(2b)} =
\left(
\begin{array}{ccc}
R_1 & * & * \\
0 & 0 & S_1 \\
\hline
R_2 & * & * \\
0 & \tilde{T}_2 & * \\
0 & 0 & S_2 \\
\hline
R_3 & * & * \\
0 & \tilde{T}_3 & * \\
0 & 0 & S_3 \\
\hline
R_4 & * & * \\
0 & 0 & S_4
\end{array}
\right)
\begin{array}{l}
\text{cluster 1} \\ \\
\text{cluster 2} \\ \\ \\
\text{cluster 3} \\ \\ \\
\text{cluster 4}
\end{array}
\tag{7.17}
$$

where $\tilde{T}_2$ and $\tilde{T}_3$ are upper triangular and nonsingular. This annihilation is organized so as to minimize intercluster communication. To illustrate this procedure, consider the annihilation of T_1 by T_2 where each is of order n_b. It is performed by elimination of diagonals as outlined in Sect. 7.6 (see Fig. 7.2). The algorithm may be described as follows:

do $k = 1 : n_b$,
    rotate $e_k^\top T_2$ and $e_1^\top T_1$ to annihilate the element of T_1 in position $(1, k)$, where e_k denotes the k-th column of the identity (requires intercluster communication).
    do $i = 1 : n_b - k$,
        rotate rows i and $i + 1$ of T_1 to annihilate the element in position $(i + 1, i + k)$ (local to the first cluster).
    end
end

Stage 3: This is similar to stage 2. Each cluster i obtains the orthogonal factorization of S_i in (7.17). Notice here, however, that the computational load is not perfectly balanced among the four clusters since S_2 and S_3 may have fewer rows that S_1 and S_4. The resulting matrix $A^{(3a)}$ is of the form,

$$A^{(3a)} = \left(\begin{array}{c|cc}
R_1 & * & * \\
0 & 0 & V_1 \\
0 & 0 & 0 \\
\hline
R_2 & * & * \\
0 & \tilde{T}_2 & * \\
0 & 0 & V_2 \\
\hline
R_3 & * & * \\
0 & \tilde{T}_2 & * \\
0 & 0 & V_3 \\
\hline
R_4 & * & * \\
0 & 0 & V_4 \\
0 & 0 & 0
\end{array}\right)
\begin{array}{l}
\text{cluster 1} \\ \\ \\
\text{cluster 2} \\ \\ \\
\text{cluster 3} \\ \\ \\
\text{cluster 4}
\end{array}
\tag{7.18}$$

where V_j is nonsingular upper triangular, $j = 1, 2, 3, 4$. This is followed by the annihilation of all except the one in *cluster2*. Here, clusters 1 and 2 cooperate (i.e. communicate) to annihilate V_1, while clusters 3 and 4 cooperate to annihilate V_4. Next, clusters 2 and 3 cooperate to annihilate V_3. Such annihilation of the V_i's is performed as described in stage 2. The resulting matrix structure is given by:

$$A^{(3b)} = \left(\begin{array}{c|cc}
R_1 & * & * \\
0 & 0 & 0 \\
\hline
R_2 & * & * \\
0 & \tilde{T}_2 & * \\
0 & 0 & \tilde{V}_2 \\
\hline
R_3 & * & * \\
0 & \tilde{T}_3 & * \\
0 & 0 & 0 \\
\hline
R_4 & * & * \\
0 & 0 & 0
\end{array}\right)
\begin{array}{l}
\text{cluster 1} \\ \\
\text{cluster 2} \\ \\ \\
\text{cluster 3} \\ \\ \\
\text{cluster 4}
\end{array}
\tag{7.19}$$

where the stars denote dense rectangular blocks.

Stage 4: If the observation vector f were appended to the block D in (7.14), the back-substitution process would follow immediately. After cluster 2 solves the upper triangular system $\widetilde{V}_2 y = g$, see (7.19), and communicating y to the other three clusters, all clusters update their portion of the right-hand side simultaneously. Next, clusters 2 and 3 simultaneously solve two triangular linear systems involving $\widetilde{T}_2$, and $\widetilde{T}_3$. Following that, clusters 1 and 2, and clusters 3 and 4 respectively cooperate in order to simultaneously update their corresponding portions of the right-hand side. Finally, each *cluster j*, $j = 1, 2, 3, 4$ independently solves a triangular system (involving R_j) to obtain the least squares solution.

7.8 Rank-Deficient Linear Least Squares Problems

When A is of rank $q < n$, there is an infinite set of solutions of (7.4): for if x is a solution, then any vector in the set $\mathscr{S} = x + \mathcal{N}(A)$ is also a solution, where $\mathcal{N}(A)$ is the $(n - q)$-dimensional null space of A. If the Singular Value Decomposition (SVD) of A is given by,

$$A = V \begin{pmatrix} \Sigma & 0_{q,n-q} \\ 0_{m-q,q} & 0_{m-q,n-q} \end{pmatrix} U^\top, \tag{7.20}$$

where the diagonal entries of the matrix $\Sigma = \mathrm{diag}(\sigma_1, \ldots, \sigma_q)$ are the singular values of A, and $U \in \mathbb{R}^{n \times n}$, $V \in \mathbb{R}^{m \times m}$ are orthogonal matrices. If b is the right-hand side of the linear least squares problem (7.4), and $c = V^\top b = \begin{pmatrix} c_1 \\ c_2 \end{pmatrix}$ where $c_1 \in \mathbb{R}^q$, then x is a solution in $\mathscr{S}$ if and only if there exists $y \in \mathbb{R}^{n-q}$ such that $x = U \begin{pmatrix} \Sigma^{-1} c_1 \\ y \end{pmatrix}$. In this case, the corresponding minimum residual norm is equal to $\|c_2\|$. In $\mathscr{S}$, the classically selected solution is the one with smallest 2-norm, i.e. when $y = 0$.

Householder Orthogonalization with Column Pivoting

In order to avoid computing the singular-value decomposition which is quite expensive, the factorization (7.20) is often replaced by the orthogonal factorization introduced in [33]. This procedure computes the QR factorization by Householder reductions with column pivoting. Factorization (7.2) is now replaced by

$$Q_q \cdots Q_2 Q_1 A P_1 P_2 \cdots P_q = \begin{pmatrix} T \\ 0 \end{pmatrix}, \tag{7.21}$$

where matrices $P_1, \ldots, P_q$ are permutations and $T \in \mathbb{R}^{q \times n}$ is upper-trapezoidal. The permutations are determined so as to insure that the absolute values of the diagonal entries of T are non increasing. To obtain the so called complete orthogonal factorization of A the process performs first the orthogonal factorization (7.21), followed by the LQ factorization of T. Here, the LQ factorization is directly obtained through the QR factorization of $T^\top$. Thus, the process yields a factorization similar to (7.20) but with Σ being a nonsingular lower-triangular matrix. In Sect. 7.8 an alternative transformation is proposed.

In floating point arithmetic, the process will not end up with exact zero blocks. Neglecting roundoff errors, we obtain a factorization of the form:

$$Q^\top A P = \begin{pmatrix} R_{11} & R_{12} \\ 0 & R_{22} \end{pmatrix},$$

where $Q = Q_1 Q_2 \cdots Q_q$ and $P_1 P_2 \cdots P_q$. The generalized inverse of A can then approximated by

$$B = P \begin{pmatrix} R_{11} & R_{12} \\ 0 & 0 \end{pmatrix}^+ Q^\top.$$

An upper bound on the Frobenius norm of the error in such an approximation of the generalized inverse is given by

$$\|B - A^+\|_F \leq 2\|R_{22}\|_F \max\{\|A^+\|^2, \|B\|^2\}, \tag{7.22}$$

e.g. see [4].

The basic algorithm is based on BLAS2 routines but a block version based on BLAS3 and its parallel implementation have been proposed in [34, 35]. Dynamic column pivoting limits potential parallelism in the QR factorization. For instance, it prevents the use of the efficient algorithm proposed for tall-narrow matrices (see Sect. 7.6). In order to enhance parallelism, an attempt to avoid global pivoting has been developed in [36]. It consists of partitioning the matrix into blocks of columns and adopting a local pivoting strategy limited to each block which is allocated to a single processor. This strategy is numerically acceptable as long as the condition number of a block remains below some threshold. In order to control numerical quality of the procedure, an incremental condition estimator (ICE) is used, e.g. see [37].

Another technique consists of delaying the pivoting procedure by first performing a regular QR factorization, and then applying the rank revealing procedure on the resulting matrix R. This is considered next.

A Rank-Revealing Post-Processing Procedure

Let $R \in \mathbb{R}^{n \times n}$ be the upper triangular matrix resulting from an orthogonal factorization without pivoting. Next, determine a permutation $P \in \mathbb{R}^{n \times n}$ and an orthogonal matrix $Q \in \mathbb{R}^{n \times n}$ such that,

$$Q^\top R P = \begin{pmatrix} R_{11} & R_{12} \\ 0 & R_{22} \end{pmatrix}, \tag{7.23}$$

is an upper triangular matrix in which $R_{11} \in \mathbb{R}^{q \times q}$, where q is the number of singular values larger than an a priori fixed threshold $\eta > 0$, and the 2-norm $\|R_{22}\| = O(\eta)$. Here, the parameter $\eta > 0$ is chosen depending on the 2-norm of R since at the end of the process $\operatorname{cond}(R_{11}) \leq \frac{\|R\|}{\eta}$.

Algorithm 7.5 Rank-revealing QR for a triangular matrix.

Input: $R \in \mathbb{R}^{n \times n}$ sparse triangular, $\eta > 0$.
Output: S : permuted upper triangular matrix, P is the permutation, q is the rank.
1: $k = 1; \ell = n ; P = I_n; S = R;$
2: $\sigma_{\min} = |S_{11}|; x_{\min} = 1;$
3: **while** $k < \ell,$
4:     **if** $\sigma_{\min} > \eta,$ **then**
5:         $k = k + 1;$
6:         Compute the smallest singular value $\sigma_{\min}$ of $S_{1:k,1:k}$ and its corresponding left
            singular vector $x_{\min};$
7:     **else**
8:         From $x_{\min}$, select the column j to be rejected;
9:         Put the rejected column in the last position;
10:        Eliminate the subdiagonal entries of $S_{k+1:\ell,k+1:\ell};$
11:        Update$(P) ; \ell = \ell - 1 ; k = k - 1;$
12:        Update $\sigma_{\min}$ and $x_{\min};$
13:    **end if**
14: **end while**
15: **if** $\sigma_{\min} > \eta,$ **then**
16:    $q = \ell;$
17: **else**
18:    $q = \ell - 1;$
19: **end if**

This post-processing procedure is given in Algorithm 7.5. The output matrix is computed column by column. At step k, the smallest singular value of the triangular matrix $S_k \in \mathbb{R}^{k \times k}$ is larger than η. If, by appending the next column, the smallest singular value passes below the threshold η, one of the columns is moved to the right end of the matrix and the remaining columns to be processed are shifted to the left by one position. This technique assumes that the smallest singular value and corresponding left singular vector can be computed. The procedure ICE [37] updates the estimate of the previous rank with $O(k)$ operations. After an updating step to restore the upper triangular form (by using Givens rotations), the process is repeated.

Proposition 7.1 *Algorithm 7.5 involves $O(Kn^2)$ arithmetic operations where K is the number of occurrences of column permutation.*

Proof Let assume that at step k, column k is rejected in the last position (Step 9 of Algorithm 7.5). Step 10 of the algorithm is performed by the sequence of Givens rotations $R_{j,j+1}^{(j)}$ for $j = k+1, \ldots, \ell$ where $R_{j,j+1}^{(j)}$ is defined in (7.7). This process involves $O((l - k)^2)$ arithmetic operations which is bounded by $O(n^2)$. This proves the proposition since the total use of ICE involves $O(n^2)$ arithmetic operations as well.

Algorithm 7.5 insures that $\|R_{22}\| = O(\eta)$ as shown in the following theorem:

Theorem 7.2 *Let us assume that the smallest singular values of blocks defined by the Eq. (7.23) satisfy the following conditions:*

$$\sigma_{\min}(R_{11}) > \eta \ and \ \sigma_{\min}\left(\begin{pmatrix} R_{11} & u_1 \\ 0 & u_2 \end{pmatrix}\right) \leq \eta$$

for any column $\begin{pmatrix} u_1 \\ u_2 \end{pmatrix}$ *of the matrix* $\begin{pmatrix} R_{12} \\ R_{22} \end{pmatrix} \in \mathbb{R}^{n \times (n-q)}$. *Therefore the following bounds hold*

$$\|u_2\| \leq \left(\frac{\sqrt{1 + \|R_{11}^{-1}u_1\|^2}}{1 - \eta\|R_{11}^{-1}\|}\right)\eta, \tag{7.24}$$

$$\|R_{22}\|_F \leq \left(\frac{\sqrt{n - q + \|R_{11}^{-1}R_{12}\|_F^2}}{1 - \eta\|R_{11}^{-1}\|}\right)\eta, \tag{7.25}$$

where $R_{11} \in \mathbb{R}^{q \times q}$ *and where* $\|.\|_F$ *denotes the Frobenius norm.*

Proof We have

$$\eta \geq \sigma_{\min}\left(\begin{pmatrix} R_{11} & u_1 \\ 0 & u_2 \end{pmatrix}\right) = \sigma_{\min}\left(\begin{pmatrix} R_{11} & u_1 \\ 0 & \rho \end{pmatrix}\right),$$

where $\rho = \|u_2\|$. Therefore,

$$\frac{1}{\eta} \leq \left\|\begin{pmatrix} R_{11}^{-1} & -\frac{1}{\rho}R_{11}^{-1}u_1 \\ 0 & \frac{1}{\rho} \end{pmatrix}\right\|,$$

$$\leq \|R_{11}^{-1}\| + \frac{1}{\rho}\sqrt{1 + \|R_{11}^{-1}u_1\|^2}$$

which implies (7.24). The bound (7.25) is obtained by summing the squares of the bounds for all the columns of R_{22}.

Remark 7.1 The hypothesis of the theorem insure that $\tau_{\text{sep}} = 1 - \eta\|R_{11}^{-1}\|$ is positive. This quantity, however, might be too small if the singular values of R are not well separated in the neighborhood of η.

Remark 7.2 The number of column permutations corresponds to the rank deficiency:

$$K = n - q,$$

where q is the η-rank of A.

Once the decomposition (7.23) is obtained, the generalized inverse of R is approximated by

$$S = P \begin{pmatrix} R_{11} & R_{12} \\ 0 & 0 \end{pmatrix}^{+}. \tag{7.26}$$

An upper bound for the approximation error is given by (7.22). To finalize the complete orthogonalization, we post-multiply $T = \begin{pmatrix} R_{11} & R_{12} \end{pmatrix} \in \mathbb{R}^{q \times n}$ by the orthogonal matrix Q obtained from the QR factorization of $T^{\top}$:

$$(L, 0) = T Q. \tag{7.27}$$

This post processing rank-revealing procedure offers only limited chances for exploiting parallelism. In fact, if the matrix A is tall and narrow or when its rank deficiency is relatively small, this post processing procedure can be used on a uniprocessor.

References

1. Halko, N., Martinsson, P., Tropp, J.: Finding structure with randomness: probabilistic algorithms for constructing approximate matrix decompositions. SIAM Rev. **53**(2), 217–288 (2011). doi:10.1137/090771806
2. Hastie, T., Tibshirani, R., Friedman, J.: The Elements of Statistical Learning. Springer Series in Statistics. Springer, New York (2001)
3. Kontoghiorghes, E.: Handbook of Parallel Computing and Statistics. Chapman & Hall/CRC, New York (2005)
4. Golub, G., Van Loan, C.: Matrix Computations, 4th edn. Johns Hopkins (2013)
5. Björck, Å.: Numerical Methods for Least Squares Problems. SIAM, Philadelphia (1996)
6. Sameh, A., Kuck, D.: On stable parallel linear system solvers. J. Assoc. Comput. Mach. **25**(1), 81–91 (1978)
7. Modi, J., Clarke, M.: An alternative givens ordering. Numerische Mathematik **43**, 83–90 (1984)
8. Cosnard, M., Muller, J.M., Robert, Y.: Parallel QR decomposition of a rectangular matrix. Numerische Mathematik **48**, 239–249 (1986)
9. Cosnard, M., Daoudi, E.: Optimal algorithms for parallel Givens factorization on a coarse-grained PRAM. J. ACM **41**(2), 399–421 (1994). doi:10.1145/174652.174660
10. Cosnard, M., Robert, Y.: Complexity of parallel QR factorization. J. ACM **33**(4), 712–723 (1986)
11. Gentleman, W.M.: Least squares computations by Givens transformations without square roots. IMA J. Appl. Math. **12**(3), 329–336 (1973)
12. Hammarling, S.: A note on modifications to the givens plane rotation. J. Inst. Math. Appl. **13**, 215–218 (1974)
13. Kontoghiorghes, E.: Parallel Algorithms for Linear Models: Numerical Methods and Estimation Problems. Advances in Computational Economics. Springer, New York (2000). http://books.google.fr/books?id=of1ghCpWOXcC
14. Lawson, C., Hanson, R., Kincaid, D., Krogh, F.: Basic linear algebra subprograms for Fortran usage. ACM Trans. Math. Softw. **5**(3), 308–323 (1979)
15. Dongarra, J., Croz, J.D., Hammarling, S., Hanson, R.: An extended set of FORTRAN basic linear algebra subprograms. ACM Trans. Math. Softw. **14**(1), 1–17 (1988)

16. Gallivan, K.A., Plemmons, R.J., Sameh, A.H.: Parallel algorithms for dense linear algebra computations. SIAM Rev. **32**(1), 54–135 (1990). doi:10.1137/1032002

17. Sameh, A.: Numerical parallel algorithms—a survey. In: Kuck, D., Lawrie, D., Sameh, A. (eds.) High Speed Computer and Algorithm Optimization, pp. 207–228. Academic Press, San Diego (1977)

18. Bischof, C., van Loan, C.: The WY representation for products of Householder matrices. SIAM J. Sci. Stat. Comput. **8**(1), 2–13 (1987). doi:10.1137/0908009

19. Schreiber, R., Parlett, B.: Block reflectors: theory and computation. SIAM J. Numer. Anal. **25**(1), 189–205 (1988). doi:10.1137/0725014

20. Anderson, E., Bai, Z., Bischof, C., Blackford, S., Demmel, J., Dongarra, J., Du Croz, J., Greenbaum, A., Hammarling, S., McKenney, A., Sorensen, D.: LAPACK Users' Guide, 3rd edn. Society for Industrial and Applied Mathematics, Philadelphia (1999)

21. Dongarra, J., Du Croz, J., Hammarling, S., Duff, I.: A set of level-3 basic linear algebra subprograms. ACM Trans. Math. Softw. **16**(1), 1–17 (1990)

22. Blackford, L., Choi, J., Cleary, A., D'Azevedo, E., Demmel, J., Dhillon, I., Dongarra, J., Hammarling, S., Henry, G., Petitet, A., Stanley, K., Walker, D., Whaley, R.: ScaLAPACK User's Guide. SIAM, Philadelphia (1997). http://www.netlib.org/scalapack

23. Björck, Å.: Solving linear least squares problems by Gram-Schmidt orthogonalization. BIT **7**, 1–21 (1967)

24. Jalby, W., Philippe, B.: Stability analysis and improvement of the block Gram-Schmidt algorithm. SIAM J. Stat. Comput. **12**(5), 1058–1073 (1991)

25. Sameh, A.: Solving the linear least-squares problem on a linear array of processors. In: Snyder, L., Gannon, D., Jamieson, L.H., Siegel, H.J. (eds.) Algorithmically Specialized Parallel Computers, pp. 191–200. Academic Press, San Diego (1985)

26. Sidje, R.B.: Alternatives for parallel Krylov subspace basis computation. Numer. Linear Algebra Appl. **4**, 305–331 (1997)

27. Demmel, J., Grigori, L., Hoemmen, M., Langou, J.: Communication-optimal parallel and sequential QR and LU factorizations. SIAM J. Sci. Comput. **34**(1), 206–239 (2012). doi:10.1137/080731992

28. Chang, X.W., Paige, C.: An algorithm for combined code and carrier phase based GPS positioning. BIT Numer. Math. **43**(5), 915–927 (2003)

29. Chang, X.W., Guo, Y.: Huber's M-estimation in relative GPS positioning: computational aspects. J. Geodesy **79**(6–7), 351–362 (2005)

30. Bomford, G.: Geodesy, 3rd edn. Clarendon Press, England (1971)

31. Golub, G., Plemmons, R.: Large scale geodetic least squares adjustment by dissection and orthogonal decomposition. Numer. Linear Algebra Appl. **35**, 3–27 (1980)

32. Wolf, H.: The Helmert block method—its origin and development. In: Proceedings of the Second Symposium on Redefinition of North American Geodetic Networks, pp. 319–325 (1978)

33. Businger, P., Golub, G.H.: Linear least squares solutions by Householder transformations. Numer. Math. **7**, 269–276 (1965)

34. Quintana-Ortí, G., Quintana-Ortí, E.: Parallel algorithms for computing rank-revealing QR factorizations. In: Cooperman, G., Michler, G., Vinck, H. (eds.) Workshop on High Performance Computing and Gigabit Local Area Networks. Lecture Notes in Control and Information Sciences, pp. 122–137. Springer, Berlin (1997). doi:10.1007/3540761691_9

35. Quintana-Ortí, G., Sun, X., Bischof, C.: A BLAS-3 version of the QR factorization with column pivoting. SIAM J. Sci. Comput. **19**, 1486–1494 (1998)

36. Bischof, C.: A parallel QR factorization algorithm using local pivoting. In: Proceedings of 1988 ACM/IEEE Conference on Supercomputing, Supercomputing'88, pp. 400–499. IEEE Computer Society Press, Los Alamitos (1988)

37. Bischof, C.H.: Incremental condition estimation. SIAM J. Matrix Anal. Appl. **11**, 312–322 (1990)

Chapter 8
The Symmetric Eigenvalue
and Singular-Value Problems

Eigenvalue problems form the second most important class of problems in numerical linear algebra. Unlike linear system solvers which could be direct, or iterative, eigensolvers can only be iterative in nature. In this chapter, we consider real symmetric eigenvalue problems (and by extension complex hermitian eigenvalue problems), as well as the problem of computing the singular value decomposition.

Given a symmetric matrix $A \in \mathbb{R}^{n \times n}$, the standard eigenvalue problem consists of computing the eigen-elements of A which are:

- *either all or a few selected eigenvalues only*: these are all the n, or $p \ll n$ selected, roots λ of the nth degree characteristic polynomial:

$$\det(A - \lambda I) = 0. \tag{8.1}$$

The set of all eigenvalues is called the spectrum of A: $\Lambda(A) = \{\lambda_1, \ldots, \lambda_n\}$, in which each λ_j is real.

- *or eigenpairs*: in addition to an eigenvalue λ, one seeks the corresponding eigenvector $x \in \mathbb{R}^n$ which is the nontrivial solution of the singular system

$$(A - \lambda I)x = 0. \tag{8.2}$$

The procedures discussed in this chapter apply with minimal changes in case the matrix A is complex Hermitian since the eigenvalues are still real and the complex matrix of eigenvectors is unitary.

Since any symmetric or Hermitian matrix can be diagonalized by an orthogonal or a unitary matrix, the eigenvalues are insensitive to small symmetric perturbations of A. In other words, when $A + E$ is a symmetric update of A, the eigenvalues of $A + E$ cannot be at a distance exceeding $\|E\|_2$ compared to those of A. Therefore, the computation of the eigenvalues of A with largest absolute values is always well conditioned. Only the relative accuracy of the eigenvalues with smallest absolute values can be affected by perturbations. Accurate computation of the eigenvectors is more difficult in case of poorly separated eigenvalues, e.g. see [1, 2].

© Springer Science+Business Media Dordrecht 2016
E. Gallopoulos et al., *Parallelism in Matrix Computations*,
Scientific Computation, DOI 10.1007/978-94-017-7188-7_8

The Singular Value Decomposition of a matrix can be expressed as a symmetric eigenvalue problem.

Theorem 8.1 *Let $A \in \mathbb{R}^{m \times n}$ ($m \geq n$) have the singular value decomposition*

$$V^\top A U = \Sigma,$$

where $U = [u_1, \ldots, u_n] \in \mathbb{R}^{n \times n}$ and $V = [v_1, \ldots, v_m] \in \mathbb{R}^{m \times m}$ are orthogonal matrices and $\Sigma \in \mathbb{R}^{m \times n}$ is a rectangular matrix which is diagonal. Then, the symmetric matrix

$$A_{nrm} = A^\top A \in \mathbb{R}^{n \times n}, \tag{8.3}$$

has eigenvalues $\sigma_1^2 \geq \cdots \geq \sigma_n^2 \geq 0$, corresponding to the eigenvectors (u_i), $(i = 1, \ldots, n)$. A_{nrm} is called the matrix of the normal equations. The symmetric matrix

$$A_{\mathrm{aug}} = \begin{pmatrix} 0 & A \\ A^\top & 0 \end{pmatrix} \tag{8.4}$$

has eigenvalues $\pm\sigma_1, \ldots, \pm\sigma_n$, corresponding to the eigenvectors

$$\frac{1}{\sqrt{2}} \begin{pmatrix} v_i \\ \pm u_i \end{pmatrix}, \; i = 1, \ldots, n.$$

A_{aug} is called the augmented matrix.

Parallel Schemes

We consider the following three classical standard eigenvalue solvers for problems (8.1) and (8.2):

- Jacobi iterations,
- QR iterations, and
- the multisectioning method.

Details of the above three methods for uniprocessors are given in many references, see e.g. [1–4].

While Jacobi's method is capable of yielding more accurate eigenvalues and perfectly orthogonal eigenvectors, in general it consumes more time on uniprocessors, or even on some parallel architectures, compared to tridiagonalization followed by QR iterations. It requires more arithmetic operations and memory references. Its high potential for parallelism, however, warrants its examination. The other two methods are often more efficient when combined with the tridiagonalization process: $T = Q^\top A Q$, where Q is orthogonal and T is tridiagonal.

Computing the full or partial singular value decomposition of a matrix is based on symmetric eigenvalue problem solvers involving either A_{nrm}, or A_{aug}, as given by

(8.3) and (8.4), respectively. We consider variants of the Jacobi methods—(one-sided [5], and two sided [6]), as well as variant of the QR method, e.g. see [3]. For the QR scheme, similar to the eigenvalue problem, an initial stage reduces A_{nrm} to the tridiagonal form,

$$T = W^\top A^\top AW, \tag{8.5}$$

where W is an orthogonal matrix. Further, if A is of maximal column rank, T is symmetric positive definite for which one can obtain the Cholesky factorization

$$T = B^\top B, \tag{8.6}$$

where B is an upper bidiagonal matrix. Therefore the matrix $V = AWB^{-1}$ is orthogonal and

$$A = VBW^\top. \tag{8.7}$$

This decomposition is called *bidiagonalization* of A. It can be directly obtained by two sequences of Householder transformations respectively applied to the left and the right sides of A without any constraints on the matrix rank.

8.1 The Jacobi Algorithms

Jacobi's algorithm consists of annihilating successively off-diagonal entries of the matrix via orthogonal similarity transformations. While the method was abandoned, on uniprocessors, due to the high computational cost compared to Householder's reduction to the tridiagonal form followed by Francis' QR iterations, it was revived in the early days of parallel computing especially when very accurate eigenvalues or perfectly orthogonal eigenvectors are needed, e.g. see [7]. From this (two-sided) method, a one-sided Jacobi scheme is directly derived for the singular value problem.

8.1.1 The Two-Sided Jacobi Scheme for the Symmetric Standard Eigenvalue Problem

Consider the eigenvalue problem

$$Ax = \lambda x \tag{8.8}$$

where $A \in \mathbb{R}^{n \times n}$ is a dense symmetric matrix. The original Jacobi's method for determining all the eigenpairs of (8.8) reduces the matrix A to the diagonal form by an infinite sequence of plane rotations.

$$A_{k+1} = U_k A_k U_k^\top, \quad k = 1, 2, \ldots,$$

where $A_1 = A$, and $U_k = R_k(i, j, \theta_{ij}^k)$ is a rotation of the (i, j)-plane in which

$$u_{ii}^k = u_{jj}^k = c_k = \cos \theta_{ij}^k \text{ and } u_{ij}^k = -u_{ji}^k = s_k = \sin \theta_{ij}^k.$$

The angle θ_{ij}^k is determined so that $\alpha_{ij}^{k+1} = \alpha_{ji}^{k+1} = 0$, i.e.

$$\tan 2\theta_{ij}^k = \frac{2\alpha_{ij}^k}{\alpha_{ii}^k - \alpha_{jj}^k},$$

where $|\theta_{ij}^k| \leq \frac{1}{4}\pi$. For numerical stability, we determine the plane rotation by

$$c_k = \frac{1}{\sqrt{1 + t_k^2}} \text{ and } s_k = c_k t_k,$$

where t_k is the smaller root (in magnitude) of the quadratic equation

$$t_k^2 + 2\tau_k t_k - 1 = 0, \text{ in which } \tau_k = \cot 2\theta_{ij}^k = \frac{\alpha_{ii}^k - \alpha_{jj}^k}{2\alpha_{ij}^k}. \tag{8.9}$$

Hence, t_k may be written as

$$t_k = \frac{\text{sign}(\tau_k)}{|\tau_k| + \sqrt{1 + \tau_k^2}} \tag{8.10}$$

Each A_{k+1} remains symmetric and differs from A_k only in the ith and jth rows and columns, with the modified elements given by

$$\begin{aligned}
\alpha_{ii}^{k+1} &= \alpha_{ii}^k + t_k \alpha_{ij}^k, \\
\alpha_{jj}^{k+1} &= \alpha_{jj}^k - t_k \alpha_{ij}^k,
\end{aligned} \tag{8.11}$$

and

$$\alpha_{ir}^{k+1} = c_k \alpha_{ir}^k + s_k \alpha_{jr}^k, \tag{8.12}$$

$$\alpha_{jr}^{k+1} = -s_k \alpha_{ir}^k + c_k \alpha_{jr}^k,$$

in which $1 \leq r \leq n$ and $r \neq i, j$. If A_k is expressed as the sum

$$A_k = D_k + E_k + E_k^\top, \tag{8.13}$$

where D_k is diagonal and E_k is strictly upper triangular, then as k increases $\|E_k\|_F$ approaches zero, and A_k approaches the diagonal matrix $D_k = \text{diag}(\lambda_1^{(k)}, \lambda_2^{(k)}, \ldots,$

$\lambda_n^{(k)}$) (here, $\|\cdot\|_F$ denotes the Frobenius norm). Similarly, the transpose of the product $(U_k \cdots U_2 U_1)$ approaches a matrix whose jth column is an eigenvector corresponding to λ_j.

Several schemes are possible for selecting the sequence of elements α_{ij}^k to be eliminated via the plane rotations U_k. Unfortunately, Jacobi's original scheme, which consists of sequentially searching for the largest off-diagonal element, is too time consuming for implementation on a multiprocessor. Instead, a simpler scheme in which the off-diagonal elements (i, j) are annihilated in the cyclic fashion $(1, 2), (1, 3), \ldots, (1, n), (2, 3), \ldots, (2, n), \ldots, (n - 1, n)$ is usually adopted as its convergence is assured [6]. We refer to each sequence of $\frac{n(n-1)}{2}$ rotations as a sweep. Furthermore, quadratic convergence for this sequential cyclic Jacobi scheme has been established (e.g. see [8, 9]). Convergence usually occurs within a small number of sweeps, typically in $O(\log n)$ sweeps.

A parallel version of this cyclic Jacobi algorithm is obtained by the simultaneous annihilation of several off-diagonal elements by a given U_k, rather than annihilating only one off-diagonal element and its symmetric counterpart as is done in the sequential version. For example, let A be of order 8 and consider the orthogonal matrix U_k as the direct sum of 4 independent plane rotations, where the c_i's and s_i's for $i = 1, 2, 3, 4$ are simultaneously determined. An example of such a matrix is

$$U_k = R_k(1, 3) \oplus R_k(2, 8) \oplus R_k(4, 7) \oplus R_k(5, 6), \qquad (8.14)$$

where $R_k(i, j)$ is that rotation which annihilates the (i, j) and (j, i) off-diagonal elements. Now, a sweep can be seen as a collection of orthogonal similarity transformations where each of them simultaneously annihilates several off-diagonal pairs and such that each of the off-diagonal entries is annihilated only once by the sweep. For a matrix of order 8, an optimal sweep will consist of 7 successive orthogonal transformations with each one annihilating distinct groups of 4 off-diagonal elements simultaneously, as shown in the left array of Table 8.1, where the similarity transformation of (8.14) is U_6. On the right array of Table 8.1, the sweep for a matrix

Table 8.1 Annihilation scheme for 2JAC

$n = 8.$

.	3	6	2	5	1	4	7
	.	2	5	1	4	7	6
		.	1	4	7	3	5
			.	7	3	6	4
				.	6	2	3
					.	5	2
						.	1
							.

$n = 9.$

.	4	8	3	7	2	6	1	5
	.	3	7	2	6	1	5	9
		.	2	6	1	5	9	4
			.	1	5	9	4	8
				.	9	4	8	3
					.	8	3	7
						.	7	2
							.	6
								.

The entries indicate the step in which these elements are annihilated of order $n = 9$ appears to be made of 9 successive orthogonal transformations of 4 independent rotations.

Although several annihilation schemes are possible, we adopt the scheme 2JAC, see [10], which is used for the two cases in Table 8.1. This scheme is optimal for any n:

Theorem 8.2 *The ordering defined in Algorithm 8.1, is optimal for any n: the* $\frac{n(n-1)}{2}$ *rotations of a sweep are partitioned into $2m - 1$ steps, each consisting of $p = \lfloor \frac{n}{2} \rfloor = n - m$ independent rotations, where $m = \lfloor \frac{n+1}{2} \rfloor$.*

Proof Obviously, the maximum number of independent rotations in one step is $\lfloor \frac{n}{2} \rfloor$. Algorithm 8.1 defines an ordering such that: (i) if n is odd, $m = \frac{n+1}{2}$ and the sweep is composed of $n = 2m - 1$ steps each consisting of $p = \frac{n-1}{2} = n - m$ independent rotations; (ii) if n is even, $m = \frac{n}{2}$ and the sweep is composed of $n - 1 = 2m - 1$ steps each consisting of $p = \frac{n}{2}$ independent rotations.

In the annihilation of a particular (i, j)-element, we update the off-diagonal elements in rows and columns i and j as given by (8.11) and (8.12). It is possible to modify only those row or column entries above the main diagonal and utilize the guaranteed symmetry of A_k. However, if one wishes to take advantage of the vectorization capability of a parallel computing platform, we may disregard the symmetry of A_k and operate with full vectors on the entirety of rows and columns i and j in (8.11) and (8.12), i.e., we may use a full matrix scheme. The product of the U_k's, which eventually yields the eigenvectors for A, is accumulated in a separate two-dimensional array by applying (8.11) and (8.12) to the identity matrix of order n.

Convergence is assumed to occur at stage k once each off-diagonal element is below a small fraction of $\|A\|_F$, see the tolerance parameter τ in line 38 of Algorithm 8.1. Moreover, in [11, 12] it has been demonstrated that various parallel Jacobi rotation ordering schemes (including 2JAC) are equivalent to the sequential row ordering scheme for which convergence is assured. Hence, the parallel schemes share the same convergence properties, e.g. quadratic convergence, see [9, 13]. An ordering used in chess and bridge tournaments that requires n to be even but which can be implemented on a ring or a grid of processors (see also [3]), has been introduced in [14].

8.1.2 The One-Sided Jacobi Scheme for the Singular Value Problem

The derivation of the *one-sided* Jacobi method is motivated by the singular value decomposition of rectangular matrices. It can be used to compute the eigenvalues of the square matrix A in (8.8) when A is symmetric positive definite. It is considerably more efficient to apply a *one-sided* Jacobi method which in effect only post-multiplies A by plane rotations. Let $A \in \mathbb{R}^{m \times n}$ with $m \geq n$ and $\mathrm{rank}(A) = r \leq n$. The singular value decomposition of A is defined by the corresponding notations:

Algorithm 8.1 2JAC: two-sided Jacobi scheme.

Input: $A = (\alpha_{ij}) \in \mathbb{R}^{n \times n}$, symmetric, $\tau > 0$
Output: $D = \mathrm{diag}(\lambda_1, \cdots, \lambda_n)$
1: $m = \lfloor \frac{n+1}{2} \rfloor$; $\mathtt{istop} = \frac{n(n-1)}{2}$;
2: **while** $\mathtt{istop} > 0$
3:    $\mathtt{istop} = \frac{n(n-1)}{2}$;
4:    **do** $k = 1 : 2m - 1$,
5:      *//Define U_k from $n - m$ pairs:*
6:      $\mathscr{P}_k = \emptyset$;
7:      **if** $k \leq m - 1$, **then**
8:        **do** $j = m - k + 1 : n - k$,
9:          **if** $j \leq 2m - 2k$, **then**
10:            $i = 2m - 2k + 1 - j$;
11:          **else if** $j \leq 2m - k - 1$, **then**
12:            $i = 4m - 2k - j$;
13:          **else**
14:            $i = n$;
15:          **end if**
16:          **if** $i > j$, **then**
17:            Exchange i and j;
18:          **end if**
19:          $\mathscr{P}_k = \mathscr{P}_k \cup \{(i, j)\}$;
20:        **end**
21:      **else**
22:        **do** $j = 4m - n - k : 3m - k - 1$,
23:          **if** $j < 2m - k + 1$, **then**
24:            $i = n$;
25:          **else if** $j \leq 4m - 2k - 1$, **then**
26:            $i = 4m - 2k - j$;
27:          **else**
28:            $i = 6m - 2k - 1 - j$;
29:          **end if**
30:          **if** $i > j$, **then**
31:            Exchange i and j;
32:          **end if**
33:          $\mathscr{P}_k = \mathscr{P}_k \cup \{(i, j)\}$;
34:        **end**
35:      **end if**
36:      *//Apply $A := U_k A U_k^\top$:*
37:      **doall** $(i, j) \in \mathscr{P}_k$,
38:        **if** $|\alpha_{ij}| \leq \|A\|_F.\tau$ **then**
39:          $\mathtt{istop} = \mathtt{istop} - 1$; $R(i, j) = I$;
40:        **else**
41:          Determine the Jacobi rotation $R(i, j)$;
42:          Apply $R(i, j)^\top$ on the right side;
43:        **end if**
44:      **end**
45:      **doall** $(i, j) \in \mathscr{P}_k$,
46:        Apply rotation $R(i, j)$ on the left side;
47:      **end**
48:    **end**
49: **end while**
50: **doall** i=1:n
51:    $\lambda_i = \alpha_{ii}$;
52: **end**

$$A = V\Sigma U^\top, \tag{8.15}$$

where $U \in \mathbb{R}^{m \times r}$ and $V \in \mathbb{R}^{n \times r}$ are orthogonal matrices and $\Sigma = \mathrm{diag}(\sigma_1, \ldots, \sigma_r)$ $\in \mathbb{R}^{r \times r}$ with $\sigma_1 \geq \sigma_2 \geq \cdots \geq \sigma_r > 0$. The r columns of U and the r columns of V yield the orthonormalized eigenvectors associated with the r non-zero eigenvalues of $A^\top A$ and $AA^\top$, respectively.

As indicated in [15], an efficient parallel scheme for computing the decomposition (8.15) on a ring of processors is realized through using a method based on the one-sided iterative orthogonalization method in [5] (see also [16, 17]). Further, this singular value decomposition scheme was first implemented in [18]. In addition, versions of the above *two-sided* Jacobi scheme have been presented in [14, 19]. Next, we consider a few modifications to the scheme discussed in [15] for the determination of the singular value decomposition (8.15) on shared memory multiprocessors, e.g., see [20].

Our main goal is to determine the orthogonal matrix $U \in \mathbb{R}^{n \times r}$, of (8.15) such that

$$AU = Q = (q_1, q_2, \ldots, q_r), \tag{8.16}$$

is a matrix of orthogonal columns, i.e.

$$q_i^\top q_j = \sigma_i^2 \delta_{ij},$$

in which δ_{ij} is the Kronecker-delta. Writing Q as

$$Q = V\Sigma \text{ where } V^\top V = I_r, \text{ and } \Sigma = \mathrm{diag}(\sigma_1, \ldots, \sigma_r),$$

then factorization (8.15) is entirely determined. We construct the matrix U via the plane rotations

$$(a_i, a_j) \begin{pmatrix} c & -s \\ s & c \end{pmatrix} = (\tilde{a}_i, \tilde{a}_j), \quad i < j,$$

so that

$$\tilde{a}_i^\top \tilde{a}_j = 0 \text{ and } \|\tilde{a}_i\| \geq \|\tilde{a}_j\|, \tag{8.17}$$

where a_i designates the ith column of the matrix A. This is accomplished by choosing

$$c = \left(\frac{\beta + \gamma}{2\gamma}\right)^{1/2} \text{ and } s = \left(\frac{\alpha}{2\gamma c}\right) \text{ if } \beta > 0, \tag{8.18}$$

or

$$s = \left(\frac{\gamma - \beta}{2\gamma}\right)^{1/2} \text{ and } c = \left(\frac{\alpha}{2\gamma s}\right) \text{ if } \beta < 0, \tag{8.19}$$

where $\alpha = 2a_i^\top a_j$, $\beta = \|a_i\|^2 - \|a_j\|^2$, and $\gamma = (\alpha^2 + \beta^2)^{1/2}$. Note that (8.17) requires the columns of Q to decrease in norm from left to right thus assuring that the resulting singular values σ_i appear in nonincreasing order. Several schemes can be used to select the order of the (i, j)-plane rotations. Following the annihilation pattern of the off-diagonal elements in the sequential Jacobi algorithm mentioned in Sect. 8.1, we could certainly orthogonalize the columns in the same cyclic fashion and thus perform the one-sided orthogonalization sequentially. This process is iterative with each sweep consisting of $\frac{1}{2}n(n - 1)$ plane rotations selected in cyclic fashion.

Adopting the parallel annihilation scheme in 2JAC outlined in Algorithm 8.1, we obtain the parallel version 1JAC of the one-sided Jacobi method for computing the singular value decomposition on a multiprocessor (see Algorithm 8.2). For example, let $n = 8$ and $m \geq n$ so that in each sweep of our one-sided Jacobi algorithm we simultaneously orthogonalize four pairs of the columns of A (see the left array in Table 8.1). More specifically, we can orthogonalize the pairs (1, 3), (2, 8) (4, 7), (5, 6) simultaneously via post-multiplication by an orthogonal transformation U_k which consists of the direct sum of 4 plane rotations (identical to $U_6^\top$ introduced in (8.14)). At the end of any particular sweep s_i we have

$$U_{s_i} = U_1 U_2 \cdots U_{2q-1},$$

where $q = \lfloor \frac{n+1}{2} \rfloor$ and hence

$$U = U_{s_1} U_{s_2} \cdots U_{s_t}, \tag{8.20}$$

where t is the number of sweeps required for convergence.

In the orthogonalization step, lines 10 and 11 of Algorithm 8.2, we are implementing the plane rotations given by (8.18) and (8.19), and hence guaranteeing the proper ordering of column norms and singular values upon termination. Whereas 2JAC must update rows and columns following each similarity transformation, 1JAC performs only post-multiplication of A_k by each U_k and hence the plane rotation (i, j) affects only columns i and j of the matrix A_k, with the updated columns given by

$$a_i^{k+1} = ca_i^k + sa_j^k, \tag{8.21}$$

$$a_j^{k+1} = -sa_i^k + ca_j^k, \tag{8.22}$$

where a_i^k denotes the ith column of A_k, and c, s are determined by either (8.18) or (8.19). On a parallel architecture with vector capability, one expects to realize high performance in computing (8.21) and (8.22). Each processor is assigned one rotation and hence orthogonalizes one pair of the n columns of matrix A_k.

Algorithm 8.2 1JAC: one-sided Jacobi for rank-revealing SVD

Input: $A = [a_1, \cdots, a_n] \in \mathbb{R}^{m \times n}$ where $m \geq n$, and $\tau > 0$.
Output: $\Sigma = \mathrm{diag}(\sigma_1, \cdots, \sigma_r)$ and $U = [u_1, \cdots, u_r]$ where r is the column-rank of A.
1: $q = \lfloor \frac{n+1}{2} \rfloor$; $\texttt{istop} = \frac{n(n-1)}{2}$; $v(A) = \|A\|_F$;
2: **while** $\texttt{istop} > 0$
3:    $\texttt{istop} = \frac{n(n-1)}{2}$;
4:    **do** $k = 1 : 2q - 1$
5:      Define the parallel list of pairs $\mathscr{P}_k$ as in Algorithm 8.1 lines 6–35;
6:      **doall** $(i, j) \in \mathscr{P}_k$
7:        **if** $\frac{a_i^\top a_j}{\|a_i\| \|a_j\|} \leq \tau$ **or** $\|a_i\| + \|a_j\| \leq \tau v(A)$ **then**
8:          $\texttt{istop} = \texttt{istop} - 1$;
9:        **else**
10:          Determine the rotation $R(i, j)$ as in (8.18) and (8.19);
11:          Apply $R(i, j)^\top$ on the right side;
12:        **end if**
13:      **end**
14:    **end**
15: **end while**
16: **doall** $i = 1 : n$
17:    $\sigma_i = \|a_i\|$;
18:    **if** $\sigma_i > \tau v(A)$ **then**
19:      $r_i = i$; $v_i = \frac{1}{\sigma_i} a_i$;
20:    **else**
21:      $r_i = 0$;
22:    **end if**
23: **end**
24: $r = \max r_i$; $\Sigma = \mathrm{diag}(\sigma_1, \cdots, \sigma_r)$; $V = [v_1, \cdots, v_r]$;

Following the convergence test used in [17], we test convergence in line 8, by counting, as in [17], the number of times the quantity

$$\frac{a_i^\top a_j}{\sqrt{(a_i^\top a_i)(a_j^\top a_j)}}, \tag{8.23}$$

falls below a given *tolerance* in any given sweep with the algorithm terminating when the counter reaches $\frac{1}{2}n(n-1)$, the total number of column pairs, after any sweep. Upon termination, the first r columns of the matrix A are overwritten by the matrix Q from (8.16) and hence the non-zero singular values σ_i can be obtained via the r square roots of the first r diagonal entries of the updated $A^\top A$. The matrix V in (8.15), which contains the leading r, left singular vectors of the original matrix A, is readily obtained by column scaling of the updated matrix A (now overwritten by $Q = V\Sigma$) by the r non-zero singular values. Similarly, the matrix U, which contains the right singular vectors of the original matrix A, is obtained as in (8.20) as the product of the orthogonal U_k's. This product is accumulated in a separate two-dimensional array by applying the rotations used in (8.21) and (8.22) to the identity matrix of order n. It is important to note that the use of the fraction in (8.23) is preferable to using $a_i^\top a_j$, since this inner product is necessarily small for relatively small singular values.

Although 1JAC concerns the singular value decomposition of rectangular matrices, it is most effective for handling the eigenvalue problem (8.8) when obtaining all the eigenpairs of square symmetric matrices. Thus, if 1JAC is applied to a symmetric matrix $A \in \mathbb{R}^{n \times n}$, the columns of the resulting $V \in \mathbb{R}^{n \times r}$ are the eigenvectors corresponding to the nonzero eigenvalues of A. The eigenvalue corresponding to v_i $(i = 1, \ldots, r)$ is obtained by the Rayleigh quotient $\lambda_i = v_i^\top A v_i$; therefore: $|\lambda_i| = \sigma_i$. The null space of A is the orthogonal complement of the subspace spanned by the columns of V.

Algorithm 1JAC has two advantages over 2JAC: (i) no need to access both rows and columns, and (ii) the matrix U need not be accumulated.

8.1.3 The Householder-Jacobi Scheme

As discussed above, 1JAC is certainly a viable parallel algorithm for computing the singular value decomposition (8.15) of a dense matrix. However, for matrices $A \in \mathbb{R}^{m \times n}$ in which $m \gg n$, the arithmetic complexity can be reduced if an initial orthogonal factorization of A is performed. One can then apply the *one-sided* Jacobi method, 1JAC, to the resulting upper-triangular matrix, which may be singular, to obtain the decomposition (8.15). In this section, we present a parallel scheme, QJAC, which can be quite effective for computing (8.15) on a variety of parallel architectures.

When $m \geq n$, the block orthogonal factorization schemes of SCALAPACK (procedure PDGEQRF) may be used for computing the orthogonal factorization

$$A = QR \tag{8.24}$$

where $Q \in \mathbb{R}^{m \times n}$ is a matrix with orthonormal columns, and $R \in \mathbb{R}^{n \times n}$ is an upper-triangular matrix. These SCALAPACK block schemes make full use of finely-tuned matrix-vector and matrix-matrix primitives (BLAS2 and BLAS3) to assure high performance. The 1JAC algorithm can then be used to obtain the singular value decomposition of the upper-triangular matrix R. The benefit that a preliminary factorization brings is to replace the scalar product of two vectors of length m by the same operation on vectors of length n for each performed rotation. In order to counterbalance the additional computation corresponding to the QR factorization, the dimension n must be much smaller than m, i.e. $m \gg n$.

If $m \gg n$, however, an alternate strategy for obtaining the orthogonal factorization (8.24) is essential for realizing high performance. In this case, the hybrid Householder—Givens orthogonal factorization scheme for tall and narrow matrices, given in Sect. 7.6, is adopted. Hence, the singular value decomposition of the matrix A $(m \gg n)$ can be efficiently determined via Algorithm QJAC, see Algorithm 8.3.

Note that in using 1JAC for computing the SVD of R, we must iterate on a full $n \times n$-matrix which is initially upper-triangular. This sacrifice in storage must be made

---

Algorithm 8.3 QJAC: SVD of a tall matrix $(m \gg n)$.

Input: $A = (\alpha_{ij}) \in \mathbb{R}^{m \times n}$, such that $m \geq np$ (p is the number of processors), $\tau > 0$.

Output: $r \leq n$: rank of A, $\Sigma = \mathrm{diag}(\sigma_1, \cdots, \sigma_r)$ the nonzero singular values, and $V \in \mathbb{R}^{m \times r}$ the corresponding matrix of left singular vectors of A.

1: On p processors, apply the hybrid orthogonal factorization scheme outlined in Sect. 7.6 to obtain the QR-factorization of A, where $Q \in \mathbb{R}^{m \times n}$ is with orthonormal columns, and $R \in \mathbb{R}^{n \times n}$ is upper triangular.

2: On p processors, apply 1JAC (Algorithm 8.2), to R: get the rank r that is determined from threshold τ, and compute the nonzero singular values $\Sigma = \mathrm{diag}(\sigma_1, \cdots, \sigma_r)$ and the matrix of left singular vectors $\tilde{V}$ of R.

3: Recover the left singular vectors of A via: $V = Q\tilde{V}$.

---

in order to take advantage of the inherent parallelism and vectorization available in 1JAC.

An implementation of Kogbetliantz algorithm for computing the singular value decomposition of upper-triangular matrices has been shown to be quite effective on systolic arrays, e.g. see [21]. Kogbetliantz method for computing the SVD of a real square matrix $A \in \mathbb{R}^{n \times n}$ mirrors the scheme 2JAC, above, in that the matrix A is reduced to the diagonal form by an infinite sequence of plane rotations,

$$A_{k+1} = V_k A_k U_k^{\top}, \quad k = 1, 2, \ldots, \tag{8.25}$$

where $A_1 \equiv A$, and $U_k = U_k(i, j, \phi_{ij}^k)$, $V_k = V_k(i, j, \theta_{ij}^k)$ are plane rotations that affect rows and columns i, and j. It follows that A_k approaches the diagonal matrix $\Sigma = \mathrm{diag}(\sigma_1, \sigma_2, \ldots, \sigma_n)$, where σ_i is the ith singular value of A, and the products $(V_k \cdots V_2 V_1)$, $(U_k \cdots U_2 U_1)$ approach matrices whose ith column is the respective left and right singular vector corresponding to σ_i. When the σ_i's are not pathologically close, it has been shown in [22] that the row (or column) cyclic Kogbetliantz method ultimately converges quadratically. For triangular matrices, it has been demonstrated in [23] that Kogbetliantz algorithm converges quadratically for those matrices having multiple or clustered singular values provided that singular values of the same cluster occupy adjacent diagonal elements of A_ν, where ν is the number of sweeps required for convergence. Even if we were to assume that R in (8.24) satisfies this condition for quadratic convergence of the parallel Kogbetliantz method in [22], the ordering of the rotations and subsequent row (or column) permutations needed to maintain the upper-triangular form is less efficient on many parallel architectures. One clear advantage of using 1JAC for obtaining the singular value decomposition of R lies in that the rotations given in (8.18) or (8.19), and applied via the parallel ordering illustrated in Table 8.1, see also Algorithm 8.1, require no processor synchronization among any set of the $\lfloor \frac{1}{2}n \rfloor$ or $\lfloor \frac{1}{2}(n-1) \rfloor$ simultaneous plane rotations. The convergence rate of 1JAC, however, does not necessarily match that of the Kogbetliantz algorithm.

Let

$$S_k = R_k^{\top} R_k = \tilde{D}_k + \tilde{E}_k + \tilde{E}_k^{\top}, \tag{8.26}$$

where $\tilde{D}_k$ is a diagonal matrices and $\tilde{E}_k$ is strictly upper-triangular. Although quadratic convergence cannot be always guaranteed for 1JAC, we can always produce clustered singular values on adjacent positions of the diagonal matrix $\tilde{D}_k$ for any A. This can be realized by monitoring the magnitudes of the elements of $\tilde{D}_k$ and $\tilde{E}_k$ in (8.27) for successive values of k in 1JAC. After a particular number of critical sweeps $k_c r$, S_k approaches a block diagonal form in which each block corresponds to a principal (diagonal) submatrix of $\tilde{D}_k$ containing a cluster of singular values of A (see [20, 21]),

$$
S_{k_{cr}} = \begin{bmatrix} T_1 & & & & & \\ & T_2 & & & & \\ & & T_3 & & & \\ & & & \ddots & & \\ & & & & \ddots & \\ & & & & & T_{n_c} \end{bmatrix} \tag{8.27}
$$

Thus, the SVD of each T_i, $i = 1, 2, \ldots, n_c$, can be computed in parallel by either a Jacobi or Kogbetliantz method. Each symmetric matrix T_i will, in general, be dense of order q_i, representing the number of singular values of A contained in the ith cluster. Since the quadratic convergence of Kogbetliantz method for upper-triangular matrices [23] mirrors the quadratic convergence of the two-sided Jacobi method, 2JAC, for symmetric matrices having clustered spectra [24], we obtain a faster global convergence for $k > k_{cr}$ if 2JAC, rather than 1JAC, were used to obtain the SVD of each block T_i. Thus, a *hybrid* method consisting of an initial phase of several 1JAC iterations followed by 2JAC on the resulting subproblems, combines the optimal parallelism of 1JAC and the faster convergence rate of the 2JAC method. Of course, optimal implementation of such a method depends on the determination of the critical number of sweeps k_{cr} required by 1JAC. We should note here that such a hybrid SVD scheme is quite suitable for implementation on multiprocessors with hierarchical memory structure and vector capability.

8.1.4 Block Jacobi Algorithms

The above algorithms are well-suited for shared memory architectures. While they can also be implemented on distributed memory systems, their efficiency on such systems will suffer due to communication costs. In order to increase the granularity of the computation (i.e., to increase the number of arithmetic operations between two successive message exchanges), block algorithms are considered.

Allocating blocks of a matrix in place of single entries or vectors and replacing the basic Jacobi rotations by more elaborate orthogonal transformations, we obtain what is known as block Jacobi schemes. Let us consider the treatment of the pair of off-diagonal blocks (i, j) in such algorithms:

- *Two-sided Jacobi for the symmetric eigenvalue problem:* the matrix $A \in \mathbb{R}^{n \times n}$ is partitioned into $p \times p$ blocks $A_{ij} \in \mathbb{R}^{q \times q}$ ($1 \leq i, j \leq p$ and $n = pq$). For any pair (i, j) with $i < j$, let $J_{ij} = \begin{pmatrix} A_{ii} & A_{ij} \\ A_{ij}^{\top} & A_{jj} \end{pmatrix} \in \mathbb{R}^{2q \times 2q}$. The matrix $U(i, j)$ is the orthogonal matrix that diagonalizes J_{ij}.
- *One-sided Jacobi for the SVD:* the matrix $A \in \mathbb{R}^{m \times n}$ ($m \geq n = pq$) is partitioned into p blocks $A_i \in \mathbb{R}^{m \times q}$ ($1 \leq i \leq p$). For any pair (i, j) with $i < j$, let $J_{ij} = (A_i, A_j) \in \mathbb{R}^{m \times 2q}$. The matrix $U(i, j)$ is the orthogonal matrix that diagonalizes $J_{ij}^{\top} J_{ij}$.
- *Two-sided Jacobi for the SVD (Kogbetliantz algorithm)* the matrix $A \in \mathbb{R}^{n \times n}$ is partitioned into $p \times p$ blocks $A_{ij} \in \mathbb{R}^{q \times q}$ ($1 \leq i, j \leq p$). For any pair (i, j) with $i < j$, let $J_{ij} = \begin{pmatrix} A_{ii} & A_{ij} \\ A_{ji} & A_{jj} \end{pmatrix} \in \mathbb{R}^{2q \times 2q}$. The matrices $U(i, j)$ and $V(i, j)$ are the orthogonal matrices defined by the SVD of J_{ij}.

For one-sided algorithms, each processor is allocated a block of columns instead of a single column. The main features of the algorithm remain the same as discussed above with the ordering of the rotations within a sweep is as given in [25].

For the two-sided version, the allocation manipulates 2-D blocks instead of single entries of the matrix. A modification of the basic algorithm in which one annihilates, in each step, two symmetrically positioned off-diagonal blocks by performing a full SVD on the smaller sized off-diagonal block has been proposed in [26, 27]. While reasonable performance can be realized on distributed memory architectures, this block strategy increases the number of sweeps needed to achieve convergence. In order to reduce the number of needed iterations, a dynamic ordering has been investigated in [28] in conjunction with a preprocessing step consisting of a preliminary QR factorization with column pivoting [29]. This procedure has also been considered for the block one-sided Jacobi algorithm for the singular value decomposition, e.g. see [30].

8.1.5 *Efficiency of Parallel Jacobi Methods*

Considering that: (a) scalable parallel schemes for tridiagonalization of a symmetric matrix A are available together with efficient parallel schemes for extracting the eigenpairs of a tridiagonal matrix, and (b) Jacobi schemes deal with the whole matrix until all the eigenvalues and vectors are extracted and hence resulting in a much higher cost of memory references and interprocessor communications; two-sided parallel Jacobi schemes are not competitive in general compared to eigensolvers that depend first on the tridiagonalization process, unless: (i) one requires all the eigenpairs with high accuracy, e.g. see [7] and Sect. 8.1.2 for handling the singular-value decomposition via Jacobi schemes, or (ii) Jacobi is implemented on a shared memory parallel architecture (multi- or many-core) in which one memory reference is almost as fast as an arithmetic operation for obtaining all the eigenpairs of a modest size matrix.

In addition to the higher cost of communications in Jacobi schemes, in general, they also incur almost the same order of overall arithmetic operations. To illustrate this, let us assume that n is even. Diagonalizing a matrix $A \in \mathbb{R}^{n \times n}$ on an architecture of $p = \frac{n}{2}$ processors is obtained by a sequence of sweeps in which each sweep requires $(n - 1)$ parallel applications of p rotations. Therefore, each sweep costs $O(n^2)$ arithmetic operations. If, in addition, we assume that the number of sweeps needed to obtain accurate eigenvalues to be $O(\log n)$, then the overall number of arithmetic operations is $O(n^2 \log n)$. This estimation can even become as low as $O(n \log n)$ by using $O(n^2)$ processors. As seen in Sect. 8.1, one sweep of rotations involves $6n^3 + O(n^2)$ arithmetic operations when taking advantage of symmetry, and including accumulation of the rotations. This estimate must be compared to the $8n^3 + O(n^2)$ arithmetic operations needed to reduce the matrix A to the tridiagonal form and to build the orthogonal matrix Q which realizes such tridiagonalization (see next section).

A favorable situation for Jacobi schemes arises when one needs to investigate the evolution of the eigenvalues of a sequence of slowly varying symmetric matrices $(A_k)_{k \geq 0} \subset \mathbb{R}^{n \times n}$. If A_k has the spectral decomposition $A_k = U_k D_k U_k^\top$, where U_k is an orthogonal matrix (full set of eigenvectors), and D_k a diagonal matrix containing the eigenvalues. Hence, if the quantity $\|A_{k+1} - A_k\| \, / \, \|A_k\|$ is small, the matrix

$$B_{k+1} = U_k^\top A_{k+1} U_k, \tag{8.28}$$

is expected to be close to a diagonal matrix. It is in such a situation that Jacobi schemes yield, after very few sweeps, accurate eigenpairs of B_{k+1},

$$V_{k+1}^\top B_{k+1} V_{k+1} = D_{k+1}, \tag{8.29}$$

and therefore yielding the spectral decomposition $A_{k+1} = U_{k+1} D_{k+1} U_{k+1}^\top$, with $U_{k+1} = U_k V_{k+1}$. When one sweep only is sufficient to get convergence, Jacobi schemes are competitive.

8.2 Tridiagonalization-Based Schemes

Obtaining all the eigenpairs of a symmetric matrix A can be achieved by the following two steps: (i) obtaining a symmetric tridiagonal matrix T which is orthogonally similar to A: $T = U^\top A U$ where U is an orthogonal matrix ; and (ii) obtaining the spectral factorization of the resulting tridiagonal matrix T, i.e. $D = V^\top T V$, and computing the eigenvectors of A by the back transformation $Q = UV$.

8.2.1 Tridiagonalization of a Symmetric Matrix

Let $A \in \mathbb{R}^{n \times n}$ be a symmetric matrix. The tridiagonalization of A will be obtained by applying successively $(n-2)$ similarity transformation via Householder reflections.

The first elementary reflector H_1 is chosen such that the bottom $(n-2)$ elements of the first column of $H_1 A$ are annihilated:

$$H_1 A = \begin{pmatrix} \star & \star & \star & \star & \star & \star \\ \star & \star & \star & \star & \star & \star \\ 0 & \star & \star & \star & \star & \star \\ 0 & \star & \star & \star & \star & \star \\ 0 & \star & \star & \star & \star & \star \\ 0 & \star & \star & \star & \star & \star \end{pmatrix}.$$

Therefore, by symmetry, applying H_1 on the right results in the following pattern:

$$A_1 = H_1 A H_1 = \begin{pmatrix} \star & \star & 0 & 0 & 0 & 0 \\ \star & \star & \star & \star & \star & \star \\ 0 & \star & \star & \star & \star & \star \\ 0 & \star & \star & \star & \star & \star \\ 0 & \star & \star & \star & \star & \star \\ 0 & \star & \star & \star & \star & \star \end{pmatrix} \tag{8.30}$$

In order to take advantage of symmetry, the reduction

$$A := H_1 A H_1 = (I - \alpha u u^\top) A (I - \alpha u u^\top)$$

can be implemented by a symmetric rank-2 update which is a BLAS2 routine [31]:

$$\begin{aligned} v &:= \alpha A u; \\ w &:= v - \tfrac{1}{2}\alpha (u^\top v) u; \\ A &:= A - (u w^\top + w u^\top). \end{aligned}$$

By exploiting symmetry, the rank-2 update (that appears in the last step) results in computational savings [32].

The tridiagonal matrix T is obtained by repeating the process successively on columns 2 to $(n-2)$. The total procedure involves $4n^3/3 + O(n^2)$ arithmetic operations. Assembling the matrix $U = H_1 \cdots H_{n-2}$ requires $4n^3/3 + O(n^2)$ additional operations. The benefit of the BLAS2 variant over that of the original BLAS1 based scheme is illustrated in [33].

As indicated in Chap. 7, successive applications of Householder reductions can be done in blocks (block Householder transformations) which allows the use of BLAS3 [34]. Such reduction is implemented in routine DSYTRD of LAPACK [35] which also takes advantage of the symmetry of the process. A parallel version of the algorithm is implemented in routine PDSYTRD of SCALAPACK [36].

8.2.2 *The QR Algorithm: A Divide-and-Conquer Approach*

The Basic QR Algorithm

The QR iteration for computing all the eigenvalues of the symmetric matrix A is given by:

$$\begin{cases} A_0 = A \text{ and } Q_0 = I \\ \text{For } k \geq 0, \\ (Q_{k+1}, R_{k+1}) = \text{ QR-Factorization of } A_k, \\ A_{k+1} = R_{k+1} Q_{k+1}, \end{cases} \tag{8.31}$$

where the QR factorization is defined in Chap. 7. Under weak assumptions, A_k approaches a diagonal matrix for sufficiently large k. A direct implementation of (8.31) would involve $O(n^3)$ arithmetic operations at each iteration. If the matrix A is tridiagonal, it is easy to see that so are the matrices A_k. Therefore, by first tridiagonalizing A, i.e. $A\colon T = Q^\top AQ$, at a cost of $O(n^3)$ arithmetic operations, the number of arithmetic operations of each QR iteration in (8.31) is reduced to only $O(n^2)$, or even $O(n)$ if only the eigenvalues are needed. To accelerate convergence, a shifted version of the QR iteration is given by

$$\begin{cases} T_0 = T \text{ and } Q_0 = I \\ \text{For } k \geq 0, \\ (Q_{k+1}, R_{k+1}) = \text{ QR-Factorization of } (T_k - \mu_k I) \\ T_{k+1} = R_{k+1} Q_{k+1} + \mu_k I, \end{cases} \tag{8.32}$$

where μ_k is usually chosen as the eigenvalue of the last 2×2 diagonal block of T_k that is the closest to the nth diagonal entry of T_k (Wilkinson shift). The effective computation of the QR-step can be either explicit as in (8.32) or implicit (e.g. see [3]).

Parallel implementation of the above iteration is modestly scalable only if the eigenvectors need to be computed as well. In the following sections, we present two techniques for achieving higher degree of parallel scalability for obtaining the eigenpairs of a tridiagonal matrix.

The Divide-and-Conquer Scheme

Since a straightforward implementation of the basic algorithm provides very limited degree of parallelism, a divide-and-conquer technique is introduced. The resulting parallel algorithm proved to yield even superior performance on uniprocessors that it has been included in the sequential version of the library LAPACKas routine xSTEDC. This method was first introduced in [37], and then modified for implementation on parallel architectures in [38].

The divide-and-conquer idea consists of "tearing" the tridiagonal matrix T in half with a rank-one perturbation. Let us denote the three nonzero entries of the kth row of the tridiagonal symmetric matrix T by β_k, α_k, β_{k+1} and let e_m, $1 < m < n$, be the mth column of the identity matrix I_n.

Thus, T can be expressed as the following block-diagonal matrix with a rank-one correction:

$$T = \begin{pmatrix} T_1 & O \\ 0 & T_2 \end{pmatrix} + \rho\, vv^{\top},$$

for some $v = e_m + \theta e_{m+1}$, with

$$\rho = \beta_m/\theta,$$
$$T_1 = T_{1:m,1:m} - \rho\, e_m e_m{}^{\top},$$
$$\text{and } T_2 = T_{m+1:n,m+1:n} - \rho\theta^2\, e_{m+1}e_{m+1}{}^{\top},$$

where the parameter θ is an arbitrary nonzero real number. A strategy for determining a safe partitioning is given in [38].

Assuming that the full spectral decompositions of T_1 and T_2 are already available: $Q_i{}^{\top} T_i Q_i = \Lambda_i$ where Λ_i is a diagonal matrix and Q_i an orthogonal matrix, for $i = 1, 2$. Let

$$\tilde{Q} = \begin{pmatrix} Q_1 & O \\ 0 & Q_2 \end{pmatrix} \text{ and } \tilde{\Lambda} = \begin{pmatrix} \Lambda_1 & O \\ 0 & \Lambda_2 \end{pmatrix}.$$

Therefore : $J = \tilde{Q}^{\top} T \tilde{Q} = \tilde{\Lambda} + \rho\, zz^{\top}$ where $z = \tilde{Q}^{\top} v$, and computing the eigenvalues of T is reduced to obtaining the eigenvalues of J.

Theorem 8.3 ([37]) *Let $\tilde{\Lambda} = \mathrm{diag}(\tilde{\lambda}_1, \dots, \tilde{\lambda}_n)$ be a diagonal matrix with distinct entries, and let $z = (\zeta_1, \dots, \zeta_n)^{\top}$ be a vector with no null entries.*

Then the eigenvalues of the matrix $J = \tilde{\Lambda} + \rho\, zz^{\top}$ are the zeros λ_i of the secular equation:

$$1 + \rho \sum_{k=1}^{n} \frac{\zeta_k{}^2}{\tilde{\lambda}_k - \lambda} = 0. \tag{8.33}$$

in which λ_i and $\tilde{\lambda}_i$ are interleaved.

Thus, $t = (\tilde{\Lambda} - \lambda_k I)^{-1} z$ is an eigenvector of J corresponding to its simple eigenvalue λ_k.

Proof The proof is based on the following identity:

$$\det(\tilde{\Lambda} + \rho\, zz^{\top}) = \det(\tilde{\Lambda} - \lambda I)\, \det(I + \rho(\tilde{\Lambda} - \lambda I)^{-1}zz^{\top}).$$

Since λ is not a diagonal entry of $\tilde{\Lambda}$, the eigenvalues of J are the zeros of the function $f(\lambda) = \det(I + \rho(\tilde{\Lambda} - \lambda I)^{-1}zz^{\top})$, or the roots of

$$1 + \rho z^{\top}(\tilde{\Lambda} - \lambda I)^{-1}z = 0. \tag{8.34}$$

which is identical to (8.33). If u is an eigenvector of J corresponding to the eigenvalue λ, then

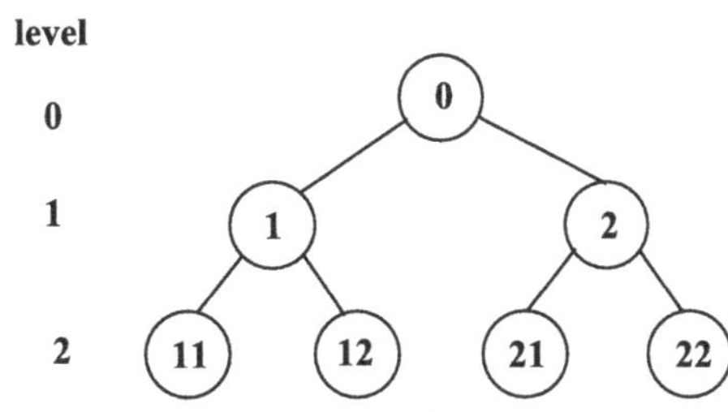

Fig. 8.1 Levels and associated tasks method Divide-and-Conquer

$$u = -\rho(z^\top u)(\tilde{\Lambda} - \lambda I)^{-1} z,$$

which concludes the proof. For further details see [3].

Parallel computation of the zeros λ_i of the secular equation (8.33) can be easily achieved since every zero is isolated in an interval defined by the set $(\tilde{\lambda}_i)_{i=1,n}$. Bisection or the secant method can be used for simultaneously obtaining these zeros with acceptable accuracy. In [38], a quadratically convergent rational approximation-based scheme is provided.

Thus, this eigensolver is organized into a tree of tasks as shown in Fig. 8.1. Using $p = 2^r$ processors ($r = 2$ in Fig. 8.1), the computation is recursively organized into an $(r + 1)$-level tree. Three types of tasks are considered:

1. Tearing (going down in the tree) involves $O(r)$ arithmetic operations.
2. Diagonalizing in parallel the p smaller tridiagonal matrices at the lowest level of the tree, each of these tasks involves $O\left(\left(\frac{n}{p}\right)^3\right)$ arithmetic operations.
3. Updating the spectrum (going up the tree): at level i, there are $q = 2^i$ tasks, consuming $O\left(\left(\frac{n}{q}\right)^3\right)$ arithmetic operations.

For $i = 0, \ldots, r$, a task of the third type is executed on 2^{r-i} processors at level i.

The algorithm computes all the eigenpairs of T. Eigenproblems corresponding to the leaves can be handled using any sequential algorithm such as the QR (or QL) algorithm. Computing the full spectrum with a Divide-and-Conquer approach consumes $O(n^3)$ arithmetic operations irrespective of the number of processors used. A more precise arithmetic operation count reveals that the divide and conquer approach results in some savings in the number of arithmetic operations compared to straightforward application of the QL algorithm. The divide and conquer algorithm is implemented in LAPACK [35] through the routine xSTEDC.

8.2.3 Sturm Sequences: A Multisectioning Approach

Following the early work in [39], or the TREPS procedure in [40], one can use the Sturm sequence properties to enable the computation of all the eigenvalues of a tridiagonal matrix T, or only those eigenvalues in a given interval of the spectrum. Once the eigenvalues are obtained, the corresponding eigenvectors can be retrieved via inverse iteration. Let T be a symmetric tridiagonal matrix of order n,

$$T = [\beta_i, \alpha_i, \beta_{i+1}],$$

and let $p_n(\lambda)$ be its characteristic polynomial:

$$p_n(\lambda) = \det(T - \lambda I).$$

Then the sequence of the principal minors of the matrix can be built using the following recursion:

$$
\begin{aligned}
p_0(\lambda) &= 1, \\
p_1(\lambda) &= \alpha_1 - \lambda, \\
p_i(\lambda) &= (\alpha_i - \lambda) p_{i-1}(\lambda) - \beta_i^2 p_{i-2}(\lambda), \qquad i = 2, \ldots, n.
\end{aligned}
\tag{8.35}
$$

Assuming that no subdiagonal element of T is zero (since if some β_i is equal to 0, the problem can be partitioned into two smaller eigenvalue problems), the sequence $\{p_i(\lambda)\}$ is called the Sturm sequence of T in λ. It is well known (e.g. see [41]) that the number of eigenvalues smaller than a given λ is equal to the number of sign variations in the Sturm sequence (8.35). Hence one can find the number of eigenvalues lying in a given interval $[\delta, \gamma]$ by computing the Sturm sequence in δ and γ. The linear recurrence (8.35), however, suffers from the possibility of over- or underflow. This is remedied by replacing the Sturm sequence $p_i(\lambda)$ by the sequence

$$q_i(\lambda) = \frac{p_i(\lambda)}{p_{i-1}(\lambda)}, \qquad i = 1, n.$$

The second order linear recurrence (8.35) is then replaced by the nonlinear recurrence

$$q_1(\lambda) = \alpha_1 - \lambda, \quad q_i(\lambda) = \alpha_i - \lambda - \frac{\beta_i^2}{q_{i-1}(\lambda)}, \qquad i = 2, \ldots, n. \tag{8.36}$$

Here, the number of eigenvalues that are smaller than λ is equal to the number of negative terms in the sequence $(q_i(\lambda))_{i=1,\ldots,n}$. It can be easily proved that $q_i(\lambda)$ is the ith diagonal element of D in the factorization $LDL^\top$ of $T - \lambda I$. Therefore, for any $i = 1, \ldots, n$,

$$p_i(\lambda) = \prod_{j=1}^{i} q_j(\lambda).$$

Given an initial interval, we can find all the eigenvalues lying in it by repeated bisection or multisection of the interval. This partitioning process can be performed until we obtain each eigenvalue to a given accuracy. On the other hand, we can stop the process once we have isolated each eigenvalue. In the latter case, the eigenvalues may be extracted using a faster method. Several methods are available for extracting an isolated eigenvalue:

- Bisection (linear convergence);
- Newton's method (quadratic convergence);
- The ZEROIN scheme [42] which is based on a combination of the secant and bisection methods (convergence of order $(\sqrt{5} + 1)/2$).

Ostrowski [43] defines an efficiency index which links the amount of computation to be done at each step and the order of the convergence. The respective indices of the three methods are 1, 1.414 and 1.618. This index, however, is not the only aspect to be considered here. Both ZEROIN and Newton methods require the use of the linear recurrence (8.35) in order to obtain the value of $\det(T - \lambda I)$, and its first derivative, for a given λ. Hence, if the possibility of over- or underflow is small, selecting the ZEROIN method is recommended, otherwise selecting the bisection method is the safer option. After the computation of an eigenvalue, the corresponding eigenvector can be found by inverse iteration [3], which is normally a very fast process that often requires no more than one iteration to achieve convergence to a low relative residual.

When some eigenvalues are computationally coincident, the isolation process actually performs "isolation of clusters", where a cluster is defined as a single eigenvalue or a number of computationally coincident eigenvalues. If such a cluster of coincident eigenvalues is isolated, the extraction stage is skipped since convergence has been achieved.

The whole computation consists of the following five steps:

1. Isolation by partitioning;
2. Extraction of a cluster by bisection or by the ZEROIN method;
3. Computation of the eigenvectors of the cluster by inverse iteration;
4. Grouping of close eigenvalues;
5. Orthogonalization of the corresponding groups of vectors by the modified Gram-Schmidt process.

The method TREPS (standing for Tridiagonal Eigenvalue Parallel Solver by Sturm sequences) is listed as Algorithm 8.4 and follows this strategy.

The Partitioning Process

Parallelism in this process is obviously achieved by performing simultaneously the computation of several Sturm sequences. However, there are several ways for achieving this. Two options are:

- Performing bisection on several intervals, or
- Partitioning of one interval into several subintervals.

A multisection of order k splits the interval $[\delta, \gamma]$ into $k+1$ subintervals $[\mu_i, \mu_{i+1}]$, where $\mu_i = \delta + i((\gamma - \delta)/(k + 1))$ for $i = 0, \ldots, k + 1$. If interval I contains only one eigenvalue, approximating it with an absolute error ε, will require

---

Algorithm 8.4 TREPS: tridiagonal eigenvalue parallel solver by Sturm sequences

---

Input: $T \in \mathbb{R}^{n \times n}$ symmetric tridiagonal; $\delta < \gamma$ (assumption: $\tau < \gamma - \delta$); $0 < \tau$; k (multisection degree).

Output: $\Lambda(T) \cap [\delta, \gamma] = \{\lambda_1, \cdots, \lambda_p\}$ and the corresponding eigenvectors $u_1, \cdots, u_p$.

1: Compute $n(\delta)$ and $n(\gamma)$; $p = n(\gamma) - n(\delta)$;   *//$n(\alpha)$ is the number of eigenvalues of T smaller than α which is computed by a Sturm sequence in α.*

2: $\mathcal{M} = \{([\delta, \gamma], n(\delta), n(\gamma))\}$; $\mathcal{S} = \emptyset$;   *//Sets of intervals including either several eigenvalues or one (simple or multiple) eigenvalue respectively.*

3: **while** $\mathcal{M} \neq \emptyset$,

4:    Select $([\alpha, \beta], n(\alpha), n(\beta)) \in \mathcal{M}$; $\mu_0 = \alpha$;

5:    **doall** $i = 1 : k$,

6:       $\mu_i = \alpha + i(\beta - \alpha)/(k + 1)$; Compute $n(\mu_i)$;

7:    **end**

8:    **do** $i = 0 : k$,

9:       $t_i = n(\mu_{i+1}) - n(\mu_i)$;

10:       **if** $t_i > 1$ and $|\mu_{i+1} - \mu_i| > \tau > 0$, **then**

11:          Store $([\mu_i, \mu_{i+1}], n(\mu_i), n(\mu_{i+1}))$ in $\mathcal{M}$;

12:       **else**

13:          **if** $n(\mu_{i+1}) - n(\mu_i) > 0$, **then**

14:             Store $([\mu_i, \mu_{i+1}], n(\mu_i), n(\mu_{i+1}))$ in $\mathcal{S}$;

15:          **end if**

16:       **end if**

17:    **end**

18: **end while**

19: **doall** $([\mu, v], n(\mu), n(v)) \in \mathcal{S}$,

20:    Extract λ from $[\mu, v]$;

21:    **if** λ is simple, **then**

22:       Compute the corresponding eigenvector u by inverse iteration ;

23:    **else**

24:       Compute a basis $V = [v_{i_1}, \cdots, v_{i_\ell}]$ of the invariant subspace by inverse iteration and combined orthogonalization ;

25:    **end if**

26: **end**

27: **doall** Grouping of close eigenvalues,

28:    Reorthogonalize the corresponding eigenvectors;

29: **end**

$$n_k = \log_2\left(\frac{\gamma - \delta}{2\varepsilon}\right) / \log_2(k + 1)$$

multisections of order k. Thus, the efficiency of the multisectioning of order k compared to bisection (multisectioning of order 1) is

$$E_f = n_1/(k\, n_k) = (\log_2(k + 1))/k.$$

Hence, for extraction of eigenvalues via parallel bisections is preferable to one multisectioning of higher order. On the other hand, during the isolation step, the efficiency of multisectioning is higher because: (i) multisectioning creates more tasks than bisection, and (ii) often, one interval contains much more than one eigenvalue. A reasonable strategy for choosing bisection or multisectioning is outlined in [44].

Computation of the Sturm Sequence

Parallelism in evaluating the recurrence (8.36) is very limited. Here, the computation of the regular Sturm sequence (8.35) is equivalent to solving a lower-bidiagonal system $Lx = c$ of order $n + 1$:

$$
L = \begin{pmatrix}
1 & & & & \\
(\lambda - \alpha_1) & 1 & & & \\
\beta_2^2 & (\lambda - \alpha_2) & 1 & & \\
& \ddots & \ddots & \ddots & \\
& & \beta_n^2 & (\lambda - \alpha_n) & 1
\end{pmatrix}, \text{ and } c = \begin{pmatrix} 1 \\ 0 \\ 0 \\ \vdots \\ 0 \end{pmatrix}.
$$

The algorithm of Sect. 3.2.2 may be used, however, to capitalize on the vectorization possible in evaluating the linear recurrence (8.35). For a vector length $k < n/2$ the total number of arithmetic operations in the parallel algorithm is roughly $10n + 11k$, compared to only $4n$ for the uniprocessor algorithm (we do not consider the operations needed to compute β_i^2 since these quantities can be provided by the user), resulting in an arithmetic redundancy which varies between 2.5 and 4. This algorithm, therefore, is efficient only when vector operations are at least 4 times faster than their sequential counterparts.

Computation of the Eigenvectors and Orthonormalization

The computation of an eigenvector can be started as soon as the corresponding eigenvalue is computed. It is obtained by inverse iteration (see Algorithm 8.5). When the

---

Algorithm 8.5 Computation of an eigenvector by inverse iteration.

Input: $T \in \mathbb{R}^{n \times n}$, $\lambda \in \mathbb{R}$ eigenvalue of T, $x_0 \in \mathbb{R}^n$ $(x_0 \neq 0)$, $\tau > 0$.
Output: x eigenvector of T corresponding to λ.
1: $x = \frac{x_0}{\|x_0\|}$; $y = (T - \lambda I)^{-1}x$; $\alpha = \|y\|$;
2: **while** $\alpha\tau < 1$,
3:     $x = \frac{y}{\alpha}$;
4:     $y = (T - \lambda I)^{-1}x$; $\alpha = \|y\|$;
5: **end while**

---

eigenvalue λ is multiple of order r, a basis of the corresponding invariant subspace can be obtained by a simultaneous iteration procedure (i.e. the iterate x in Algorithm 8.5 is replaced by a block $X \in \mathbb{R}^{n \times r}$ in which the columns are maintained orthogonal by means of a Gram-Schmidt procedure).

So, we consider the extraction of an eigenvalue and the computation of its eigenvector as two parts of the same task, with the degree of potential parallelism being dependent on the number of desired eigenvalues.

The computed eigenvectors of distinct eigenvalues are necessarily orthogonal. However, for close eigenvalues, this property is poorly satisfied and an orthogonalization step must be performed. Orthonormalization is only performed on eigenvectors whose corresponding eigenvalues meet a predefined grouping criterion (e.g.; in the procedure TINVIT of the library EISPACK [45], two eigenvalues λ_i and λ_{i+1} are in the same group when $|\lambda_i - \lambda_{i+1}| \leq 10^{-3} \|T\|_R$, where $\|T\|_R = \max_{i=1:n}(\alpha_i + \beta_i)$). The modified Gram-Schmidt method is used to orthonormalize each group.

If we consider processing several groups of vectors in parallel, then we have two levels of parallelism: (i) concurrent orthogonalization of the groups; and (ii) parallel orthogonalization of a group (as discussed in Sect. 7.4). Selection of the appropriate algorithm will depend on: (a) the levels of parallelism in the architecture, and (b) whether we have many groups each containing a small number of eigenvalues, or only a few groups each containing many more eigenvalues.

Reliability of the Sturm Sequences

The reliability of Sturm sequence computation in floating point arithmetic where the sequence is no longer monotonic has been considered in [46]. It states that in very rare situations, it is possible to obtain incorrect eigenvalues with regular bisection. A robust algorithm, called MRRR in [47], is implemented in LAPACK via routine DSTEMR for the computation of high quality eigenvectors. In addition, several approaches are considered for its parallel implementation.

8.3 Bidiagonalization via Householder Reduction

The classical algorithm for reducing a rectangular matrix which reduces A to an upper bidiagonal matrix B, via the orthogonal transformation (8.7) (e.g. see [3]) requires $4(mn^2 - n^3/3) + O(mn)$ arithmetic operations. V and U can be assembled in $4(mn^2 - n^3/3) + O(mn)$ and $4n^3/3 + O(n^2)$ additional operations, respectively. The corresponding routine in LAPACK is based on BLAS3 procedures and a parallel version of this algorithm is implemented in routine PDGEBRD of SCALAPACK.

A bidiagonalization scheme that is more suited for parallel processing has been proposed in [48]. This one-sided reduction consists of determining, as a first step, the orthogonal matrix W which appears in the tridiagonalization (8.5). The matrix W can be obtained by using Householder or Givens transformations, without forming $A^\top A$ explicitly. The second step performs a QR factorization of $F = AW$. From (8.5) and (8.6), we can see that R is upper bidiagonal. It is therefore possible to set $V = Q$ and $B = R$. This allows a simplified QR factorization. Two adaptations of the approach in [48] have been considered in [49]: one for obtaining a numerically more robust algorithm, and a second parallel block version which allows the use of BLAS3.

Once the bidiagonalization is performed, all that remains is to compute the singular value decomposition of B. When only the singular values are sought, these can be achieved via the uniprocessor scheme in [3] in which the number of arithmetic operations per iteration is $O(n)$, see routine PDGESVD in SCALAPACK .

References

1. Higham, N.: Accuracy and Stability of Numerical Algorithms, 2nd edn. SIAM, Philadelphia (2002)
2. Stewart, G.W., Sun, J.: Matrix Perturbation Theory. Academic Press, Boston (1990)
3. Golub, G., Van Loan, C.: Matrix Computations, 4th edn. Johns Hopkins (2013)
4. Parlett, B.: The Symmetric Eigenvalue Problem. SIAM (1998)
5. Hestenes, M.: Inversion of matrices by biorthogonalization and related results. J. Soc. Ind. Appl. Math. **6**(1), 51–90 (1958). doi:10.1137/0106005, http://epubs.siam.org/doi/abs/10.1137/0106005
6. Forsythe, G.E., Henrici, P.: The cyclic Jacobi method for computing the principal values of a complex matrix (January 1960)
7. Demmel, J., Veselic, K., Physik, L.M., Hagen, F.: Jacobi's method is more accurate than QR. SIAM J. Matrix Anal. Appl **13**, 1204–1245 (1992)
8. Schönhage, A.: Zur konvergenz des Jacobi-verfahrens. Numer. Math. **3**, 374–380 (1961)
9. Wilkinson, J.H.: Note on the quadratic convergence of the cyclic Jacobi process. Numer. Math. **4**, 296–300 (1962)
10. Sameh, A.: On Jacobi and Jacobi-like algorithms for a parallel computer. Math. Comput. **25**, 579–590 (1971)
11. Luk, F., Park, H.: A proof of convergence for two parallel Jacobi SVD algorithms. IEEE Trans. Comp. (to appear)
12. Luk, F., Park, H.: On the equivalence and convergence of parallel Jacobi SVD algorithms. IEEE Trans. Comp. **38**(6), 806–811 (1989)
13. Henrici, P.: On the speed of convergence of cyclic and quasicyclic Jacobi methods for computing eigenvalues of Hermitian matrices. Soc. Ind. Appl. Math. **6**, 144–162 (1958)
14. Brent, R., Luk, F.: The solution of singular-value and symmetric eigenvalue problems on multiprocessor arrays. SIAM J. Sci. Stat. Comput. **6**(1), 69–84 (1985)
15. Sameh, A.: Solving the linear least-squares problem on a linear array of processors. In: L. Snyder, D. Gannon, L.H. Jamieson, H.J. Siegel (eds.) Algorithmically Specialized Parallel Computers, pp. 191–200. Academic Press (1985)
16. Kaiser, H.: The JK method: a procedure for finding the eigenvectors and eigenvalues of a real symmetric matrix. Computers **15**, 271–273 (1972)
17. Nash, J.: A one-sided transformation method for the singular value decomposition and algebraic eigenproblem. Computers **18**(1), 74–76 (1975)
18. Luk, F.: Computing the singular value decomposition on the Illiac IV. ACM Trans. Math. Sftw. **6**(4), 524–539 (1980)
19. Brent, R., Luk, F., van Loan, C.: Computation of the singular value decomposition using mesh connected processors. VLSI Comput. Syst. **1**(3), 242–270 (1985)
20. Berry, M., Sameh, A.: An overview of parallel algorithms for the singular value and symmetric eigenvalue problems. J. Comp. Appl. Math. **27**, 191–213 (1989)
21. Charlier, J., Vanbegin, M., van Dooren, P.: On efficient implementations of Kogbetlianz's algorithm for computing the singular value decomposition. Numer. Math. **52**, 279–300 (1988)
22. Paige, C., van Dooren, P.: On the quadratic convergence of Kogbetliantz's algorithm for computing the singular value decomposition. Numer. Linear Algebra Appl. **77**, 301–313 (1986)
23. Charlier, J., van Dooren, P.: On Kogbetliantz's SVD algorithm in the presence of clusters. Numer. Linear Algebra Appl. **95**, 135–160 (1987)
24. Wilkinson, J.: Almost diagonal matrices with multiple or close eigenvalues. Numer. Linear Algebra Appl. **1**, 1–12 (1968)
25. Luk, F.T., Park, H.: On parallel Jacobi orderings. SIAM J. Sci. Stat. Comput. **10**(1), 18–26 (1989)
26. Bečka, M., Vajteršic, M.: Block-Jacobi SVD algorithms for distributed memory systems I: hypercubes and rings. Parallel Algorithms Appl. **13**, 265–287 (1999)
27. Bečka, M., Vajteršic, M.: Block-Jacobi SVD algorithms for distributed memory systems II: meshes. Parallel Algorithms Appl. **14**, 37–56 (1999)

28. Bečka, M., Okša, G., Vajteršic, M.: Dynamic ordering for a parallel block-Jacobi SVD algorithm. Parallel Comput. **28**(2), 243–262 (2002). doi:10.1016/S0167-8191(01)00138-7, http://dx.doi.org/10.1016/S0167-8191(01)00138-7
29. Okša, G., Vajteršic, M.: Efficient pre-processing in the parallel block-Jacobi SVD algorithm. Parallel Comput. **32**(2), 166–176 (2006). doi:10.1016/j.parco.2005.06.006, http://www.sciencedirect.com/science/article/pii/S0167819105001341
30. Bečka, M., Okša, G., Vajteršic, M.: Parallel Block-Jacobi SVD methods. In: M. Berry, K. Gallivan, E. Gallopoulos, A. Grama, B. Philippe, Y. Saad, F. Saied (eds.) High-Performance Scientific Computing, pp. 185–197. Springer, London (2012), http://dx.doi.org/10.1007/978-1-4471-2437-5_1
31. Dongarra, J., Croz, J.D., Hammarling, S., Hanson, R.: An extended set of FORTRAN basic linear algebra subprograms. ACM Trans. Math. Softw. **14**(1), 1–17 (1988)
32. Dongarra, J.J., Kaufman, L., Hammarling, S.: Squeezing the most out of eigenvalue solvers on high-performance computers. Linear Algebra Appl. **77**, 113–136 (1986)
33. Gallivan, K.A., Plemmons, R.J., Sameh, A.H.: Parallel algorithms for dense linear algebra computations. SIAM Rev. **32**(1), 54–135 (1990). http://dx.doi.org/10.1137/1032002
34. Dongarra, J., Du Croz, J., Hammarling, S., Duff, I.: A set of level-3 basic linear algebra subprograms. ACM Trans. Math. Softw. **16**(1), 1–17 (1990)
35. Anderson, E., Bai, Z., Bischof, C., Blackford, S., Demmel, J., Dongarra, J., Du Croz, J., Greenbaum, A., Hammarling, S., McKenney, A., Sorensen, D.: LAPACK Users' Guide, 3rd edn. Society for Industrial and Applied Mathematics, Philadelphia (1999)
36. Blackford, L., Choi, J., Cleary, A., D'Azevedo, E., Demmel, J., Dhillon, I., Dongarra, J., Hammarling, S., Henry, G., Petitet, A., Stanley, K., Walker, D., Whaley, R.: ScaLAPACK User's Guide. SIAM, Philadelphia (1997). http://www.netlib.org/scalapack
37. Cuppen, J.: A divide and conquer method for the symmetric tridiagonal eigenproblem. Numer. Math. **36**, 177–195 (1981)
38. Dongarra, J.J., Sorensen, D.C.: A fully parallel algorithm for the symmetric eigenvalue problem. SIAM J. Sci. Stat. Comput. **8**(2), s139–s154 (1987)
39. Kuck, D., Sameh, A.: Parallel computation of eigenvalues of real matrices. In: Information Processing '71, pp. 1266–1272. North-Holland (1972)
40. Lo, S.S., Philippe, B., Sameh, A.: A multiprocessor algorithm for the symmetric tridiagonal eigenvalue problem. SIAM J. Sci. Stat. Comput. **8**, S155–S165 (1987)
41. Wilkinson, J.H.: The Algebraic Eigenvalue Problem. Oxford University Press, New York (1965)
42. Forsythe, G., Malcom, M., Moler, C.: Computer Methods for Mathematical Computation. Prentice-Hall, New Jersey (1977)
43. Ostrowski, A.: Solution of Equations and Systems of Equations. Academic Press, New York (1966)
44. Bernstein, H., Goldstein, M.: Parallel implementation of bisection for the calculation of eigenvalues of tridiagonal symmetric matrices. Computing **37**, 85–91 (1986)
45. Garbow, B.S., Boyle, J.M., Dongarra, J.J., Moler, C.B.: Matrix Eigensystem Routines—EISPACK Guide Extension. Springer, Heidelberg (1977)
46. Demmel, J.W., Dhillon, I., Ren, H.: On the correctness of some bisection-like parallel eigenvalue algorithms in floating point arithmetic. Electron. Trans. Numer. Anal. pp. 116–149 (1995)
47. Dhillon, D., Parlett, B., Vömel, C.: The design and implementation of the MRRR algorithm. ACM Trans. Math. Softw. **32**, 533–560 (2006)
48. Rui, Ralha: One-sided reduction to bidiagonal form. Linear Algebra Appl. **358**(1–3), 219–238 (2003)
49. Bosner, N., Barlow, J.L.: Block and parallel versions of one-sided bidiagonalization. SIAM J. Matrix Anal. Appl. **29**, 927–953 (2007)

Part III
Sparse Matrix Computations

Chapter 9
Iterative Schemes for Large Linear Systems

Sparse linear systems occur in a multitude of applications in computational science and engineering. While the need for solving such linear systems is most prevalent in numerical simulations using mathematical models based on differential equations, it also arises in other areas. For example, the PageRank vector that was used by Google to order the nodes of a network based on its link structure can be interpreted as the solution of a linear system of order equal to the number of nodes [1, 2]. The system can be very large and sparse so that parallel iterative methods become necessary; see e.g. [3, 4]. We note that we will not describe any parallel asynchronous iterative methods that are sometimes proposed for solving such systems when they are very large, possibly distributed across several computers that are connected by means of a relatively slow network; cf. [5–8].

As we mentioned in the preface of this book, iterative methods for the parallel solution of the sparse linear systems that arise in partial differential equations have been proposed since the very early days of parallel processing. In fact, some basic ideas behind parallel architectures have been inspired by such methods. For instance, SOLOMON, the first parallel SIMD system, was designed in order to facilitate the communication necessary for the fast implementation of iterative schemes and for solving partial differential equations with marching schemes. The Illiac IV followed similar principles while with proper design of the numerical model, the SIMD Massively Parallel Processor (MPP) built for NASA's Goddard Space Flight Center by Goodyear Aerospace between 1980 and 1985 was also shown to lend itself to the fast solution of partial differential equations for numerical weather prediction with marching schemes in the spirit of "Richardson's forecast factory" [9–12].

In this chapter, we explore some methods for the solution of nonsingular linear systems of the form $Ax = f$. We explore some of the basic iterative schemes for obtaining an approximation y of the solution vector x with a specified relative residual $\|r\|/\|f\| \leq \tau$, where $r = f - Ay$, with τ being a given tolerance. We divide our presentation into two parts: (i) splitting, and (ii) polynomial methods. We illustrate these parallel iterative schemes through the use of the linear algebraic systems resulting from the finite-difference discretization of linear elliptic partial differential equations.

© Springer Science+Business Media Dordrecht 2016
E. Gallopoulos et al., *Parallelism in Matrix Computations*,
Scientific Computation, DOI 10.1007/978-94-017-7188-7_9

9.1 An Example

Let us consider the two-dimensional linear elliptic boundary value problem

$$
\frac{-\partial}{\partial x}\left(a(x, y)\frac{\partial u}{\partial x}\right) - \frac{\partial}{\partial y}\left(c(x, y)\frac{\partial u}{\partial y}\right) + d(x, y)\frac{\partial u}{\partial x},
$$

$$
+ e(x, y)\frac{\partial u}{\partial y} + f(x, y)u = g(x, y), \tag{9.1}
$$

defined on the unit square $0 \le x, y \le 1$, with the Dirichlet boundary conditions $u(x, y) = 0$. Using the five-point central finite-difference discretization on a uniform mesh of width h, we obtain the linear system of equations

$$
Au = g, \tag{9.2}
$$

in which u represents an approximation of the solution at the points (ih, jh), $i, j = 1, 2, \ldots, n$, where n is assumed to be even in what follows. Here A is a block-tridiagonal matrix of order n^2,

$$
A = \begin{pmatrix}
A_1 & B_1 & & & & \\
C_2 & A_2 & B_2 & & & \\
& \cdot & \cdot & \cdot & & \\
& & \cdot & \cdot & \cdot & \\
& & & C_{n-1} & A_{n-1} & B_{n-1} \\
& & & & C_n & A_n
\end{pmatrix}, \tag{9.3}
$$

in which $A_j = [\gamma_i^{(j)}, \alpha_i^{(j)}, \beta_i^{(j)}]$, $j = 1, 2, \ldots, n$, is tridiagonal of order n, $B_j = \mathrm{diag}(\mu_1^{(j)}, \mu_2^{(j)}, \ldots, \mu_n^{(j)})$, and $C_j = \mathrm{diag}(\nu_1^{(j)}, \nu_2^{(j)}, \ldots, \nu_n^{(j)})$. Correspondingly, we write

$$
u^\top = (u_1^\top, u_2^\top, \ldots, u_n^\top),
$$

and

$$
g^\top = (g_1^\top, g_2^\top, \ldots, g_n^\top),
$$

If we assume that $a(x, y), c(x, y) > 0$, and $f(x, y) \ge 0$ on the unit square, then $\alpha_i^{(j)} > 0$, and provided that

$$
0 < h < \min_{i,j}\left(\frac{2a_{i\pm\frac{1}{2},j}}{|d_{ij}|}, \frac{2c_{i\pm\frac{1}{2},j}}{|e_{ij}|}\right),
$$

where $d_{i}j$ and $e_{i}j$ are the values of $d(ih, jh)$ and $e(ih, jh)$, respectively, while $a_{i\pm\frac{1}{2},j}$ and $c_{i\pm\frac{1}{2},j}$ denote the values of $a(x, y)$ and $c(x, y)$ on the staggered grid. Thus, we have

$$\beta_i^{(j)}, \gamma_i^{(j)}, \mu_i^{(j)}, \text{ and } \nu_i^{(j)} < 0. \tag{9.4}$$

Furthermore,

$$\alpha_i^{(j)} \geq |\beta_i^{(j)} + \gamma_i^{(j)} + \mu_i^{(j)} + \nu_i^{(j)}|. \tag{9.5}$$

With the above assumptions it can be shown that the linear system (9.2) has a unique solution [13]. Before we show this, however, we would like to point out that the above block-tridiagonal is a special one with particular properties. Nevertheless, we will use this system to illustrate some of the basic iterative methods for solving sparse linear systems. We will explore the properties of such a system by first presenting the following preliminaries. These and their proofs can be found in many textbooks, see for instance the classical treatises [13–15].

Definition 9.1 A square matrix B is irreducible if there exists no permutation matrix Q for which

$$QBQ^\top = \begin{pmatrix} B_{11} & B_{12} \\ 0 & B_{22} \end{pmatrix}$$

where B_{11} and B_{22} are square matrices.

Definition 9.2 A matrix $B = (\beta_{ij}) \in \mathbb{R}^{m \times m}$ is irreducibly diagonally dominant if it is irreducible, and

$$|\beta_{ii}| \geq \sum_{\substack{j=1 \\ j \neq 1}} |\beta_{ij}|, \qquad i = 1, 2, \ldots, m,$$

with strict inequality holding at least for one i.

Lemma 9.1 *The matrix* $A \in \mathbb{R}^{n^2 \times n^2}$ *in (9.2) is irreducibly diagonally dominant.*

Theorem 9.1 (Perron-Frobenius) *Let* $B \in \mathbb{R}^{m \times m}$ *be nonnegative, i.e.,* $B \geq 0$ *or* $\beta_{ij} \geq 0$ *for all* i, j, *and irreducible. Then*

(i) B has a real, simple, positive eigenvalue equal to its spectral radius, $\rho(B) = \max_{1 \leq i \leq m} |\lambda_i(B)|$, with a corresponding nonnegative eigenvector $x \geq 0$.

(ii) $\rho(B)$ increases when any entry in B increases, i.e., if $C \geq B$ with $C \neq B$, then $\rho(B) < \rho(C)$.

Theorem 9.2 *The matrix A in (9.2) is nonsingular. In fact A is an M-matrix and* $A^{-1} > 0.$ $\qquad\qquad\qquad\qquad\qquad\qquad\qquad\qquad\qquad\qquad\qquad\qquad\qquad\qquad\square$

Corollary 9.1 *The tridiagonal matrices* $A_j \in \mathbb{R}^{n \times n}$, $j = 1, 2, \ldots, n$ *in (9.3) are nonsingular, and* $A_j^{-1} > 0.$

It is clear then from Theorem 9.2 that the linear system (9.2) has a unique solution u, which can be obtained by an iterative or a direct linear system solver. We explore first some basic iterative methods based on classical options for matrix splitting.

9.2 Classical Splitting Methods

If the matrix A in (9.2) can be written as

$$A = M - N, \tag{9.6}$$

where M is nonsingular, then the system (9.2) can be expressed as

$$Mu = Nu + g. \tag{9.7}$$

This, in turn, suggests the iterative scheme

$$Mu_{k+1} = Nu_k + g, \quad k \geq 0, \tag{9.8}$$

with u_0 chosen arbitrarily. In order to determine the condition necessary for the convergence of the iteration (9.8), let $\delta u_k = u_k - u$ and subtract (9.7) from (9.8) to obtain the relation

$$M \ \delta u_{k+1} = N \ \delta u_k, \quad k \geq 0. \tag{9.9}$$

Thus,

$$\delta u_k = H^k \delta u_0, \tag{9.10}$$

where $H = M^{-1}N$. Consequently, $\lim_{k \to \infty} \delta u_k = 0$ for any initial vector u_0, if and only if the spectral radius of H is strictly less than 1, i.e. $\rho(H) < 1$. For linear systems in which the coefficient matrix have the same properties as those of (9.3) an important splitting of the matrix A is given by the following.

Definition 9.3 $A = M - N$ is a regular splitting of A if M is nonsingular with $M^{-1} \geq 0$ and $N \geq 0.$

Theorem 9.3 *If* $A = M - N$ *is a regular splitting, then the iterative scheme (9.8) converges for any initial iterate* u_0 *[13].*

Simple examples of regular splitting of A are the classical Jacobi and Gauss-Seidel iterative methods. For example, if we express A as

$$A = D - L - U,$$

where $D = \text{diag}(\alpha_1^{(1)}, \ldots, \alpha_n^{(1)}; \ldots; \alpha_1^{(n)}, \ldots, \alpha_n^{(n)})$, and $-L$, $-U$ are the strictly lower and upper triangular parts of A, respectively, then the iteration matrices of the Jacobi and Gauss-Seidel schemes are respectively given by,

$$H_J = D^{-1}(L + U)$$

and

$$H_{G.S.} = (D - L)^{-1}U. \qquad (9.11)$$

$\square$

9.2.1 Point Jacobi

Here, the iteration is given by

$$u_{k+1} = H_J\, u_k + b, \qquad k \geq 0, \qquad (9.12)$$

where u_0 is arbitrary, $H_J = D^{-1}(L+U)$, or $(I - D^{-1}A)$, and $b = D^{-1}g$. Assuming that H_J and b are obtained at a preprocessing stage, parallelism in each iteration is realized by an effective sparse matrix-vector multiplication as outlined in Sect. 2.4.

If instead of using the natural ordering of the mesh points (i.e. ordering the mesh points row by row) that yields the coefficient matrix in (9.3), we use the so called red-black ordering [15], the linear system (9.2) is replaced by $A'u' = g'$, or

$$\begin{pmatrix} D_R & E_R \\ E_B & D_B \end{pmatrix} \begin{pmatrix} u^{(R)} \\ u^{(B)} \end{pmatrix} = \begin{pmatrix} g_R \\ g_B \end{pmatrix}, \qquad (9.13)$$

where D_R and D_B are diagonal matrices each of order $n^2/2$, and each row of E_R (or E_B) contains no more than 4 nonzero elements. The subscripts (or superscripts) B, R denote the quantities associated with the black and red points, respectively. One can also show that

$$A' = \begin{pmatrix} D_R & E_R \\ E_B & D_B \end{pmatrix} = P^\top A P,$$

where A is given by (9.3), and P is a permutation matrix. Hence, the iterative point-Jacobi scheme can be written as

$$\begin{pmatrix} D_R & 0 \\ 0 & D_B \end{pmatrix} \begin{pmatrix} u_{k+1}^{(R)} \\ u_{k+1}^{(B)} \end{pmatrix} = \begin{pmatrix} 0 & -E_R \\ -E_B & 0 \end{pmatrix} \begin{pmatrix} u_k^{(R)} \\ u_k^{(B)} \end{pmatrix} + \begin{pmatrix} g_R \\ g_B \end{pmatrix}, \qquad k \geq 0.$$

Observing that $u_{k+1}^{(R)}$ depends only on $u_k^{(B)}$, and $u_{k+1}^{(B)}$ depends only on $u_k^{(R)}$, it is clear we need to compute only half of each iterate u_j. Thus, choosing $u_0^{(R)}$ arbitrarily, we generate the sequences

$$\begin{aligned} u_{2k+1}^{(B)} &= -E_B' u_{2k}^{(R)} + g_B', \\ u_{2k+2}^{(R)} &= -E_R' u_{2k+1}^{(B)} + g_R' \end{aligned} \qquad k = 0, 1, 2, \ldots \qquad (9.14)$$

where

$$\begin{aligned} E_B' &= D_B^{-1} E_B, \quad E_R' = D_R^{-1} E_R \\ g_B' &= D_B^{-1} g_B, \quad g_R' = D_R^{-1} g_R. \end{aligned} \qquad (9.15)$$

Evaluating E_R', E_B', g_R', and g_B', in parallel, in a pre-processing stage, the degree of parallelism in each iteration is again governed by the effectiveness of the matrix-vector multiplication involving the matrices E_B' and E_R'. Note that each iteration in (9.14) is equivalent to two iterations of (9.12). In other words we update roughly half the unknowns in each half step of (9.14). Once $u_j^{(B)}$ (say) is accepted as a reasonable estimate of $u^{(B)}$, i.e., convergence has taken place, $u_j^{(R)}$ is obtained directly from (9.13) as

$$u^{(R)} = -E_R' u_j^{(B)} + g_R'.$$

Clearly, the point Jacobi iteration using the red-black ordering of the mesh points exhibits a higher degree of parallelism than using the natural ordering.

9.2.2 Point Gauss-Seidel

Employing the natural ordering of the uniform mesh, the iterative scheme is given by

$$u_{k+1} = (D - L)^{-1}(U u_k + g), \qquad k \geq 0. \qquad (9.16)$$

Here again the triangular system to be solved is of a special structure, and obtaining an approximation of the action of $(D - L)^{-1}$ on a column vector can be obtained effectively on a parallel architecture, e.g. via use of the Neumann series. Using the point red-black ordering of the mesh, and applying the point Gauss-Seidel splitting to (9.13), we obtain the iteration

$$\begin{pmatrix} u_{k+1}^{(R)} \\ u_{k+1}^{(B)} \end{pmatrix} = \begin{pmatrix} D_R^{-1} & 0 \\ D_B^{-1}E_B D_R^{-1} & D_B^{-1} \end{pmatrix} \left[\begin{pmatrix} 0 & -E_R \\ 0 & 0 \end{pmatrix} \begin{pmatrix} u_k^{(R)} \\ u_k^{(B)} \end{pmatrix} + \begin{pmatrix} g_R \\ g_B \end{pmatrix} \right], \qquad k \geq 0.$$

Simplifying, we get

$$u_{k+1}^{(R)} = E_R' u_k^{(B)} + g_R',$$

and

$$u_{k+1}^{(B)} = -E_B' u_{k+1}^{(R)} + g_B' \tag{9.17}$$

where $u_0^{(B)}$ is chosen arbitrarily, and E_R', E_B', g_B', g_R' are obtained in parallel via a preprocessing stage as given in (9.15). Taking in consideration, however, that [13]

$$\rho(H_{G.S.}') = \rho^2(H_J'), \tag{9.18}$$

where,

$$H_J' = \begin{pmatrix} D_R^{-1} & 0 \\ 0 & D_B^{-1} \end{pmatrix} \begin{pmatrix} 0 & -E_R \\ -E_B & 0 \end{pmatrix},$$

$$H_{G.S.}' = \begin{pmatrix} D_R & 0 \\ E_B & D_B \end{pmatrix}^{-1} \begin{pmatrix} 0 & -E_R \\ 0 & 0 \end{pmatrix},$$

we see that the red-black point-Jacobi scheme (9.14) and the red-black point-Gauss-Seidel scheme (9.17) are virtually equivalent regarding degree of parallelism and for solving (9.13) to a given relative residual.

9.2.3 Line Jacobi

Here we consider the so called line red-black ordering [15], where one row of mesh points is colored red while the one below and above it is colored black. In this case the linear system (9.2) is replaced by $A''u'' = g''$, or

$$\begin{pmatrix} T_R & F_R \\ F_B & T_B \end{pmatrix} \begin{pmatrix} v^{(R)} \\ v^{(B)} \end{pmatrix} = \begin{pmatrix} f_R \\ f_B \end{pmatrix}, \tag{9.19}$$

where $T_R = \mathrm{diag}(A_1, A_3, A_5, \ldots, A_{n-1})$, $T_B = \mathrm{diag}(A_2, A_4, A_6, \ldots, A_n)$,

$$
F_R = \begin{pmatrix}
B_1 & & & & \\
C_3 & B_3 & & & \\
& C_5 & B_5 & & \\
& & \ddots & & \\
& & & C_{n-1} & B_{n-1}
\end{pmatrix},
$$

$$
F_B = \begin{pmatrix}
C_2 & B_2 & & & \\
& C_4 & B_4 & & \\
& & \ddots & & \\
& & & C_{n-2} & B_{n-2} \\
& & & & C_n
\end{pmatrix}
$$

and A_j, B_j, and C_j are as given by (9.3). In fact, it can be shown that

$$
A'' = Q^\top A Q,
$$

where Q is a permutation. Now,

$$
A'' = \begin{pmatrix} T_R & 0 \\ 0 & T_B \end{pmatrix} - \begin{pmatrix} 0 & -F_R \\ -F_B & 0 \end{pmatrix}
$$

is a regular splitting of A'' which gives rise to the line-Jacobi iterative scheme

$$
\begin{aligned}
v_{k+1}^{(R)} &= T_R^{-1}(-F_R v_k^{(B)} + f_R), \\
v_{k+1}^{(B)} &= T_B^{-1}(-F_B v_k^{(R)} + f_B),
\end{aligned} \qquad k \geq 0 \qquad (9.20)
$$

which can be written more economically as,

$$
\begin{aligned}
v_{2k+1}^{(B)} &= T_B^{-1}(-F_B v_{2k}^{(R)} + f_B), \\
v_{2k+2}^{(R)} &= T_R^{-1}(-F_R v_{2k+1}^{(B)} + f_R),
\end{aligned} \qquad k \geq 0 \qquad (9.21)
$$

with $v_0^{(R)}$ chosen arbitrarily. For efficient implementation of (9.21), we start the pre-processing stage by obtaining the factorizations $A_j = D_j L_j U_j$, $j = 1, 2, \ldots, n$, where D_j is diagonal, and L_j, U_j are unit lower and unit upper bidiagonal matrices, respectively. Since each A_j is diagonally dominant, the above factorizations are obtained by Gaussian elimination without pivoting. Note that processor i handles the factorization of A_{2i-1} and A_{2i}. Using an obvious notation, we write

$$
T_R = D_R L_R U_R, \text{ and } T_B = D_B L_B U_B, \qquad (9.22)
$$

where, for example, $L_R = \mathrm{diag}(L_1, L_3, \ldots, L_{n-1})$, and $L_B = \mathrm{diag}(L_2, L_4, \ldots, L_n)$. Now, the iteration (9.21) is reduced to solving the following systems

$$L_B w_{2k+1}^{(B)} = (-G_B v_{2k}^{(R)} + h_B) \ ; \ U_B v_{2k+1}^{(B)} = w_{2k+1}^{(B)},$$
$$\text{and} \tag{9.23}$$
$$L_R w_{2k+2}^{(R)} = (-G_R v_{2k+1}^{(B)} + h_R) \ ; \ U_R v_{2k+2}^{(R)} = w_{2k+2}^{(R)}$$

where

$$G_R = D_R^{-1} F_R, \quad G_B = D_B^{-1} F_B, \tag{9.24}$$

$$h_R = D_R^{-1} f_R, \quad h_B = D_B^{-1} f_B, \tag{9.25}$$

with $v_0^{(R)}$ chosen arbitrarily. Thus, if we also compute G_R, h_R, and G_B, h_B in the pre-processing stage, each iteration (9.23) has ample degree of parallelism that can be further enhanced if solving triangular systems involving each L_j is achieved via any of the parallel schemes outlined in Chap. 3. Considering the ample parallelism inherent in each iteration of the red-black point Jacobi scheme using $n/2$ multicore nodes, it is natural to ask why should one consider the line red-black ordering. The answer is essentially provided by the following theorem.

Theorem 9.4 ([13]) *Let A be that matrix given in (9.2). Also let* $A = M_1 - N_1 = M_2 - N_2$ *be two regular splittings of A. If* $N_2 \geq N_1 \geq 0$, *with neither* N_1 *nor* $(N_2 - N_1)$ *being the null matrix, then*

$$\rho(H_1) < \rho(H_2) < 1,$$

where $H_i = M_i^{-1} N_i$, $i = 1,2$.

Note that the regular splittings of A' (red-black point-Jacobi) and A'' (red-black line-Jacobi) are related. In fact, if we have the regular splittings:

$$A' = \begin{pmatrix} D_R & 0 \\ 0 & D_B \end{pmatrix} - \begin{pmatrix} 0 & -E_R \\ -E_B & 0 \end{pmatrix},$$
$$= M'_J - N'_J,$$

and

$$A'' = \begin{pmatrix} T_R & 0 \\ 0 & T_B \end{pmatrix} - \begin{pmatrix} 0 & -F_R \\ -F_B & 0 \end{pmatrix},$$
$$= M''_J - N''_J$$

Then A'' has the two regular splittings,

$$A'' = M_J'' - N_J'' = S^{\mathsf{T}} M_J' S - S^{\mathsf{T}} N_J' S,$$

where $S = P^{\mathsf{T}} Q$ is a permutation matrix, and $S^{\mathsf{T}} N_J' S \geq N'' \geq 0$, equality excluded. Consequently, $\rho(H_J'') < \rho(H_J') < 1$, where

$$H_J' = {M_J'}^{-1} N_J' \quad \text{and} \quad H_J'' = {M_J''}^{-1} N_J'', \tag{9.26}$$

i.e., the line red-black Jacobi converges in fewer iterations than the point red-black Jacobi scheme. Such advantage, however, quickly diminishes as n increases, i.e., as $h = 1/(n+1) \to 0$, $\rho(H_J'') \to \rho(H_J')$. For example, in the case of the Poisson equation, i.e., when $a(x, y) = c(x, y) = 1$, $d(x, y) = e(x, y) = 0$, and $f(x, y) = f$, we have

$$\frac{\rho(H_J'')}{\rho(H_J')} = \frac{1}{2 - \cos \pi h}.$$

For n as small as 100, the above ratio is roughly 0.9995.

9.2.4 Line Gauss-Seidel

Applying the Gauss-Seidel splitting to (9.19)—linered-black ordering—we obtain the iteration

$$\begin{aligned} v_{k+1}^{(R)} &= T_R^{-1}(-F_R v_k^{(B)} + f_R), \\ v_{k+1}^{(B)} &= T_B^{-1}(-F_B v_{k+1}^{(R)} + f_B), \end{aligned} \qquad k \geq 0 \tag{9.27}$$

with $v_0^{(B)}$ chosen arbitrarily. If, in the pre-processing stage, T_R and T_B are factored as shown in (9.22), the iteration (9.27) reduces to solving the following linear systems

$$L_R x_{k+1}^{(R)} = (-G_R v_k^{(B)} + h_R); \quad U_R v_{k+1}^{(R)} = x_{k+1}^{(R)},$$
$$\text{and} \tag{9.28}$$
$$L_B x_{k+1}^{(B)} = (-G_B v_{k+1}^{(R)} + h_B); \quad U_B v_{k+1}^{(B)} = x_{k+1}^{(B)},$$

where G and h are as given in (9.24 and 9.25). Consequently, each iteration (9.28) can be performed with high degree of parallelism using $(n/2)$ nodes assuming that G_R, B_B, h_R and h_B have already been computed in a preprocessing stage. Similarly, such parallelism can be further enhanced if we use one of the parallel schemes in Chap. 3 for solving each triangular system involving $L_j, j = 1, 2, \ldots, n$.

We can also show that

$$\frac{\rho(H''_{G.S.})}{\rho(H'_{G.S.})} = \left(\frac{\rho(H''_J)}{\rho(H'_J)}\right)^2.$$

9.2.5 The Symmetric Positive Definite Case

If $d(x, y) = e(x, y) = 0$ in (9.1), then the linear system (9.2) is symmetric. Since A is nonsingular with $\alpha_i^{(j)} > 0, i, j = 1, 2, \ldots, n$, then by virtue of Gerschgorin's theorem, e.g., see [16] and (9.5), the system (9.2) is positive-definite. Consequently, the linear systems in (9.13) and (9.19) are positive-definite. In particular, the matrices T_R and T_B in (9.19) are also positive-definite. In what follows, we consider two classical acceleration schemes: (i) the cyclic Chebyshev semi-iterative method e.g. see [13, 17] for accelerating the convergence of line-Jacobi, and (ii) Young's successive over-relaxation, e.g. see [13, 15] for accelerating the convergence of line Gauss-Seidel, without adversely affecting the parallelism inherent in these two original iterations.

In this section, we do not discuss the notion of multisplitting where the coefficient matrix A has, say k possible splittings of the form $M_j - N_j, j = 1, 2, \ldots, k$. On a parallel architecture k iterations as in Eq. (9.8) can proceed in parallel but, under some conditions, certain combinations of such iterations can converge faster than any based on one splitting. For example, see [5, 18–20].

The Cyclic Chebyshev Semi-Iterative Scheme

For the symmetric positive definite case, each A_j in (9.2) is symmetric positive definite tridiagonal matrix, and $C_{j+1} = B_j^\top$. Hence the red-black line-Jacobi iteration (9.20) reduces to,

$$\begin{pmatrix} T_R & 0 \\ 0 & T_B \end{pmatrix}\begin{pmatrix} v_{k+1}^{(R)} \\ v_{k+1}^{(B)} \end{pmatrix} = \begin{pmatrix} 0 & -F_R \\ -F_R^\top & 0 \end{pmatrix}\begin{pmatrix} v_k^{(R)} \\ v_k^{(B)} \end{pmatrix} + \begin{pmatrix} f_R \\ f_B \end{pmatrix}.$$

Since T_R and T_B are positive-definite, the above iteration may be written as

$$\begin{pmatrix} x_{k+1}^{(R)} \\ x_{k+1}^{(B)} \end{pmatrix} = \begin{pmatrix} 0 & -K \\ -K^\top & 0 \end{pmatrix}\begin{pmatrix} x_k^{(R)} \\ x_k^{(B)} \end{pmatrix} + \begin{pmatrix} b_R \\ b_B \end{pmatrix}, \tag{9.29}$$

where $K = T_R^{-1/2} F_R T_B^{-1/2}, b_R = T_R^{-1/2} f_R, b_B = T_B^{-1/2} f_B$, and

$$\begin{pmatrix} x_k^{(R)} \\ x_k^{(B)} \end{pmatrix} = \begin{pmatrix} T_R^{1/2} & 0 \\ 0 & T_B^{1/2} \end{pmatrix}\begin{pmatrix} v_k^{(R)} \\ v_k^{(B)} \end{pmatrix}$$

converges to the true solution

$$x = \begin{pmatrix} I & K \\ K^\top & I \end{pmatrix}^{-1} \begin{pmatrix} b_R \\ b_B \end{pmatrix}. \tag{9.30}$$

The main idea of the semi-iterative method is to generate iterates

$$y_k = \sum_{j=0}^{k} \psi_j(K) x_j, \qquad k \geq 0 \tag{9.31}$$

with the coefficients $\psi_j(K)$ chosen such that y_k approaches the solution x faster than the iterates $x_j^\top = (x_j^{(R)^\top}, x_j^{(B)^\top})$ produced by (9.29). Again from (9.29) it is clear that if x_0 is chosen as x, then $x_j = x$ for $j > 0$. Hence, if y_k is to converge to x we must require that

$$\sum_{j=0}^{k} \psi_j(K) = 1, \qquad k \geq 0.$$

Let $\delta y_k = y_k - x$ and $\delta x_k = x_k - x$, then from (9.29) to (9.31), we have

$$\delta y_k = q_k(H) \delta y_0, \qquad k \geq 0$$

where

$$H = \begin{pmatrix} 0 & -K \\ -K^\top & 0 \end{pmatrix}, \qquad q_k(H) = \sum_{j=0}^{k} \psi_j(K) H^j,$$

and $\delta y_0 = \delta x_0$. Note that $q_k(I) = I$. If all the eigenvalues of H were known beforehand, it would have been possible to construct the polynomials $q_k(H), k \geq 1$, such that the exact solution (ignoring roundoff errors) is obtained after a finite number of iterations. Since this is rarely the case, and since

$$\|\delta y_k\| \leq \|q_k(H)\| \, \|\delta y_0\| = q_k(\rho)\|\delta y_0\|,$$

where $\rho = \rho(H)$ is the spectral radius of H, we seek to minimize $q_k(\rho)$ for $k \geq 1$. Observing that the eigenvalues of H are given by $\pm\lambda_i, i = 1, 2, \ldots, n^2/2$, with $0 < \lambda_i < 1$, we construct the polynomial $\hat{q}_k(\xi)$ characterized by

$$\min_{q_k(\xi)\varepsilon Q} \ \max_{-\rho \leq \xi \leq \rho} |q_k(\xi)| = \max_{-\rho \leq \xi \leq \rho} |\hat{q}_k(\xi)|,$$

where Q is the set of all polynomials $q_k(\xi)$ of degree k such that $q_0(\xi) = q_k(1) = 1$. The unique solution of this problem is well known [13]

$$\hat{q}_k(\xi) = \frac{\tau_k(\xi/\rho)}{\tau_k(1/\rho)}, \qquad -\rho \leq \xi \leq \rho \tag{9.32}$$

where $\tau_k(\xi)$ is the Chebyshev polynomial of the first kind of degree k,

$$\begin{aligned}
\tau_k(\xi) &= \cos(k\cos^{-1}\xi), \; |\xi| \leq 1, \\
&= \cosh(k\cosh^{-1}\xi), \; |\xi| > 1
\end{aligned}$$

which satisfy the 3-term recurrence relation

$$\tau_{j+1}(\xi) = 2\xi\tau_j(\xi) - \tau_{j-1}(\xi), \qquad j \geq 1, \tag{9.33}$$

with $\tau_0(\xi) = 1$, and $\tau_1(\xi) = \xi$. Hence, $\delta y_k = \hat{q}_k(H)\delta y_0$, and from (9.29), (9.32), and (9.33), we obtain the iteration

$$y_{k+1} = \omega_{k+1}[b + Hy_k] + (1 - \omega_{k+1})y_{k-1}, \qquad k \geq 1 \tag{9.34}$$

where $y_1 = Hy_0 + b$, i.e. $\omega_1 = 1$, with y_0 chosen arbitrarily, and

$$\begin{aligned}
\omega_{k+1} &= \frac{2}{\rho}\frac{\tau_k(1/\rho)}{\tau_{k+1}(1/\rho)} \\
&= 1/\left(1 - \frac{\rho^2}{4}\omega_k\right), \qquad k = 2, 3, \ldots
\end{aligned} \tag{9.35}$$

with $\omega_2 = 2/(2 - \rho^2)$. Consequently, the sequence y_k converges to the solution x with the errors $\delta y_k = y_k - x$ satisfying the relation

$$\begin{aligned}
\|\delta y_k\| &\leq \|\delta x_0\|/\tau_k(\rho^{-1}), \\
&= 2\|\delta x_0\|e^{-k\theta}(1 + e^{-2k\theta})^{-1},
\end{aligned}$$

where

$$\theta = \log_e(\rho^{-1} + \sqrt{\rho^{-2} - 1}).$$

Such reduction in errors is quite superior to the line-Jacobi iterative scheme (9.21) in which

$$\|\delta v_k\| \leq \rho^k\|\delta v_0\|.$$

Note that H is similar to H_j'', hence $\rho \equiv \rho(H) = \rho(H_j'')$ where H_j'' is defined in (9.26). Similar to the Jacobi iteration (9.21), (9.34) can be written more efficiently as

$$\begin{aligned}
z_{2k}^{(R)} &= z_{2k-2}^{(R)} + \omega_{2k}\Delta z_{2k}^{(R)}, \\
z_{2k+1}^{(B)} &= z_{2k-1}^{(B)} + \omega_{2k+1}\Delta z_{2k+1}^{(B)},
\end{aligned} \qquad k \geq 1 \tag{9.36}$$

where $z_0^{(R)}$ is chosen arbitrarily, $z_j^{(R)} = T_R^{-1/2} y_j^{(R)}$, $z_j^{(B)} = T_B^{-1/2} y_j^{(B)}$, with

$$
\begin{aligned}
z_1^{(B)} &= T_B^{-1}(f_B - F_R^\top z_0^{(R)}), \\
\Delta z_{2k}^{(R)} &= T_R^{-1}(f_R - F_R z_{2k-1}^{(B)}) - z_{2k-2}^{(R)},
\end{aligned}
\tag{9.37}
$$

and

$$
\Delta z_{2k+1}^{(B)} = T_B^{-1}(f_B - F_R^\top z_{2k}^{(R)}) - z_{2k-1}.
$$

Assuming that we have a good estimate of ρ, the spectral radius of H_j'', and hence have available the array ω_j, in a preprocessing stage, together with a Cholesky factorization of each tridiagonal matrix $A_j = L_j D_j L_j^\top$, then each iteration (9.37) can be performed with almost perfect parallelism. This is assured if each pair A_{2i-1} and A_{2i}, as well as B_{2i-1} and B_{2i} are stored in the local memory of the ith multicore node i, $i = 1, 2, \ldots, n/2$. Note that each matrix-vector multiplication involving F_R or $F_R^\top$, as well as solving tridiagonal systems involving T_R or T_B in each iteration (9.37), will be performed with very low communication overhead. If we have no prior knowledge of ρ, then we either have to deal with evaluating the largest eigenvalue in modulus of the generalized eigenvalue problem

$$
\begin{pmatrix} 0 & F_R \\ F_R^\top & 0 \end{pmatrix} u = \lambda \begin{pmatrix} T_R & 0 \\ 0 & T_B \end{pmatrix} u
$$

before the beginning of the iterations, or use an adaptive procedure for estimating ρ during the iterations (9.36). In the latter, early iterations can be performed with nonoptimal acceleration parameters ω_j in which ρ^2 is replaced by

$$
\rho_{2k}^2 = \frac{\Delta z_{2k}^{(R)^\top} F_R T_B^{-1} F_R^\top \Delta z_{2k}^{(R)}}{\Delta z_{2k}^{(R)^\top} T_R \Delta z_{2k}^{(R)}},
$$

and as k increases, ρ_{2k} approaches ρ, see [21].

Note that this method may be regarded as a polynomial method in which $r_k = (I - AM^{-1})^k r_0$, where we have the matrix splitting $A = M - N$.

The Line-Successive Overrelaxation Method

The line Gauss-Seidel iteration for the symmetric positive definite case is given by,

$$
\begin{pmatrix} T_R & 0 \\ F_R^\top & T_B \end{pmatrix} \begin{pmatrix} v_{k+1}^{(R)} \\ v_{k+1}^{(B)} \end{pmatrix} = \begin{pmatrix} 0 & -F_R \\ 0 & 0 \end{pmatrix} \begin{pmatrix} v_k^{(R)} \\ v_k^{(B)} \end{pmatrix} + \begin{pmatrix} f_R \\ f_B \end{pmatrix}.
$$

Consider, instead, the modified iteration

$$\begin{pmatrix} \frac{1}{\omega} T_R & 0 \\ F_R^{\top} & \frac{1}{\omega} T_B \end{pmatrix} \begin{pmatrix} v_{k+1}^{(R)} \\ v_{k+1}^{(B)} \end{pmatrix} = \begin{pmatrix} \left(\frac{1}{\omega} - 1\right) T_R & -F_R \\ 0 & \left(\frac{1}{\omega} - 1\right) T_B \end{pmatrix} \begin{pmatrix} v_k^{(R)} \\ v_k^{(B)} \end{pmatrix} + \begin{pmatrix} f_R \\ f_B \end{pmatrix}, \quad (9.38)$$

in which ω is a real parameter. If $\omega = 1$, we have the line Gauss-Seidel iteration, and for $\omega > 1$ ($\omega < 1$), we have the line-successive overrelaxation (underrelaxation). It is worthwhile mentioning that for $0 < \omega \leq 1$, we have a regular splitting of

$$C = \begin{pmatrix} T_R & F_R \\ F_R^{\top} & T_B \end{pmatrix} \quad (9.39)$$

and convergence of (9.38) is assured. The following theorem shows how large can ω be and still assuring that $\rho(H_{\omega}'') < 1$, where

$$H_{\omega}'' = M_{\omega}^{-1} N_{\omega},$$

in which

$$M_{\omega} = \begin{pmatrix} \frac{1}{\omega} T_R & 0 \\ F_R^{\top} & \frac{1}{\omega} T_B \end{pmatrix}, \quad (9.40)$$

and

$$N_{\omega} = \begin{pmatrix} \left(\frac{1}{\omega} - 1\right) T_R & -F_R \\ 0 & \left(\frac{1}{\omega} - 1\right) T_B \end{pmatrix}. \quad (9.41)$$

i.e.,

$$H_{\omega}'' = \begin{pmatrix} I & 0 \\ -\omega T_B^{-1} F_R^{\top} & I \end{pmatrix} \begin{pmatrix} (1-\omega)I & -\omega T_R^{-1} F_R \\ 0 & (1-\omega)I \end{pmatrix}. \quad (9.42)$$

Theorem 9.5 $\rho(H_{\omega}'') < 1$ *if and only if* $0 < \omega < 2$.

Proof The proof follows directly from the fact that the matrix

$$S = M_{\omega}^{\top} + N_{\omega}$$
$$= \begin{pmatrix} \left(\frac{2}{\omega} - 1\right) T_R & 0 \\ 0 & \left(\frac{2}{\omega} - 1\right) T_B \end{pmatrix}$$

is symmetric positive definite, see [22], for $0 < \omega < 2$. $\square$

Theorem 9.6 *Let*

$$\omega_0 = \frac{2}{1 + \sqrt{1 - \rho^2(H_J'')}},$$

where

$$H_J'' = \begin{pmatrix} T_R^{-1} & 0 \\ 0 & T_B^{-1} \end{pmatrix} \begin{pmatrix} 0 & -F_R \\ -F_R^\top & 0 \end{pmatrix}.$$

Then

$$\rho(H_{\omega_0}'') = \min_{0 < \omega < 2} \rho(H_\omega'').$$

Proof Consider the algebraic eigenvalue problem $H_\omega'' u = \lambda u$, or

$$\begin{pmatrix} (1-\omega)I & -\omega T_R^{-1} F_R \\ 0 & (1-\omega)I \end{pmatrix} \begin{pmatrix} u_1 \\ u_2 \end{pmatrix} = \lambda \begin{pmatrix} I & 0 \\ \omega T_B^{-1} F_R^\top & I \end{pmatrix} \begin{pmatrix} u_1 \\ u_2 \end{pmatrix}.$$

This can be reduced to the eigenvalue problem

$$T_B^{-1} F_R^\top T_R^{-1} F_R \, u_2 = \frac{(1 - \omega - \lambda)^2}{\lambda \omega^2} u_2.$$

If we let $H_J'' w = \mu w$, with $w^\top = (w_1^\top, w_2^\top)$, we can similarly verify that

$$T_B^{-1} F_R^\top T_R^{-1} F_R w_2 = \mu^2 w_2.$$

Thus, if λ is an eigenvalue of H_ω'', then

$$\mu^2 = \frac{(\lambda + \omega - 1)^2}{\lambda \omega^2} \tag{9.43}$$

is the square of an eigenvalue of H_J'', or

$$\lambda^2 + [2(\omega - 1) - \omega^2 \mu^2]\lambda + (\omega - 1)^2 = 0.$$

Differentiating w.r.t. λ, we get

$$\lambda + \omega - 1 = \frac{1}{2} \mu^2 \omega^2. \tag{9.44}$$

From (9.43) and (9.44) $\rho(H_\omega)$ is a minimum when

$$\rho^2(H_J'') = \frac{\rho^4(H_J'')\omega^4}{4\omega^2\left(\frac{1}{2}\rho^2(H_J'')\omega^2 - \omega + 1\right)},$$

i.e.,

$$\rho^2(H_J'')\omega^2 - 4\omega + 4 = 0.$$

Taking the smaller root of the above quadratic, we obtain the optimal value of ω,

$$\omega_0 = \frac{2}{1 + \sqrt{1 - \rho^2(H_J'')}},$$

which minimizes $\rho(H_\omega)$. Hence,

$$\begin{aligned}
\rho(H_{\omega_0}'') &= \omega_0 - 1 \\
&= \frac{\rho^2(H_J'')}{[1 + \sqrt{1 - \rho^2(H_J'')}]^2},
\end{aligned} \tag{9.45}$$

note that $\rho(H_J'') < 1$. $\qquad\qquad\qquad\qquad\qquad\qquad\qquad\qquad\square$

Algorithm 9.1 Line SOR iteration

1: Solve $\begin{array}{l}(i)\ \ L_R x_{k+1}^{(R)} = -G_R v_k^{(B)} + h_R \\ (ii)\ U_R y_{k+1}^{(R)} = x_{k+1}^{(R)}\end{array}$

2: $v_{k+1}^{(R)} = \omega_0 y_{k+1}^{(R)} + (1 - \omega_0)v_k^{(R)}$

3: Solve
$\begin{array}{l}(i)\ \ L_B x_{k+1}^{(B)} = -G_B v_{k+1}^{(R)} + h_B \\ (ii)\ U_B y_{k+1}^{(B)} = x_{k+1}^{(B)}\end{array}$

4: $v_{k+1}^{(B)} = \omega_0 y_{k+1}^{(B)} + (1 - \omega_0)v_k^{(B)}$

Using the factorization (9.22) of T_R and T_B, then similar to the line Gauss-Seidel iteration (9.23), the line-SOR iteration is given by Algorithm 9.2.5 where ω_0 is the optimal acceleration parameter, $v_0^{(B)}$ is chosen arbitrarily, and G, h are as given in (9.23). Assuming that G_R, G_B, and h_R, h_B are computed in a preprocessing stage each iteration exhibits ample parallelism as before provided ω_0 is known. If the optimal parameter ω_0 is not known beforehand, early iterations may be performed with the non-optimal parameters

$$\omega_j = \frac{2}{1 + \sqrt{1 - \mu_j^2}},$$

where

$$\mu_j^2 = \frac{s_j^\top F_R^\top T_R^{-1} F_R s_j}{s_j^\top T_B s_j}$$

in which $s_j = v_j^{(B)} - v_{j-1}^{(B)}$, and as j increases μ_j approaches $\rho(H_j'')$, see [21].

9.3 Polynomial Methods

Given an initial guess $x_0 \in \mathbb{R}^n$, the exact solution of $Ax = f$, can be expressed as

$$x = x_0 + A^{-1} r_0, \tag{9.46}$$

where $r_0 = f - Ax_0$ is the initial residual. A polynomial method is one which approximates $A^{-1} r_0$ for each iteration $k = 1, 2, \ldots$ by some polynomial expression $p_{k-1}(A) r_0$ so that the kth iterate is expressed as:

$$x_k = x_0 + p_{k-1}(A) r_0, \tag{9.47}$$

where $p_{k-1} \in \mathscr{P}_{k-1}$ is the set of all polynomials of degree no greater than $k - 1$.

In the following two sections, we consider two such methods. In Sect. 9.3.1 we consider the use of an explicit polynomial for symmetric positive definite linear systems. In Sect. 9.3.2 we consider the more general case where the linear system could be nonsymmetric and the polynomial p_{k-1} is defined implicitly. This results in an iterative scheme characterized by the way the vector $p_{k-1}(A) r_0$ is selected. This class of methods is referred to as Krylov subspace methods.

9.3.1 Chebyshev Acceleration

In this section we consider the classical Chebyshev method (or Stiefel iteration [23]), and one of its generalizations—the block Stiefel algorithm [24]—for solving symmetric positive definite linear systems of the form,

$$Ax = f, \tag{9.48}$$

where A is of order n. Further, for the sake of illustration, let us assume that A has a spectral radius less than 1. Also, let

$$0 < v = \mu_1 \leqq \mu_2 \leqq \cdots \mu_n = \mu < 1,$$

where ν and μ are the smallest and largest eigenvalues of A, respectively. Then, the classical Chebyshev iteration (or Stiefel iteration [23]) for solving the above linear system is given by,

---

1: (a) Initial step

$$x_0 : \text{arbitrary}$$
$$r_0 = f - Ax_0$$
$$x_1 = x_0 + \gamma^{-1}r_0 \; ; \; r_1 = f - Ax_1.$$

2: (b) **For** $j = 0, 1, 2, 3, \cdots$, obtain the following:

$$
\begin{aligned}
&1. \; \Delta x_j = \omega_j r_j + (\gamma\omega_j - 1)\Delta x_{j-1}, \\
&2. \; x_{j+1} = x_j + \Delta x_j, \\
&3. \; r_{j+1} = f - Ax_{j+1}.
\end{aligned}
\tag{9.49}
$$

Here, $\gamma = \beta/\alpha$, in which

$$\alpha = \frac{2}{\mu - \nu}, \qquad \beta = \frac{\mu + \nu}{\mu - \nu}$$

and

$$\omega_j = \left(\gamma - \frac{1}{4\alpha^2}\omega_{j-1}\right)^{-1}, \qquad j \geq 1$$

with $\omega_0 = 2/\gamma$.

This iterative scheme produces residuals r_j that satisfy the relation

$$r_j = P_j(A)r_0 \tag{9.50}$$

where $P_j(\lambda)$ is a polynomial of degree j given by,

$$P_j(\lambda) = \frac{\tau_j(\beta - \alpha\lambda)}{\tau_j(\beta)}, \qquad \nu \leq \lambda \leq \mu, \tag{9.51}$$

where, as defined earlier, $\tau_j(\xi)$ is the Chebyshev polynomial of the first kind:

$$\tau_j(\xi) = \begin{cases} \cos(j\cos^{-1}\xi), & |\xi| \leq 1, \\ \cos h(j\cos h^{-1}\xi), & \xi \geq 1. \end{cases}$$

As a result

$$\frac{\|r_j\|_2}{\|r_0\|_2} \leq [\tau_j(\beta)]^{-1}.$$

Suppose now that v is just an estimate of an interior eigenvalue μ_{s+1}, where s is a small integer

$$0 < \mu_1 \leqq \cdots \leqq \mu_s < v \leqq \mu_{s+1} \leqq \cdots \leqq \mu_n \leq \mu < 1.$$

Thus, if $r_0 = \sum_{i=1}^{n} \eta_i z_i$, where z_i is the eigenvector of A corresponding to μ_i, then r_j can be expressed as

$$r_j = \sum_{i=1}^{s} \eta_i P_j(\mu_i) z_i + \sum_{i=s+1}^{n} \eta_i P_j(\mu_i) z_i = r_j' + r_j''. \tag{9.52}$$

While r_j'' is damped out quickly as j increases, i.e., for $\beta = (\mu + v)/(\mu - v)$

$$\frac{\|r_j''\|_2}{\|r_0''\|_2} \leqq [\tau_j(\beta)]^{-1}, \tag{9.53}$$

the term r_j' is damped out at a much slower rate,

$$\frac{\|r_j'\|_2}{\|r_0'\|_2} \leqq \frac{\tau_j(\beta - \alpha\mu_1)}{\tau_j(\beta)}.$$

The basic strategy of the block Stiefel algorithm is to annihilate the contributions of the eigenvectors $z_1, z_2, \ldots, z_s$ to the residuals r_j so that eventually $\|r_j\|_2$ approaches zero as $\zeta_j = 1/\tau_j[(\mu_n + \mu_{s+1})/(\mu_n - \mu_{s+1})]$ rather than $\psi_j = 1/\tau_j[(\mu_n + \mu_1)/(\mu_n - \mu_1)]$ as in the classical Stiefel iteration [25]. Let $Z = [z_1, z_2, \ldots, z_s]$ be the orthonormal matrix consisting of the s-smallest eigenvectors. Then, from the fact that $r_j = -A(x_j - x)$ a projection process [26] produces the improved iterate

$$\hat{x}_j = x_j + Z(Z^\top A Z)^{-1} Z^\top r_j,$$

for which the corresponding residual $\hat{r}_j = b - A\hat{x}_j$ has zero projection onto z_i, $1 \leqq i \leqq s$, i.e., $Z^\top \hat{r}_j = 0$. Note that $Z^\top A Z = \mathrm{diag}(\mu_1, \ldots, \mu_s)$. This procedure is essentially a deflation technique.

Assuming, for the time being, that we also have the optimal parameters $\mu = \mu_n$, $v = \mu_{s+1}$ where s is 2 or 3 such that $\mu_s < \mu_{s+1}$, and $\zeta_k \ll \psi_k$ for k not too large, together with a reasonable approximation of the eigenpairs μ_i and z_i, $1 \leqq i \leqq s$. In order to avoid the computation of inner products in each iteration, we adopt an unconventional stopping criterion. Once ζ_i, the right-hand side of (9.53), drops below a given tolerance, we consider $\|r_\ell''\|_2$ to be sufficiently damped and start the projection step. Therefore, once $x_{\ell+1}$ and $r_{\ell+1}$ are obtained, the improved iterate $x_{\ell+1}$ is computed by

$$\hat{x}_{\ell+1} = x_{\ell+1} + \sum_{i=1}^{s} \mu_i (z_i^{\top} r_{\ell+1}) z_i. \tag{9.54}$$

While it is reasonable to expect that the optimal parameters μ and $v (\mu_n \leq \mu < 1, \mu_s \leq v < \mu_{s+1})$ are known, for example, as a result of previously solving problem (9.48) with a different right-hand side using the CG algorithm, it may not be reasonable to assume that Z is known a priori. In this case, the projection step (9.54) may be performed as follows.

Let $k = \ell - s + 1$, where ℓ is determined as before, i.e., so that $\tau_k(\beta)$ is large enough to assure that $\|r_k''\|_2$ is negligible compared to $\|r_k'\|_2$. Now, from (9.50) and (9.52)

$$r_k \simeq P_k(A) Z y,$$

where $y^{\top} = (\eta_1, \eta_2, \ldots, \eta_s)$, or

$$r_k \simeq Z w_k$$

in which

$$w_k^{\top} = (\eta_1 P_k(\mu_1), \ldots, \eta_s P_k(\mu_s)).$$

Consequently,

$$R_\ell = [r_{\ell-s+1}, r_{\ell-s+2}, \ldots, r_\ell] \simeq Z[w_{\ell-s+1}, w_{\ell-s+2}, \ldots, w_\ell] = Z W_\ell.$$

Let

$$R_\ell = Q_\ell U_\ell \tag{9.55}$$

be the orthogonal factorization of R_ℓ where Q_ℓ has orthonormal columns and U_ℓ is upper triangular of order s. As a result

$$Q_\ell \simeq Z \Theta_\ell$$

in which Θ_ℓ is an orthogonal matrix or order s, and

$$\hat{x}_{\ell+1} = x_{\ell+1} + Q_\ell (Q_\ell^{\top} A Q_\ell)^{-1} Q_\ell^{\top} r_{\ell+1} \tag{9.56}$$

has the desired property that $Z^{\top} \hat{r}_{\ell+1} \simeq 0$. Note that the eigenvalues of $Q_\ell^{\top} A Q_\ell$ are good approximations of $\mu_1, \ldots, \mu_s$.

The projection stage consists of the six steps shown below.

1. The modified Gram-Schmidt factorization $R_\ell = Q_\ell U_\ell$, as described in Algorithm 7.3, where

$$R_\ell = [r^{(1)}_{\ell-s+1}, \ldots, r^{(1)}_\ell], \quad Q_\ell = [q_{\ell-s+1}, \ldots, q_\ell], \text{ and } U_\ell = \begin{bmatrix} p_{11} & p_{12} & \cdots & p_{1s} \\ & p_{22} & \cdots & p_{2s} \\ & & & p_{ss} \end{bmatrix}.$$

2. Obtain the s inner-products $\gamma_i = q_i^\top r_{\ell+1}$, $\ell - s + 1 \leq i \leq \ell$.
3. Compute the upper triangular part of the symmetric matrix $S = Q_\ell^\top A Q_\ell$.
4. Solve the linear system $Sd = c$, where $c^\top = (\gamma_{\ell-s+1}, \ldots, \gamma_\ell)$.
5. Compute $d' = Q_\ell d = \sum_{i=\ell-s+1}^{\ell} q_i \delta_i$, where $\delta_i = e_i^\top d$ in which e_i is the ith column of the identity.
6. Finally, obtain $\hat{x}_{\ell+1} = x_{\ell+1} + d'$.

For small s, e.g. $s = 2$ or 3, the cost of this projection stage is modest.

Algorithm 9.2 Block Stiefel iterations.

Input: $A \in \mathbb{R}^{n \times n}$ (A SPD), f, $x_0 \in \mathbb{R}^n$, μ, $\nu \in \mathbb{R}$, $tol > 0$.
Output: $\ell \in \mathbb{N}$, $x \in \mathbb{R}^n$, $\frac{\|r_{\ell+1}\|}{\|f\|}$.
1: $\alpha = \frac{2}{\mu-\nu}$; $\beta = \frac{\mu+\nu}{\mu-\nu}$; $\gamma = \frac{\beta}{\alpha}$; $r_0 = f - A x_0$; $x_1 = x_0 + \gamma_0^{-1}$; $r_1 = f - A x_1$; $\Delta x_0 = 0$; $\omega_{-1} = 0$;
2: Determine ℓ for which $\tau_\ell(\beta)$;
3: **for** $j = 1 : \ell$, **do**
4:    $\omega_j = \left(\gamma - \frac{1}{4\alpha^2}\omega_{j-1}\right)^{-1}$;
5:    $\theta_j = \gamma\omega_j - 1$;
6:    $\Delta x_j = \omega_j r_j + \theta_j \Delta x_{j-1}$;
7:    $x_{j+1} = x_j + \Delta x_j$;
8:    $r_{j+1} = f - A x_{j+1}$;
9:    $R_j = [r_{j-s+1}, \cdots, r_{j-1}, r_j]$;
10: **end**
   //Projection step: Obtain the orthogonal factorization of R_ℓ via MGS:
11: $R_\ell = Q_\ell U_\ell$ (Q_ℓ orthonormal) ;
12: Form $S = Q_\ell^\top A Q_\ell$ and $c = Q_\ell r_{\ell+1}$;
13: Solve the system : $Sd = c$;
14: $\hat{x}_{\ell+1} = x_\ell + Q_\ell d$;
15: $x = \hat{x}_{\ell+1}$;
16: $r = f - A x$;
17: Compute rel. res. $\frac{\|r_{\ell+1}\|}{\|f\|}$;

Remark 9.1 On a parallel computing platform the block Stiefel iteration takes more advantage of parallelism than the classical Chebyshev scheme. Further, on a platform of many multicore nodes (peta- or exa-scale architectures), the block Stiefel iteration could be quite scalable and consume far less time than the conjugate gradient algorithm (CG) for obtaining a solution with a given level of the relative residual. This is due to the fact that the block Stiefel scheme avoids the repeated fan-in and fan-out operations (inner products) needed in each CG iteration.

9.3.2 Krylov Methods

Modern algorithms for the iterative solution of linear systems involving large sparse matrices are frequently based on approximations in Krylov subspaces. They allow for implicit polynomial approximations in classical linear algebra problems, namely: (i) computation of eigenvalues and corresponding eigenvectors, (ii) solving linear systems, and (iii) matrix function evaluation. To make such a claim more precise, we consider the problem of solving the linear system:

$$Ax = f, \tag{9.57}$$

where $A \in \mathbb{R}^{n \times n}$ is a large nonsingular matrix and $f \in \mathbb{R}^n$.

In this section, we consider the Arnoldi process, which is at the core of many such iterative schemes, and explore the potential of high performance on parallel architectures. For general presentations of Krylov subspace methods, one is referred to some of these references [27–30].

Krylov Subspaces

Definition 9.4 Given a matrix $A \in \mathbb{R}^{n \times n}$ and a vector r_0, for $k \geq 1$, the Krylov subspace $\mathcal{K}_k(A, r_0)$ is the subspace spanned by the set of vectors $\{r_0, Ar_0, A^2 r_0, \ldots, A^{k-1} r_0 \}$.

A Krylov method is characterized by how the vector $y_k \in \mathcal{K}_k(A, r_0)$ is selected to define the kth iterate as $x_k = x_0 + y_k$. In general, the method implicitly defines the polynomial p_{k-1} such that $y_k = p_{k-1}(A)r_0$. For instance, the GMRES method [31] consists of selecting y_k such that the 2-norm of the residual $r_k = f - Ax_k = r_0 - Ay_k$ is minimized over $\mathcal{K}_k(A, r_0)$; this defines a unique polynomial but it is not made explicit by the method.

Proposition 9.1 *With the previous notations, we have the following assertions:*

- *The sequence of the Krylov subspaces is nested:*

$$\mathcal{K}_k(A, r_0) \subseteq \mathcal{K}_{k+1}(A, r_0), \text{ for } k \geq 0.$$

- *There exists $\eta \geq 0$ such that the previous sequence is increasing for $k \leq \eta$ (i.e. $\dim(\mathcal{K}_k(A, r_0)) = k$), and is constant for $k \geq \eta$ (i.e. $\mathcal{K}_k(A, r_0) = \mathcal{K}_\eta(A, r_0)$). The subspace $\mathcal{K}_\eta(A, r_0)$ is invariant with respect to A. When A is nonsingular, the solution of (9.57) satisfies $x \in x_0 + \mathcal{K}_\eta(A, r_0)$.*
- *If $A^{k+1} r_0 \in \mathcal{K}_k(A, r_0)$, then $\mathcal{K}_k(A, r_0)$ is an invariant subspace of A.*

Proof The proof is straightforward.

Working with the Krylov subspace $\mathcal{K}_k(A, r_0)$ usually requires the knowledge of a basis. As we will point out later, the canonical basis $\{r_0, Ar_0, A^2 r_0, \ldots, A^{k-1} r_0 \}$ is ill-conditioned and hence not appropriate for direct use. The most robust approach consists of obtaining an orthogonal version of this canonical basis. Such a procedure is called the Arnoldi process.

The Arnoldi Process

Let us assume that, for $k \geq 0$, the columns of the matrix $V_k = [v_1, v_2, \ldots, v_k]$ form an orthonormal basis of $\mathcal{K}_k(A, r_0)$. An algorithm can be derived inductively as long as the sequence of Krylov subspaces is not constant:

$k = 1$: by normalizing r_0, one gets $v_1 = \frac{r_0}{\|r_0\|}$.

induction: let $V_k = [v_1, v_2, \ldots, v_k]$ be an orthonormal basis of $\mathcal{K}_k(A, r_0)$. The vector $w_k = Av_k$ obviously lies in $\mathcal{K}_{k+1}(A, r_0)$. If $w_k \in \mathcal{K}_k(A, r_0)$ then $\mathcal{K}_k(A, r_0)$ is invariant and the sequence is constant from that point on and the process stops. Else, $w_k \in \mathcal{K}_{k+1}(A, r_0) \backslash \mathcal{K}_k(A, r_0)$.[1] Then the vector v_{k+1} can be defined by $v_{k+1} = \frac{P_k w_k}{\|P_k w_k\|}$ where P_k is the orthogonal projector onto $\mathcal{K}_k(A, r_0)^\perp$.

With this procedure, one realizes a Gram-Schmidt process as implemented in Algorithm 7.2 of Chap. 7. As a result, it performs (step-by-step) a QR-factorization of the matrix $[r_0, Av_1, Av_2, \ldots, Av_k] = V_{k+1} R_{k+1}$, where the upper-triangular matrix R_{k+1} can be partitioned into

$$R_{k+1} = \begin{bmatrix} \|r_0\| & \\ & \hat{H}_k \\ 0 & \end{bmatrix},$$

where $\hat{H}_k \in \mathbb{R}^{(k+1) \times k}$ is an augmented upper-Hessenberg matrix. By considering the modified version MGS of the Gram-Schmidt process, the Arnoldi process is then expressed by Algorithm 9.3.

Algorithm 9.3 Arnoldi procedure.

Input: $A \in \mathbb{R}^{n \times n}$, $r_0 \in \mathbb{R}^n$ and $m > 0$.
Output: $V = [v_1, \cdots, v_{m+1}] \in \mathbb{R}^{n \times (m+1)}$, $\hat{H}_k = (\mu_{ij}) \in \mathbb{R}^{(m+1) \times m}$.
1: $v_1 = r_0/\|r_0\|$;
2: **for** $k = 1 : m$, **do**
3:	$w = Av_k$;
4:	**for** $j = 1 : k$, **do**
5:		$\mu_{j,k} = v_j^\top w$;
6:		$w = w - \mu_{j,k} v_j$;
7:	**end**
8:	$\mu_{k+1,k} = \|w\|$; $v_{k+1} = w/\mu_{k+1,k}$;
9: **end**

We can now claim the main result:

---

[1] Given two sets $\mathcal{A}$ and $\mathcal{B}$, we define $\mathcal{A} \backslash \mathcal{B} = \{x \in \mathcal{A} \mid x \notin \mathcal{B}\}$.

Theorem 9.7 (Arnoldi identities) *The matrices V_{m+1} and $\hat{H}_m$ satisfy the following relation:*

$$AV_m = V_{m+1}\hat{H}_m, \tag{9.58}$$

that can also be expressed by

$$AV_m = V_m H_m + \mu_{m+1,m} v_{m+1} e_m^\top, \tag{9.59}$$

where

- $e_m \in \mathbb{R}^m$ *is the last canonical vector of $\mathbb{R}^m$,*
- $H_m \in \mathbb{R}^{m \times m}$ *is the matrix obtained by deleting the last row of $\hat{H}_m$.*

The upper Hessenberg matrix H_m corresponds to the projection of the restriction of the operator A on the Krylov subspace with basis V_m:

$$H_m = V_m^\top A V_m. \tag{9.60}$$

The Arnoldi process involves m sparse matrix-vector multiplications and requires $O(m^2 n)$ arithmetic operations for the orthogonalization process. Note that we need to store the sparse matrix A and the columns of V_{m+1}.

Lack of Efficiency in Parallel Implementation

A straightforward parallel implementation consists of creating parallel procedures for $v \to Av$ and the BLAS-1 kernels as well (combinations of _DOT and _AXPY). As outlined in Sect. 2.1, inner products impede achieving high efficiency on parallel computing platforms. It is not possible to replace the Modified Gram-Schmidt MGS procedure by its classical counterpart CGS, not only due to the lack of numerical stability of the latter, but also because of a dependency in the computation of v_{k+1} which requires availability of v_k.

The situation would be much better if a basis of $\mathcal{K}_k(A, r_0)$ were available before the orthogonalization process. This is addressed below.

Nonorthogonal Bases for a Krylov Subspace

Consider $Z_{m+1} = [z_1, \ldots, z_{m+1}] \in \mathbb{R}^{n \times m+1}$ such that $\{z_1, \ldots, z_k\}$ is a basis of $\mathcal{K}_k(A, r_0)$ for any $k \leq m + 1$, with $\alpha_0 z_1 = r_0$, for $\alpha_0 \neq 0$. A natural model of recursion which builds bases of this type is given by,

$$\alpha_k z_{k+1} = (A - \beta_k I)z_k - \gamma_k z_{k-1}, \qquad k \geq 1, \tag{9.61}$$

where $\alpha_k \neq 0$, β_k, and γ_k are scalars with $\gamma_1 = 0$ (and $z_0 = 0$). For instance, the canonical basis of $\mathcal{K}_{m+1}(A, r_0)$ corresponds to the special choice: $\alpha_k = 1$, $\beta_k = 0$, and $\gamma_k = 0$ for any $k \geq 1$. Conversely, such a recursion insures that, for $1 \leq k \leq m + 1$, $z_k = p_{k-1}(A)r_0$, where p_{k-1} is a polynomial of degree $k - 1$.

By defining the augmented tridiagonal matrix

$$
\hat{T}_m =
\begin{bmatrix}
\beta_1 & \gamma_2 & & & & \\
\alpha_1 & \beta_2 & \gamma_3 & & & \\
& \alpha_2 & \beta_3 & \gamma_4 & & \\
& & \alpha_3 & \ddots & \ddots & \\
& & & & \ddots & \beta_{m-1} & \gamma_m \\
& & & & & \alpha_{m-1} & \beta_m \\
& & & & & & \alpha_m
\end{bmatrix}
\in \mathbb{R}^{(m+1)\times m},
\tag{9.62}
$$

the relations (9.61), can be expressed as

$$
A Z_m = Z_{m+1} \hat{T}_m.
\tag{9.63}
$$

Given the QR factorization,

$$
Z_{m+1} = W_{m+1} R_{m+1},
\tag{9.64}
$$

where the matrix $W_{m+1} \in \mathbb{R}^{n \times (m+1)}$ has orthonormal columns with $W_{m+1} e_1 = r_0 / \|r_0\|$, and where $R_{m+1} \in \mathbb{R}^{(m+1)\times(m+1)}$ is upper triangular, then $Z_m = W_m R_m$, where R_m is the leading $m \times m$ principal submatrix of R_{m+1}, and W_m consists of the first m columns of W_{m+1}. Consequently, $Z_m = W_m R_m$ is an orthogonal factorization. Substituting the QR factorizations of Z_{m+1} and Z_m into (9.63) yields

$$
A W_m = W_{m+1} \hat{G}_m, \qquad \hat{G}_m = R_{m+1} \hat{T}_m R_m^{-1},
\tag{9.65}
$$

where we note that the $(m + 1) \times m$ matrix $\hat{G}_m$ is an augmented upper Hessenberg matrix. We are interested in relating the matrices W_{m+1} and $\hat{G}_m$ above to the corresponding matrices V_{m+1} and $\hat{H}_m$ determined by the Arnoldi process.

Proposition 9.2 *Assume that m steps of the Arnoldi process can be applied to the matrix $A \in \mathbb{R}^{n \times n}$ with an initial vector $r_0 \in \mathbb{R}^n$ without breakdown yielding the decomposition (9.58). Let*

$$
A W_m = W_{m+1} \hat{G}_m
$$

be another decomposition, such that $W_{m+1} \in \mathbb{R}^{n \times (m+1)}$ has orthonormal columns, $W_m e_1 = V_m e_1$, and $\hat{G}_m \in \mathbb{R}^{(m+1)\times m}$ is an augmented upper Hessenberg matrix with nonvanishing subdiagonal entries. Then $W_{m+1} = V_{m+1} D_{m+1}$ and $\hat{G}_m = D_{m+1}^{-1} \hat{H}_m D_m$, where $D_{m+1} \in \mathbb{C}^{(m+1)\times(m+1)}$ is diagonal with ± 1 diagonal entries, and D_m is the leading $m \times m$ principal submatrix of D_{m+1}. In particular, the matrices $\hat{H}_m$ and $\hat{G}_m$ are orthogonally equivalent, i.e., they have the same singular values. The matrices H_m and G_m obtained by removing the last row of $\hat{H}_m$ and $\hat{G}_m$, respectively, are also orthogonally similar, i.e., they have the same eigenvalues.

Proof The proposition is a consequence of the Implicit Q Theorem (e.g. see [16]).

Suitability for Parallel Implementation

In implementing the Arnoldi process through a non-orthonormal Krylov basis, the two critical computational steps that need to be implemented as efficiently as possible are those expressed in (9.63) and (9.64):

(a) Step 1—The Krylov basis generation in (9.63) consists of two levels of computations. The first involves sparse matrix-vector multiplication while the second involves the tridiagonalization process through a three-term recurrence. By an adequate memory allocation of the sparse matrix A, the computational kernel $v \longrightarrow Av$ can usually be efficiently implemented on a parallel architecture (see Sect. 2.4.1). This procedure, however, involves a global sum (see Algorithms 2.9 and 2.10) which, in addition to the three-term recurrence limit parallel scalability. To avoid such limitations, it is necessary to consider a second level of parallelism at the three-term recursion level which must be pipelined. The pipelining pattern depends on the matrix sparsity structure.

(b) Step 2—The QR factorization in (9.64) can be implemented on a parallel architecture as outlined in Sect. 7.6 since the basis dimension m is much smaller than the matrix size, i.e. $m \ll n$).

The combination of these two steps (a) and (b) leads to an implementation which is fairly scalable unlike a straightforward parallel implementation of the Arnoldi procedure. This is illustrated in [32, 33] where the matrix consists of overlapped sparse diagonal blocks. In this case, it is shown that the number of interprocessor communications is linear with respect to the number of diagonal blocks. This is not true for the classical Arnoldi approach which involves inner products. An example is also given in the GPREMS procedure, e.g. see [34] and Sect. 10.3. Unfortunately, there is a limit on the size m of the basis generated this way as we explain below.

Ill-Conditioned Bases

From the relation (9.65), the regular Arnoldi relation (9.58) can be recovered by considering the Hessenberg matrix $\hat{G}_m$. Clearly this computation can be suspect when R_{m+1} is ill-conditioned. If we define the condition number of Z_{m+1} as,

$$\text{cond}(Z_{m+1}) = \frac{\displaystyle\max_{\|y\|=1} \|Z_{m+1}y\|}{\displaystyle\min_{\|y\|=1} \|Z_{m+1}y\|}, \tag{9.66}$$

then, when the basis Z_{m+1} is ill-conditioned the recovery fails since $\text{cond}(Z_{m+1}) = \text{cond}(R_{m+1})$. This is exactly the situation with the canonical basis. In the next section, we propose techniques for choosing the scalars α_k, β_k, and γ_k in (9.62) in order to limit the growth of the condition number of Z_k as k increases.

Limiting the Growth of the Condition Number of Krylov Subspace Bases

The goal is therefore to define more appropriate recurrences of the type (9.61). We explore two options: (i) using orthogonal polynomials, and (ii) defining a sequence of shifts in the recursion (9.61). A comparison of the two approaches with respect to controling the condition number is given in [35]. Specifically, both options are compared by examining their effect on the convergence speed of GMRES.

The algorithms corresponding to these options require some knowledge of the spectrum of the matrix A, i.e. $\Lambda(A)$. Note that, $\Lambda(A)$ need not be determined to high accuracy. Applying, Algorithm 9.3 with a small basis dimension $m_0 \leq m$, the convex hull of $\Lambda(A)$ is estimated by the eigenvalues of the Hessenberg matrix H_{m_0}. Since the Krylov bases are repeatedly built from a sequence of initial vectors $r_0^{(j)}$, the whole computation is not made much more expensive by such an estimation of $\Lambda(A)$. Moreover, at each restart, the convex hull may be updated by considering the convex hull of the union of the previous estimates and the Ritz values obtained from the last basis generation.

In the remainder of this section, we assume that $A \in \mathbb{C}^{n \times n}$. We shall indicate how to maintain real arithmetic operations when the matrix A is real.

Chebyshev Polynomial Bases

The short generation recurrences of Chebyshev polynomials of the first kind are ideal for our objective of creating Krylov subspace bases. As outlined before, such recurrence is given by

$$T_k(t) := \cosh(k \cosh^{-1}(t)), \qquad k \geq 0, \qquad |t| \geq 1. \tag{9.67}$$

Generating such Krylov bases has been discussed in several papers, e.g. see [35–37]. In what follows, we adopt the presentation given in [35]. Introducing the family of ellipses in $\mathbb{C}$,

$$\mathscr{E}^{(\rho)} := \left\{ e^{i\theta} + \rho^{-2}e^{-i\theta} \; : \; -\pi < \theta \leq \pi \right\}, \qquad \rho \geq 1,$$

where $\mathscr{E}^{(\rho)}$ has foci at $\pm c$ with $c = 2\rho^{-1}$, with semi-major and semi-minor axes given, respectively, by $\alpha = 1 + \frac{1}{\rho^2}$ and $\beta = 1 - \frac{1}{\rho^2}$, in which

$$\rho = \frac{\alpha + \beta}{c}. \tag{9.68}$$

When ρ grows from 1 to ∞, the ellipse $\mathscr{E}^{(\rho)}$ evolves from the interval $[-2, +2]$ to the unit circle.

Consider the scaled Chebyshev polynomials

$$C_k^{(\rho)}(z) := \frac{1}{\rho^k} T_k(\frac{\rho}{2} z), \qquad k = 0, 1, 2, \dots .$$

then rom the fact that

$$T_k\left(\frac{1}{2} \left(\rho e^{i\theta} + \rho^{-1}e^{-i\theta} \right) \right) = \frac{1}{2} \left(\rho^k e^{ik\theta} + \rho^{-k}e^{-ik\theta} \right),$$

and

$$C_k^{(\rho)}(e^{i\theta} + \rho^{-2}e^{-i\theta}) = \frac{1}{2}\left(e^{ik\theta} + \rho^{-2k}e^{-ik\theta}\right), \tag{9.69}$$

it follows that when $z \in \mathcal{E}^{(\rho)}$ we have $C_k^{(\rho)}(z) \in \frac{1}{2}\mathcal{E}^{(\rho^{2k})}$. Consequently, it can be proved that, for any ρ, the scaled Chebyshev polynomials $(C_k^{(\rho)})$ are quite well conditioned on $\mathcal{E}^{(\rho)}$ for the uniform norm [35, Theorem 3.1]. This result is independent of translation, rotation, and scaling of the ellipse, provided that the standard Chebyshev polynomials T_k are translated, rotated, and scaled accordingly. Let $\mathcal{E}(c_1, c_2, \tau)$ denote the ellipse with foci c_1 and c_2 and semi-major axis of length τ. This ellipse can then be mapped by a similarity transformation $\phi(z) = \mu z + \nu$ onto the ellipse $\mathcal{E}^{(\rho)} = \mathcal{E}(-\frac{1}{2\rho}, \frac{1}{2\rho}, 1 + \frac{1}{\rho^2})$ with a suitable value of $\rho \geq 1$ by translation, rotation, and scaling:

$$\begin{bmatrix} m = \frac{c_1+c_2}{2}, & \text{is the center point of } \mathcal{E}(c_1, c_2, \tau), \\ c = \frac{|c_2-c_1|}{2}, & \text{is the semi-focal distance of } \mathcal{E}(c_1, c_2, \tau), \\ s = \sqrt{\tau^2 - c^2}, & \text{is the semi-minor axis of } \mathcal{E}(c_1, c_2, \tau), \\ \rho = \frac{\tau+s}{c}, & \\ \mu = \frac{2}{\rho(c_2-m)}, & \\ \nu = -\mu m. & \end{bmatrix} \tag{9.70}$$

The three last expressions are obtained from (9.68) and from the equalities $\phi(m) = 0$ and $\phi(c_2) = \frac{2}{\rho}$.

It is now easy to define polynomials $S_k(z)$ which are well conditioned on the ellipse $\mathcal{E}(c_1, c_2, \tau)$, by defining for $k \geq 0$

$$S_k(z) = C_k^{(\rho)}(\phi(z)). \tag{9.71}$$

Before using the polynomials S_k to generate Krylov subspace bases, however, we outline first the recurrence which generates them, e.g. see, [38]):

$$\begin{cases} T_0(z) & = 1, \\ T_1(z) & = z, \\ T_{k+1}(z) = 2z\, T_k(z) - T_{k-1}(z), & \text{for } k \geq 1. \end{cases} \tag{9.72}$$

Since

$$\begin{aligned} S_{k+1}(z) &= C_{k+1}^{(\rho)}(\phi(z)) \\ &= \frac{1}{\rho^{k+1}} T_{k+1}(\frac{\rho}{2}\,\phi(z)) \\ &= \frac{1}{\rho^{k+1}}\left(\rho\phi(z)\, T_k(\frac{\rho}{2}\,\phi(z)) - T_{k-1}(\frac{\rho}{2}\,\phi(z))\right), \end{aligned}$$

we obtain the following recurrence for $S_j(z)$,

$$\begin{cases} S_0(z) & = 1, \\ S_1(z) & = \frac{1}{2}\phi(z), \\ S_{k+1}(z) = \phi(z)\,S_k(z) - \frac{1}{\rho^2}\,S_{k-1}(z), & \text{for } k \geq 1, \end{cases} \tag{9.73}$$

where ρ and $\phi(z) = \mu z + v$ are obtained from (9.70).

Let,

$$Z_{m+1} = [r_0, S_1(A)r_0, \ldots, S_m(A)r_0] \in \mathbb{C}^{n \times (m+1)} \tag{9.74}$$

form a basis for the Krylov subspace $\mathcal{K}_{m+1}(A, r_0)$, with m being a small integer.
Then the matrix $\hat{T}_m$, which satisfies (9.63), is given by,

$$\hat{T}_m = \begin{pmatrix} -\frac{2v}{\mu} & \frac{2}{\mu} & & & & \\ \frac{1}{\mu\rho^2} & -\frac{v}{\mu} & \frac{1}{\mu} & & & \\ & \frac{1}{\mu\rho^2} & -\frac{v}{\mu} & \frac{1}{\mu} & & \\ & & \frac{1}{\mu\rho^2} & \ddots & \ddots & \\ & & & \ddots & -\frac{v}{\mu} & \frac{1}{\mu} \\ & & & & \frac{1}{\mu\rho^2} & -\frac{v}{\mu} \\ & & & & & \frac{1}{\mu\rho^2} \end{pmatrix} \in \mathbb{C}^{(m+1)\times m}. \tag{9.75}$$

The remaining question concerns how the ellipse is selected for a given operator $A \in \mathbb{C}^{n \times n}$. By extension of the special situation when A is normal, one selects the smallest ellipse which contains the spectrum $\Lambda(A)$ of A, see [35] and Sect. 9.3.2.

A straightforward application of the technique presented here is described in Algorithm 9.4.

Algorithm 9.4 Chebyshev-Krylov procedure.

Input: $A \in \mathbb{C}^{n \times n}$, $m > 1$, $r_0 \in \mathbb{C}^n$, c_1, c_2, and $\tau > 0$ such that the ellipse $\mathcal{E}(c_1, c_2, \tau) \supset \Lambda(A)$.
Output: $Z_{m+1} \in \mathbb{C}^{n \times (m+1)}$ basis of $\mathcal{K}_{m+1}(A, r_0)$, $\hat{T}_m \in \mathbb{C}^{(m+1)\times m}$ tridiagonal matrix such that $AZ_m = Z_{m+1}\hat{T}_m$.
1: Compute ρ, μ, and v by (9.70) ;
2: Generate the Krylov subspace basis matrix Z_{m+1} using the recurrence (9.73).
3: Build the matrix $\hat{T}_m$; see (9.75).

When A and r_0 are real ($A \in \mathbb{R}^{n \times n}$ and $r_0 \in \mathbb{R}^n$), the computations in Algorithm 9.4 are real if the ellipse is correctly chosen: since the spectrum $\Lambda(A)$ is symmetric with respect to the real axis, so must be the chosen ellipse. By considering only ellipses where the major axis is the real axis, the foci as well as the function ϕ are real.

In order to avoid the possibility of overflow or underflow, the columns $(z_1, \ldots, z_{m+1})$ of Z_{m+1} are normalized to obtain the new basis $W_{m+1} = Z_{m+1} D_{m+1}^{-1}$ where $D_{m+1} = \mathrm{diag}(\|z_1\|, \ldots, \|z_{m+1}\|)$ is diagonal. The recurrence (9.73) must then be modified accordingly, with the new basis W_{m+1} satisfying the relation

$$A\, W_m = W_{m+1} \check{T}_m, \tag{9.76}$$

in which $\check{T}_m = D_{m+1} \hat{T}_m D_m^{-1}$.

Newton Polynomial Bases

This approach has been developed in [39] from an earlier result [40], and its implementation on parallel architectures is given in [32]. More recently, an improvement of this approach has been proposed in [35].

Recalling the notations of Sect. 9.3.2, we consider the recurrence (9.61) with the additional condition that $\gamma_k = 0$ for $k \geq 1$. This condition reduces the matrix $\hat{T}_m$ (introduced in (9.62)) to the bidiagonal form. Denoting this bidiagonal matrix by $\hat{B}_m$, the corresponding polynomials $p_k(z)$ can then be generated by the following recurrence

$$p_k(z) = \eta_k (z - \beta_k) p_{k-1}(z), \qquad k = 1, 2, \ldots, m, \tag{9.77}$$

where $z_k = p_{k-1}(A) r_0$, with $p_0 = 1$, $\eta_k = \frac{1}{\alpha_k}$ is a scaling factor, and β_k is a zero of the polynomial $p_k(z)$. The corresponding basis Z_{m+1} is called a scaled Newton polynomial basis and is built from $z_1 = \alpha_0 r_0$ for any nonzero α_0, by the recurrence

$$z_{k+1} = \eta_k (A - \beta_k I) z_k, \qquad k = 1, 2, \ldots, m. \tag{9.78}$$

Therefore

$$A\, Z_m = Z_{m+1} \hat{B}_m, \tag{9.79}$$

with

$$\hat{B}_m = \begin{pmatrix} \beta_1 & & & & \\ \alpha_1 & \beta_2 & & & \\ & \alpha_2 & \beta_3 & & \\ & & \alpha_3 & \ddots & \\ & & & \ddots & \beta_{m-1} \\ & & & & \alpha_{m-1} & \beta_m \\ & & & & & \alpha_m \end{pmatrix} \in \mathbb{C}^{(m+1) \times m}. \tag{9.80}$$

The objective here is to mimic the characteristic polynomial of A with a special order for enumerating the eigenvalues of A.

Definition 9.5 (*Leja points*) Let $\mathscr{S}$ be a compact set in $\mathbb{C}$, such that $(\mathbb{C} \cup \{\infty\}) \backslash \mathscr{S}$ is connected and possesses a Green's function. Let $\zeta_1 \in \mathscr{S}$ be arbitrary and let ζ_j for $j = 2, 3, 4, \ldots$, satisfy

$$\prod_{j=1}^{k} |\zeta_{k+1} - \zeta_j| = \max_{z \in \mathscr{S}} \prod_{j=1}^{k} |z - \zeta_j|, \qquad \zeta_{k+1} \in \mathscr{S}, \qquad k = 1, 2, 3, \ldots . \tag{9.81}$$

Any sequence of points $\zeta_1, \zeta_2, \zeta_3, \ldots$ satisfying (9.81) is said to be a sequence of Leja points for $\mathscr{S}$.

In [39], the set $\mathscr{S}$ is chosen to be $\Lambda(H_m)$, the spectrum of the upper Hessenberg matrix H_m generated in Algorithm 9.3. These eigenvalues of $\Lambda(H_m)$ are the Ritz values of A corresponding to the Krylov subspace $\mathscr{K}_m(A, r_0)$. They are sorted with respect to the Leja ordering, i.e., they are ordered to satisfy (9.81) with $\mathscr{S} = \Lambda(H_m)$, and are used as the nodes β_k in the Newton polynomials (9.78).

In [35], this idea is extended to handle more elaborate convex sets $\mathscr{S}$ containing the Ritz values. This allows starting the process with a modest integer m_0 to determine the convex hull of $\Lambda(H_{m_0})$. From this set, an infinite Leja sequence for which $\mathscr{S} = \Lambda(H_m)$. Note that the sequence has at most m terms. Moreover, when an algorithm builds a sequence of Krylov subspaces, the compact set $\mathscr{S}$ can be updated by $\mathscr{S} := \mathrm{co}\,(\mathscr{S} \cup \Lambda(H_m))$ at every restart.

The general pattern of a Newton-Krylov procedure is given by Algorithm 9.5.

Algorithm 9.5 Newton-Krylov procedure.

Input: $A \in \mathbb{C}^{n \times n}$, $m > 1$, $r_0 \in \mathbb{C}^n$, a set $\mathscr{S} \supset \Lambda(A)$.
Output: $Z_{m+1} \in \mathbb{C}^{n \times (m+1)}$ basis of $\mathscr{K}_{m+1}(A, r_0)$, $\hat{B}_m \in \mathbb{C}^{(m+1) \times m}$ bidiagonal matrix such that $AZ_m = Z_{m+1}\hat{B}_m$.
1: From $\mathscr{S}$, build a Leja ordered sequence of points $\{\beta_k\}_{k=1,\cdots,m}$;
2: **for** $k = 1 : m$, **do**
3: $\quad w = A\, z_k - \beta_{k+1} z_k$;
4: $\quad \alpha_{k+1} = 1/\|w\|$; $z_{k+1} = \alpha_{k+1}\, w$;
5: **end**
6: Build the matrix $\hat{B}_m$ (9.80).

When the matrix A is real, we can choose the set $\mathscr{S}$ to be symmetric with respect to the real axis. In such a situation, the nonreal shifts β_k are supposed to appear in conjugate complex pairs. The Leja ordering is adapted to keep consecutive elements of the same pair. Under this assumption, the recurrence (which appears in Algorithm 9.6) involves real operations. The above bidiagonal matrix $\hat{B}_m$ becomes the tridiagonal matrix $\hat{T}_m$ since at each conjugate pair of shifts an entry appears on the superdiagonal.

Algorithm 9.6 Real Newton-Krylov procedure.

Input: $A \in \mathbb{R}^{n \times n}$, $m > 1$, $r_0 \in \mathbb{R}^n$, a set $\mathscr{S} \supset \Lambda(A)$.
Output: $Z_{m+1} \in \mathbb{R}^{n \times (m+1)}$ basis of $\mathscr{K}_{m+1}(A, r_0)$, $\hat{T}_m \in \mathbb{R}^{(m+1) \times m}$ tridiagonal matrix such that $A Z_m = Z_{m+1} \hat{T}_m$.

1: From $\mathscr{S}$, build a Leja ordered sequence of points $\{\beta_k\}_{k=1,\cdots,m}$;
    //conjugate values appear consecutively, the first value of the pair is the one with positive imaginary part.
2: **for** $k = 1 : m$, **do**
3:   **if** $\mathrm{Im}(\beta_k) == 0$ **then**
4:     $z_{k+1} = A z_k - \beta_k z_k$;
5:     $\eta_{k+1} = \|w\|$; $z_{k+1} = w/\eta_{k+1}$;
6:   **else**
7:     **if** $\mathrm{Im}(\beta_k) > 0$ **then**
8:       $w^{(1)} = A z_k - \mathrm{Re}(\beta_k) z_k$;
9:       $w^{(2)} = A w^{(1)} - \mathrm{Re}(\beta_k) w^{(1)} + \mathrm{Im}(\beta_k)^2 z_k$;
10:       $\alpha_{k+1} = 1/\|w^{(1)}\|$; $z_{k+1} = \alpha_{k+1} w^{(1)}$;
11:       $\alpha_{k+2} = 1/\|w^{(2)}\|$; $z_{k+2} = \alpha_{k+1} w^{(2)}$;
12:     **end if**
13:   **end if**
14: **end**
15: Build the matrix $\hat{T}_m$.

References

1. Brin, S., Page, L.: The anatomy of a large-scale hypertextual web search engine. In: Proceedings of 7th International Conference World Wide Web, pp. 107–117. Elsevier Science Publishers B.V, Brisbane (1998)
2. Langville, A., Meyer, C.: Google's PageRank and Beyond: The Science of Search Engine Rankings. Princeton University Press, Princeton (2006)
3. Gleich, D., Zhukov, L., Berkhin., P.: Fast parallel PageRank: a linear system approach. Technical report, Yahoo Corporate (2004)
4. Gleich, D., Gray, A., Greif, C., Lau, T.: An inner-outer iteration for computing PageRank. SIAM J. Sci. Comput. **32**(1), 349–371 (2010)
5. Bahi, J., Contassot-Vivier, S., Couturier, R.: Parallel Iterative Algorithms. Chapman & Hall/CRC, Boca Raton (2008)
6. Bertsekas, D.P., Tsitsiklis, J.N.: Parallel and Distributed Computation. Prentice Hall, Englewood Cliffs (1989)
7. Kollias, G., Gallopoulos, E., Szyld, D.: Asynchronous iterative computations with web information retrieval structures: The PageRank case. In: PARCO, pp. 309–316 (2005)
8. Ishii, H., Tempo, R.: Distributed randomized algorithms for the PageRank computation. IEEE Trans. Autom. Control **55**(9), 1987–2002 (2010)
9. Kalnay, E., Takacs, L.: A simple atmospheric model on the sphere with 100% parallelism. Advances in Computer Methods for Partial Differential Equations IV (1981). http://ntrs.nasa.gov/archive/nasa/casi.ntrs.nasa.gov/19820017675.pdf. Also published as NASA Technical Memorandum No. 83907, Laboratory for Atmospheric Sciences, Research Review 1980–1981, pp. 89–95, Goddard Sapce Flight Center, Maryland, December 1981
10. Gallopoulos, E.: Fluid dynamics modeling. In: Potter, J.L. (ed.) The Massively Parallel Processor, pp. 85–101. The MIT Press, Cambridge (1985)
11. Gallopoulos, E., McEwan, S.D.: Numerical experiments with the massively parallel processor. In: Proceedings of the 1983 International Conference on Parallel Processing, August 1983, pp. 29–35 (1983)

12. Potter, J. (ed.): The Massively Parallel Processor. MIT Press, Cambridge (1985)
13. Varga, R.S.: Matrix Iterative Analysis. Prentice Hall Inc., Englewood Cliffs (1962)
14. Wachspress, E.L.: Iterative Solution of Elliptic Systems. Prentice-Hall Inc., Englewood Cliffs (1966)
15. Young, D.: Iterative Solution of Large Linear Systems. Academic Press, New York (1971)
16. Golub, G., Van Loan, C.: Matrix Computations, 4th edn. Johns Hopkins (2013)
17. Golub, G.H., Varga, R.S.: Chebychev semi-iterative methods, successive overrelaxation iterative methods, and second order Richardson iterative methods: part I. Numer. Math. **3**, 147–156 (1961)
18. O'Leary, D., White, R.: Multi-splittings of matrices and parallel solution of linear systems. SIAM J. Algebra Discret. Method **6**, 630–640 (1985)
19. Neumann, M., Plemmons, R.: Convergence of parallel multisplitting iterative methods for M-matrices. Linear Algebra Appl. **88–89**, 559–573 (1987)
20. Szyld, D.B., Jones, M.T.: Two-stage and multisplitting methods for the parallel solution of linear systems. SIAM J. Matrix Anal. Appl. **13**, 671–679 (1992)
21. Hageman, L., Young, D.: Applied Iterative Methods. Academic Press, New York (1981)
22. Keller, H.: On the solution of singular and semidefinite linear systems by iteration. J. Soc. Indus. Appl. Math. **2**(2), 281–290 (1965)
23. Stiefel, E.L.: Kernel polynomials in linear algebra and their numerical approximations. U.S. Natl. Bur. Stand. Appl. Math. Ser. **49**, 1–22 (1958)
24. Saad, Y., Sameh, A., Saylor, P.: Solving elliptic difference equations on a linear array of processors. SIAM J. Sci. Stat. Comput. **6**(4), 1049–1063 (1985)
25. Rutishauser, H.: Refined iterative methods for computation of the solution and the eigenvalues of self-adjoint boundary value problems. In: Engli, M., Ginsburg, T., Rutishauser, H., Seidel, E. (eds.) Theory of Gradient Methods. Springer (1959)
26. Householder, A.S.: The Theory of Matrices in Numerical Analysis. Dover Publications, New York (1964)
27. Dongarra, J., Duff, I., Sorensen, D., van der Vorst, H.: Numerical Linear Algebra for High-Performance Computers. SIAM, Philadelphia (1998)
28. Meurant, G.: Computer Solution of Large Linear Systems. Studies in Mathematics and its Applications. Elsevier Science (1999). http://books.google.fr/books?id=fSqfb5a3WrwC
29. Saad, Y.: Iterative Methods for Sparse Linear Systems. SIAM, Philadelphia (2003)
30. van der Vorst, H.A.: Iterative Krylov Methods for Large Linear Systems. Cambridge University Press, Cambridge (2003). http://dx.doi.org/10.1017/CBO9780511615115
31. Saad, Y., Schultz, M.H.: GMRES: A generalized minimal residual algorithm for solving nonsymmetric linear systems. SIAM J. Sci. Stat. Comput. **7**(3), 856–869 (1986)
32. Sidje, R.B.: Alternatives for parallel Krylov subspace basis computation. Numer. Linear Algebra Appl. 305–331 (1997)
33. Nuentsa Wakam, D., Erhel, J.: Parallelism and robustness in GMRES with the Newton basis and the deflated restarting. Electron. Trans. Linear Algebra (ETNA) **40**, 381–406 (2013)
34. Nuentsa Wakam, D., Atenekeng-Kahou, G.A.: Parallel GMRES with a multiplicative Schwarz preconditioner. J. ARIMA **14**, 81–88 (2010)
35. Philippe, B., Reichel, L.: On the generation of Krylov subspace bases. Appl. Numer. Math. (APNUM) **62**(9), 1171–1186 (2012)
36. Joubert, W.D., Carey, G.F.: Parallelizable restarted iterative methods for nonsymmetric linear systems. Part I: theory. Int. J. Comput. Math. **44**, 243–267 (1992)
37. Joubert, W.D., Carey, G.F.: Parallelizable restarted iterative methods for nonsymmetric linear systems. Part II: parallel implementation. Int. J. Comput. Math. **44**, 269–290 (1992)
38. Parlett, B.: The Symmetric Eigenvalue Problem. SIAM, Philadelphia (1998)
39. Bai, Z., Hu, D., Reichel, L.: A Newton basis GMRES implementation. IMA J. Numer. Anal. **14**, 563–581 (1994)
40. Reichel, L.: Newton interpolation at Leja points. BIT **30**, 332–346 (1990)

Chapter 10
Preconditioners

In order to make iterative methods effective or even convergent it is frequently necessary to combine them with an appropriate preconditioning scheme; cf. [1–5]. A large number of preconditioning techniques have been proposed in the literature and have been implemented on parallel processing systems. We described earlier in this book methods for solving systems involving banded preconditioners, for example the Spike algorithm discussed in Chap. 5.

In this chapter we consider some preconditioners that are particularly amenable to parallel implementation. If the coefficient matrix A is available explicitly, in solving the linear system $Ax = f$, then we seek to partition it, either before or after reordering depending on its structure, into two or more submatrices that can be handled simultaneously in parallel. For example, A can be partitioned into several (non-overlapped or overlapped) block rows in order to solve the linear system via an accelerated block row projection scheme, see Sect. 10.2. If the reordered matrix A were to result in a banded or block-tridiagonal matrix, then the block row projection method will exhibit two or more levels of parallelism. Alternatively, let A be such that it can be rewritten, after reordering, as the sum of a "generalized" banded matrix M whose Frobenius norm is almost equal to that of A, and a general sparse matrix E of a much lower rank than that of A and containing a much lower number of nonzeros than A. Then this matrix M that encapsulates as many of the nonzeros as possible may be used as a preconditioner. Hence, in the context of Krylov subspace methods, in each outer iteration for solving $Ax = f$, there is a need to solve systems of the for $Mz = r$, involving the preconditioner M. Often, M can be represented as overlapped diagonal blocks to maximize the number of nonzero elements encapsulated by M. Solving systems with such coefficient matrices M on parallel architectures can be realized by: (i) Tearing—that we first encountered in Sect. 5.4, or (ii) via a multiplicative Schwarz approach. Preconditioners of this type are discussed in the following Sects. 10.1 and 10.3.

© Springer Science+Business Media Dordrecht 2016
E. Gallopoulos et al., *Parallelism in Matrix Computations*,
Scientific Computation, DOI 10.1007/978-94-017-7188-7_10

10.1 A Tearing-Based Solver for Generalized Banded Preconditioners

The tearing-based solver described in Sect. 5.4 is applicable for handling systems involving such sparse preconditioner M. The only exception here is that each linear system corresponding to an overlapped diagonal block M_j can be solved using a sparse direct solver. Such a direct solver, however, should have the following capability: given a system $M_j Y_j = G_j$ in which G_j has only nonzero elements at the top and bottom m rows, then the solver should be capable of obtaining the top and bottom m rows of the solution Y_j much faster than computing all of Y_j. The sparse direct solver PARDISO possesses such a feature [6]. This will allow solving the balance system, and hence the original system, much faster and with a high degree of parallel scalability.

10.2 Row Projection Methods for Large Nonsymmetric Linear Systems

Most sparse nonsymmetric linear system solvers either require: (i) storage and computation that grow excessively as the number of iterations increases, (ii) special spectral properties of the coefficient matrix A to assure convergence or, (iii) a symmetrization process that could result in potentially disastrously ill-conditioned problems. One group of methods which avoids these difficulties is accelerated row projection (RP) algorithms. This class of sparse system solvers start by partitioning the coefficient matrix A of the linear system $Ax = f$ of order n, into m block rows:

$$A^\top = (A_1, A_2, \ldots, A_m), \tag{10.1}$$

and partition the vector f accordingly. A *row projection* (RP) method is any algorithm which requires the computation of the orthogonal projections $P_i\, x = A_i (A_i^\top A_i)^{-1} A_i^\top x$ of a vector x onto $\mathscr{R}(A_i)$, $i = 1, 2, \ldots, m$. Note that the nonsingularity of A implies that A_i has full column rank and so $(A_i^\top A_i)^{-1}$ exists.

In this section we present two such methods, bearing the names of their inventors, and describe their properties. The first (Kaczmarz) has an iteration matrix formed as the product of orthogonal projectors, while the second RP method (Cimmino) has an iteration matrix formed as the sum of orthogonal projectors. Conjugate gradient (CG) acceleration is used for both. Most importantly, we show the underlying relationship between RP methods and the CG scheme applied to the normal equations. This, in turn, provides an explanation for the behavior of RP methods, a basis for comparing them, and a guide for their effective use.

Possibly the most important implementation issue for RP methods is that of choosing the row partitioning which defines the projectors. An approach for banded systems yields scalable parallel algorithms that require only a few extra vectors of storage, and allows for accurate computations involving the applications of the necessary projections. Numerous numerical experiments show that these algorithms have superior robustness and can be quite competitive with other solvers of sparse nonsymmetric linear systems.

RP schemes have also been attracting the attention of researchers because of their robustness in solving overdetermined and difficult systems that appear in applications and because of the many interesting variations in their implementation, including parallel asynchronous versions; see e.g. [7–12].

10.2.1 The Kaczmarz Scheme

As the name suggests, a projection method can be considered as a method of solution which involves the projection of a vector onto a subspace. This method, which was first proposed by Kaczmarz considers each equation as a hyperplane; (i.e. partition using $m = n$); thus reducing the problem of finding the solution of a set of equations to the equivalent problem of finding the coordinates of the point of intersection of those hyperplanes. The method of solution is to project an initial iterate onto the first hyperplane, project the resulting point onto the second, and continue the projections on the hyperplanes in a cyclic order thus approaching the solution more and more closely with each projection step [13–16].

Algorithm 10.1 Kaczmarz method (classical version)

1: Choose x_0; $r_0 = f - Ax_0$; set $k = 0$. ρ_i^k
2: **do** $i = 1 : n$,
3: $\quad \alpha_k^i = \dfrac{r_k^i}{\|a_i\|_2^2}$;
4: $\quad x_k = x_k + \alpha_k^i a_i$;
5: $\quad r_k = r_k - \alpha_k^i A\, a_i$;
6: **end**
7: If a convergence criterion is satisfied, terminate the iterations; else set $k = k+1$ and go to Step 2.

Here ρ_i^k is the ith component of the residual $r_k = f - Ax_k$ and $a_i^\top$ is the ith row of the matrix A. For each k, Step 2 consists of n projections, one for each row of A. Kaczmarz's method was recognized as a projection in [17, 18]. It converges for any system of linear equations with nonzero rows, even when it is singular and inconsistent, e.g. see [19, 20], as well as [21–23]. The method has been used under the name (unconstrained) ART (Algebraic Reconstruction Techniques) in the area of image reconstruction [24–28].

The idea of projection methods was further generalized by in [29, 30] to include the method of steepest descent, Gauss-Seidel, and other relaxation schemes.

In order to illustrate this idea, let the error and the residual at the kth step be defined, respectively, as

$$\delta x_k = x - x_k, \tag{10.2}$$

and

$$r_k = f - Ax_k. \tag{10.3}$$

Then a method of projection is one in which at each step k, the error δx_k is resolved into two components, one of which is required to lie in a subspace selected at that step, and the other is δx_{k+1}, which is required to be less than δx_k in some norm. The subspace is selected by choosing a matrix Y_k whose columns are linearly independent. Equivalently,

$$\delta x_{k+1} = \delta x_k - Y_k u_k \tag{10.4}$$

where u_k is a vector (or a scalar if Y_k has one column) to be selected at the kth step so that

$$\|\delta x_{k+1}\| < \|\delta x_k\| \tag{10.5}$$

where $\|\cdot\|$ is some vector norm. The method of projection depends on the choice of the matrix Y_k in (10.4) and the vector norm in (10.5).

If we consider ellipsoidal norms, i.e.

$$\|s\|^2 = s^\top G s, \tag{10.6}$$

where G is a positive definite matrix, then u_k is selected such that $\|\delta x_{k+1}\|$ is minimized yielding,

$$u_k = (Y_k^\top G Y_k)^{-1} Y_k^\top G \delta x_k \tag{10.7}$$

and

$$\|\delta x_k\|^2 - \|\delta x_{k+1}\|^2 = \delta x_k^\top G Y_k u_k. \tag{10.8}$$

Since we do not know δx_k, the process can be made feasible if we require that,

$$Y_k^\top G = V_k^\top A \tag{10.9}$$

in which the matrix V_k will have to be determined at each iteration. This will allow u_k to be expressed in terms of r_k,

$$u_k = (Y_k^\top G Y_k)^{-1} V_k^\top r_k. \tag{10.10}$$

The 2-Partitions Case

Various choices of G in (10.6) give rise to various projection methods. In this section, we consider the case when

$$G = I. \tag{10.11}$$

This gives from (10.7) and (10.9)

$$u_k = (V_k^\top A A^\top V_k)^{-1} V_k^\top A \delta x_k \tag{10.12}$$

and

$$x_{k+1} = x_k + A^\top V_k (V_k^\top A A^\top V_k)^{-1} V_k^\top r_k. \tag{10.13}$$

Now let n be even, and let A be partitioned as

$$A = \begin{pmatrix} A_1^\top \\ A_2^\top \end{pmatrix} \tag{10.14}$$

where $A_i^\top$, $i = 1, 2$, is an $n/2 \times n$ matrix. Let

$$V_1^\top = (I_{n/2}, 0), \qquad V_2^\top = (0, I_{n/2}), \quad \text{and} \quad f^\top = (f_1^\top, f_2^\top) \tag{10.15}$$

where f_i, $i = 1, 2$, is an $n/2$ vector.

Then, one iteration of (10.13) can be written as

$$\begin{aligned} z_1 &= x_k, & z_2 &= z_1 + A_1(A_1^\top A_1)^{-1}(f_1 - A_1^\top z_1), \\ z_3 &= z_2 + A_2(A_2^\top A_2)^{-1}(f_2 - A_2^\top z_2), & x_{k+1} &= z_3. \end{aligned} \tag{10.16}$$

This is essentially the same as applying the block Gauss-Seidel method to

$$AA^\top y = f, \qquad A^\top y = x. \tag{10.17}$$

Symmetrized m-Partition

Variations of (10.16) include the block-row Jacobi, the block row JOR and the block-row SOR methods [31–34]. For each of these methods, we also have the corresponding block-column method, obtained by partitioning the matrix in (10.14) by columns instead of rows. Note that if A had been partitioned into n parts, each part being a row of A, then (10.16) would be equivalent to the Kaczmarz method. In general, for m partitions, the method of successive projections yields the iteration

$$x_{k+1} = Q_u x_k + f_u = (I - P_m)(I - P_{m-1}) \cdots (I - P_1)x_k + f_u \tag{10.18}$$

where

$$f_u = \hat{f}_m + (I - P_m)\hat{f}_{m-1} + (I - P_m)(I - P_{m-1})\hat{f}_{m-2} + \cdots + (I - P_m)\cdots(I - P_2)\hat{f}_1,$$

and $\hat{f}_i = A_i(A_i^\top A_i)^{-1} f_i$.

While the robustness of (10.18) is remarkable and the iteration converges even when A is singular or rectangular, as with any linear stationary process, the rate of convergence is determined by the spectral radius of Q_u and can be arbitrarily slow. For this reason, it is proposed in [31, 35], to symmetrize Q_u by following a forward sweep through the rows with a backward sweep, and introduce an acceleration parameter to get the iteration

$$x_{k+1} = Q(\omega)x_k + c \tag{10.19}$$

Here, c is some vector to be defined later, and

$$Q(\omega) = (I - \omega P_1)(I - \omega P_2)\cdots(I - \omega P_m)^2 \cdots (I - \omega P_2)(I - \omega P_1), \tag{10.20}$$

When A is nonsingular and $0 < \omega < 2$, the eigenvalues of the symmetric matrix $(I - Q(\omega))$ lie in the interval $(0,1]$ and so the conjugate gradient (CG) may be used to solve

$$(I - Q(\omega))x = c. \tag{10.21}$$

Note that iteration, (10.19) is equivalent to that of using the block symmetric successive overrelaxation (SSOR) method to solve the linear system

$$\begin{cases} AA^\top y = f \\ \quad\; x = A^\top y \end{cases} \tag{10.22}$$

in which the blocking is that induced by the row partitioning of A. This gives a simple expression for the right-hand side $c = T(\omega)f$ where,

$$T(\omega) = A^\top (D + \omega L)^{-T} D(D + \omega L)^{-1}, \tag{10.23}$$

in which $AA^\top = L + D + L^\top$ is the usual splitting into strictly block-lower triangular, block-diagonal, and striclty block-upper triangular parts.

Exploration of the effectiveness of accelerated row projection methods has been made in [35] for the single row ($m = n$), and in [14, 36] for the block case ($m \geq 2$). Using sample problems drawn from nonself-adjoint elliptic partial differential equations, the numerical experiments in [14, 36] examined the issues of suitable block row partitioning and methods for the evaluation of the actions of the induced projections. Comparisons with preconditioned Krylov subspace methods, and preconditioned CG applied to the normal equations show that RP algorithms are more robust but not necessarily the fastest on uniprocessors.

The first implementation issue is the choice of ω in (10.21). Normally, the 'optimal' ω is defined as the ω_{min} that minimizes the spectral radius of $Q(\omega)$. Later, we show that $\omega_{min} = 1$ for the case in which A is partitioned into two block rows, i.e., $m = 2$, see also [14]. This is no longer true for $m > 3$, as can be seen by considering

$$A = \begin{pmatrix} 1 & 0 & 0 \\ 1 & 1 & 0 \\ 1 & 0 & 1 \end{pmatrix} = \begin{pmatrix} A_1^\top \\ A_2^\top \\ A_3^\top \end{pmatrix}.$$

For $Q(\omega)$ defined in (10.20), it can be shown that the spectral radii of $Q(1)$, and $Q(0.9)$ satisfy,

$$\rho(Q(1)) = (7 + \sqrt{17})/16 = 0.69519 \pm 10^{-5},$$
$$\rho(Q(0.9)) \leq 0.68611 \pm 10^{-5} < \rho(Q(1)).$$

Hence, $\omega_{min} \neq 1$.

However, taking $\omega = 1$ is recommended as we explain it in the following. First, however, we state two important facts.

Proposition 10.1 *At least* rank(A_1) *of the eigenvalues of $Q(1)$ are zero.*

Proof $(I - P_1)x = 0$ for $x \in \mathscr{R}(P_1) = \mathscr{R}(A_1)$. From the definition of $Q(\omega)$, $\mathscr{N}(I - P_1) \subseteq \mathscr{N}(Q(1))$.

Proposition 10.2 *When $\omega = 1$ and $x_0 = 0$, $A_1^\top x_k = f_1$ holds in exact arithmetic for every iteration of (10.19).*

Proof Using the definition $P_i = A_i(A_i^\top A_i)^{-1}A_i^\top = A_i A_i^+$, (10.23) can be expanded to show that the ith block column of $T(1)$ is given by

$$\prod_{j=1}^{i-1}(I - P_j)\left[I + \prod_{j=i}^{m}(I - P_j)\prod_{j=m-1}^{i+1}(I - P_j)\right](A_i^\top)^+. \tag{10.24}$$

The first product above should be interpreted as I when $i = 1$ and the third product should be interpreted as I when $i = m - 1$ and 0 when $i = m$, so that the first summand in forming $T(1)f$ is

$$[I + (I - P_1)\cdots(I - P_m)\cdots(I - P_2)](A_1^\top)^+ f_1.$$

Since $A_1^\top(I - P_1) = 0$, then $A_1^\top x_1 = A_1^\top(A_1^\top)^+ f_1 = f_1$. The succeeding iterates x_k are obtained by adding elements in $\mathscr{R}(Q(1)) \subseteq \mathscr{R}(I - P_1) = \mathscr{N}(A_1^\top)$ to x_1.

Taking $\omega = 1$ is recommended for the following three reasons:

- Since CG acceleration is to be applied, the entire distribution of the spectrum must be considered, not simply the spectral radius. Thus, from Proposition 10.1, when $\omega = 1$, many of the eigenvalues of the coefficient matrix in (10.21) are exactly equal to 1. Moreover, since the number of needed CG iterations (in exact arithmetic) is equal to the number of distinct eigenvalues, this suggests that numerically fewer iterations are needed as compared to when $\omega \neq 1$.
- Numerical experience shows that $\rho(Q(\omega))$ is not sensitive to changes in ω. This matches classical results for the symmetric successive overrelaxation SSOR iterations, which are not as sensitive to minor changes in ω compared to the method of successive overrelaxation SOR. Hence, the small improvement that does result from choosing $\omega_{\min}$ is more than offset by the introduction of extra nonzero eigenvalues.
- From Proposition 10.2, $A_1^\top x_k = f_1$ is satisfied for all k, and as we will show later, remains so even after the application of CG acceleration. This feature means that, when $\omega = 1$, those equations deemed more important than others can be placed into the first block and kept satisfied to machine precision throughout this iterative procedure.

Definition 10.1 Considering the previous notations, for solving the system $Ax = f$, the system resulting from the choice $Q = Q(1)$ and $T = T(1)$ is :

$$(I - Q)x = c, \tag{10.25}$$

Where Q and T are as given by (10.20) and (10.23), respectively, with $\omega = 1$, and $c = Tf$. The corresponding solver given by Algorithm 10.2 is referred to as KACZ, the symmetrized version of the Kaczmarz method.

Algorithm 10.2 KACZ: Kaczmarz method (symmetrized version)

1: $c = Tf$;
2: Choose x_0; set $k = 0$.
3: **do** $i = 1 : m$,
4:     $x_k = (I - P_i)x_k$;
5: **end**
6: **do** $i = m - 1 : 1$,
7:     $x_k = (I - P_i)x_k$;
8: **end**
9: $x_k = x_k + c$;
10: If a convergence criterion is satisfied, terminate the iterations; else set $k = k + 1$ and go to Step 3.

10.2.2 The Cimmino Scheme

This RP method can be derived as a preconditioner for the CG algorithm. Premultiplying the system $Ax = f$ by

$$\tilde{A} = (A_1(A_1^\top A_1)^{-1}, A_2(A_2^\top A_2)^{-1}, \ldots, A_m(A_m^\top A_m)^{-1}) \tag{10.26}$$

we obtain

$$(P_1 + P_2 + \cdots + P_m)x = \tilde{A}f. \tag{10.27}$$

This system can also be derived as a block Jacobi method applied to the system (10.22); see [37]. For nonsingular A, this system is symmetric positive definite and can be solved via the CG algorithm. The advantage of this approach is that the projections can be computed in parallel and then added.

In 1938, Cimmino [22, 38] first proposed an iteration related to (10.27), and since then it has been examined by several others [21, 31, 32, 37, 39–42]. Later, we will show how each individual projection can, for a wide class of problems, be computed in parallel creating a solver with two levels of parallelism.

10.2.3 Connection Between RP Systems and the Normal Equations

Although KACZ can be derived as a block SSOR, and the Cimmino scheme as a block Jacobi method for (10.22), a more instructive comparison can be made with CGNE—the conjugate gradient method applied to the normal equations $A^\top Ax = A^\top f$. All three methods consist of the CG method applied to a system with coefficient matrix $W^\top W$, where $W^\top$ is shown for each of the three methods in Table 10.1. Intuitively an ill-conditioned matrix A is one in which some linear combination of rows yields approximately the zero vector. For a block row partitioned matrix near linear dependence may occur *within* a block, that is, some linear combination of the rows within a particular block is approximately zero, or across the blocks, that is, the

Table 10.1 Comparison of system matrices for three row projection methods

Method	$W^\top$
CGNE	$(A_1, A_2, \ldots, A_m)$
Cimmino	$(Q_1, Q_2, \ldots, Q_m)$
KACZ	$(P_1, (I - P_1)P_2, (I - P_1)(I - P_2)P_3, \ldots, \prod_{i=1}^{m-1}(I - P_i)P_m)$

linear combination must draw on rows from more than one block row $A_i^\top$. Now let $A_i = Q_i U_i$ be the orthogonal decomposition of A_i in which the columns of Q_i are orthonormal. Examining the matrices W shows that CGNE acts on $A^\top A$, in which near linear dependence could occur both from within and across blocks. Cimmino, however, replaces each A_i with the orthonormal matrix Q_i. In other words, Cimmino avoids forming linear dependence within each block, but remains subject to linear dependence formed across the blocks.

Similar to Cimmino, KACZ also replaces each A_i with the orthonormal matrix Q_i, but goes a step further since $P_i^\top (I - P_i) = 0$.

Several implications follow from this heuristic argument. We make two practical observations. First, we note that the KACZ system matrix has a more favorable eigenvalue distribution than that of Cimmino in the sense that KACZ has fewer small eigenvalues and many more near the maximal one. Similarly, the Cimmino system matrix is better conditioned than that of CGNE. Second, we note that RP methods will require fewer iterations for matrices A where the near linear dependence arises primarily from within a block row rather than across block rows. A third observation is that one should keep the number of block rows small. The reason is twofold: (i) partial orthogonalization across blocks in the matrix W of Table 10.1 becomes less effective as more block rows appear; (ii) the preconditioner becomes less effective (i.e. the condition number of the preconditioned system increases) when the number of blocks increases. Further explanation of the benefit of keeping the number of block rows small is seen from the case in which $m = n$, i.e. having n blocks. In such a case, the ability to form a near linear dependence occurs only across rows where the outer CG acceleration method has to deal with it.

10.2.4 CG Acceleration

Although the CG algorithm can be applied directly to the RP systems, special properties allow a reduction in the amount of work required by KACZ. CG acceleration for RP methods was proposed in [35], and considered in [14, 36]. The reason that a reduction in work is possible, and assuring that $A_1^\top x_k = f_1$ is satisfied in every CG outer iteration for accelerating KACZ follows from:

Theorem 10.1 *Suppose that the CG algorithm is applied to the KACZ system (10.25). Also, let $r_k = c - (I - Q)x_k$ be the residual, and d_k the search direction. If $x_0 = c$ is chosen as the starting vector, then $r_k, d_k \in \mathscr{R}(I - P_1)$ for all k.*

Proof

$$r_0 = c - (I - Q)c = Qc$$
$$= (I - P_1)(I - P_2)\cdots(I - P_m)\cdots(I - P_2)(I - P_1)c \in \mathscr{R}(I - P_1).$$

Since $d_0 = r_0$, the same is true for d_0. Suppose now that the theorem holds for step $(k - 1)$. Then $d_{k-1} = (I - P_1)d_{k-1}$ and so

$$w_k \equiv (I - Q)d_{k-1}$$
$$= (I - P_1)d_{k-1} - (I - P_1)(I - P_2)\cdots(I - P_m)\cdots(I - P_2)(I - P_1)d_{k-1}$$
$$\in \mathscr{R}(I - P_1).$$

Since r_k is a linear combination of r_{k-1} and w_k, then $r_k \in \mathscr{R}(I - P_1)$. Further, since d_k is a linear combination of d_{k-1} and r_k, then $d_k \in \mathscr{R}(I - P_1)$. The result follows by induction.

This reduces the requisite number of projections from $2m - 1$ to $2m - 2$ because the first multiplication by $(I - P_1)$ when forming $(I - Q)d_k$ can be omitted. Also, using $x_0 = c$ for KACZ keeps the first block of equations satisfied in exact arithmetic.

Corollary 10.1 *If $x_0 = c$ is chosen in the CG algorithm applied to the KACZ system, then $A_1^\top x_k = f_1$ for all $k \geq 0$.*

Proof The proof of Theorem 10.1 shows that $c \in (A_1^\top)^+ f_1 + \mathscr{R}(I - P_1)$. Since $A_1^\top(I - P_1) = A_1^\top(I - A_1 A_1^+) = 0$, $A_1^\top x_0 = A_1^\top(A_1^\top)^+ f_1 = f_1$ because A_1 has full column rank. For $k > 0$, $d_{k-1} \in \mathscr{R}(I - P_1)$, so

$$A_1^\top x_k = A_1^\top(x_{k-1} + \alpha_k d_{k-1}) = A_1^\top x_{k-1} = \cdots = A_1^\top x_0 = f_1.$$

In summary, CG acceleration for KACZ allows one projection per iteration to be omitted and one block of equations to be kept exactly satisfied, provided that $x_0 = c$ is used.

10.2.5 The 2-Partitions Case

If the matrix A is partitioned into two block rows (i.e. $m = 2$), a complete eigenanalysis of RP methods is possible using the concept of the angles θ_k between the two subspaces $L_i = \mathscr{R}(A_i)$, $i = 1, 2$. The definition presented here follows [43], but for convenience L_1 and L_2 are assumed to have the same dimension. The smallest angle $\theta_1 \in [0, \pi/2]$ between L_1 and L_2 is defined by

$$\cos\theta_1 = \max_{u \in L_1} \max_{v \in L_2} u^\top v$$
$$\text{subject to } \|u\| = \|v\| = 1.$$

Let u_1 and v_1 be the attainment vectors; then for $k = 2, 3, \ldots, n/2$ the remaining angles between the two subspaces are defined as

$$\cos\theta_k = \max_{u \in L_1} \max_{v \in L_2} u^\top v$$
$$\text{subject to } \begin{cases} \|u\| = \|v\| = 1 \\ u_i^\top u = v_i^\top v = 0, \ i = 1, 2, \ldots, k - 1. \end{cases}$$

Furthermore, when u_i and v_j are defined as above, $u_i^T v_j = 0$, $i \neq j$ also holds. From this, one can obtain the CS decomposition [44, 45], which is stated below in terms of the projectors P_i.

Theorem 10.2 (CS Decomposition). *Let $P_i \in \mathbb{R}^{n \times n}$ be the orthogonal projectors onto the subspaces L_i, for $i = 1, 2$. Then there are orthogonal matrices U_1 and U_2 such that*

$$P_1 = U_1 \begin{pmatrix} I & 0 \\ 0 & 0 \end{pmatrix} U_1^T, \quad P_2 = U_2 \begin{pmatrix} I & 0 \\ 0 & 0 \end{pmatrix} U_2^T, \quad U_1^T U_2 = \begin{pmatrix} C & -S \\ S & C \end{pmatrix},$$

$$\begin{aligned} C &= \mathrm{diag}(c_1, c_2, \ldots, c_{n/2}) \\ S &= \mathrm{diag}(s_1, s_2, \ldots, s_{n/2}) \\ I &= C^2 + S^2 \\ 1 &\geq c_1 \geq c_2 \geq \cdots \geq c_{n/2} \geq 0. \end{aligned} \tag{10.28}$$

In the above theorem $c_k = \cos \theta_k$ and $s_k = \sin \theta_k$, where the angles θ_k are as defined above. Now consider the nonsymmetric RP iteration matrix $Q_u = (I - \omega P_1)(I - \omega P_2)$. Letting $\alpha = 1 - \omega$, using the above expressions of P_1 and P_2, we get

$$Q_u = U_1 \begin{pmatrix} \alpha^2 C & -\alpha S \\ \alpha S & C \end{pmatrix} U_2^T.$$

Hence,

$$U_2^T Q_u U_2 = \begin{pmatrix} \alpha^2 C^2 + \alpha S^2 & (1 - \alpha)CS \\ \alpha(1 - \alpha)CS & C^2 + \alpha S^2 \end{pmatrix}. \tag{10.29}$$

Since each of the four blocks is diagonal, $U_2^T Q_u U_2$ has the same eigenvalues as the scalar 2×2 principal submatrices of the permuted matrix (inverse odd-even permutation on the two sides). The eigenvalues are given by,

$$\frac{1}{2} \left((1 - \alpha)^2 c_i^2 + 2\alpha \pm |1 - \alpha| c_i \sqrt{(1 - \alpha)^2 c_i^2 + 4\alpha} \right), \tag{10.30}$$

for $i = 1, 2, \ldots, n/2$. For a given α the modulus of this expression is a maximum when c_i is largest, i.e., when $c_i = c_1$. The spectral radius of Q_u can then be found by taking $c_i = c_1$ and choosing the positive sign in (10.30). Doing so and minimizing with respect to α yields $\omega_{\min} = 1 - \alpha_{\min} = 2/(1 + s_1)$ and a spectral radius of $(1 - s_1)/(1 + s_1)$. The same result was derived in [31] using the classical SOR theory. The benefit of the CS decomposition is that the full spectrum of Q_u is given and not simply only its spectral radius. In particular, when $\omega = 1$, the eigenvalues of the nonsymmetric RP iteration matrix Q_u become $\{c_1^2, c_2^2, \ldots, c_{n/2}^2, 0\}$ with the zero eigenvalue being of multiplicity $n/2$.

Now, applying the CS decomposition to the symmetrized RP iteration, the matrix $Q(\omega)$ is given by,

$$Q(\omega) = (I - \omega P_1)(I - \omega P_2)^2(I - \omega P_1)$$
$$= U_1 \begin{pmatrix} \alpha^2(\alpha C^2 + S^2) & \alpha(\alpha - 1)CS \\ \alpha(\alpha - 1)CS & \alpha S^2 + C^2 \end{pmatrix} U_1^\top \qquad (10.31)$$

where again $\alpha = 1 - \omega$. When $\omega = 1$, the eigenvalues of the symmetrized matrix Q are identical to those of the unsymmetrized matrix Q_u. One objection to the symmetrization process is that $Q(\omega)$ requires three projections while Q_u only needs two. In the previous section, however, we showed that when $\omega = 1$, KACZ can be implemented with only two projections per iteration.

When $m = 2$, $\omega = 1$ minimizes the spectral radius of $Q(\omega)$ as was shown in [14]. This result can be obtained from the representation (10.31) in the same way as (10.29) is obtained.

The CS decomposition also allows the construction of an example showing that no direct relationship need to exist between the singular value distribution of A and its related RP matrices. Define

$$A = \begin{pmatrix} A_1^\top \\ A_2^\top \end{pmatrix} = \begin{pmatrix} 0 & D \\ S & C \end{pmatrix} \qquad (10.32)$$

where each block is $n/2 \times n/2$, with C and S satisfying (10.28) and $D = \mathrm{diag}(d_1, d_2, \ldots, d_{n/2})$. Then $P_1 = \begin{pmatrix} I & 0 \\ 0 & 0 \end{pmatrix}$ and $P_2 = \begin{pmatrix} C \\ S \end{pmatrix}(C\ S)$, and the eigenvalues of $[I - Q(1)]$ are $\{s_1^2, s_2^2, \ldots, s_{n/2}^2, 1\}$ while those of A are $[c_i \pm \sqrt{c_i^2 + 4s_i d_i}]/2$. Clearly, A is nonsingular provided that each s_i and d_i is nonzero. If the s_i's are close to 1 while the d_i's are close to 0, then $[I - Q(1)]$ has eigenvalues that are clustered near 1, while A has singular values close to both 0 and 1. Hence, A is badly conditioned while $[I - Q(1)]$ is well-conditioned. Conversely if the d_i's are large while the s_i's are close to 0 in such a way that $d_i s_i$ is near 1, then A is well-conditioned while $[I - Q(1)]$ is badly conditioned. Hence, the conditioning of A and its induced RP system matrix may be significantly different.

We have already stated that the eigenvalue distribution for the KACZ systems is better for CG acceleration than that of Cimmino, which in turn is better than that of CGNE (CG applied to $A^\top A x = A^\top f$). For $m = 2$, the eigenvalues of KACZ are $\{s_1^2, s_2^2, \ldots, s_{n/2}^2, 1\}$ while those of Cimmino are easily seen to be $\{1 - c_1, \ldots, 1 - c_{n/2}, 1 + c_{n/2}, \ldots, 1 + c_1\}$, verifying that KACZ has better eigenvalue distribution than that of Cimmino. The next theorem shows that in terms of condition numbers this heuristic argument is valid when $m = 2$.

Theorem 10.3 *When $m = 2$, $\kappa(A^\top A) \geq \kappa(P_1 + P_2) \geq \kappa(I - (I - P_1)(I - P_2)(I - P_1))$. Furthermore, if $c_1 = \cos\theta_1$ is the canonical cosine corresponding to the* *smallest angle between* $\mathcal{R}(A_1)$ *and* $\mathcal{R}(A_2)$, *then* $\kappa(P_1 + P_2) = (1+c_1)^2\kappa(I - (I - P_1)(I - P_2)(I - P_1))$.

Proof Without loss of generality, suppose that $\mathcal{R}(A_1)$ and $\mathcal{R}(A_2)$ have dimension $n/2$. Let $U_1 = (G_1, G_2)$ and $U_2 = (H_1, H_2)$ be the matrices defined in Theorem 10.2 so that $P_1 = G_1 G_1^\top$, $P_2 = H_1 H_1^\top$, and $G_1^\top H_1 = C$. Set $X = G_1^\top A_1$ and $Y = H_1^\top A_2$ so that $A_1 = P_1 A_1 = G_1 G_1^\top A_1 = G_1 X$ and $A_2 = H_1 Y$. It is easily verified that the eigenvalues of $(P_1 + P_2)$ are $1 \pm c_i$, corresponding to the eigenvectors $g_i \pm h_i$ where $G_1 = (g_1, g_2, \ldots, g_{n/2})$ and $H_1 = (h_1, h_2, \ldots, h_{n/2})$. Furthermore, $G_1 g_1 = H_1 h_1 = e_1$ and $G_1 h_1 = H_1 g_1 = c_1 e_1$, where e_1 is the first unit vector. Then $A^\top A = A_1 A_1^\top + A_2 A_2^\top = G_1 X X^\top G_1^\top + H_1 Y Y^\top H_1^\top$, so $(g_1 + h_1)^\top (A^\top A)(g_1 + h_1) = (1+c_1)^2 e_1^\top (A^\top A) e_1$, and $(g_1 - h_1)^\top (A^\top A)(g_1 - h_1) = (1-c_1)^2 e_1^\top (A^\top A) e_1$. Thus, the minimax characterization of eigenvalues yields,

$$
\begin{aligned}
\lambda_{\max}(A^\top A) &\geq (1 + c_1)^2 (e_1^\top (A^\top A) e_1)/(g_1 + h_1)^\top (g_1 + h_1) \\
&= (1 + c_1)^2 (e_1^\top (A^\top A) e_1)/2(1 + c_1) = (1 + c_1)(e_1^\top (A^\top A) e_1)/2
\end{aligned}
$$

and $\lambda_{\min}(A^\top A) \leq (1 - c_1)(e_1^\top (A^\top A) e_1)/2$. Hence $\kappa(A^\top A) \geq \dfrac{1 + c_1}{1 - c_1} = \kappa(P_1 + P_2)$, proving the first inequality.

To prove the second inequality, we see that from the CS Decomposition the eigenvalues of $(I - (I - P_1)(I - P_2)(I - P_1))$ are given by: $\{s_1^2, s_2^2, \ldots, s_{n/2}^2, 1\}$, where $s_i^2 = 1 - c_i^2$ are the square of the canonical sines with the eigenvalue 1 being of multiplicity $n/2$. Consequently,

$$
\kappa(I - (I - P_1)(I - P_2)(I - P_1)) = \frac{1}{s_1^2},
$$

and

$$
\frac{\kappa(P_1 + P_2)}{\kappa(I - (I - P_1)(I - P_2)(I - P_1))} = \frac{1 + c_1}{1 - c_1} \times s_1^2 = (1 + c_1)^2.
$$

Note that $(1 + c_1)^2$ is a measure of the lack of orthogonality between $\mathcal{R}(A_1)$ and $\mathcal{R}(A_2)$, and so measures the partial orthogonalization effect described above.

10.2.6 Row Partitioning Goals

The first criterion for a row partitioning strategy is that the projections $P_i x = A_i (A_i^\top A_i)^{-1} A_i^\top x$ must be efficiently computable. One way to achieve this is through parallelism: if $A_i^\top$ is the direct sum of blocks C_j for $j \in S_i$, then P_i is block-diagonal [14]. The computation of $P_i x$ can then be done by assigning each block of P_i to a different multicore node of a multiprocessor. The second criterion is storage efficiency.

The additional storage should not exceed $\mathcal{O}(n)$, that is, a few extra vectors, the number of which must not grow with increasing problem size n. The third criterion is that the condition number of the subproblems induced by the projections should be kept under control, i.e. monitored. The need for this is made clear by considering the case when $m = 1$, i.e. when A is partitioned into a single block of rows. In this case, KACZ simply solves the normal equations $A^\top A x = A^\top b$, an approach that can fail if A is severely ill-conditioned. More generally when $m > 1$, computing $y = P_i x$ requires solving a system of the form $A_i^\top A_i v = w$. Hence the accuracy with which the action of P_i can be computed depends on the condition number of $A_i^\top A_i$ (i.e., $\kappa(A_i^\top A_i)$), requiring the estimation of an upper bound of $\kappa(A_i^\top A_i)$. The fourth criterion is that the number of partitions m, i.e. the number of projectors should be kept as small as possible, and should not depend on n. One reason has already been outlined in Proposition 10.1.

In summary, row partitioning should allow parallelism in the computations, require at most $\mathcal{O}(n)$ storage, should yield well conditioned subproblems, with the number of partitions m being a small constant. We should note that all four goals can be achieved simultaneously for an important class of linear systems, namely banded systems.

10.2.7 Row Projection Methods and Banded Systems

For general linear systems, algorithm KACZ (Algorithm 10.2) can exploit parallelism only at one level, that within each projection: given a vector u obtain $v = (I - P_j)u$, where $P_j = A_j(A_j^\top A_j)^{-1}A_j^\top$. This matrix-vector multiplication is essentially the solution of the least-squares problem

$$\min_v \|u - A_j w\|_2$$

where one seeks only the minimum residual v.

If A were block-tridiagonal of the form

$$A = \begin{pmatrix} G_1 & H_1 & & & & & & \\ J_2 & G_2 & H_2 & & & & & \\ & J_3 & G_3 & H_3 & & & & \\ & & J_4 & G_4 & H_4 & & & \\ & & & J_5 & G_5 & H_5 & & \\ & & & & J_6 & G_6 & H_6 & \\ & & & & & J_7 & G_7 & H_7 \\ & & & & & & J_8 & G_8 \end{pmatrix}$$

where each G_i, H_i, J_i is square of order q, for example, then using suitable row permutations, one can enhance parallelism within each projection by extracting block rows in which each block consists of independent submatrices.

Case 1: $m = 2$,

$$
\pi_1 A =
\left(
\begin{array}{ccc|ccc|ccc}
G_1 & H_1 &   &   &   &   &   &   &   \\
J_2 & G_2 & H_2 &   &   &   &   &   &   \\
    &   &   & J_5 & G_5 & H_5 &   &   &   \\
    &   &   &   & J_6 & G_6 & H_6 &   &   \\
\hline
    & J_3 & G_3 & H_3 &   &   &   &   &   \\
    &   & J_4 & G_4 & H_4 &   &   &   &   \\
    &   &   &   &   & J_7 & G_7 & H_7 &   \\
    &   &   &   &   &   & J_8 & G_8 &   \\
\end{array}
\right)
$$

Case 2: $m = 3$,

$$
\pi_2 A =
\left(
\begin{array}{ccc|ccc|ccc}
G_1 & H_1 &   &   &   &   &   &   &   \\
    &   & J_4 & G_4 & H_4 &   &   &   &   \\
    &   &   &   &   & J_7 & G_7 & H_7 &   \\
\hline
J_2 & G_2 & H_2 &   &   &   &   &   &   \\
    &   &   & J_5 & G_5 & H_5 &   &   &   \\
    &   &   &   &   &   & J_8 & G_8 &   \\
\hline
    & J_3 & G_3 & H_3 &   &   &   &   &   \\
    &   &   &   & J_6 & G_6 & H_6 &   &   \\
\end{array}
\right)
$$

Such permutations have two benefits. First, they introduce an outer level of parallelism in each single projection. Second, the size of each independent linear least-squares problem being solved is much smaller, leading to reduction in time and memory requirements.

The Cimmino row projection scheme (Cimmino) is capable of exploiting parallelism at two levels. For m = 3 in the above example, we have

(i)   An outer level of parallel projections on all block rows using 8 nodes, one node for each independent block, and
(ii) An inner level of parallelism in which one performs each projection (i.e., solving a linear least-squares problem) as efficiently as possible on the many cores available on each node of the parallel architecture.

10.3 Multiplicative Schwarz Preconditioner with GMRES

We describe in this section an approach for creating a parallel Krylov subspace method, namely GMRES, for solving systems of the form $Ax = f$ preconditioned via a Block Multiplicative Schwarz iteration, which is related to block-row projection via the Kaczmarz procedure—see Sect. 10.2.1. Here, A consists of overlapped diagonal blocks, see Fig. 10.1. While multiplicative Schwarz is not amenable for parallel computing as it introduces a recursion between the local solves, it is often more

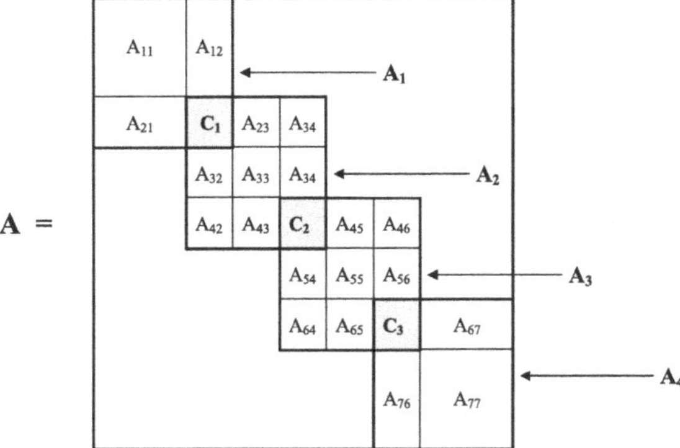

Fig. 10.1 A matrix domain decomposition with block overlaps

robust than its additive counterpart. By using Newton Krylov bases, as introduced in Sect. 9.3.2, the resulting method does not suffer from the classical bottleneck of excessive internode communications.

10.3.1 Algebraic Domain Decomposition of a Sparse Matrix

Let us consider a sparse matrix $A \in \mathbb{R}^{n \times n}$. The pattern of A is the set $\mathbb{P} = \{(k, l) | a_{k,l} \neq 0\}$ which is the set of the edges of the graph $G = (W, \mathbb{P})$ where $W = \{1, \ldots, n\} = [1 : n]$ is the set of vertices.

Definition 10.2 A domain decomposition of matrix A into p subdomains is defined by a collection of sets of integers $W_i \subset W = [1 : n], i = 1, \ldots, p$ such that:

$$\begin{cases} |i - j| > 1 \implies W_i \cap W_j = \emptyset \\ \mathbb{P} \subset \bigcup_{i=1}^{p} (W_i \times W_i) \end{cases}$$

In order to simplify the notation, the sets W_i are assumed to be intervals of integers. This is not a restriction, since it is always possible to define a new numbering of the unknowns which satisfies this constraint. Following this definition, a "domain" decomposition can be considered as resulting from a graph partitioner but with potential overlap between domains. It can be observed that such a decomposition does not necessarily exist (e.g. when A is dense matrix). For the rest of our discussion, we will assume that a graph partitioner has been applied resulting in p intervals $W_i = w_i + [1 : m_i]$ whose union is W, $W = [1 : n]$. The submatrix of A corresponding to $W_i \times W_i$ is denoted by A_i. We shall denote by $I_i \in \mathbb{R}^{n \times n}$ the diagonal matrix, or the sub-identity matrix, whose diagonal elements are set to one if the corresponding node belongs to W_i and set to zero otherwise. In effect, I_i is the orthogonal projector onto the subspace L_i corresponding to the unknowns numbered by W_i. We still denote by A_i the extension of block A_i to the whole space, in other words,

$$A_i = I_i A I_i, \tag{10.33}$$

For example, from Fig. 10.1 , we see that unlike the *tearing* method, the whole overlap block C_j belongs to A_j as well as A_{j+1}.

Also, let

$$\bar{A}_i = A_i + \bar{I}_i, \tag{10.34}$$

where $\bar{I}_i = I - I_i$ is the complement sub-identity matrix. For the sake of simplifying the presentation of what follows, we assume that all the matrices $\bar{A}_i$, for $i = 1, \dots, p$ are nonsingular. Hence, the generalized inverse A_i^+ of A_i is given by $A_i^+ = I_i \bar{A}_i^{-1} = \bar{A}_i^{-1} I_i$.

Proposition 10.3 *For any domain decomposition as given in Definition 10.2 the following property holds:*

$$|i - j| > 2 \Rightarrow I_i A I_j = 0, \quad \forall i, j \in \{1, \dots, p\}.$$

Proof Let $(k, l) \in W_i \times W_j$ such that $a_{k,l} \neq 0$. Since $(k, l) \in \mathbb{P}$, there exists $m \in \{1 \dots n\}$ such that $k \in W_m$ and $l \in W_m$; therefore $W_i \cap W_m \neq \emptyset$ and $W_j \cap W_m \neq \emptyset$. Consequently, from Definition 10.2, $|i - m| \leq 1$ and $|j - m| \leq 1$, which implies $|i - j| \leq 2$.

Next, we consider a special case which arises often in practice.

Definition 10.3 The domain decomposition is with a *weak overlap* if and only if the following is true:

$$|i - j| > 1 \Rightarrow I_i A I_j = 0, \quad \forall i, j \in \{1, \dots, p\}.$$

The set of unknowns which represents the overlap is defined by the set of integers $J_i = W_i \cap W_{i+1}$, $i = 1, \dots, p - 1$, with the size of the overlap being s_i. Similar to (10.33) and (10.34), we define

$$C_i = O_i A O_i, \tag{10.35}$$

and

$$\bar{C}_i = C_i + \bar{O}_i, \tag{10.36}$$

where the diagonal matrix $O_i \in \mathbb{R}^{n \times n}$ is a sub-identity matrix whose diagonal elements are set to one if the corresponding node belongs to J_i and set to zero otherwise, and $\bar{O}_i = I - O_i$.

Example 10.1 In the following matrix, it can be seen that the decomposition is of a weak overlap if $A_2^{b,t} = 0$, and $A_2^{t,b} = 0$, i.e. when A is block-tridiagonal.

$$A = \begin{pmatrix} A_1^{m,m} & A_1^{m,b} & & & \\ A_1^{b,m} & C_1 & A_2^{t,m} & A_2^{t,b} & \\ & A_2^{m,t} & A_2^{m,m} & A_2^{m,b} & \\ & A_2^{b,t} & A_2^{b,m} & C_2 & A_3^{t,m} \\ & & & A_3^{t,m} & A_3^{m,m} \end{pmatrix} \tag{10.37}$$

10.3.2 Block Multiplicative Schwarz

The goal of Multiplicative Schwarz methods is to iteratively solve the linear system

$$A\,x = f \tag{10.38}$$

where matrix A is decomposed into overlapping subdomains as described in the previous section. The iteration consists of solving the original system in sequence on each subdomain. This is a well-known method; for more details, see for instance [2, 4, 46, 47]. In this section, we present the main properties of the iteration and derive an explicit formulation of the corresponding matrix splitting that allows a parallel expression of the preconditionner when used with Krylov methods.

Classical Formulation

This method corresponds to a relaxation iteration defined by the splitting $A = P - N$. The existence of the preconditioner P is not made explicit in the classical formulation of the multiplicative Schwarz procedure. However, the existence of P is a direct consequence of the coming expression (10.39). Let x_k be the current iterate and $r_k = f - Ax_k$ the corresponding residual. The two sequences of vectors can be generated via $x_{k+1} = x_k + P^{-1}r_k$ and $r_{k+1} = f - A\,x_{k+1}$, as given by Algorithm 10.3 for building p sub-iterates together their corresponding residuals.

Algorithm 10.3 Single block multiplicative Schwarz step

Input: $A \in \mathbb{R}^{n \times n}$, a partition of A, x_k and $r_k = f - Ax_k \in \mathbb{R}^n$.
Output: $x_{k+1} = x_k + P^{-1}r_k$ and $r_{k+1} = f - Ax_{k+1}$, where P corresponds to one Multiplicative Schwarz step.
1: $x_{k,0} = x_k$ and $r_{k,0} = r_k$;
2: **do** $i = 1 : p$,
3:    $z = A_i^+ r_{k,i-1}$;
4:    $r_{k,i} := r_{k,i-1} - A\,z$;
5:    $x_{k,i} := x_{k,i-1} + z$;
6: **end**
7: $x_{k+1} = x_{k,p}$ and $r_{k+1} = r_{k,p}$;

Hence, it follows that,

$$r_{k+1} = (I - AA_p^+) \cdots (I - AA_1^+) r_k. \tag{10.39}$$

Convergence of this iteration to the solution x of the system (10.38) is proven for M-matrices and SPD matrices (eg. see [46]).

Embedding in a System of Larger Dimension

If the subdomains do not overlap, it can be shown [47] that the Multiplicative Schwarz is equivalent to a Block Gauss-Seidel method applied on an extended system. In this section, following [48], we present an extended system which embeds the original system (10.38) into a larger one with no overlapping between subdomains.

For that purpose, we define the prolongation mapping and the restriction mapping. We assume for the whole section that the set of indices defining the domains are intervals. As mentioned before, this does not limit the scope of the study since a preliminary symmetric permutation of the matrix, corresponding to the same renumbering of the unknowns and the equations, can always end up with such a system.

For any vector $x \in \mathbb{R}^n$, we consider the set of overlapping subvectors $x^{(i)} \in \mathbb{R}^{m_i}$ for $i = 1, \ldots, p$, where $x^{(i)}$ is the subvector of x corresponding to the indices W_i. This vector can also be partitioned into $x^{(i)} = \begin{pmatrix} x^{(i,t)} \\ x^{(i,m)} \\ x^{(i,b)} \end{pmatrix}$ accordingly to the indices of the overlapping blocks C_{i-1} and C_i (with the obvious convention that $x^{(1,t)}$ and $x^{(p,b)}$ are zero-length vectors).

Definition 10.4 The prolongation mapping which injects $\mathbb{R}^n$ into a space $\mathbb{R}^m$ where $m = \sum_{i=1}^{p} m_i = n + \sum_{i=1}^{p-1} s_i$ is defined as follows :

$$\begin{aligned} \mathscr{D} : \mathbb{R}^n &\to \mathbb{R}^m \\ x &\mapsto \widetilde{x}, \end{aligned}$$

where $\widetilde{x}$ is obtained from vector x by duplicating all the blocks of entries corresponding to overlapping blocks: therefore, according to the previous notation,

$$\widetilde{x} = \begin{pmatrix} x^{(1)} \\ \vdots \\ x^{(p)} \end{pmatrix}.$$

The restriction mapping consists of projecting a vector $\widetilde{x} \in \mathbb{R}^m$ onto $\mathbb{R}^n$, which consists of deleting the subvectors corresponding to the first appearance of each overlapping blocks

$$\begin{aligned} \mathscr{P} : \mathbb{R}^m &\to \mathbb{R}^n \\ \widetilde{x} &\mapsto x. \end{aligned}$$

Embedding the original system in a larger one is done for instance in [47, 48]. We present here a special case. In order to avoid a tedious formal presentation of the augmented system, we present its construction on an example which is generic enough to understand the definition of $\widetilde{A}$. In (10.37), is displayed an example with three domains. Mapping $\mathscr{D}$ builds $\tilde{x}$ by duplicating some entries in vector x; mapping $x \to \tilde{x} = \mathscr{D}x$ expands vector x to include subvectors $x^{(1,b)}$ and $x^{(2,b)}$:

$$
x = \begin{pmatrix} x^{(1,m)} \\ x^{(2,t)} \\ x^{(2,m)} \\ x^{(3,t)} \\ x^{(3,m)} \end{pmatrix} \xrightarrow{\;\mathscr{D}\;} \tilde{x} = \begin{pmatrix} x^{(1,m)} \\ x^{(1,b)} \\ x^{(2,t)} \\ x^{(2,m)} \\ x^{(2,b)} \\ x^{(3,t)} \\ x^{(3,m)} \end{pmatrix}
\tag{10.40}
$$

$$
\text{with} \quad x^{(1,b)} = x^{(2,t)} \text{ and } x^{(2,b)} = x^{(3,t)}.
\tag{10.41}
$$

The matrix $\widetilde{A}$ corresponding to the matrix given in (10.37) is

$$
\widetilde{A} = \left(\begin{array}{ccc|ccc|cc}
A_1^{m,m} & A_1^{m,b} & & & & & & \\
A_1^{b,m} & C_1 & & A_2^{t,m} & A_2^{t,b} & & & \\
\hline
A_1^{b,m} & & C_1 & A_2^{t,m} & A_2^{t,b} & & & \\
 & & A_2^{m,t} & A_2^{m,m} & A_2^{m,b} & & & \\
 & & A_2^{b,t} & A_2^{b,m} & C_2 & & A_3^{t,m} & \\
\hline
 & & A_2^{b,t} & A_2^{b,m} & & C_2 & A_3^{t,m} & \\
 & & & & & A_3^{t,m} & A_3^{m,m} &
\end{array}\right)
\tag{10.42}
$$

The equalities (10.41) define a subspace $\mathscr{J}$ of $\mathbb{R}^m$. This subspace is the range of mapping $\mathscr{D}$. These equalities combined with the definition of matrix $\widetilde{A}$ show that $\mathscr{J}$ is an invariant subspace of $\widetilde{A} : \widetilde{A}\,\mathscr{J} \subset \mathscr{J}$. Therefore solving system $Ax = f$ is equivalent to solving system $\widetilde{A}\tilde{x} = \tilde{f}$ where $\tilde{f} = \mathscr{D}f$. Operator $\mathscr{P}$ deletes entries $x^{(1,b)}$ and $x^{(2,b)}$ from vector $\tilde{x}$.

Remark 10.1 The following properties are straightforward consequences of the previous definitions:

1. $Ax = \mathscr{P}\widetilde{A}\mathscr{D}x$,
2. The subspace $\mathscr{J} = \mathscr{R}(\mathscr{D}) \subset \mathbb{R}^m$ is an invariant subspace of $\widetilde{A}$,
3. $\mathscr{P}\mathscr{D} = I_n$ and $\mathscr{D}\mathscr{P}$ is a projection onto $\mathscr{J}$,
4. $\forall x,\ y \in \mathbb{R}^n,\ (y = Ax \Leftrightarrow \mathscr{D}y = \widetilde{A}\mathscr{D}x)$.

This can be illustrated by diagram (10.43):

$$
\begin{array}{ccc}
\mathbb{R}^n & \xrightarrow{A} & \mathbb{R}^n \\
\mathscr{D}\downarrow & & \uparrow\mathscr{P} \\
\mathbb{R}^m & \xrightarrow{\widetilde{A}} & \mathbb{R}^m\,.
\end{array}
\tag{10.43}
$$

One iteration of the Block Multiplicative Schwarz method on the original system (10.38) corresponds to one Block-Seidel iteration on the enhanced system

$$\widetilde{A}\widetilde{x} = \mathscr{D}f, \tag{10.44}$$

where the diagonal blocks are the blocks defined by the p subdomains. More precisely, denoting by $\widetilde{P}$ the block lower triangular part of $\widetilde{A}$, the iteration defined in Algorithm 10.3 can be expressed as follows :

$$\begin{cases} \widetilde{x}_k = \mathscr{D}x_k, \\ \widetilde{r}_k = \mathscr{D}r_k, \\ \widetilde{x}_{k+1} = \widetilde{x}_k + \widetilde{P}^{-1}\widetilde{r}_k, \\ x_{k+1} = \mathscr{P}\widetilde{x}_{k+1}. \end{cases} \tag{10.45}$$

To prove it, let us partition $\widetilde{A} = \widetilde{P} - \widetilde{N}$, where $\widetilde{N}$ is the strictly upper block-triangular part of $(-\widetilde{A})$. Matrices $\widetilde{P}$ and $\widetilde{N}$ are partitioned by blocks accordingly to the domain definition. One iteration of the Block Gauss-Seidel method can then be expressed by

$$\widetilde{x}_{k+1} = \widetilde{x}_k + \widetilde{P}^{-1}\widetilde{r}_k.$$

The resulting block-triangular system is solved successively for each diagonal block. To derive the iteration, we partition $\widetilde{x}_k$ and $\widetilde{x}_{k+1}$ accordingly. At the first step, and assuming $\widetilde{x}_{k,0} = \widetilde{x}_k$ and $\widetilde{r}_{k,0} = \widetilde{r}_k$, we obtain

$$\widetilde{x}_{k+1,1} = \widetilde{x}_{k,0} + A_1^{-1}\widetilde{r}_{k,0}, \tag{10.46}$$

which is identical to the first step of the Multiplicative Schwarz $x_{k,1} = x_{k,0} + A_1^{-1} + r_{k,0}$. The ith step $(i = 2, \ldots, p)$

$$\widetilde{x}_{k+1,i} = \widetilde{x}_{k,i-1} + A_i^{-1}\left(\widetilde{f}_i - \widetilde{P}_{i,1:i-1}\,\widetilde{x}_{k+1,1:i-1} - A_i\widetilde{x}_{k,i} + \widetilde{N}_{i,i+1:p}\,\widetilde{x}_{k,i+1:p}\right), \tag{10.47}$$

is equivalent to its counterpart $x_{k,i+1} = x_{k,i} + A_i^+ r_{k,i}$ in the Multiplicative Schwarz algorithm.

Therefore, we have the following diagram

$$\begin{array}{ccc} \mathbb{R}^n & \xrightarrow{P^{-1}} & \mathbb{R}^n \\ \mathscr{D}\downarrow & & \uparrow\mathscr{P} \\ \mathbb{R}^m & \xrightarrow{\widetilde{P}^{-1}} & \mathbb{R}^m \end{array}$$

and we conclude that $P^{-1} = \mathscr{P}\widetilde{P}^{-1}\mathscr{D}$. Similarly, we claim that

$$N = \mathscr{P}\widetilde{N}\mathscr{D}, \tag{10.48}$$

and $\mathscr{J}$ is an invariant subspace of $\mathbb{R}^m$ for $\widetilde{N}$. We must remark that there is an abuse in the notation, in the sense that the matrix denoted by P^{-1} can be singular even when $\widetilde{P}^{-1}$ is non singular.

Explicit Formulation of the Multiplicative Schwarz Preconditioner

A lemma must be first given to then express the main result.

Notation: For $1 \leq i \leq j \leq p$, $I_{i:j}$ is the identity on the union of the domains W_k, $(i \leq k \leq j)$ and $\bar{I}_{i:j} = I - I_{i:j}$.

Lemma 10.1 *For any $i \in \{1, \ldots, p-1\}$,*

$$\bar{A}_{i+1} I_i + I_{i+1} \bar{A}_i - I_{i+1} A I_i = \bar{C}_i I_{i:i+1}, \tag{10.49}$$

and for any $i \in \{1, \ldots, p\}$,

$$A_i^+ = \bar{A}_i^{-1} I_i = I_i \bar{A}_i^{-1}.$$

Straightforward (see Fig. 10.2).

Theorem 10.4 ([49]) *Let $A \in \mathbb{R}^{n \times n}$ be algebraically decomposed into p subdomains as described in Sect. 10.3.1 such that all the matrices $\bar{A}_i$, and the matrix C_i for $i = 1, \ldots, p$ are non singular. The inverse of the Multiplicative Schwarz preconditioner, i.e. the matrix P^{-1}, can be explicitly expressed by:*

$$P^{-1} = \bar{A}_p^{-1} \bar{C}_{p-1} \bar{A}_{p-1}^{-1} \bar{C}_{p-2} \cdots \bar{A}_2^{-1} \bar{C}_1 \bar{A}_1^{-1} \tag{10.50}$$

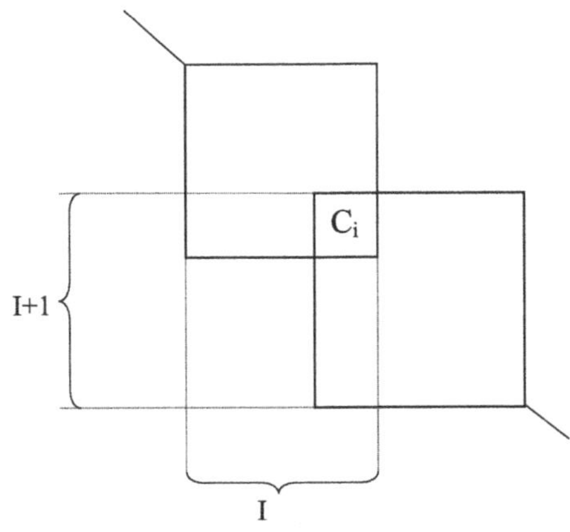

Fig. 10.2 Illustration of (10.49). *Legend* $\mathsf{I} = I_i$ and $\mathsf{I}+1 = I_{i+1}$

where for $i = 1, \ldots, p$, *matrix* $\bar{A}_i$ *corresponds to the block* A_i *and for* $i = 1, \ldots, p - 1$, *matrix* $\bar{C}_i$ *to the overlap* C_i *as given by (10.34), and (10.36), respectively.*

Proof In [49], two proofs are proposed. We present here the proof by induction.

Since $x_{k+1} = x_k + P^{-1} r_k$ and since x_{k+1} is obtained as result of successive steps as defined in Algorithm 10.3, for $i = 1, \ldots, p$: $x_{k,i+1} = x_{k,i} + A_i^+ r_{k,i}$, we shall prove by reverse induction on $i = p - 1, \ldots, 0$ that:

$$x_{k,p} = x_{k,i} + \bar{A}_p^{-1} \bar{C}_{p-1} \cdots \bar{C}_{i+1} \bar{A}_{i+1}^{-1} I_{i+1:p}\, r_{k,i} \tag{10.51}$$

For $i = p - 1$, the relation $x_{k,p} = x_{k,p-1} + \bar{A}_p^{-1} I_p r_{k,p-1}$ is obviously true. Let us assume that (10.51) is valid for i and let us prove it for $i - 1$.

$$x_{k,p} = x_{k,i-1} + A_i^+ r_{k,i-1} + A_p^{-1} \bar{C}_{p-1} \cdots \bar{C}_{i+1} \bar{A}_{i+1}^{-1} I_{i+1:p} (I - A A_i^+) r_{k,i-1}$$

$$= x_{k,i-1} + \bar{A}_p^{-1} \cdots \bar{C}_{i+1} (A_i^+ + A_{i+1}^+ I_{i+1:p} - \bar{A}_{i+1}^{-1} I_{i+1:p} A A_i^+) r_{k,i-1}.$$

The last transformation was possible since the supports of A_p, and of C_{p-1}, C_{p-1}, $\ldots, C_{i+1}$ are disjoint from domain i. Let us transform the matrix expression:

$$B = A_i^+ + A_{i+1}^+ I_{i+1:p} - \bar{A}_{i+1}^{-1} I_{i+1:p} A A_i^+$$

$$= \bar{A}_{i+1}^{-1} (A_{i+1}^+ I_i + I_{i+1:p} \bar{A}_i - I_{i+1:p} A I_i) \bar{A}_i^{-1}$$

Lemma 10.1 and elementary calculations imply that:

$$B = \bar{A}_{i+1}^{-1} (\bar{C}_i I_{i:i+1} + I_{i+2:p} - O_{i+1}) \bar{A}_i^{-1}$$

$$= \bar{A}_{i+1}^{-1} \bar{C}_i \bar{A}_i^{-1} I_{i:p}$$

which proves that relation (10.51) is valid for $i - 1$. This ends the proof.

Once the inverse of P is explicitly known, the matrix $N = P - A$ can be expressed explicitly as well. For this purpose, we assume that every block A_i, for $i = 1, \ldots, p - 1$ is partitioned as follows:

$$A_i = \begin{pmatrix} B_i & F_i \\ E_i & C_i \end{pmatrix},$$

and $A_p = B_p$.

Proposition 10.4 *The matrix* N, *is defined by the multiplicative Schwarz splitting* $A = P - N$. *By embedding it in* $\widetilde{N}$ *as expressed in (10.48), block* $N_{i,j}$ *is the upper part of block* $\widetilde{N}_{i,j}$ *as referred by the block structure of* $\widetilde{N}$. *The blocks can be expressed as follows:*

$$\begin{cases} N_{i,j} &= G_i \cdots G_{j-1} B_j, \text{ when } j > i+1 \\ N_{i,i+1} &= G_i B_{i+1} - [F_i,\ 0], \\ N_{i,j} &= 0 \ otherwise, \end{cases} \tag{10.52}$$

for $i = 1, \ldots, p-1$ and $j = 2, \ldots, p$ and where $G_i = [F_i C_i^{-1},\ 0]$. When the algebraic domain decomposition is with a weak overlap, expression (10.52) becomes:

$$\begin{cases} N_{i,i+1} = G_i B_{i+1} - [F_i,\ 0], & for\ i = 1, \ldots, p-1, \\ N_{i,j} = 0 \ otherwise. \end{cases} \tag{10.53}$$

The matrix N is of rank $r \leq \sum_{i=1}^{p-1} s_i$, where s_i is the size of C_i.

Proof The proof of the expression (10.52) is based on a constructive proof of Theorem 10.4 as explicited in [49].

The structure of row block $N_i = [N_{i,1}, \ldots, N_{i,p}]$, for $i = 1, \ldots, p-1$, of matrix N is:

$$N_i = \left[0,\ \cdots 0,\ [F_i\ 0] \begin{bmatrix} 0 & C_i^{-1} B_{i+1}(1,2) \\ 0 & -I \end{bmatrix},\ G_i G_{i+1} B_{i+2},\ \ldots,\ G_i \cdots G_{p-1} B_p \right].$$

Therefore, the rank of row block of N_i is limited by the rank of factor $[F_i\ 0]$ which cannot exceed s_i. This implies $r \leq \sum_{i=1}^{p-1} s_i$.

Example 10.2 In Fig. 10.3, a block-triagonal matrix A is considered. Three overlapping blocks are defined on it. There is a weak overlap. The pattern of the matrix N as defined in Proposition 10.4 has only two nonzero blocks as shown in Fig. 10.3.

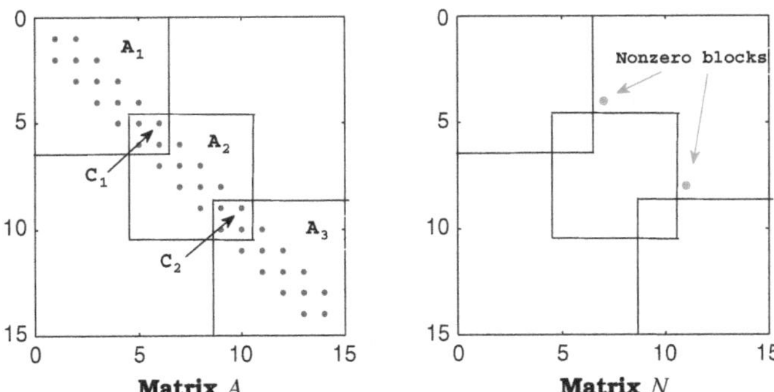

Fig. 10.3 Expression of the block multiplicative Schwarz splitting for a block-tridiagonal matrix. *Left* Pattern of A; *Right* Pattern of N where $A = P - N$ is the corresponding splitting

10.3.3 Block Multiplicative Schwarz as a Preconditioner
for Krylov Methods

Left and Right Preconditioner of a Krylov Method

The goal is to solve the system (10.38) by using a preconditioned Krylov method from an initial guess x_0 and therefore with $r_0 = f - Ax_0$ as initial residual. Let $P \in \mathbb{R}^{n \times n}$ be a preconditioner of the matrix A. Two options are possible:

Left preconditioning: the system to be solved is $P^{-1}Ax = P^{-1}f$ which defines the Krylov subspaces $\mathscr{K}_k(P^{-1}A, P^{-1}r_0)$.

Right preconditioning: the system to be solved is $AP^{-1}y = f$ and the solution is recovered from $x = P^{-1}y$. The corresponding sequence of Krylov subspaces is therefore $\mathscr{K}_k(AP^{-1}, r_0)$.

Because $\mathscr{K}_k(P^{-1}A, P^{-1}r_0) = P^{-1}\mathscr{K}_k(AP^{-1}, r_0)$, the two preconditioners exhibit similar behavior since they involve two related sequences of Krylov subspaces.

Early Termination

We consider the advantage of preconditioning a Krylov method, via the splitting $A = P - N$ in which N is rank deficient—the case for the Multiplicative Schwarz preconditioning. For solving the original system $Ax = f$, we define a Krylov method, as being an iterative method which builds, from an initial guess x_0, a sequence of iterates $x_k = x_0 + y_k$ such that $y_k \in \mathscr{K}_k(A, r_0)$ where $\mathscr{K}_k(A, r_0)$ is the Krylov subspace of degree k, built from the residual r_0 of the initial guess: $\mathscr{K}_k(A, r_0) = \mathbb{P}_{k-1}(A)r_0$, where $\mathbb{P}_{k-1}(\mathbb{R})$ is the set of polynomials of degree $k - 1$ or less. The vector y_k is obtained by minimizing a given norm of the error $x_k - x$, or by projecting the initial error onto the subspace $\mathscr{K}_k(A, r_0)$ in a given direction. Here, we consider that, for a given k, when the property $x \in x_0 + \mathscr{K}_k(A, r_0)$ holds, it implies that $x_k = x$. This condition is satisfied when the Krylov subspace sequence becomes stationary as given in Proposition 9.1.

Proposition 10.5 *With the previous notations, if* $rank(N) = r < n$, *then any Krylov method, as it has just been defined, reaches the exact solution in at most* $r + 1$ *iterations.*

Proof In $P^{-1}A = I - P^{-1}N$ the matrix $P^{-1}N$ is of rank r. For any degree k, the following inclusion $\mathscr{K}_k(P^{-1}A, P^{-1}r_0) \subset \mathrm{Span}(r_0) + \mathscr{R}(P^{-1}N)$ guarantees that the dimension of $\mathscr{K}_k(P^{-1}A, P^{-1}r_0)$ is at most $r + 1$. Therefore, the method is stationary from $k = r + 1$ at the latest. The proof is identical for the right preconditioning.

For a general nonsingular matrix, this result is applicable to the methods BiCG and QMR, preconditioned by the Multiplicative Schwarz method. In exact arithmetic, the number of iterations cannot exceed the total dimension s of the overlap by more than 1. The same result applies to GMRES(m) when m is greater than s.

Explicit Formulation and Nonorthogonal Krylov Bases

In the classical formulation of the Multiplicative Schwarz iteration (Algorithm 10.3) the computation of the two steps

$$x_{k+1} = x_k + P^{-1}r_k \text{ and } r_{k+1} = f - Ax_{k+1},$$

is carried out recursively through the domains, whereas our explicit formulation (see Theorem 10.4) decouples the two computations. The computation of the residual is therefore more easily implemented in parallel since it is included in the recursion. Another advantage of the explicit formulation arises when it is used as a preconditioner of a Krylov method. In such a case, the user is supposed to provide a code for the procedure $x \rightarrow P^{-1}x$. Since the method computes the residual, the classical algorithm implies calculation of the residual twice.

The advantage is even higher when considering the a priori construction of a nonorthogonal basis of the Krylov subspace. For this purpose, the basis is built by a recursion of the type of (9.61) but where the operator A is now replaced by either $P^{-1}A$ or AP^{-1}. The corresponding algorithm is one of the following: 9.4, 9.5, or 9.6. Here, we assume that we are using the Newton-Arnoldi iteration (i.e. the Algorithm 9.5) with a left preconditionner.

Next, we consider the parallel implementation of the above scheme on a linear array of processors $P(q)$, $q = 1, \ldots, p$. We assume that the matrix A and the vectors involved are distributed as indicated in Sect. 2.4.1. In order to get rid of any global communication, the normalization of w, must be postponed. For a clearer presentation, we skip for now the normalizing factors but we will show later how to incorporate these normalization factors without harming parallel scalability.

Given a vector z_1, building a Newton-Krylov basis can be expressed by the loop:

do $k = 1 : m$,
$\quad z_{k+1} = P^{-1}A\, z_k - \lambda_{k+1}z_k;$
end

At each iteration, this loop involves multiplying A by a vector (MV as defined in Sect. 2.4.1), followed by solving a linear system involving the preconditioner P. As shown earlier, Sect. 2.4.1, the MV kernel is implemented via Algorithm 2.11. Solving systems involving the preconditioner P corresponds to the application of one Block Multiplicative Schwarz iteration, which can be expressed from the explicit formulation of P^{-1} as given by Theorem 10.4, and implemented by Algorithm 10.4.

Algorithms 2.11 and 10.4 can be concatenated into one. Running the resulting program on each processor defines the flow illustrated in Fig. 10.4. The efficiency of this approach is analyzed in [50]. It shows that if there is a good load balance across all the subdomains and if τ denotes the maximum number of steps necessary to compute a subvector v_q, then the number of steps to compute one entire vector is given by $T = p\tau$, and consequently, to compute m vectors of the basis the number of parallel steps is

$$\mathbf{T}_p = (p - 3 + 3m)\tau. \tag{10.54}$$

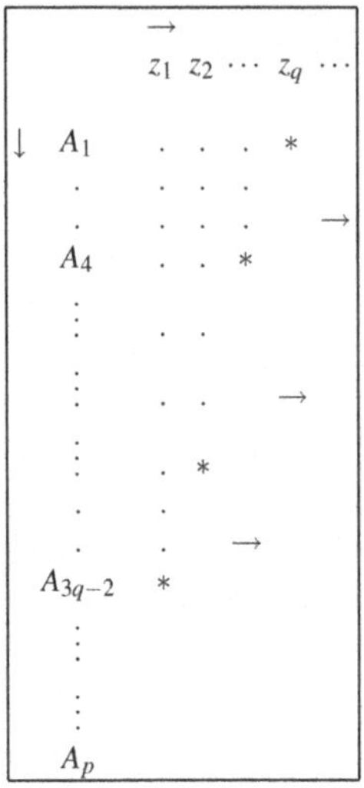

Horizontal direction:

> **do** $k = 1 : m$
> $\quad z_{k+1} = \alpha_k (P^{-1} A - \lambda_k I) z_k;$
> **end**

Vertical direction:

Recursion across the domains.

Asymptotic efficiency :

The stars illustrate the wavefront of the flow of the computation.

Efficiency : $1/3$ for 1 domain / processor.

Fig. 10.4 Pipelined costruction of the Newton-Krylov basis corresponding to the block multiplicative Schwarz preconditioning

---

Algorithm 10.4 Pipelined multiplicative Schwarz iteration $w := P^{-1} v$: program for processor $p(q)$

Input: In the local memory: A_q, $v_q^T = [(v_q^1)^\top, (v_q^2)^\top (v_q^3)^\top]$ and $w_q^T = [(w_q^1)^\top, (w_q^2)^\top, (w_q^3)^\top]$.
Output: $w := P^{-1} v$.
1: **if** $q > 1$, **then**
2: $\quad$ receive z_q^1 from processor p_{q-1};
3: $\quad w_q^1 = v_q^1$;
4: **end if**
5: solve $A_q w_q = v_q$;
6: **if** $q < p$, **then**
7: $\quad w_q^3 := C_q w_q^3$;
8: $\quad$ send w_q^3 to processor p_{q+1};
9: **end if**
10: **if** $q > 1$, **then**
11: $\quad$ send w_q^1 to processor p_{q-1};
12: **end if**
13: **if** $q < p$, **then**
14: $\quad$ receive t from processor p_{q+1};
15: $\quad w_q^3 := t$;
16: **end if**

Thus, the speedup of the basis construction is given by,

$$\mathbf{S}_p = \frac{p}{3 + (p-3)/m}. \tag{10.55}$$

While efficiency $\mathbf{E}_p = \frac{1}{3+(p-3)/m}$ grows with m, it is asymptotically limited to $\frac{1}{3}$. Assuming that the block-diagonal decomposition is with weak overlap as given by Definition 10.3, then by using (10.53) we can show that the asymptotic efficiency can reach $\frac{1}{2}$.

If we do not skip the computation of the normalizing coefficients α_k, the actual computation should be as follows:

$$\textbf{do } k = 1 : m,$$
$$z_{k+1} = \alpha_k (P^{-1} A z_k - \lambda_{k+1} z_k);$$
$$\textbf{end}$$

where $\alpha_k = \frac{1}{\|P^{-1} A z_k - \lambda_{k+1} z_k\|}$. When a slice of the vector z_{k+1} has been computed on processor P_q, the corresponding norm can be computed and added to the norms of the previous slices which have been received from the previous processor P_{q-1}. Then the result is sent to the next processor $p(q + 1)$. When reaching the last processor P_p, the norm of the entire vector is obtained, and can be sent back to all processors in order to update the subvectors. This procedure avoids global communication which could harm parallel scalability. The entries of the tridiagonal matrix $\hat{T}_m$, introduced in (9.62), must be updated accordingly.

A full implementation of the GMRES method preconditioned with the block multiplicative Schwarz iteration is coded in GPREMS following the PETSc formulations. A description of the method is given in [50] where the pipeline for building the Newton basis is described. The flow of the computation is illustrated by Fig. 10.5. A new version is available in a more complete set where deflation is also incorporated [51]. Deflation is of importance here as it limits the increase of the number of GMRES iterations for large number of subdomains.

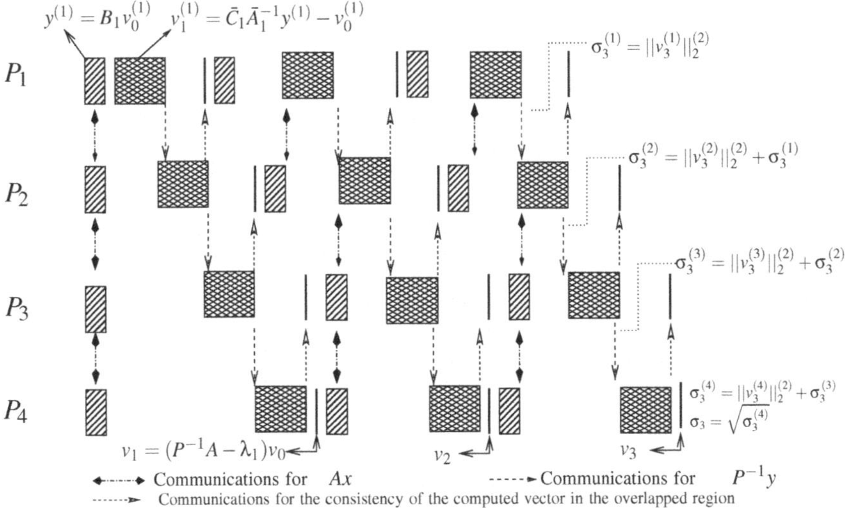

Fig. 10.5 Flow of the computation $v_{k+1} = \sigma_k P^{-1}(A - \lambda_k I) v_k$ (*courtesy* of the authors of [50])

References

1. Axelsson, O., Barker, V.A.: Finite Element Solution of Boundary Value Problems. Academic Press Inc., Orlando (1984)
2. Meurant, G.: Computer Solution of Large Linear Systems. Studies in Mathematics and its Applications. Elsevier Science, North-Holland (1999). http://books.google.fr/books?id=fSqfb5a3WrwC
3. Chen, K.: Matrix Preconditioning Techniques and Applications. Cambridge University Press, Cambridge (2005)
4. Saad, Y.: Iterative Methods for Sparse Linear Systems. SIAM, Philadelphia (2003)
5. van der Vorst, H.A.: Iterative Krylov Methods for Large Linear Systems. Cambridge University Press, Cambridge (2003). http://dx.doi.org/10.1017/CBO9780511615115
6. Schenk, O., Gärtner, K.: Solving unsymmetric sparse systems of linear equations with pardiso. Future Gener. Comput. Syst. **20**(3), 475–487 (2004)
7. Censor, Y., Gordon, D., Gordon, R.: Component averaging: an efficient iterative parallel algorithm for large and sparse unstructured problems. Parallel Comput. **27**(6), 777–808 (2001)
8. Gordon, D., Gordon, R.: Component-averaged row projections: a robust, block-parallel scheme for sparse linear systems. SIAM J. Sci. Stat. Comput. **27**(3), 1092–1117 (2005)
9. Zouzias, A., Freris, N.: Randomized extended Kaczmarz for solving least squares. SIAM J. Matrix Anal. Appl. **34**(2), 773–793 (2013)
10. Liu, J., Wright, S., Sridhar, S.: An Asynchronous Parallel Randomized Kaczmarz Algorithm. CoRR (2014). arXiv:abs/1201.3120 math.NA
11. Popa, C.: Least-squares solution of overdetermined inconsistent linear systems using Kaczmarz's relaxation. Int. J. Comput. Math. **55**(1–2), 79–89 (1995)
12. Popa, C.: Extensions of block-projections methods with relaxation parameters to inconsistent and rank-deficient least-squares problems. BIT Numer. Math. **38**(1), 151–176 (1998)
13. Bodewig, E.: Matrix Calculus. North-Holland, Amsterdam (1959)
14. Kamath, C., Sameh, A.: A projection method for solving nonsymmetric linear systems on multiprocessors. Parallel Comput. **9**, 291–312 (1988/1989)
15. Tewarson, R.: Projection methods for solving space linear systems. Comput. J. **12**, 77–80 (1969)
16. Tompkins, R.: Methods of steep descent. In: E. Beckenbach (ed.) Modern Mathematics for the Engineer. McGraw-Hill, New York (1956). Chapter 18
17. Gastinel, N.: Procédé itératif pour la résolution numérique d'un système d'équations linéaires. Comptes Rendus Hebd. Séances Acad. Sci. (CRAS) **246**, 2571–2574 (1958)
18. Gastinel, N.: Linear Numerical Analysis. Academic Press, Paris (1966). Translated from the original French text Analyse Numerique Lineaire
19. Tanabe, K.: A projection method for solving a singular system of linear equations. Numer. Math. **17**, 203–214 (1971)
20. Tanabe, K.: Characterization of linear stationary iterative processes for solving a singular system of linear equations. Numer. Math. **22**, 349–359 (1974)
21. Ansorge, R.: Connections between the Cimmino-methods and the Kaczmarz-methods for the solution of singular and regular systems of equations. Computing **33**, 367–375 (1984)
22. Cimmino, G.: Calcolo approssimato per le soluzioni dei sistemi di equazioni lineari. Ric. Sci. Progr. Tech. Econ. Naz. **9**, 326–333 (1938)
23. Dyer, J.: Acceleration of the convergence of the Kaczmarz method and iterated homogeneous transformations. Ph.D. thesis, University of California, Los Angeles (1965)
24. Gordon, R., Bender, R., Herman, G.: Algebraic reconstruction photography. J. Theor. Biol. **29**, 471–481 (1970)
25. Gordon, R., Herman, G.: Three-dimensional reconstruction from projections, a review of algorithms. Int. Rev. Cytol. **38**, 111–115 (1974)
26. Trummer, M.: A note on the ART of relaxation. Computing **33**, 349–352 (1984)
27. Natterer, F.: Numerical methods in tomography. Acta Numerica **8**, 107–141 (1999)

28. Byrne, C.: Applied Iterative Methods. A.K. Peters (2008)
29. Householder, A., Bauer, F.: On certain iterative methods for solving linear systems. Numer. Math. **2**, 55–59 (1960)
30. Householder, A.S.: The Theory of Matrices in Numerical Analysis. Dover Publications, New York (1964)
31. Elfving, T.: Block iterative methods for consistent and inconsistent linear equations. Numer. Math. **35**, 1–12 (1980)
32. Kydes, A., Tewarson, R.: An iterative method for solving partitioned linear equations. Computing **15**, 357–363 (1975)
33. Peters, W.: Lösung linear Gleichungeneichungssysteme durch Projektion auf Schnitträume von Hyperebenen and Berechnung einer verallgemeinerten Inversen. Beit. Numer. Math. **5**, 129–146 (1976)
34. Wainwright, R., Keller, R.: Algorithms for projection methods for solving linear systems of equations. Comput. Math. Appl. **3**, 235–245 (1977)
35. Björck, Å., Elfving, T.: Accelerated projection methods for computing pseudoinverse solutions of systems for linear equations. BIT **19**, 145–163 (1979)
36. Bramley, R., Sameh, A.: Row projection methods for large nonsymmetric linear systems. SIAM J. Sci. Stat. Comput. **13**, 168–193 (1992)
37. Elfving, T.: Group iterative methods for consistent and inconsistent linear equations. Technical Report LITH-MAT-R-1977-11, Linkoping University (1977)
38. Benzi, M.: Gianfranco cimmino's contribution to numerical mathematics. In: Atti del Seminario di Analisi Matematica dell'Università di Bologna, pp. 87–109. Technoprint (2005)
39. Gilbert, P.: Iterative methods for the three-dimensional reconstruction of an object from projections. J. Theor. Biol. **36**, 105–117 (1972)
40. Lakshminarayanan, A., Lent, A.: Methods of least squares and SIRT in reconstruction. J. Theor. Biol. **76**, 267–295 (1979)
41. Whitney, T., Meany, R.: Two algorithms related to the method of steepest descent. SIAM J. Numer. Anal. **4**, 109–118 (1967)
42. Arioli, M., Duff, I., Noailles, J., Ruiz, D.: A block projection method for general sparse matrices. SIAM J. Sci. Stat. Comput. **13**, 47–70 (1990)
43. Bjorck, A., Golub, G.: Numerical methods for computing angles between linear subspaces. Math. Comput. **27**, 579–594 (1973)
44. Golub, G., Van Loan, C.: Matrix Computations, 4th edn. Johns Hopkins (2013)
45. Stewart, G.: On the perturbation of pseudo-inverse, projections and linear least squares problems. SIAM Rev. **19**, 634–662 (1977)
46. Benzi, M., Frommer, A., Nabben, R., Szyld, D.B.: Algebraic theory of multiplicative Schwarz methods. Numerische Mathematik **89**, 605–639 (2001)
47. Hackbusch, W.: Iterative Solution of Large Sparse Systems of Equations. Springer, New York (1999)
48. Tang, W.: Generalized Schwarz splittings. SIAM J. Sci. Stat. Comput. **13**, 573–595 (1992)
49. Atenekeng-Kahou, G.A., Kamgnia, E., Philippe, B.: An explicit formulation of the multiplicative Schwarz preconditionner. Appl. Numer. Math. **57**(11–12), 1197–1213 (2007)
50. Nuentsa Wakam, D., Atenekeng-Kahou, G.A.: Parallel GMRES with a multiplicative Schwarz preconditioner. J. ARIMA **14**, 81–88 (2010)
51. Nuentsa Wakam, D., Erhel, J.: Parallelism and robustness in GMRES with the newton basis and the deflated restarting. Electron. Trans. Linear Algebra (ETNA) **40**, 381–406 (2013)

Chapter 11
Large Symmetric Eigenvalue Problems

In this chapter we consider the following problems:

Eigenvalue problem: Given symmetric $A \in \mathbb{R}^{n \times n}$ compute eigenpairs $\{\lambda, x\}$ that satisfy $Ax = \lambda x$.

Generalized eigenvalue problem: Given symmetric $A \in \mathbb{R}^{n \times n}$ and $B \in \mathbb{R}^{n \times n}$ that is s.p.d., compute generalized eigenpairs $\{\lambda, x\}$ that satisfy $Ax = \lambda Bx$.

Singular value problem: Given $A \in \mathbb{R}^{m \times n}$, compute normalized vectors $u \in \mathbb{R}^m$, $v \in \mathbb{R}^n$ and scalars $\sigma \geq 0$ such that $Av = \sigma u$ and $A^{\top} u = \sigma v$.

We assume here that the underlying matrix is large and sparse so that it is impractical to form a full spectral or singular-value decomposition as was the strategy in Chap. 8. Instead, we present methods that seek only a few eigenpairs or singular triplets. The methods can easily be adapted for complex Hermitian matrices.

11.1 Computing Dominant Eigenpairs and Spectral Transformations

The simplest method for computing the dominant eigenvalue (i.e. the eigenvalue of largest modulus) and its corresponding eigenvector is the Power Method, listed in Algorithm 11.1. The power method converges to the leading eigenpair for almost all initial iterates x_0 if the matrix has a unique eigenvalue of maximum modulus. For symmetric matrices, this implies that the dominant eigenvalue λ is simple and $-\lambda$ is not an eigenvalue. This method can be easily implemented on parallel architectures given an efficient parallel sparse matrix-vector multiplication kernel MV (see Sect. 2.4.1).

Theorem 11.1 *Let $\lambda_1 > \lambda_2 \geq \cdots \geq \lambda_n$ be the eigenvalues of the symmetric matrix $A \in \mathbb{R}^{n \times n}$ and $u_1, \ldots u_n$ the corresponding normalized vectors. If $\lambda_1 > 0$ is the unique dominant eigenvalue and if $x_0^{\top} u_1 \neq 0$, then the sequence of vectors x_k*

© Springer Science+Business Media Dordrecht 2016

E. Gallopoulos et al., *Parallelism in Matrix Computations*,
Scientific Computation, DOI 10.1007/978-94-017-7188-7_11

Algorithm 11.1 Power method.

Input: $A \in \mathbb{R}^{n \times n}$, $x_0 \in \mathbb{R}^n$ and $\tau > 0$.
Output: $x \in \mathbb{R}^n$ and $\lambda \in \mathbb{R}$ such that $\|Ax - \lambda x\| \le \tau$.
1: $x_0 = x_0/\|x_0\|$; $k = 0$;
2: $y = A\,x_0$;
3: $\lambda = x_0^\top y$; $r = y - \lambda\,x_0$;
4: **while** $\|r\| > \tau$,
5:   $k := k + 1$; $x_k = y/\|y\|$;
6:   $y = A\,x_k$;
7:   $\lambda = x_k^\top y$; $r = y - \lambda\,x_k$;
8: **end while**
9: $x = x_k$;

---

created by Algorithm 11.1 converges to $\pm u_1$. *The method converges linearly with the convergence factor given by:* $\max\{\frac{|\lambda_2|}{\lambda_1}, \frac{|\lambda_n|}{\lambda_1}\}$.

Proof See e.g. [1]. $\qquad\blacksquare$

Note, however, that the power method could exhibit very slow convergence. As an illustration, we consider the Poisson matrix introduced in Eq. (6.64) of Sect. 6.64. For order $n = 900$ (5-point discretization of the Laplace operator on the unit square with a 30×30 uniform grid), the power method has a convergence factor of 0.996.

The most straightforward way for simultaneously improving the convergence of the power method and enhancing parallel scalability is to generalize it by iterating on a block of vectors instead of a single vector. A direct implementation of the power method on each vector of the block will not work since every column of the iterated block will converge to the same dominant eigenvector. It is therefore necessary to maintain strong linear independence across the columns. For that purpose, the block will need to be orthonormalized in every iteration. The simplest implementation of this method, called Simultaneous Iteration, for the standard symmetric eigenvalue problem is given by Algorithm 11.2. Simultaneous Iteration was originally introduced as Treppen-iteration (cf. [2–4]) and later extended for non-Hermitian matrices (cf. [5, 6]).

For $s = 1$, this algorithm mimics Algorithm 11.1 exactly.

Theorem 11.2 *Let* λ_i *($i = 1, \ldots, n$) be the eigenvalues of the symmetric matrix* $A \in \mathbb{R}^{n \times n}$ *and* $u_1, \ldots u_n$ *the corresponding normalized vectors, with the eigenvalues ordered such that :* $|\lambda_1| \ge \cdots \ge |\lambda_s| > |\lambda_{s+1}| \ge \cdots \ge |\lambda_n|$. *Let* P *be the spectral projector associated with the invariant subspace belonging to the eigenvalues* $\lambda_1, \ldots, \lambda_s$. *Also, let* $X_0 = [x_1, \ldots, x_s] \in \mathbb{R}^{n \times s}$ *be a basis of this subspace such that* $P X_0$ *is of full column rank. Then,*

$$\|(I - P_k)u_i\| = O\left(\left|\frac{\lambda_{s+1}}{\lambda_i}\right|^k\right), \; \text{for } i = 1, \ldots, s, \tag{11.1}$$

Algorithm 11.2 Simultaneous iteration method.

Input: $A \in \mathbb{R}^{n \times n}$, $X_0 \in \mathbb{R}^{n \times s}$ and $\tau > 0$.
Output: $X \in \mathbb{R}^{n \times s}$ orthonormal and $\Lambda \in \mathbb{R}^{s \times s}$ diagonal such that $\| AX - X\Lambda \| \leq \tau$.
1: $X_0 = \text{MGS}(X_0)$;
2: $Y = AX_0$;
3: $H = X_0^{\top} Y$;
4: $H = Q\Lambda Q^{\top}$;   //Diagonalization of H
5: $R = Y - X_0 H$;
6: **while** $\| R \| > \tau$,
7:    $X_k = \text{MGS}(YQ)$;
8:    $Y = A X_k$;
9:    $H = X_k^{\top} Y$;
10:    $H = Q\Lambda Q^{\top}$;   //Diagonalization of H
11:    $R = Y - X_k H$; ;
12: **end while**
13: $X = X_k$;

---

where P_k is the orthogonal projector onto the subspace spanned by the columns of X_k which is created by Algorithm 11.2.

Proof See e.g. [7] for the special case of a diagonalizable matrix.

Algorithms which implement simultaneous iteration schemes are inherently parallel. For example, in Algorithm 11.2, steps 7–11, are performed using dense matrix kernels whose parallel implementations that guarantee high performance have been discussed in earlier chapters. Most of the time consumed in each iteration of Algorithm 11.2 is due to the sparse matrix-dense matrix multiplication, e.g. step 8. We should point out that the eigendecomposition of $H = X_k^{\top} Y$ may be performed on one multicore node when s is small relative to the size of the original matrix.

11.1.1 Spectral Transformations

In the previous section, we explored methods for computing a few dominant eigenpairs. In this section, we outline techniques that allow the computation of those eigenpairs belonging to the interior of the spectrum of a symmetric matrix (or operator) $A \in \mathbb{R}^{n \times n}$. These techniques consist of considering a transformed operator which has the same invariant subspaces as the original operator but with a rearranged spectrum.

Deflation

Let us order the eigenvalues of A as: $|\lambda_1| \geq \cdots \geq |\lambda_p| > |\lambda_{p+1}| \geq \cdots \geq |\lambda_n|$. We assume that an orthonormal basis $V = (v_1, \ldots, v_p) \in \mathbb{R}^{n \times p}$ of the invariant subspace $\mathscr{V}$ corresponding to the eigenvalues $\{\lambda_1, \cdots, \lambda_p\}$ is already known. Let $P = I - VV^{\top}$ be the orthogonal projector onto $\mathscr{V}^{\perp}$ (the orthogonal complement of $\mathscr{V}$).

Proposition 11.1 *The invariant subspaces of the operator* $B = PAP$ *are the same as those of* A, *and the spectrum of* B *is given by,*

$$\Lambda(B) = \{0\} \cup \{\lambda_{p+1}, \ldots, \lambda_n\}. \tag{11.2}$$

Proof Let u_i be the eigenvector of A corresponding to λ_i. If $i \leq p$, then $Pu_i = 0$ and $Bu_i = 0$, else $Pu_i = u_i$ and $Bu_i = \lambda_i u_i$.

Using the power method or simultaneous iteration on the operator B provides the dominant eigenvalues of B which are now $\lambda_{p+1}, \lambda_{p+2}, \ldots$. To implement the matrix-vector multiplication involving the matrix B, it is necessary to perform the operation Pv for any vector v. We need to state here that it is not recommended to use the expression $Px = x - V(V^\top x)$ since this corresponds to the Classical Gram-Schmidt method (CGS) which is not numerically reliable unless applied twice (see Sect. 7.4). It is better to implement the Modified Gram-Schmidt version (MGS) via the operator: $Pv = \prod_{i=1}^{p} (I - u_i u_i^\top)\, v$. Since $B = PAP$, a multiplication of B by a vector implies two applications of the projector P to this vector. However, we can ignore the pre-multiplication of A by P since the iterates x_k in Algorithm 11.1, and X_k in Algorithm 11.2 satisfy, respectively, the relationships $Px_k = x_k$ and $PX_k = X_k$. Therefore we can use only the operator $\tilde{B} = PA$ in both the power and simultaneous iteration methods. Hence, this procedure—known as the *deflation process*—increases the number of arithmetic operations in each iteration by $4pn$ or $4spn$, respectively, on a uniprocessor.

In principle, by applying the above technique, it should be possible to determine any part of the spectrum. There are two obstacles, however, which make this approach not viable for determining eigenvalues that are very small in magnitude: (i) the number of arithmetic operations increases significantly; and (ii) the storage required for the eigenvectors $u_1, \ldots, u_p$ could become too large. Next, we consider two alternate strategies for a spectral transformation that could avoid these two disadvantages.

Shift-and-Invert Transformation

This shift-and-invert strategy transforms the eigenvalues of A which lie near zero into the largest eigenvalues of A^{-1} and therefore allow efficient use of the power or the simultaneous iteration methods for determining eigenvalues of A lying close to the origin. This can be easily extended for the computation of the eigenvalues in the neighborhood of any given value μ.

Lemma 11.1 (Shift-and-Invert) *Let* $A \in \mathbb{R}^{n \times n}$ *and* $\mu \in \mathbb{R}$. *For any eigenpair* (v, u) *of* $(A - \mu I)^{-1}$, $(\lambda = \mu + \frac{1}{v}$, $u)$ *is an eigenpair of* A. *Using the power method (Algorithm 11.1) for obtaining the dominant eigenpair of the matrix* $B = (A - \mu I)^{-1}$, *one builds a sequence of vectors which converges to the eigenvector* u *of the eigenvalue* λ *closest to* μ, *as long as the initial vector used to start the power method is not chosen orthogonal to* u. *The convergence factor of the method is given by,* $\tau = \left| \frac{\lambda - \mu}{\tilde{\lambda} - \mu} \right|$, *where* $\tilde{\lambda}$ *is the second closest eigenvalue to* μ.

Algorithm 11.3 Shift-and-invert method .

Input: $A \in \mathbb{R}^{n \times n}$, $\mu \in \mathbb{R}$, $x_0 \in \mathbb{R}^n$ and $\tau > 0$.
Output: $x \in \mathbb{R}^n$ and $\lambda \in \mathbb{R}$ such that $\| Ax - \lambda x \| \leq \tau$.
1: Perform an LU-factorization of $(A - \mu I)$.
2: $x_0 := x_0 / \|x_0\|$; $k = 0$;
3: Solve $(A - \mu I)y = x_0$; $\rho = \|y\|$;
4: $x_1 = y/\rho$; $t = x_0/\rho$;
5: $\lambda_1 = x_1^\top t$; $r_1 = t - \lambda_1 x_1$;
6: **while** $\|r_{k+1}\| > \tau$,
7:    $k := k + 1$;
8:    Solve $(A - \mu I)y = x_k$; $\rho = \|y\|$;
9:    $x_{k+1} = y/\rho$; $t = x_k/\rho$;
10:    $\lambda_{k+1} = x_{k+1}^\top t$; $r_{k+1} = t - \lambda_{k+1} x_k$;
11: **end while**
12: $\lambda := \lambda_{k+1} + \mu$; $x = x_{k+1}$;

The resulting method is illustrated in Algorithm 11.3.

Solving the linear systems (*step* 8) in each iteration k is not as costly since the LU-factorization of $(A - \mu I)$ is performed only once (*step* 1). In order to increase the order of convergence to an asymptotic cubic rate, it is possible to update the shift μ at each iteration (see the Rayleigh Quotient Iteration [1]) requiring an LU-factorization of $(A - \mu_k I)$ at each iteration k. As a compromise, one can re-evaluate the shift μ only periodically, e.g. every $q > 1$ iterations.

The Shift-and-Invert transformation may also be combined with the Simultaneous Iteration method (Algorithm 11.2), resulting in algorithms which are better suited for parallel architectures than the single vector iteration which could have better rates of convergence.

Polynomial Transformations for Intermediate Eigenvalues

The Shift-and-Invert transformation is efficient for targetting any region of the spectrum of a matrix but it involves solving linear systems. A more easily computable transformation is via the use of polynomials: for any polynomial p and any eigenpair (λ, u) of A, the pair $(p(\lambda), u)$ is an eigenpair of $p(A)$.

Although polynomial transformations usually give rise to a considerably higher number of iterations than Shift-and-Invert transformations, they are still useful because of their ability to cope with extremely large matrices and realizing high efficiency on parallel architectures.

As a simple illustration, we consider the quadratic transformation $p(A) = I - c(A - aI)(A - bI)$ combined with the method of simultaneous iterations for computing all the eigenvalues of A lying in a given interval $[a, b]$ as introduced in [8]. The scaling factor c is chosen so that $\min(p(c_1), p(c_2)) = -1$ where the interval $[c_1, c_2]$ includes, as closely as possible, the spectrum of A. This yields

$$c = \min_{k=1,2} \frac{2}{(c_k - a)(c_k - b)}, \tag{11.3}$$

which implies that for any eigenvalue λ of A,

$$\text{if } \lambda \in [a, b], \text{ then }\quad 1 \le p(\lambda) \le 1 + g, \text{ where } g = \frac{c(b-a)^2}{4},$$

$$\text{else }\quad -1 \le p(\lambda) \le 1.$$

Therefore the polynomial transformation makes dominant the eigenvalues lying in $[a, b]$. The eigenvalues of $p(A)$, however, can be densely concentrated around 1. For that reason, the authors of [8] apply the simultaneous iteration not to $B = p(A)$ but to $\mathscr{T}_m(B)$ where $\mathscr{T}_m$ is the Chebyshev polynomial of the first kind of order m. The resulting procedure is given in Algorithm 11.4.

Algorithm 11.4 Computing intermediate eigenvalues.

Input: $A \in \mathbb{R}^{n \times n}$, $[a, b]$: interval of the eigenvalues to compute, $[c_1, c_2]$: interval including all the eigenvalues, $X_0 \in \mathbb{R}^{n \times s}$, m, and $\tau > 0$.
Output: $X = [x_1, \cdots, x_s] \in \mathbb{R}^{n \times s}$ orthonormal and $\Lambda = (\lambda_1, \cdots, \lambda_s)$ such that $\|Ax_i - \lambda_i x_i\| \le \tau$ for $i = 1, \cdots, s$.
1: $c = \min_{k=1,2} 2/[(c_k - a)(c_k - b)]$; $B \equiv I - c(A - aI)(A - bI)$;
2: $X_0 = \mathrm{MGS}(X_0)$;
3: **while** True ,
4:     $Y_k^0 = X_k$;
5:     $Y_k^1 = \mathrm{multiply}(B, X_k)$;
6:     **do** $t = 1 : m - 1$,
7:       $Y_k^{t+1} = 2 \times \mathrm{multiply}(B, Y_k^t) - Y_k^{t-1}$;
8:     **end**
9:     $Y_k = Y_k^m$;
10:    **do** $i = 1 : s$,
11:      $\lambda_i = X_k(:, i)^\top Y_k(:, i)$;
12:      $\rho_i = \|Y_k(:, i) - \lambda_i Y_k(:, i)\|$;
13:    **end**
14:    **if** $\max_{i=1:s} \rho_i \le \tau$, **then**
15:      EXIT ;
16:    **end if**
17:    $[U_k, R_k] = qr(Y_k)$;
18:    $H = R_k R_k^\top$;
19:    $H_k = Q_k \Lambda_k Q_k^\top$;   //Diagonalization of H
20:    $X_{k+1} = U_k Q_k$;
21: **end while**
22: $X = X_k$;

Theorem 11.3 *The ith column x_i of X_k as generated by Algorithm 11.4 converges to an eigenvector u corresponding to an eigenvalue $\lambda \in [a, b]$ of A such that*

$$\|u - x_i^{(k)}\| = O(q^k), \text{ where} \tag{11.4}$$

$$q = \max_{\substack{\mu \in \Lambda(A) \\ \mu \notin [a,b]}} \left\{ \frac{|\mathscr{T}_m(p(\mu))|}{|\mathscr{T}_m(p(\lambda))|} \right\}. \tag{11.5}$$

Proof This result is a straightforward application of Theorem 11.2 to the matrix $\mathscr{T}_m(p(A))$.

Note that the intervals $[a, b]$ and $[c_1, c_2]$ should be chosen so as to avoid obtaining a very small g which could lead to very slow convergence.

11.1.2 Use of Sturm Sequences

The scheme for obtaining the eigenpairs of a tridiagonal matrix that lie in a given interval, TREPS (Algorithm 8.4), introduced in Sect. 8.2.3, can be applied to any sparse symmetric matrix A with minor modifications as described below.

Let $\mu \in \mathbb{R}$ be located somewhere inside the spectrum of the sparse symmetric matrix A, but not one of its eigenvalues. Hence, $A - \mu I$ is indefinite . Computing the symmetric factorization $P^\top (A - \mu I) P = LDL^\top$, via the stable algorithm introduced in [9], where D is a symmetric block-diagonal matrix with blocks of order 1 or 2, P is a permutation matrix, and L is lower triangular matrix with a unit diagonal, then according to the Sylvester Law of Inertia, e.g. see [10], the number of negative eigenvalues of D is equal to the number of eigenvalues of A which are smaller than μ. This allows one to iteratively partition a given interval for extracting the eigenvalues belonging to this interval. On a parallel computer, the strategy of TREPS can be followed as described in Sect. 8.2.3. The eigenvector corresponding to a computed eigenvalue λ can be obtained by inverse iteration using the last $LDL^\top$ factorization utilized in determining λ.

An efficient sparse $LDL^\top$ factorization scheme by Iain Duff et al. [11], is implemented as routine `MA57` in Harwell Subroutine Library [12] (it also exists in `MATLAB` as procedure `ldl`).

If the matrix $A \in \mathbb{R}^{n \times n}$ is banded with a narrow semi-bandwidth d, two options are viable for applying Sturm sequences:

1. The matrix A is first tridiagonalized via the orthogonal similarity transformation $T = Q^\top A Q$ using parallel schemes such as those outlined in either [13] or [14], which consume $O(n^2 d)$ arithmetic operations on a uniprocessor.
2. The TREPS strategy is used through the band-LU factorization of $A - \mu I$ with no pivoting to determine the Sturm sequences, e.g. see [15]. Note that the calculation of one sequence on a uniprocessor consumes $O(nd^2)$ arithmetic operations.

The first option is usually more efficient on parallel architectures since the reduction to the tridiagonal form is performed only once, while in the second option we will need to obtain the LU-factorization of $A - \mu I$ for different values of μ.

In handling very large symmetric eigenvalue problems on uniprocessors, Sturm sequences are often abandoned as inefficient compared to the Lanczos method. However, on parallel architectures the Sturm sequence approach offers great advantages. This is due mainly to the three factors: (i) recent improvements in parallel factorization schemes of large sparse matrices on a multicore node, and (ii) the ability of the Sturm sequence approach in determining exactly the number of eigenvalues lying in a given subinterval, and (iii) the use of many multicore nodes, one node per each subinterval.

11.2 The Lanczos Method

The Lanczos process is equivalent to the Arnoldi process which is described in Sect. 9.3.2 but adapted to symmetric matrices.

When A is symmetric, Eq. (9.60) implies that H_m is symmetric and upper Hessenberg, i.e. $H_m = T_m$ is a tridiagonal matrix. Later, we show that for sufficiently large values of m, the eigenvalues of T_m could be used as good approximations of the extreme eigenvalues of A. Denoting the matrix $\hat{T}_m$ by

$$
\hat{T}_m =
\begin{pmatrix}
\alpha_1 & \beta_2 & & & & \\
\beta_2 & \cdot & \cdot & & & \\
 & \cdot & \cdot & \cdot & & \\
 & & \beta_k & \alpha_k & \beta_{k+1} & \\
 & & & \cdot & \cdot & \cdot \\
 & & & & \cdot & \cdot & \beta_m \\
 & & & & & \beta_m & \alpha_m \\
 & & & & & & \beta_{m+1}
\end{pmatrix}
\in \mathbb{R}^{(m+1)\times m},
\tag{11.6}
$$

we present the following theorem.

Theorem 11.4 (Lanczos identities) *The matrices V_{m+1} and $\hat{T}_m$ satisfy the following relation:*

$$
V_{m+1}^{\top} V_{m+1} = I_{m+1},
\tag{11.7}
$$

$$
\text{and} \quad A V_m = V_{m+1}\hat{T}_m,
\tag{11.8}
$$

The equality (11.8) can also be written as

$$
A V_m = V_m T_m + \beta_{m+1} v_{m+1} e_m^{\top},
\tag{11.9}
$$

where $e_m \in \mathbb{R}^m$ is the last canonical vector of $\mathbb{R}^m$, and $T_m \in \mathbb{R}^{m\times m}$ is the symmetric tridiagonal matrix obtained by deleting the last row of $\hat{T}_m$.

11.2.1 The Lanczos Tridiagonalization

Many arithmetic operations can be skipped in the Lanczos process when compared to that of Arnoldi: since the entries of the matrix located above the first superdiagonal are zeros, orthogonality between most of the pairs of vectors is mathematically guaranteed. The procedure is implemented as shown in Algorithm 11.5. The matrix $\hat{T}_m$

Algorithm 11.5 Lanczos procedure (no reorthogonalization).

Input: $A \in \mathbb{R}^{n \times n}$ symmetric, $v \in \mathbb{R}^n$ and $m > 0$.
Output: $V = [v_1, \cdots, v_{m+1}] \in \mathbb{R}^{n \times (m+1)}$, $\hat{T}_m \in \mathbb{R}^{(m+1) \times m}$ as given in (11.6).
1: $v_1 = v/\|v\|$;
2: **do** $k = 1 : m$,
3:    $w := A v_k$;
4:    **if** $k > 1$, **then**
5:       $w := w - \beta_k \, v_{k-1}$;
6:    **end if**
7:    $\alpha_k = v_k^\top w$;
8:    $w := w - \alpha_k \, v_k$;
9:    $\beta_{k+1} = \|w\|$;
10:    **if** $\beta_{k+1} = 0$, **then**
11:       Break ;
12:    **end if**
13:    $v_{k+1} = w/\beta_{k+1}$;
14: **end**

---

is usually stored in the form of the two vectors $(\alpha_1, \ldots, \alpha_m)$ and $(0, \beta_2, \ldots, \beta_{m+1})$. In this algorithm, it is easy to see that it is possible to proceed without storing the basis V_m since only v_{k-1} and v_k are needed to perform iteration k. This interesting feature allows us to use large values for m. Theoretically (i.e. in exact arithmetic), at some $m \le n$ the entry β_{m+1} becomes zero. This implies that V_m is an orthonormal basis of an invariant subspace and that all the eigenvalues of T_m are eigenvalues of A. Thus, the Lanczos process terminates with T_m being irreducible (since $(\beta_k)_{k=2:m}$ are nonzero) with all its eigenvalues simple.

The Effect of Roundoff Errors

Unfortunately, in floating-point arithmetic the picture is not so as straightforward: the orthonormality expressed in (11.7) is no longer assured. Instead, the relation (11.8) is now replaced by

$$AV_m = V_{m+1}\hat{T}_m + E_m, \tag{11.10}$$

where $E_m \in \mathbb{R}^{n \times m}$, with $\|E_m\| = O(\mathbf{u}\|A\|)$.

Enforcing Orthogonality

An alternate version of the Lanczos method that maintains orthogonality during the tridiagonalization process can be realized as done in the Arnoldi process (see algorithm 9.3). The resulting procedure is given by Algorithm 11.6. Since in this version the basis V_{m+1} must be stored, the parameter m cannot be chosen as large as in Algorithm 11.5. Hence, a restarting strategy must be considered (see Sect. 11.2.2).

A Block Version of the Lanczos Method with Reorthogonalization

There exists a straightforward generalization of the Lanczos method which operates with blocks of r vectors instead of a single vector (see [16–18]). Starting with an

---

Algorithm 11.6 Lanczos procedure (with reorthogonalization).

Input: $A \in \mathbb{R}^{n \times n}$ symmetric, $v \in \mathbb{R}^n$ and $m > 0$.
Output: $V = [v_1, \ldots, v_{m+1}] \in \mathbb{R}^{n \times (m+1)}$, $\hat{T}_m \in \mathbb{R}^{(m+1) \times m}$ as given in (11.6).
1: $v_1 = v/\|v\|$;
2: **do** $k = 1 : m$,
3:    $w := Av_k$;
4:    **do** $j = 1 : k$,
5:      $\mu = v_j^\top w$;
6:      $w := w - \mu\, v_j$;
7:    **end**
8:    $\alpha_k = v_k^\top w$;
9:    $w := w - \alpha_k\, v_k$;
10:   $\beta_{k+1} = \|w\|$;
11:   **if** $\beta_{k+1} = 0$, **then**
12:      Break ;
13:   **end if**
14:   $v_{k+1} = w/\beta_{k+1}$;
15: **end**

---

initial matrix $V_1 \in \mathbb{R}^{n \times p}$ with orthonormal columns, the block Lanczos algorithm for reducing A to the block-tridiagonal matrix

$$
T = \begin{bmatrix}
G_1 & R_1^\top & & & \\
R_1 & G_2 & R_2^\top & & \\
 & \ddots & \ddots & \ddots & \\
 & & R_{k-2} & G_{k-1} & R_{k-1}^\top \\
 & & & R_{k-1} & G_k
\end{bmatrix}
\tag{11.11}
$$

is given by Algorithm 11.7. In other words,

$$
AV = VT + Z(0, \ldots, 0, I_p)
\tag{11.12}
$$

where

$$
V = (V_1, V_2, \ldots, V_k).
\tag{11.13}
$$

Since the blocks R_j are upper-triangular, the matrix T is a symmetric block-tridiagonal matrix of half-bandwidth p.

11.2.2 The Lanczos Eigensolver

Let us assume that the tridiagonalization is given by (11.9). For any eigenpair (μ, y) of T_m, where $y = (\gamma_1, \ldots, \gamma_m)^\top$ is of unit 2-norm, the vector $x = V_m y$ is called the Ritz vector which satisfies

---

Algorithm 11.7 Block-Lanczos procedure (with full reorth.)

Input: $A \in \mathbb{R}^{n \times n}$ symmetric, $Y \in \mathbb{R}^{n \times p}$ and $k > 0$, and $p > 0$.
Output: $V = [v_1, \cdots, v_m] \in \mathbb{R}^{n \times m}$, $T \in \mathbb{R}^{m \times m}$ as given in (11.11) and where $m \leq kp$, and
 $Z \in \mathbb{R}^{n \times p}$ which satisfies (11.12).
1: Orthogonalize Y into V_1 ; $V = []$;
2: **do** $j = 1 : k$,
3:   $V := [V, V_j]$;
4:   $W := A V_j$;
5:   **do** $i = 1 : j$,
6:     $F = V_i^{\mathsf{T}} W$;
7:     $W := W - V_i F$;
8:   **end**
9:   $G_j = V_j^{\mathsf{T}} W$;
10:   $W := W - V_j G_j$;
11:   $[V_{j+1}, R_j]=\mathrm{qr}(W)$;
12:   **if** R_j singular, **then**
13:     Break ;
14:   **end if**
15: **end**
16: $Z = W$;

$$Ax - \mu x = \beta_{m+1} \gamma_m v_{m+1}, \tag{11.14}$$

$$\text{and } \|Ax - \mu x\| = \beta_{m+1} |\gamma_m|. \tag{11.15}$$

Clearly, there exists an eigenvalue $\lambda \in \Lambda(A)$ such that

$$|\mu - \lambda| \leq \frac{\|Ax - \mu x\|}{\|x\|}, \tag{11.16}$$

(e.g. see [7, p. 79]). Therefore, in exact arithmetic, we have

$$|\mu - \lambda| \leq \beta_{m+1} |\gamma_m|. \tag{11.17}$$

The convergence of the Ritz values to the eigenvalues of A has been investigated extensively (e.g. see the early references [19, 20]), and can be characterized by the following theorem.

Theorem 11.5 (Eigenvalue convergence) *Let T_m be the symmetric tridiagonal matrix built by Algorithms 11.5 from the inital vector v. Let the eigenvalues of A and T_m be respectively denoted by $(\lambda_i)_{i=1:n}$ and $(\mu_i^{(m)})_{i=1:m}$ and labeled in decreasing order. The difference between the ith exact and approximate eigenvalues, λ_i and $\mu_i^{(m)}$, respectively, satisfies the following inequality,*

$$0 \leq \lambda_i - \mu_i^{(m)} \leq (\lambda_1 - \lambda_n) \left(\frac{\kappa_i^{(m)} \tan \angle(v, u_i)}{\mathscr{T}_{m-i}(1 + 2\gamma_i)} \right)^2 \tag{11.18}$$

where u_i is the eigenvector corresponding to λ_i, $\gamma_i = \frac{\lambda_i - \lambda_{i+1}}{\lambda_{i+1} - \lambda_n}$, and $\kappa_i^{(m)}$ is given by

$$\kappa_1^{(m)} = 1, \ and \ \kappa_i^{(m)} = \prod_{j=1}^{i-1} \frac{\mu_j^{(m)} - \lambda_n}{\mu_j^{(m)} - \lambda_i}, \ for \ i > 1, \tag{11.19}$$

with $\mathscr{T}_k$ being the Chebyshev polynomial of first kind of order k.

Proof See [7] or [21].

The Lanczos Eigensolver without Reorthogonalization

Let us consider the effect of the loss of orthogonality on any Ritz pair $(\mu, x = V_m y)$. Recalling that the orthonormality of V_m could be completely lost, then from (11.10) (similar to (11.15) and (11.16)), there exists an eigenvalue $\lambda \in \Lambda(A)$ such that

$$|\mu - \lambda| \le \frac{\beta_{m+1} |\gamma_m| + \|F_m\|}{\|x\|}, \tag{11.20}$$

where $\|x\| \ge \sigma_{\min}(V_m)$ can be very small. Several authors discussed how accurately can the eigenvalues of T_m approximate the eigenvalues of A in finite precision, for example, see the historical overview in [21]. The framework of the theory proving that loss of orthogonality appears when V_m includes a good approximation of an eigenvector of A and, by continuing the process, new copies of the same eigenvalue could be regenerated was established in [20, 22].

A scheme for discarding "spurious" eigenvalues by computing the eigenvalues of the tridiagonal matrix $\tilde{T}_m$ obtained from T_m by deleting its first row and first column was developed in [23]. As a result, any common eigenvalue of T_m and $\tilde{T}_m$ is deemed spurious.

It is therefore possible to compute a given part of the spectrum $\Lambda(A)$. For instance, if the sought after eigenvalues are those that belong to an interval $[a, b]$, Algorithm 11.8 provides those estimates with their corresponding eigenvectors without storing the basis V_m. Once an eigenvector y of T_m is computed, the Ritz vector $x = V_m y$ needs to be computed. To overcome the lack of availability of V_m, however, a second pass of the Lanczos process is restarted with the same initial vector v and accumulating products in Step 6 of Algorithm 11.8 on the fly.

Algorithm 11.8 exhibits a high level of parallelism as outlined below:

- *Steps 2 and 3*: Computing the eigenvalues of a tridiagonal matrix via multisectioning and bisections using Sturm sequences (TREPS method),
- *Step 5*: Inverse iterations generate q independent tasks, and
- *Steps 6 and 7*: Using parallel sparse matrix—vector (and multivector) multiplication kernels, as well as accumulating dense matrix-vector products on the fly, and
- *Step 7*: Computing simultaneously the norms of the q columns of W.

Algorithm 11.8 LANCZOS1: Two-pass Lanczos eigensolver.

Input: $A \in \mathbb{R}^{n \times n}$ symmetric, $[a, b] \subset \mathbb{R}$, and $v \in \mathbb{R}^n$, and $m \geq n$.
Output: $\{\lambda_1, \cdots, \lambda_p\} \subset [a, b] \cap \Lambda(A)$ and their corresponding eigenvectors $U = [u_1, \ldots, u_p] \in \mathbb{R}^{n \times p}$.

1: Perform the Lanczos procedure from v to generate T_m without storing the basis V_m;
2: Compute the eigenvalues of T_m in $[a, b]$ by TREPS (Algorithm 8.4);
3: Compute the eigenvalues of $\tilde{T}_m$ in $[a, b]$ by TREPS;
4: Discard the spurious eigenvalues of T_m to get a list of q eigenvalues: $\{\lambda_1, \cdots, \lambda_q\} \subset [a, b]$;
5: Compute the corresponding eigenvectors of T_m by inverse iteration : $Y = [y_1, \ldots, y_q]$;
6: Redo the Lanczos procedure from v and perform $W = V_m Y$ without storing the basis V_m;
7: Using (11.16), check for convergence of the columns of W as eigenvectors of A;
8: List the converged eigenvalues $\{\lambda_1, \cdots, \lambda_p\} \subset [a, b]$ and corresponding normalized eigenvectors $U = [u_1, \ldots, u_p] \in \mathbb{R}^{n \times p}$.

The Lanczos Eigensolver with Reorthogonalization

When Algorithm 11.6 is used, the dimension m is usually too small to obtain good approximations of the extremal eigenvalues of A from the Ritz values $(\mu_i^{(m)})_{1:m}$, i.e. the eigenvalues of T_m. Therefore, a restarting strategy becomes mandatory.

Assuming that we are seeking the p largest eigenvalues of A, the simplest technique consists of computing the p largest eigenpairs (μ_i, y_i) of T_m and restarting the Lanczos process with the vector v which is the linear combination $v = V_m(\sum_{i=1}^{p} \mu_i y_i)$. This approach of creating the restarting vector is not as effective as the Implicitly Restarted Arnoldi Method (IRAM) given in [24] and implemented in ARPACK [25]. Earlier considerations of implicit restarting appeared in [26, 27]. The technique can be seen as a special polynomial filter applied to the initial starting vector v_1, i.e. the first vector used in the next restart is given by, $\tilde{v}_1 = p_{m-1}(A)v_1 \in \mathcal{K}_m(A, v_1)$ where the polynomial p_{m-1} is implicitly built.

Algorithm 11.9 LANCZOS2: Iterative Lanczos eigensolver.

Input: $A \in \mathbb{R}^{n \times n}$ symmetric, and $v \in \mathbb{R}^n$, and $m \geq n$ and $p < m$.
Output: The p largest eigenvalues $\{\lambda_1, \cdots, \lambda_p\} \subset \Lambda(A)$ and their corresponding eigenvectors $U = [u_1, \cdots, u_p] \in \mathbb{R}^{n \times p}$.

1: $k = 0$;
2: Perform the Lanczos procedure from v to generate T_m^1 and the basis V_m^1 (Algorithm 11.6);
3: **repeat**
4:    $k = k + 1$;
5:    **if** $k > 1$, **then**
6:       Apply the restarting procedure IRAM to generate V_p^k and T_p^k from V_m^{k-1} and T_m^{k-1};
7:       Perform the Lanczos procedure to generate V_m^k and the basis T_m^k from V_p^k and T_p^k;
8:    **end if**
9:    Compute the p largest Ritz pairs and the corresponding residuals;
10: **until** Convergence

Since all the desired eigenpairs do not converge at the same iteration, Algorithm 11.9 can be accelerated by incorporating a deflating procedure that allows the storing of the converged vectors and applying the rest of the algorithm on PA, instead of A, where P is the orthogonal projector that maintains the orthogonality of the next basis with respect to the converged eigenvectors.

Parallel scalability of Algorithm 11.9 can be further enhanced by using a block version of the Lanczos eigensolver based on Algorithm 11.7. Once the block-tridiagonal matrix T_m is obtained, any standard algorithm can be employed to find all its eigenpairs. For convergence results for the block scheme analogous Theorem 11.5 see [17, 18]. An alternative which involves fewer arithmetic operations consists of orthogonalizing the newly computed Lanczos block against the (typically few) converged Ritz vectors. This scheme is known as the block Lanczos scheme with *selective orthogonalization*. The practical aspects of enforcing orthogonality in either of these ways are discussed in [28–30]. The IRAM strategy was adapted for the block Lanczos scheme in [31].

11.3 A Block Lanczos Approach for Solving Symmetric Perturbed Standard Eigenvalue Problems

In this chapter, we have presented several algorithms for obtaining approximations of a few of the extreme eigenpairs of the $n \times n$ sparse symmetric eigenvalue problem,

$$Ax = \lambda x, \tag{11.21}$$

which are suitable for implementation on parallel architectures with high efficiency. In several computational science and engineering applications there is the need for developing efficient parallel algorithms for approximating the extreme eigenpairs of the series of slightly perturbed eigenvalue problems,

$$A(S_i)y = \lambda(S_i)y, \qquad i = 1, 2, .., m \tag{11.22}$$

where

$$A(S_i) = A + B S_i B^\top \tag{11.23}$$

in which $B \in \mathbb{R}^{n \times p}$, $S_i = S_i^\top \in \mathbb{R}^{p \times p}$ with B being of full-column rank $p \ll n$, and m is a small positive integer.

While any of the above symmetric eigenvalue problem solvers can be adapted to handle the series of problems in (11.22), we consider here an approach based on the block Lanczos algorithm as defined in Sect. 11.2.1, [16].

11.3.1 Starting Vectors for $A(S_i)x = \lambda X$

We are now in a position to discuss how to choose the starting block V_1 for the block Lanczos reduction of $A(S_i)$ in order to take advantage of the fact that A has already been reduced to block tridiagonal form via the algorithm described above. Recalling that

$$A(S_i) = A + B S_i B^\top,$$

where $1 \le i \le m$, $p \ll n$, and the matrix B has full column rank, the idea of the approach is rather simple. Taking as a starting block the matrix V_1 given by the orthogonal factorization of B, $B = V_1 R_0$ where $R_0 \in \mathbb{R}^{p \times p}$ is upper triangular. Then, the original matrix A is reduced to the block tridiagonal matrix T via the block Lanczos algorithm.

In the following, we show that this choice of the starting Lanczos block leads to a reduced block tridiagonal form of the perturbed matrices $A(S_i)$ which is only a rank p perturbation of the original block tridiagonal matrix T. Let V, contain the Lanczos blocks $V_1, V_2, \ldots, V_k$ generated by the algorithm. From the orthogonality property of these matrices, we have

$$V_i^\top V_j = \begin{cases} I_p & \text{for } i = j \\ 0 & \text{for } i \neq j \end{cases} \tag{11.24}$$

where I_p is the identity matrix of order p. Now let $E_i = B S_i B^\top$, hence

$$V^\top E_i V = V^\top B S_i B^\top V = V^\top V_1 R_0 S_i R_0^\top V_1^\top V, \quad i = 1, 2, \ldots, m. \tag{11.25}$$

Since

$$V_1 = V(I_p, 0, \ldots, 0), \tag{11.26}$$

then,

$$V^\top E_i V = \begin{pmatrix} R_0 S_i R_0^\top & & & \\ & 0 & & \\ & & \ddots & \\ & & & 0 \end{pmatrix}, \tag{11.27}$$

and,

$$V^\top A(S_i) V = V^\top (A + B S_i B^\top) V = V^\top A V + V^\top E_i V = T(S_i) \tag{11.28}$$

where

$$T(S_i) = \begin{pmatrix} \tilde{G}_1 & R_1^\top & & & \\ R_1 & G_2 & R_2^\top & & \\ & \ddots & \ddots & \ddots & \\ & & R_{k-2} & G_{k-1} & R_{k-1}^\top \\ & & & R_{k-1} & G_k \end{pmatrix} \tag{11.29}$$

with

$$\tilde{G}_1 = G_1 + R_0 S_i R_0^\top,$$

i.e. the matrix $V^\top A(S_i)V$ is of the same structure as $V^\top AV$. It is then clear that the advantage of choosing such a set of starting vectors is that the block Lanczos algorithm needs to be applied only to the matrix A to yield the block tridiagonal matrix T. Once T is formed, all block tridiagonal matrices $T(S_i)$ can be easily obtained by the addition of terms $R_0 S_i R_0^\top$ to the first diagonal block G_1 of T. The matrix V is independent of S_i and, hence, remains the same for all $A(S_i)$. Consequently, the computational savings will be significant for large-scale engineering computations which require many small modifications (or reanalyses) of the structure.

11.3.2 Starting Vectors for $A(S_i)^{-1}x = \mu x$

The starting set of vectors discussed in the previous section are useful for handling $A(S_i)x = \lambda x$ when the largest eigenvalues of $A(S_i)$ are required. For many engineering or scientific applications, however, only a few of the smallest (in magnitude) eigenvalues and their corresponding eigenvectors are desired. In such a case, one usually considers the shift and invert procedure instead of working with $A(S_i)x = \lambda x$ directly. In other words, to seek the eigenvalue near zero, one considers instead the problem $A(S_i)^{-1}x = \dfrac{1}{\lambda}x$. Accordingly, to be able to take advantage of the nature of the perturbations, we must choose an appropriate set of starting vectors.

It is not clear at this point how the starting vectors for $(A + BS_i B^\top)^{-1}x = \mu x$, where $\mu = \dfrac{1}{\lambda}$, can be properly chosen so as to yield a block tridiagonal structure analogous to that shown in (11.29). However, if we assume that both A and S_i are nonsingular, the Woodbury formula yields,

$$(A + BS_i B^\top)^{-1} = A^{-1} - A^{-1}B(S_i^{-1} + B^\top A^{-1}B)^{-1}B^\top A^{-1}. \qquad (11.30)$$

Thus, if in reducing A^{-1} to the block tridiagonal form

$$\tilde{V}^\top A^{-1}\tilde{V} = \tilde{T} \qquad (11.31)$$

via the block Lanczos scheme, we choose the first block, $\tilde{V}_1$ of $\tilde{V}$ as that orthonormal matrix resulting from the orthogonal factorization

$$A^{-1}B = \tilde{V}_1 \tilde{R}_0 \qquad (11.32)$$

where $\tilde{R}_0$ is upper triangular of order p, then

$$\tilde{V}^\top (A + B S_i B^\top)^{-1} \tilde{V} = \tilde{T} + \binom{I_p}{0} \tilde{E}(S_i)(I_p, 0) = \tilde{T}(S_i) \tag{11.33}$$

in which $\tilde{E}(S_i)$ is a $p \times p$ matrix given by

$$\tilde{E}(S_i) = \tilde{R}_0 (S_i^{-1} + B^\top A^{-1} B)^{-1} \tilde{R}_0^\top . \tag{11.34}$$

Note that $\tilde{T}(S_i)$ is a block tridiagonal matrix identical to $\tilde{T}$ in (11.31), except for the first diagonal block. Furthermore, note that as before, $\tilde{V}$ is independent of S_i and hence, remains constant for $1 \le i \le m$.

11.3.3 Extension to the Perturbed Symmetric Generalized Eigenvalue Problems

In this section, we address the extension of our approach to the perturbed generalized eigenvalue problems of type

$$K(S_i)x = \lambda M x, \quad K(S_i) = (K + B S_i B^\top) \tag{11.35}$$

and

$$Kx = \lambda M(S_i)x, \quad M(S_i) = (M + B S_i B^\top) \tag{11.36}$$

where $K(S_i)$ and $M(S_i)$ are assumed to be symmetric positive definite for all i, $1 \le i \le m$. In structural mechanics, $K(S_i)$ and $M(S_i)$ are referred to as stiffness and mass matrices, respectively.

Let $K = L_K L_K{}^\top$ and $M = L_M L_M{}^\top$ be the Cholesky factorization of K and M, respectively. Then the generalized eigenvalue problems (11.35) and (11.36) are reduced to the standard form

$$(\tilde{K} + \tilde{B} S_i \tilde{B}^\top)y = \lambda y \tag{11.37}$$

and

$$(\hat{M} + \hat{B} S_i \hat{B}^\top)z = \lambda^{-1} z \tag{11.38}$$

where

$$\tilde{K} = L_M^{-1} K L_M^{-\top}, \quad \tilde{B} = L_M^{-1} B, \quad \text{and} \quad y = L_M^\top x \tag{11.39}$$

and

$$\hat{M} = L_K^{-1} M L_K^{-\top}, \quad \hat{B} = L_K^{-1} B, \quad \text{and} \quad z = L_K^\top x. \tag{11.40}$$

Now, both problems can be treated as discussed above. If one seeks those eigenpairs closest to zero in (11.35), then we need only obtain the block tridiagonal form associated with $\tilde{K}^{-1}$ once the relevant information about the starting orthonormal block is obtained from the orthogonal factorization for $\tilde{B}$. Similarly, in (11.36), one needs the block tridiagonal form associated with $\hat{M}$ based on the orthogonal factorization of $\hat{B}$.

11.3.4 Remarks

Typical examples of the class of problems described in the preceding sections arise in the dynamic analysis of modified structures. A frequently encountered problem is how to take into account, in analysis and design, changes introduced after the initial structural dynamic analysis has been completed. Typically, the solution process is of an iterative nature and consists of repeated modifications to either the stiffness or the mass of the structure in order to fine-tune the constraint conditions. Clearly, the number of iterations depends on the complexity of the problem, together with the nature and number of constraints. Even though these modifications may be only slight, a complete reanalysis of the new eigenvalues and eigenvectors of the modified eigensystem is often necessary. This can drive the computational cost of the entire process up dramatically especially for large scale structures.

The question then is how information obtained from the initial/previous analysis can be readily exploited to derive the response of the new modified structure without extensive additional computations. To illustrate the usefulness of our approach in this respect, we present some numerical applications from the free vibrations analysis of an undamped cantilever beam using finite elements. Without loss of generality, we consider only modifications to the system stiffness matrix.

We assume that the beam is uniform along its span and that it is composed of a linear, homogeneous, isotropic, elastic material. Further, the beam is assumed to be slender, i.e. deformation perpendicular to the beam axis is due primarily to bending (flexing), and shear deformation perpendicular to the beam axis can be neglected; shear deformation and rotational inertia effects only become important when analyzing deep beams at low frequencies or slender beams at high frequencies. Subsequently, we consider only deformations normal to the undeformed beam axis and we are only interested in the fewest lowest natural frequencies.

The beam possesses an additional support at its free end by a spring (assumed to be massless) with various stiffness coefficients α_i, $i = 1, 2, \ldots, m$, as shown in Fig. 11.1. The beam is assumed to have length $L = 3.0$, a distributed mass $\bar{m}$ per unit length, and a flexural rigidity EI.

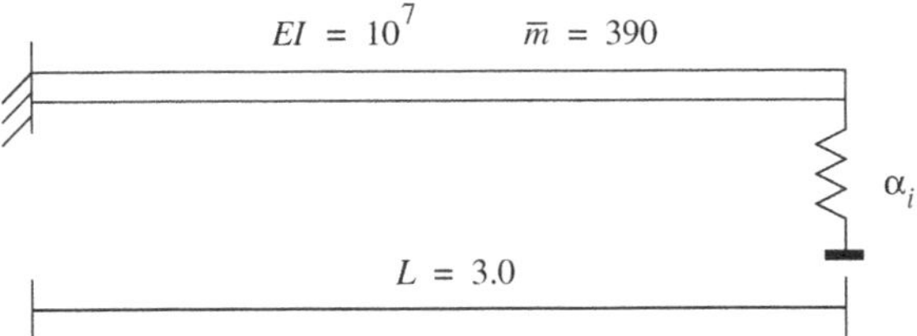

Fig. 11.1 A cantilever beam with additional spring support

First, we discretize the beam, without spring support using one-dimensional solid beam finite elements, each of length $\ell = 0.3$. Using a lumped mass approach, we get a diagonal mass matrix M, while the stiffness matrix K is block-tridiagonal of the form

$$K = \frac{EI}{\ell^3} = \begin{pmatrix} A_s & C^\top & & & \\ C & A_s & C^\top & & \\ \vdots & \vdots & \vdots & & \\ & & C & A_s & C^\top \\ & & & C & A_r \end{pmatrix}$$

in which each A_s, A_r, and C is of order 2. Note that both M and K are symmetric positive definite and each is of order 20. Including the spring with stiffness α_i, we obtain the perturbed stiffness matrix

$$K(\alpha_i) = K + \alpha_i bb^\top \tag{11.41}$$

where $b = e_{19}$ (the 19th column of I_{20}). In other words, $K(\alpha_i)$ is a rank-1 perturbation of K. Hence, the generalized eigenvalue problems for this discretization are given by

$$K(\alpha_i)x = (K + \alpha_i bb^\top)x = \lambda M x, \quad i = 1, 2, \ldots, m \tag{11.42}$$

which can be reduced to the standard form by symmetrically scaling $K(\alpha_i)$ using the diagonal mass matrix M, i.e.

$$M^{-1/2}K(\alpha_i)M^{-1/2}y = \lambda y$$

in which $y = M^{1/2}x$.

Second, we consider modeling the beam as a three-dimensional object using regular hexahedral elements, see Fig. 11.2. This element possesses eight nodes, one at each corner, with each having three degrees of freedom, namely, the components of displacement u, v, and w along the directions of the x, y, and z axes, respectively.

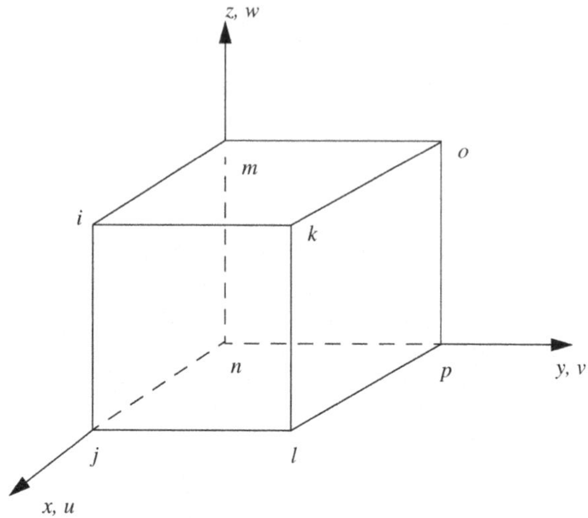

Fig. 11.2 The regular hexahedral finite element

In the case of the hexahedral element, [32, 33], the element stiffness and mass matrices possess the SAS property (see Chap. 6) with respect to some reflection matrix and, hence, they can each be recursively decomposed into eight submatrices.

Because of these properties of the hexahedral element (i.e. three levels of symmetric and antisymmetric decomposability), if SAS ordering of the nodes is employed, then the system stiffness matrix K, of order n, satisfies the relation

$$PKP = K \tag{11.43}$$

where $P = P^\top$ is a permutation matrix ($P^2 = I$). Therefore, depending on the number of planes of symmetry, the problem can be decomposed into up to eight subproblems. In the case of the three-dimensional cantilever beam studied here, see Fig. 11.3, having only two axes of symmetry with the springs removed, then the problem can be decomposed into only 4 independent subproblems,

$$Q^\top K Q = \mathrm{diag}(K_1, \ldots, K_4) \tag{11.44}$$

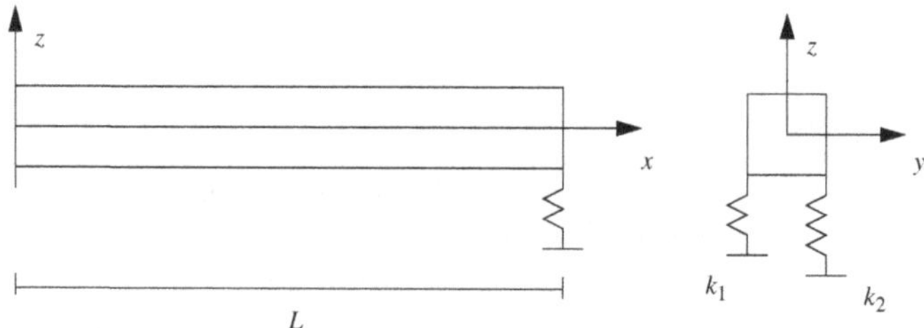

Fig. 11.3 A 3-D cantilever beam with additional spring supports

in which Q is an orthogonal matrix that can be easily constructed from the permutation submatrices that constitute P. The matrices K_i are each of order $n/4$ and of a much smaller bandwidth. Recalling that in obtaining the smallest eigenpairs, we need to solve systems of the form

$$Kz = g. \tag{11.45}$$

It is evident that in each step of the block Lanczos algorithm, we have four independent systems that can be solved in parallel, with each system of a much smaller bandwidth than that of the stiffness matrix K.

11.4 The Davidson Methods

In 1975, an algorithm for solving the symmetric eigenvalue problem, called Davidson's method [34], emerged from the computational chemistry community. This successful eigensolver was later generalized and its convergence proved in [35, 36]. The Davidson method can be viewed as a modification of Newton's method applied to the system that arises from treating the symmetric eigenvalue problem as a constrained optimization problem involving the Rayleigh quotient. This method is a precursor to other methods that appeared later such as Trace Minimization [37, 38], and Jacobi-Davidson [39] which are discussed in Sect. 11.5. These approaches essentially take the viewpoint that the eigenvalue problem is a nonlinear system of equations and attempt to find a good way to correct a given approximate eigenpair. In practice, this requires solving a correction equation which updates the current approximate eigenvector, in a subspace that is orthogonal to it.

11.4.1 General Framework

All versions of the Davidson method may be regarded as various forms of preconditioning the basic Lanczos method. In order to illustrate this point, let us consider a symmetric matrix $A \in \mathbb{R}^{n \times n}$. Both the Lanczos and Davidson algorithms generate, at some iteration k, an orthonormal basis $V_k = (v_1, \ldots, v_k)$ of a k-dimensional subspace $\mathcal{V}_k$ of $\mathbb{R}^{n \times k}$ and the symmetric interaction matrix $H_k = V_k^{\top} A V_k \in \mathbb{R}^{k \times k}$. In the Lanczos algorithm, $\mathcal{V}_k$ is the Krylov subspace $\mathcal{V}_k = \mathcal{K}_k(A, v)$, but for Davidson methods, this is not the case. In the Lanczos method, the goal is to obtain V_k such that some eigenvalues of H_k are good approximations of some eigenvalues of A. In other words, if (λ, y) is an eigenpair of H_k, then it is expected that the Ritz pair (λ, x), where $x = V_k y$, is a good approximation of an eigenpair of A. Note that this occurs only for some eigenpairs of H_k and only at convergence.

Davidson methods differ from the Lanczos scheme in the definition of the new direction which will be added to the subspace $\mathcal{V}_k$ to obtain V_{k+1}. In Lanczos schemes, V_{k+1} is the basis of the Krylov subspace $\mathcal{K}_{k+1}(A, v)$. In other words, if reorthogonalization is not considered (see Sect. 11.2), the vector v_{k+1} is computed by a three term recurrence. In Davidson methods, however, the local improvement of the direction of the Ritz vector towards the sought after eigenvector is obtained by a quasi-Newton step in which the next vector is obtained by reorthogonalization with respect to $\mathcal{V}_k$. Moreover, in this case, the matrix H_k is no longer tridiagonal. Therefore, one iteration of Davidson methods involves more arithmetic operations than the basic Lanczos scheme, and at least as expensive as the Lanczos scheme with full reorthogonalization. Note also, that Davidson methods require the storage of the basis V_k thus limiting the maximum value $k_{\max}$ in order to control storage requirements. Consequently, Davidson methods are implemented with periodic restarts.

To compute the smallest eigenvalue of a symmetric matrix A, the template of the basic Davidson method is given by Algorithm 11.10. In this template, the operator C_k of Step 12 represents any type of correction, to be specified later in Sect. 11.4.3. Each Davidson method is characterized by this correction step.

Algorithm 11.10 Generic Davidson method.

Input: $A \in \mathbb{R}^{n \times n}$ symmetric, and $v \in \mathbb{R}^n$, and $m < n$.
Output: The smallest eigenvalue $\lambda \in \Lambda(A)$ and its corresponding eigenvector $x \in \mathbb{R}^n$.
1: $v_1 = v/\|v\|$;
2: **repeat**
3:    **do** $k = 1 : m$,
4:       compute $W_k = A V_k$;
5:       compute the interaction matrix $H_k = V_k^\top A V_k$;
6:       compute the smallest eigenpair (λ_k, y_k) of H_k ;
7:       compute the Ritz vector $x_k = V_k y_k$;
8:       compute the residual $r_k = W_k y_k - \lambda_k x_k$;
9:       **if** convergence **then**
10:          Exit from the loop k;
11:       **end if**
12:       compute the new direction $t_k = C_k r_k$;   //Davidson correction
13:       $V_{k+1} = \text{MGS}([V_k, t_k])$;
14:    **end**
15:    $V_1 = \text{MGS}([x_k, t_k])$;
16: **until** Convergence
17: $x = x_k$;

11.4.2 Convergence

In this section, we present a general convergence result for the Davidson methods as well as the Lanczos eigensolver with reorthogonalization (Algorithm 11.9). For a more general context, we consider the block version of these algorithms. Creating Algorithm 11.11 as the block version of Algorithm 11.10, we generalize Step 11 so that the corrections $C_{k,i}$ can be independently chosen. Further, when some, but not all, eigenpairs have converged, a deflation process is considered by appending the converged eigenvectors as the leading columns of V_k.

Algorithm 11.11 Generic block Davidson method.

Input: $A \in \mathbb{R}^{n \times n}$ symmetric, $p \geq 1$, $V \in \mathbb{R}^{n \times p}$, and $p \leq m \leq n$.
Output: The p smallest eigenvalue $\{\lambda_1, \cdots, \lambda_p\} \subset \Lambda(A)$ and their corresponding eigenvector
  $X = [x_1, \cdots, x_p] \in \mathbb{R}^{n \times p}$.
1: $V_1 = \mathrm{MGS}(V)$; $k = 1$;
2: **repeat**
3:   compute $W_k = AV_k$;
4:   compute the interaction matrix $H_k = V_k{}^\top AV_k$;
5:   compute the p smallest eigenpairs $(\lambda_{k,i}, y_{k,i})$ of H_k, for $i = 1, \cdots, p$;
6:   compute the p corresponding Ritz vectors $X_k = V_k Y_k$ where $Y_k = [y_{k,1}, \cdots y_{k,p}]$;
7:   compute the residuals $R_k = [r_{k,1}, \cdots, r_{k,p}] = W_k Y_k - X_k \, \mathrm{diag}(\lambda_{k,1}, \cdots, \lambda_{k,p})$;
8:   **if** convergence **then**
9:     Exit from this loop k;
10: **end if**
11: compute the new block $T_k = [t_{k,1}, \ldots, t_{k,p}]$ where $t_{k,i} = C_{k,i} r_{k,i}$;   //Davidson correction
12: $V_{k+1} = \mathrm{MGS}([V_k, T_k])$;
13: **if** $\dim(V_k) \leq m - p$, **then**
14:    $V_{k+1} = \mathrm{MGS}([V_k, T_k])$;
15: **else**
16:    $V_{k+1} = \mathrm{MGS}([X_k, T_k])$;
17: **end if**
18: $k = k + 1$;
19: **until** Convergence
20: $X = X_k$; $\lambda_i = \lambda_{k,i}$ for $i = 1, \cdots, p$.

Theorem 11.6 (Convergence of the Block Davidson methods) *In Algorithm 11.11, let $\mathcal{V}_k$ be the subspace spanned by the columns of V_k. Under the assumption that*

$$x_{k,i} \in V_{k+1}, \quad for \; i = 1, \ldots, p, \; and \; k \in \mathbb{N}, \tag{11.46}$$

the sequence $\{\lambda_{k,i}\}_{k \geq 1}$ is nondecreasing and convergent for $i = 1, \ldots, p$. Moreover, if in addition,

1. there exist K_1, $K_2 > 0$ such that for any $k \geq 1$ and for any vector $v \in \mathcal{V}_k^\perp$, the set of matrices $\{C_{k,i}\}_{k \geq 1}$ satisfies,

$$K_1 \|v\|^2 \leq v^\top C_{k,i} v \leq K_2 \|v\|^2, and \tag{11.47}$$

2. for any $i = 1, \ldots, p$ and $k \geq 1$,

$$(I - V_k V_k^\top) C_{k,i} (A - \lambda_{k,i} I) x_{k,i} \in V_{k+1}, \tag{11.48}$$

then $\lim_{k \to \infty} \lambda_{k,i}$ is an eigenvalue of A and the elements in $\{x_{k,i}\}_{k \geq 1}$ yield the corresponding eigenvectors.

Proof See [36].

This theorem provides a convergence proof of the single vector version of the Davidson method expressed in Algorithm 11.10, as well as the LANCZOS2 method since in this case $C_{k,i} = I$, even for the block version.

11.4.3 Types of Correction Steps

To motivate the distinct approaches for the correction Step 12 in Algorithm 11.10, let us assume that a vector x of unit norm approximates an unknown eigenvector $(x + y)$ of A, where y is chosen orthogonal to x. The quantity $\lambda = \rho(x)$ (where $\rho(x) = x^\top A x$ denotes the Rayleigh quotient of x) approximates the eigenvalue $\lambda + \delta$ which corresponds to the eigenvector $x + y : \lambda + \delta = \rho(x + y) = \frac{(x+y)^\top A(x+y)}{\|x+y\|^2}$. The quality of the initial approximation is measured by the norm of the residual $r = Ax - \lambda x$. Since the residual $r = (I - xx^\top)Ax$ is orthogonal to the vector x, then if we denote by θ the angle $\angle(x, x + y)$, and t is the orthogonal projection of x onto $x + y$ and $z = x - t$, then we have:

$$\|x+y\|^2 = 1 + \tan^2 \theta,$$
$$\|y\| = \tan \theta,$$
$$\|z\| = \sin \theta,$$
$$\|t\| = \cos \theta,$$

where x is of unit 2-norm.

Proposition 11.2 *Using the previous notations, the correction δ to the approximation λ of an eigenvalue, and the orthogonal correction y for the corresponding approximate eigenvector x, satisfy:*

$$\begin{cases} (A - \lambda I)y = -r + \delta(x + y), \\ y \perp x, \end{cases} \tag{11.49}$$

for which the following inequalities hold :

$$|\delta| \leq 2\|A\| \tan^2 \theta, \tag{11.50}$$

$$\|r\| \leq 2\|A\| \tan \theta \left(\frac{\sin \theta}{\cos^2 \theta} + 1 \right). \tag{11.51}$$

Proof Equation (11.49) is directly obtained from

$$A(x + y) = (\lambda + \delta)(x + y).$$

Moreover

$$\begin{aligned}
\lambda + \delta &= \rho(t), \\
&= (1 + \tan^2 \theta)\, t^\top A t, \\
&= (1 + \tan^2 \theta)(x - z)^\top A(x - z), \\
&= (1 + \tan^2 \theta)(\lambda - 2x^\top A z + z^\top A z).
\end{aligned}$$

Since z is orthogonal to the eigenvector t, we obtain $x^\top A z = (t + z)^\top A z = z^\top A z$ and therefore

$$\begin{aligned}
\delta &= -z^\top A z + (\tan^2 \theta)(\lambda - z^\top A z), \\
&= (\tan^2 \theta)(\lambda - \rho(z)), \quad \text{or} \\
|\delta| &\le 2(\tan^2 \theta)\, \|A\|,
\end{aligned}$$

which proves (11.50). Inequality (11.51) is a straightforward consequence of (11.49) and (11.50).

Since the system (11.49) is not easy to solve, it is replaced by an approximate one obtained by deleting the nonlinear term $\delta(x + y)$ whose norm is of order $O(\theta^2)$ whereas $\|r\| = O(\theta)$. Thus, the problem reduces to solving a linear system under constraints:

$$\begin{cases}
(A - \lambda I)y = -r, \\
y \perp x,
\end{cases} \tag{11.52}$$

Unfortunately, this system has no solution except the trivial solution when λ is an eigenvalue. Three options are now considered to overcome this difficulty.

First Option: Using a Preconditioner:

The first approach consists of replacing Problem (11.52) by

$$(M - \lambda I)y = -r, \tag{11.53}$$

where M is chosen as a preconditioner of A. Then, the solution y is projected onto $\mathcal{V}_k^\perp$. Originally in [34], Davidson considered $M = \mathrm{diag}(A)$ to solve (11.53). This method is called the *Davidson method*.

Second Approach: Using an Iterative Solver

When λ is close to an eigenvalue, solve approximately

$$(A - \lambda I)y = -r, \tag{11.54}$$

by an iterative method with a fixed or variable number of iterations. In such a situation, the method is similar to inverse iteration in which the ill-conditioning of the system provokes an error which is in the direction of the desired eigenvector. The iterative solver, however, must be capable of handling symmetric indefinite systems.

Third Approach: Using Projections

By considering the orthogonal projector $P = I - V_k V_k^\top$ with respect to $\mathcal{V}_k^\perp$, the system to be solved becomes

$$\begin{cases} P(A - \lambda I)y = -r, \\ Py = y, \end{cases} \tag{11.55}$$

Note that the system matrix can be also expressed as $P(A - \lambda I)P$ since $Py = y$. This system is then solved via an iterative scheme that accommodates symmetric indefiniteness. The Jacobi-Davidson method [39] follows this approach.

The study [40] compares the second approach which can be seen as the Rayleigh Quotient method, and the Newton Grassmann method which corresponds to this approach. The study concludes that the two correction schemes have comparable behavior. In [40], the authors also provide a stopping criterion for controlling the inner iterations of an iterative solver for the correction vectors.

This approach is studied in the section devoted to the Trace Minimization method as an eigensolver of the generalized symmetric eigenvalue problem (see Sect. 11.5). Also, in this section we describe the similarity between the Jacobi-Davidson method and the method that preceded it by almost two decades: the Trace Minimization method.

11.5 The Trace Minimization Method for the Symmetric Generalized Eigenvalue Problem

The generalized eigenvalue problem

$$Ax = \lambda Bx, \tag{11.56}$$

where A and B are $n \times n$ real symmetric matrices with B being positive definite, arises in many applications, most notably in structural mechanics [41, 42] and plasma physics [43, 44]. Usually, A and B are large, sparse, and only a few of the eigenvalues and the associated eigenvectors are desired. Because of the size of the problem, methods that rely only on operations like matrix-vector multiplications, inner products, and vector updates are usually considered.

Many methods fall into this category (see, for example [45, 46]). The basic idea in all of these methods is building a sequence of subspaces that, in the limit, contain the desired eigenvectors. Most of the early methods iterate on a single vector, i.e. using one-dimensional subspaces, to compute one eigenpair at a time. If, however, several eigenpairs are needed, a deflation technique is frequently used. Another alternative is to use block analogs of the single vector methods to obtain several eigenpairs simultaneously. The well-known *simultaneous iteration* [47], or *subspace iteration* [1] (originally described in [2]) has been extensively studied in the 1960s and the 1970s [45, 47–50].

In this chapter we do not include practical eigensolvers based on contour integration. The first of such methods was proposed in [51], and later it was enhanced as a subspace iteration similar to the trace minimization scheme, or Davidson-type trace minimization, resulting in the Hermitian eigensolver FEAST, e.g. see [52].

Let A be symmetric positive definite and assume that we seek the smallest p eigenpairs, where $p \ll n$. In simultaneous iteration, the sequence of subspaces of dimension p is generated by the following recurrence:

$$X_{k+1} = A^{-1}BX_k, \qquad k = 0, 1, \ldots, \tag{11.57}$$

where X_0 is an $n \times p$ matrix of full rank. The eigenvectors of interest are magnified at each iteration step, and will eventually dominate X_k. The downside of simultaneous iteration is that linear systems of the form $Ax = b$ have to be solved repeatedly which is a significant challenge for large problems. Solving these linear systems inexactly often compromises global convergence. A variant of simultaneous iteration that avoids this difficulty is called the *trace minimization method*. This method was proposed in 1982 [37]; see also [38]. Let X_k be the current approximation to the eigenvectors corresponding to the p smallest eigenvalues where $X_k^\top BX_k = I_p$. The idea of the trace minimization scheme is to find a correction term Δ_k that is B-orthogonal to X_k such that

$$\operatorname{tr}(X_k - \Delta_k)^\top A(X_k - \Delta_k) < \operatorname{tr}(X_k^\top AX_k).$$

It follows that, for any B-orthonormal basis X_{k+1} of the new subspace $\operatorname{span}\{X_k - \Delta_k\}$, we have

$$\operatorname{tr}(X_{k+1}^\top AX_{k+1}) < \operatorname{tr}(X_k^\top AX_k),$$

i.e. $\operatorname{span}\{X_k - \Delta_k\}$ gives rise to a better approximation of the desired eigenspace than $\operatorname{span}\{X_k\}$. This trace reduction property can be maintained without solving any linear system exactly.

The trace minimization method can be accelerated via shifting strategies. The introduction of shifts, however, may compromise the robustness of the trace minimization scheme. Various techniques have been developed to prevent *unstable convergence* (see the section on "Randomization" for details). A simple way to get around this difficulty is to utilize expanding subspaces. This, in turn, places the trace minimization method into a class of methods that includes the Lanczos method [53], Davidson's method [34], and the Jacobi-Davidson method [39, 54, 55].

The Lanczos method has become popular after the ground-breaking analysis by Paige [20], resulting in many practical algorithms [17, 56–58] (see [10, 59] for an overview). The original Lanczos algorithm was developed for handling the standard eigenvalue problem only, i.e. $B = I$. Extensions to the generalized eigenvalue problem [60–62] require solving a linear system of the form $Bx = b$ at each iteration step, or factorizing matrices of the form $(A - \sigma B)$ during each iteration. Davidson's method, which can be regarded as a preconditioned Lanczos method, was intended to be a practical method for standard eigenvalue problems in quantum chemistry where the matrices involved are diagonally dominant. In the past two decades, Davidson's method has gone through a series of significant improvements [35, 36, 63–65]. A development is the Jacobi-Davidson method [39], published in 1996, which is a variant of Davidson's original scheme and the well-known Newton's method. The Jacobi-Davidson algorithm for the symmetric eigenvalue problem may be regarded as a generalization of the trace minimization scheme (which was published 15 years earlier) that uses expanding subspaces. Both utilize an idea that dates back to Jacobi [66]. As we will see later, the current Jacobi-Davidson scheme can be further improved by the techniques developed in the trace minimization method.

Throughout this section, the eigenpairs of (11.56) are denoted by (x_i, λ_i), $1 \leqslant i \leqslant n$, with the eigenvalues arranged in ascending order.

11.5.1 Derivation of the Trace Minimization Algorithm

Here, we derive the trace minimization method originally presented in [37]. We assume that A is positive definite, otherwise problem (11.56) can be replaced by

$$(A - \mu B)x = (\lambda - \mu)Bx$$

with $\mu < \lambda_1 < 0$, that ensures a positive definite $(A - \mu B)$.

The trace minimization method is motivated by the following theorem.

Theorem 11.7 (Beckenbach and Bellman [67], Sameh and Wisniewski [37]). *Let A and B be as given in Problem (11.56), and let X^* be the set of all $n \times p$ matrices X for which $X^\top BX = I_p$, $1 \leqslant p \leqslant n$. Then*

$$\min_{X \in X^*} \operatorname{tr}(X^\top AX) = \sum_{i=1}^{p} \lambda_i. \tag{11.58}$$

where $\lambda_1 \leqslant \lambda_2 \leqslant \cdots \leqslant \lambda_n$ are the eigenvalues of Problem (11.56). The equality holds if and only if the columns of the matrix X, which achieves the minimum, span the eigenspace corresponding to the smallest p eigenvalues.

If we denote by E/F the matrix EF^{-1} and $\mathscr{X}$ the set of all $n \times p$ matrices of full rank, then (11.58) is equivalent to

$$\min_{X \in \mathscr{X}} \operatorname{tr}\left(\frac{X^\top AX}{X^\top BX}\right) = \sum_{k=1}^{p} \lambda_i.$$

$X^\top AX/X^\top BX$ is called the generalized Rayleigh quotient. Most of the early methods that compute a few of the smallest eigenvalues are devised explicitly or implicitly by reducing the generalized Rayleigh quotient step by step. A simple example is the simultaneous iteration scheme for a positive definite matrix A where the current approximation X_k is updated by (11.58). It can be shown by the Courant-Fischer theorem [1] and the Kantorovič inequality [68, 69] that

$$\lambda_i\left(\frac{X_{k+1}^\top AX_{k+1}}{X_{k+1}^\top BX_{k+1}}\right) \leqslant \lambda_i\left(\frac{X_k^\top AX_k^\top}{X_k^\top BX_k}\right), \qquad 1 \leqslant i \leqslant p. \tag{11.59}$$

The equality holds only when X_k is already an eigenspace of Problem (11.56).

Algorithm 11.12 Simultaneous iteration for the generalized eigenvalue problem.

1: Choose a block size $s \geqslant p$ and an $n \times s$ matrix V_1 of full rank such that $V_1^\top BV_1 = I_s$.
2: **do** $k = 1, 2, \ldots$ until convergence,
3:   Compute $W_k = AV_k$ and the interaction matrix $H_k = V_k^\top W_k$.
4:   Compute the eigenpairs (Y_k, Θ_k) of H_k. The eigenvalues are arranged in ascending order and the eigenvectors are chosen to be orthogonal.
5:   Compute the corresponding Ritz vectors $X_k = V_k Y_k$.
6:   Compute the residuals $R_k = W_k Y_k - BX_k \Theta_k$.
7:   Test for convergence.
8:   Solve the linear system

$$AZ_{k+1} = BX_k, \tag{11.60}$$

   by an iterative method.
9:   B-orthonormalize Z_{k+1} into V_{k+1}.
10: **end**

In [37], simultaneous iteration is derived in a way that allows exploration of the trace minimization property explicitly. At each iteration step, the previous approximation X_k, which satisfies $X_k^\top BX_k = I_s$ and $X_k^\top AX_k = \Theta_k$, where Θ_k is diagonal (with the ith element denoted by $\theta_{k,i}$), is corrected with Δ_k that is obtained by

$$\begin{aligned} \text{minimizing} \quad & \operatorname{tr}(X_k - \Delta_k)^\top A(X_k - \Delta_k), \\ \text{subject to} \quad & X_k^\top B\Delta_k = 0. \end{aligned} \tag{11.61}$$

As a result, the matrix $Z_{k+1} = X_k - \Delta_k$ always satisfies

$$\operatorname{tr}(Z_{k+1}^\top AZ_{k+1}) \leqslant \operatorname{tr}(X_k^\top AX_k), \tag{11.62}$$

and

$$Z_{k+1}^\top BZ_{k+1} = I_s + \Delta_k^\top B\Delta_k, \tag{11.63}$$

which guarantee that

$$\operatorname{tr}(X_{k+1}^\top A X_{k+1}) \leqslant \operatorname{tr}(X_k^\top A X_k) \tag{11.64}$$

for any B-orthonormal basis X_{k+1} of the subspace $\operatorname{span}\{Z_{k+1}\}$. The equality in (11.64) holds only when $\Delta_k = 0$, i.e. X_k spans an eigenspace of (11.56) (see Theorem 11.10 for details).

Using Lagrange multipliers, the solution of the minimization problem (11.61) can be obtained by solving the saddle-point problem

$$\begin{pmatrix} A & BX_k \\ X_k^\top B & 0 \end{pmatrix} \begin{pmatrix} \Delta_k \\ L_k \end{pmatrix} = \begin{pmatrix} AX_k \\ 0 \end{pmatrix}, \tag{11.65}$$

where L_k represents the Lagrange multipliers. Several methods may be used to solve (11.65) using either direct methods via the Schur complement or via preconditioned iterative schemes, e.g. see the detailed survey [70]. In [37], (11.65) is further reduced to solving the following positive-semidefinite system

$$(PAP)\Delta_k = PAX_k, \tag{11.66}$$

where P is the orthogonal projector $P = I - BX_k(X_k^\top B^2 X_k)^{-1} X_k^\top B$. This system is solved by the conjugate gradient method (CG) in which zero is chosen as the initial iterate so that the linear constraint $X_k^\top B \Delta_k^{(\ell)} = 0$ is automatically satisfied for any intermediate $\Delta_k^{(\ell)}$. This results in the following basic trace minimization algorithm.

Algorithm 11.13 The basic trace minimization algorithm.

1: Choose a block size $s \geqslant p$ and an $n \times s$ matrix V_1 of full rank such that $V_1^\top B V_1 = I_s$.
2: **do** $k = 1, 2, \ldots$ until convergence,
3:　　Compute $W_k = A V_k$ and the interaction matrix $H_k = V_k^\top W_k$.
4:　　Compute the eigenpairs (Y_k, Θ_k) of H_k. The eigenvalues are arranged in ascending order and the eigenvectors are chosen to be orthogonal.
5:　　Compute the corresponding Ritz vectors $X_k = V_k Y_k$.
6:　　Compute the residuals $R_k = A x_k - B X_k \Theta_k = W_k Y_k - B X_k \Theta_k$.
7:　　Test for convergence.
8:　　Solve the positive-semidefinite linear system (11.66) approximately via the CG scheme.
9:　　B-orthonormalize $X_k - \Delta_k$ into V_{k+1}.
10: **end**

From now on, we will refer to the linear system (11.66) in step 8 as the *inner system(s)*. It is easy to see that the exact solution of the inner system is

$$\Delta_k = X_k - A^{-1} B X_k (X_k^\top B A^{-1} B X_k)^{-1}, \tag{11.67}$$

thus the subspace spanned by $X_k - \Delta_k$ is the same subspace spanned by $A^{-1} B X_k$. In other words, if the inner system (11.66) is solved exactly at each iteration step, the trace minimization algorithm above is mathematically equivalent to simultaneous iteration. As a consequence, global convergence of the basic trace minimization algorithm follows exactly from that of simultaneous iteration.

Theorem 11.8 ([1, 4, 37]) *Let A and B be positive definite and let $s \geqslant p$ be the block size such that the eigenvalues of Problem (11.56) satisfy $0 < \lambda_1 \leqslant \lambda_2 \leqslant \cdots \leqslant \lambda_s < \lambda_{s+1} \leqslant \cdots \leqslant \lambda_n$. Let also the initial iterate X_0 be chosen such that it has linearly independent columns and is not deficient in any eigen-component associated with the p smallest eigenvalues. Then the ith column of X_k, denoted by $x_{k,i}$, converges to the eigenvector x_i corresponding to γ_i for $i = 1, 2, \ldots, p$ with an asymptotic rate of convergence bounded by $\lambda_i / \lambda_{s+1}$. More specifically, at each step, the error*

$$\phi_i = (x_{k,i} - x_i)^\top A (x_{k,i} - x_i) \tag{11.68}$$

is reduced asymptotically be a factor of $(\lambda_i / \lambda_{s+1})^2$.

The main difference between the trace minimization algorithm and simultaneous iteration is in step 8. If both (11.60) and (11.66) are solved via the CG scheme exactly, the performance of either algorithm is comparable in terms of time consumed, as observed in practice. The additional cost in performing the projection P at each CG step (once rather than twice) is not high because the block size s is usually small, i.e. $s \ll n$. This additional cost is sometimes compensated for by the fact that PAP, when it is restricted to the subspace $\{v \in R^n | Pv = v\}$, is better conditioned than A as will be seen in the following theorem.

Theorem 11.9 *Let A and B be as given in Theorem 11.8 and P be given as in (11.66), and let v_i, μ_i, $1 \leqslant i \leqslant n$ be the eigenvalues of A and PAP arranged in ascending order, respectively. Then, we have*

$$0 = \mu_1 = \mu_2 = \cdots = \mu_s < v_1 \leqslant \mu_{s+1} \leqslant \mu_{s+2} \leqslant \cdots \leqslant \mu_n \leqslant v_n.$$

Proof The proof is a straightforward consequence of the Courant-Fischer theorem [1], and hence omitted.

11.5.2 Practical Considerations

In practice, however, the inner systems (11.66) are always solved approximately, particularly for large problems. Note that the error (11.68) in the ith column of X_k is reduced asymptotically by a factor of $(\lambda_i / \lambda_{s+1})^2$ at each iteration step. Thus, we should not expect high accuracy in the early Ritz vectors even if the inner systems are solved to machine precision. Further, convergence of the trace minimization scheme is guaranteed if a constant relative residual tolerance is used for the inner system (11.66) in each outer iteration.

A Convergence Result

We prove convergence of the trace minimization algorithm under the assumption that the inner systems in (11.66) are solved inexactly. We assume that, for each i, $1 \leqslant i \leqslant s$, the ith inner system in (11.66) is solved approximately by the CG scheme with zero as the initial iterate such that the 2-norm of the residual is reduced by a factor $\gamma < 1$. The computed correction matrix will be denoted by $\Delta_k^c = (d_{k,1}^c d_{k,2}^c, \ldots, d_{k,s}^c)$ to distinguish it from the exact solution $\Delta_k = (d_{k,1}, d_{k,2}, \ldots, d_{k,s})$ of (11.66).

We begin the convergence proof with two lemmas. We first show that, in each iteration, the columns of $X_k - \Delta_k^c$ are linearly independent, and the sequence $\{X_k\}_0^\infty$ in the trace minimization algorithm is well-defined. In the second, we show that the computed correction matrix Δ_k^c satisfies

$$\mathrm{tr}(X_k - \Delta_k^c)^\top A(X_k - \Delta_k^c) \leqslant \mathrm{tr}(X_k^\top A X_k).$$

This assures that, no matter how prematurely the CG process is terminated, $\mathrm{tr}(X_k^\top A X_k)$ always forms a decreasing sequence bounded from below by $\Sigma_{i=1}^s \lambda_i$.

Lemma 11.2 *For each $k = 0, 1, 2, \ldots$, $Z_{k+1} = X_k - \Delta_k^c$ is of full column rank.*

Proof Since $d_{k,i}^c$ is an intermediate approximation obtained from the CG process, there exists a polynomial $p(t)$ such that

$$d_{k,i}^c = p(PAP)(PAx_{k,i}),$$

where $x_{k,i}$ is the ith column of X_k and P is the projector in (11.66). As a consequence, for each i, $d_{k,i}^c$ is B-orthogonal to X_k, i.e. $X_k^\top B d_{k,i}^c = 0$. Thus the matrix

$$Z_{k+1}^\top B Z_{k+1} = I_s + (\Delta_k^c)^\top B \Delta_k^c$$

is nonsingular, and Z_{k+1} is of full column rank. $\square$

Lemma 11.3 *Suppose that the inner systems in (11.66) are solved by the CG scheme with zero as the initial iterate. Then, for each i, $(x_{k,i} - d_{k,i}^{(\ell)})^\top A(x_{k,i} - d_{k,i}^{(\ell)})$ decreases monotonically with respect to step ℓ of the CG scheme.*

Proof The exact solution of the inner system (11.66) is given by

$$\Delta_k = X_k - A^{-1}BX_k(X_k^\top BA^{-1}BX_k)^{-1}$$

for which $P\Delta_k = \Delta_k$. For each i, $1 \leqslant i \leqslant s$, the intermediate $d_{k,i}^{(\ell)}$ in the CG process also satisfies $Pd_{k,i}^{(\ell)} = d_{k,i}^{(\ell)}$. Thus, it follows that

$$\begin{aligned}
(d_{k,i}^{(\ell)} - d_{k,i})^\top PAP(d_{k,i}^{(\ell)} - d_{k,i}) &= (d_{k,i}^{(\ell)} - d_{k,i})^\top A(d_{k,i}^{(\ell)} - d_{k,i}) \\
&= (x_{k,i} - d_{k,i}^{(\ell)})^\top A(x_{k,i} - d_{k,i}^{(\ell)}) - e_i^\top(X_k^\top BA^{-1}BX_k)^{-1}e_i.
\end{aligned}$$

Since the CG process minimizes the *PAP*-norm of the error $\tilde{d}_{k,i}^{(\ell)} = d_{k,i}^{(\ell)} - d_{k,i}$ on the expanding Krylov subspace [37], both $(d_{k,i}^{(\ell)} - d_{k,i})^\top PAP(d_{k,i}^{(\ell)} - d_{k,i})$ and $(x_{k,i} - d_{k,i}^{(\ell)})^\top A(x_{k,i} - d_{k,i}^{(\ell)})$ decrease monotonically. $\qquad\square$

Theorem 11.10 *Let X_k, Δ_k^c, and Z_{k+1} be as given in Lemma 11.2. Then $\lim_{k\to\infty} \Delta_k^c = 0$.*

Proof By the definition of Δ_k^c, we have

$$Z_{k+1}^\top B Z_{k+1} = I_s + (\Delta_k^c)^\top B \Delta_k^c \triangleq I_s + T_k.$$

From the spectral decomposition,

$$Z_{k+1}^\top B Z_{k+1} = U_{k+1} D_{k+1}^2 U_{k+1}^\top,$$

where U_{k+1} is an $s \times s$ orthogonal matrix and $D_{k+1}^2 = \mathrm{diag}(\delta_1^{(k+1)}, \delta_2^{(k+1)}, \ldots, \delta_s^{(k+1)})$, we see that $\delta_i^{(k+1)} = 1 + \lambda_i(T_k) \geqslant 1$. Further, from the definition of X_{k+1}, there exists an orthogonal matrix V_{k+1} for which

$$X_{k+1} = Z_{k+1} \cdot U_{k+1} D_{k+1}^{-1} V_{k+1}.$$

Denoting by $z_i^{(k+1)}$ the diagonal elements of the matrix $U_{k+1}^\top Z_{k+1}^\top A Z_{k+1} U_{k+1}$, it follows that

$$
\begin{aligned}
\mathrm{tr}\left(X_{k+1}^\top A X_{k+1}\right) &= \mathrm{tr}\left(D_{k+1}^{-1}\left(U_{k+1}^\top Z_{k+1}^\top A Z_{k+1} U_{k+1}\right) D_{k+1}^{-1}\right), \\
&= \frac{z_1^{(k+1)}}{\delta_1^{(k+1)}} + \frac{z_2^{(k+1)}}{\delta_2^{(k+1)}} + \cdots + \frac{z_s^{(k+1)}}{\delta_s^{(k+1)}}, \\
&\leqslant z_1^{(k+1)} + z_2^{(k+1)} + \cdots + z_s^{(k+1)}, \\
&= \mathrm{tr}\left(Z_{k+1}^\top A Z_{k+1}\right), \\
&\leqslant \mathrm{tr}\left(X_k^\top A X_k\right),
\end{aligned}
$$

which implies that

$$\cdots \geqslant \mathrm{tr}\left(X_k^\top A X_k\right) \geqslant \mathrm{tr}\left(Z_{k+1}^\top A Z_{k+1}\right) \geqslant \mathrm{tr}\left(X_{k+1}^\top A X_{k+1}\right) \geqslant \cdots.$$

Since the sequence is bounded from below by $\Sigma_{i=1}^s \lambda_i$, it converges to a positive number $t \geqslant \Sigma_{i=1}^s \lambda_i$. Moreover, the two sequences

$$\frac{z_1^{(k+1)}}{\delta_1^{(k+1)}} + \frac{z_2^{(k+1)}}{\delta_2^{(k+1)}} + \cdots + \frac{z_s^{(k+1)}}{\delta_s^{(k+1)}}, \qquad k = 1, 2, \ldots$$

and

$$z_1^{(k+1)} + z_2^{(k+1)} + \cdots + z_s^{(k+1)}, \qquad k = 1, 2 \ldots$$

also converge to t. Therefore,

$$\left(\frac{z_1^{(k+1)} \lambda_1(T_k)}{1 + \lambda_1(T_k)} \right) + \left(\frac{z_2^{(k+1)} \lambda_2(T_k)}{1 + \lambda_2(T_k)} \right) + \cdots + \left(\frac{z_s^{(k+1)} \lambda_s(T_k)}{1 + \lambda_s(T_k)} \right) \to 0.$$

Observing that for any i, $1 \leqslant i \leqslant s$,

$$z_i^{(k+1)} \geqslant \lambda_1 \left(U_{k+1}^\top Z_{k+1}^\top A Z_{k+1} U_{k+1} \right),$$

$$= \lambda_1 \left(Z_{k+1}^\top A Z_{k+1} \right),$$

$$= \min_{y \neq 0} \frac{y^\top Z_{k+1}^\top A Z_{k+1} y}{y^\top y},$$

$$= \min_{y \neq 0} \left(\frac{y^\top Z_{k+1}^\top A Z_{k+1} y}{y^\top Z_{k+1}^\top B Z_{k+1} y} \right) \cdot \left(\frac{y^\top Z_{k+1}^\top B Z_{k+1} y}{y^\top y} \right),$$

$$\geqslant \left(\frac{y^\top Z_{k+1}^\top A Z_{k+1} y}{y^\top Z_{k+1}^\top B Z_{k+1} y} \right),$$

$$\geqslant \lambda_1(A, B),$$

$$> 0.$$

Hence, we have

$$\lambda_1(T_k) \to 0, \qquad i = 1, 2, \ldots, s,$$

i.e. $\lim_{k \to \infty} \Delta_k^c = 0.$ $\qquad\qquad\qquad\qquad\qquad\qquad\qquad\qquad\qquad\qquad\qquad\qquad\qquad$ $\square$

Theorem 11.11 *If for each $1 \leqslant i \leqslant s$, the CG process for the ith inner system* (11.66)

$$(PAP)d_{k,i} = PAx_{k,i}, \qquad d_{k,i}^\top BX_k = 0,$$

is terminated such that the 2-norm of the residual is reduced by a factor $\gamma < 1$, i.e.

$$\|PAx_{k,i} - (PAP)d_{k,i}^c\|_2 \leqslant \gamma \|PAx_{k,i}\|_2, \tag{11.69}$$

then the columns of X_k converge to the s eigenvectors of Problem 11.56.

Proof Condition (11.69) implies that

$$\|PAx_{k,i}\|_2 - \|PAd^c_{k,i}\|_2 \leqslant \gamma \|PAx_{k,i}\|_2,$$

and consequently

$$\|PAx_{k,i}\|_2 \leqslant \frac{1}{1-\gamma}\|PAd^c_{k,i}\|_2.$$

It follows from Theorem 11.10 that $\lim_{k\to\infty} PAX_k = 0$, i.e.

$$\lim_{k\to\infty}\left(AX_k - BX_k\left[\left(X_k^\top B^2 X_k\right)^{-1} X_k^\top BAX_k\right]\right) = 0.$$

In other words, $span\{X_k\}$ converges to an eigenspace of Problem 11.56.   □

Randomization

Condition (11.69) in Theorem 11.11 is not essential because the constant γ can
be arbitrarily close to 1. The only deficiency in Theorem 11.11) is that it does not
establish ordered convergence in the sense that the ith column of X_k converges to the
ith eigenvector of the problem. This is called *unstable convergence* in [4]. In practice,
roundoff errors turn unstable convergence into delayed stable convergence. In [4], a
randomization technique to prevent unstable convergence in simultaneous iteration
was introduced. Such an approach can be incorporated into the trace minimization
algorithm as well: *After step 8 of Algorithm 11.13, we append a random vector to
X_k and perform the Ritz processes 3 and 4 on the augmented subspace of dimension
$s + 1$. The extra Ritz pair is discarded after step 4.*

Randomization slightly improves the convergence of the first s Ritz pairs [47].
Since it incurs additional cost, it should be used only in the first few steps when a
Ritz pair is about to converge.

Terminating the CG Process

Theorem 11.11 gives a sufficient condition for the convergence of the trace minimization algorithm. However, the asymptotic rate of convergence of the trace minimization algorithm will be affected by the premature termination of the CG processes.
The algorithm behaves differently when the inner systems are solved inexactly. It
is not clear how the parameter γ should be chosen to avoid performing excessive
CG iterations while maintaining the asymptotic rate of convergence. In [37], the CG
process is terminated by a heuristic stopping strategy.

Let $d^{(\ell)}_{k,i}$ be the approximate solution at the ith step of the CG process for the ith
column of X_k and $d_{k,i}$ the exact solution, then the heuristic stopping strategy in [37]
can be outlined as follows:

1. From Theorem 11.8, it is reasonable to terminate the CG process for the ith column
   of Δ_k when the error

$$\varepsilon_{k,i}^{(\ell)} = \left[\left(d_{k,i}^{(\ell)} - d_{k,i} \right)^{\top} A \left(d_{k,i}^{(\ell)} - d_{k,i} \right) \right]^{1/2},$$

   is reduced by a factor of $\tau_i = \lambda_i/\lambda_{s+1}$, called the *error reduction factor*.
2. The quantity $\varepsilon_{k,i}^{(\ell)}$ can be estimated by

$$\left[\left(d_{k,i}^{(\ell)} - d_{k,i}^{(l+1)} \right)^{\top} A \left(d_{k,i}^{(\ell)} - d_{k,i}^{(l+1)} \right) \right]^{1/2},$$

   which is readily available from the CG process.
3. The error reduction factor $\tau_i = \lambda_i/\lambda_{s+1}$, $1 \leqslant i \leqslant s$, can be estimated by the
   ratio of the Ritz values $\tilde{\tau}_{k,i} = \theta_{k,i}/\theta_{k,s+1}$. Since $\theta_{k,s+1}$ is not available, $\theta_{k-1,s}$ is
   used instead and is fixed after a few steps because it will eventually converge to
   λ_s rather than λ_{s+1}.

11.5.3 Acceleration Techniques

For problems in which the desired eigenvalues are poorly separated from the remaining part of the spectrum, the algorithm converges slowly. Like other inverse iteration schemes, the trace minimization algorithm can be accelerated by shifting. Actually, the formulation of the trace minimization algorithm makes it easier to incorporate shifts. For example, if the Ritz pairs (x_i, θ_i), $1 \leqslant i \leqslant i_0$, have been accepted as eigenpairs and $\theta_{i_0} < \theta_{i_0+1}$, then θ_{i_0} can be used as a shift parameter for computing subsequent eigenpairs. Due to the deflation effect, the linear systems

$$[P(A - \theta_{i_0} B)P]d_{k,i} = PAx_{k,i}, \qquad X_k^{\top} Bd_{k,i} = 0, \qquad i_0 + 1 \leqslant i \leqslant s,$$

are consistent and can still be solved by the CG scheme. Moreover, the trace reduction property still holds. In the following, we introduce two more efficient shifting techniques which improve further the performance of the trace minimization algorithm.

Single Shift

We know from Sect. 11.5.1 that global convergence of the trace minimization algorithm follows from the monotonic reduction of the trace, which in turn depends on the positive definiteness of A. A simple and robust shifting strategy would be finding a scalar σ close to λ_1 from below and replace A by $(A - \sigma B)$ in step 8 of Algorithm 11.13. After the first eigenvector has converged, find another σ close to λ_2 from below and continue until all the desired eigenvectors are obtained. If both A and B are explicitly available, it is not hard to find a σ satisfying $\sigma \leqslant \lambda_1$.

In the trace minimization algorithm, the subspace spanned by X_k converges to the invariant subspace V_s corresponding to the s smallest eigenvalues. If the subspace spanned by X_k is close enough to V_s, a reasonable bound for the smallest eigenvalue can be obtained. More specifically, let Q be a B-orthonormal matrix obtained by appending $n - s$ columns to X_k, i.e. $Q = (X_k, Y_k)$ and $Q^\top B Q = I_n$. Then Problem (11.56) is reduced to the standard eigenvalue problem

$$(Q^\top A Q)u = \lambda u. \tag{11.70}$$

Since

$$Q^\top A Q = \begin{pmatrix} \Theta_k & X_k^\top A Y_k \\ Y_k^\top A X_k & Y_k^\top A Y_k \end{pmatrix} = \begin{pmatrix} \Theta_k & C_k^\top \\ C_k & Y_k^\top A Y_k \end{pmatrix}, \tag{11.71}$$

by the Courant-Fischer theorem, we have

$$\lambda_1 \geq \lambda_{\min} \begin{pmatrix} \Theta_k & 0 \\ 0 & Y_k^\top A Y_k \end{pmatrix} + \lambda_{\min} \begin{pmatrix} 0 & C_k^\top \\ C_k & 0 \end{pmatrix}$$

$$\geq \min \left\{ \theta_1, \lambda_1(Y_k^\top A Y_k) \right\} - \|C_k\|_2.$$

Similarly [1], it is easy to show that $\|C_k\|_2 = \|R_k\|_{B^{-1}}$, in which $R_k = AX_k - BX\Theta_k$ is the residual matrix. If

$$\theta_{k,1} \leq \lambda_1 \left(Y_k^\top A Y_k \right), \tag{11.72}$$

we get

$$\lambda_1 \geq \theta_{k,1} - \|R_k\|_{B^{-1}}. \tag{11.73}$$

In particular, if (11.71) holds for the orthonormal complement of x_{k+1}, we have

$$\lambda_1 \geq \theta_{k,1} - \|r_{k,1}\|_{B^{-1}}. \tag{11.74}$$

This heuristic bound for the smallest eigenvalue suggests the following shifting strategy (we denote $-\infty$ by λ_0). If the first $i_0 \geq 0$ eigenvalues have converged, use $\sigma = \max\{\lambda_{i_0}, \theta_{k,i_0+1} - \|r_{k,i_0+1}\|_{B^{-1}}\}$ as the shift parameter. If θ_{k,i_0+1} lies in a cluster, replace the B^{-1}-norm of r_{k,i_0+1} by the B^{-1}-norm of the residual matrix corresponding to the cluster containing θ_{k,i_0+1}.

Multiple Dynamic Shifts

In [37], the trace minimization algorithm is accelerated with a more aggressive shifting strategy. At the beginning of the algorithm, a single shift is used for all the columns of X_k. As the algorithm proceeds, multiple shifts are introduced dynamically and the CG process is modified to handle possible breakdown. This shifting strategy is motivated by the following theorem.

Theorem 11.12 [1] *For an arbitrary nonzero vector u and scalar σ, there is an eigenvalue λ of (11.56) such that*

$$[\lambda - \sigma] \leqslant \|(A - \sigma B)u\|_{B^{-1}}/\|Bu\|_{B^{-1}}.$$

We know from the Courant-Fischer theorem that the targeted eigenvalue λ_i is always below the Ritz value $\theta_{k,i}$. Further, from Theorem 11.12, if $\theta_{k,i}$ is already very close to the targeted eigenvalue λ_i, then λ_i must lie in the interval $[\theta_{k,i} - \|r_{k,i}\|_{B^{-1}}, \theta_{k,i}]$. This observation leads to the following shifting strategy for the trace minimization algorithm. At step k of the outer iteration, the shift parameters $\sigma_{k,i}$, $1 \leqslant i \leqslant s$, are determined by the following rules (here, $\lambda_0 = -\infty$ and the subscript k is dropped for the sake of simplicity):

1. *If the first i_0, $i_0 \geqslant 0$, eigenvalues have converged, choose*

$$\sigma_{k,i_0+1} = \begin{cases} \theta_{i_0+1} & \text{if } \theta_{i_0+1} + \|r_{i_0+1}\|_{B^{-1}} \leqslant \theta_{i_0+2} - \|r_{i_0+2}\|_{B^{-1}}, \\ \max\{\theta_{i_0+1} - \|r_{i_0+1}\|_{B^{-1}}, \lambda_{i_0}\} & \text{otherwise.} \end{cases}$$

2. *For any other column j, $i_0 + 1 < j \leqslant p$, choose the largest θ_ℓ such that*

$$\theta_l < \theta_j - \|r_j\|_{B^{-1}}$$

   as the shift parameter σ_j. If no such θ_l exists, use θ_{i_0+1} instead.
3. *Choose $\sigma_i = \theta_i$ if θ_{i-1} has been used as the shift parameter for column $(i-1)$ and*

$$\theta_i < \theta_{i+1} - \|r_{i+1}\|_{B^{-1}}.$$

4. *Use σ_{i_0+1} as the shift parameter for other columns if any.*

Solving the Inner Systems

With the multiple shifts, the inner systems in (11.66) become

$$[P(A - \sigma_{k,i} B)P]d_{k,i} = PAx_{k,i}, \qquad X_k^\top B d_{k,i} = 0, \qquad 1 \leqslant i \leqslant s \qquad (11.75)$$

with $P = I - BX_k(X_k^\top B^2 X_k)^{-1}X_k^\top B$. Clearly, the linear systems could be indefinite, and the CG process for such systems may break down. A simple way to get around this problem is either by using another solver such as MINRES (e.g. see [71] or [72]), or terminating the CG process when a near breakdown is detected. In [37], the CG process is also terminated when the error $(x_{k,i}^{(\ell)} - d_{k,i}^{(\ell)})^\top A(x_{k,i}^{(\ell)} - d_{k,i}^{(\ell)})$, increases by a small factor. This helps maintain global convergence which is not guaranteed in the presence of shifting.

11.5.4 A Davidson-Type Extension

The shifting strategies described in the previous section, improve the performance of the trace minimization algorithm considerably. Although the randomization technique, the shifting strategy, and the roundoff error actually make the algorithm surprisingly robust for a variety of problems, further measures to guard against unstable convergence are necessary for problems in which the desired eigenvalues are clustered. A natural way to maintain stable convergence is by using expanding subspaces with which the trace reduction property is automatically maintained.

The best-known method that utilizes expanding subspaces is that of Lanczos. It uses the Krylov subspaces to compute an approximation of the desired eigenpairs, usually the largest. This idea was adopted by Davidson, in combination with the simultaneous coordinate relaxation method, to obtain what he called the "compromise method" [34]) known as Davidson's method today. In this section, we generalize the trace minimization algorithm described in the previous sections by casting it into the framework of the Davidson method. We start by the Jacobi-Davidson method, explore its connections to the trace minimization method, and develop a *Davidson-type trace minimization algorithm*.

The Jacobi-Davidson Method

As mentioned in Sect. 11.5, the Jacobi-Davidson scheme is a modification of the Davidson method. It uses the same ideas presented in the trace minimization method to compute a correction term to a previously computed Ritz pair, but with a different objective. In the Jacobi-Davidson method, for a given Ritz pair (x_i, θ_i) with $x_i^\top B x_i = 1$, a correction vector d_i is sought such that

$$A(x_i + d_i) = \lambda_i B(x_i + d_i), \quad x_i^\top B d_i = 0, \tag{11.76}$$

where λ_i is the eigenvalue targeted by θ_i. Since the targeted eigenvalue λ_i is not available during the iteration, it is replaced by an approximation σ_i. Ignoring high-order terms in (11.76), we get

$$\begin{pmatrix} A - \sigma_i B & B x_i \\ x_i^\top B & 0 \end{pmatrix} \begin{pmatrix} d_i \\ l_i \end{pmatrix} = \begin{pmatrix} -r_i \\ 0 \end{pmatrix}, \tag{11.77}$$

where $r_i = A x_i - \theta_i B x_i$ is the residual vector associated with the Ritz pair (x_i, θ_i). Note that replacing r_i with $A x_i$ does not affect d_i. In [39, 55], the Ritz value θ_i is used in place of σ_i at each step.

The block Jacobi-Davidson algorithm, described in [55], may be outlined as shown in Algorithm 11.14 which can be regarded as a trace minimization scheme with expanding subspaces.

The performance of the block Jacobi-Davidson algorithm depends on how good the initial guess is and how efficiently and accurately the inner system (11.78) is solved. If the right-hand side of (11.78) is taken as the approximate solution to the inner system (11.78), the algorithm is reduced to the Lanczos method. If the inner

---

Algorithm 11.14 The block Jacobi-Davidson algorithm.

Choose a block size $s \geqslant p$ and an $n \times s$ matrix V_1 such that $V_1^\top B V_1 = I_s$.

For $k = 1, 2, \ldots$ until convergence, do

1: Compute $W_k = A V_k$ and the interaction matrix $H_k = V_k^\top W_k$.
2: Compute the s smallest eigenpairs (Y_k, Θ_k) of H_k. The eigenvalues are arranged in ascending order and the eigenvectors are chosen to be orthogonal.
3: Compute the corresponding Ritz vectors $X_k = V_k Y_k$.
4: Compute the residuals $R_k = W_k Y_k - B X_k \Theta_k$.
5: Test for convergence.
6: for $1 \leqslant i \leqslant s$, solve the indefinite system

$$\begin{pmatrix} A - \theta_i B & B x_{k,i} \\ x_{k,i}^\top B & 0 \end{pmatrix} \begin{pmatrix} d_{k,i} \\ l_{k,i} \end{pmatrix} = \begin{pmatrix} r_{k,i} \\ 0 \end{pmatrix}, \tag{11.78}$$

or preferably its projected form

$$[P_i(A - \theta_{k,i} B) P_i] d_{k,i} = P_i r_{k,i}, \qquad x_{k,i}^\top B d_{k,i} = 0, \tag{11.79}$$

approximately, where $P_i = I - B x_{k,i} (x_{k,i}^\top B^2 x_{k,i})^{-1} x_{k,i}^\top B$ is an orthogonal projector, and $r_{k,i} = A x_{k,i} - \theta_{k,i} B x_{k,i}$ is the residual corresponding to the Ritz pair $(x_{k,i}, \theta_{k,i})$.
7: If $\dim(V_k) \leqslant m - s$, then

$$V_{k+1} = \mathrm{Mod}GS_B(V_k, \Delta_k),$$

else

$$V_{k+1} = \mathrm{Mod}GS_B(X_k, \Delta_k).$$

Here, $\mathrm{Mod}GS_B$ stands for the Gram-Schmidt process with reorthogonalization [73] with respect to B-inner products, i.e. $(x, y) = x^\top B y$.

End for

---

system (11.78) is solved to a low relative residual, it is reduced to the simultaneous Rayleigh quotient iteration (RQI, see [1]) with expanding subspaces, which converges cubically. If the inner system (11.78) is solved to a modest relative residual, the performance of the algorithm is in-between. In practice, however, the stage of cubic convergence is often reached after many iterations. The algorithm almost always "stagnates" at the beginning. Increasing the number of iterations for the inner systems makes little difference or, in some cases, even derails convergence to the desired eigenpairs. Note that the Ritz shifting strategy in the block Jacobi-Davidson algorithm forces convergence to eigenpairs far away from the desired ones. However, since the subspace is expanding, the Ritz values are decreasing and the algorithm is forced to converge to the smallest eigenpairs.

Another issue with the block Jacobi-Davidson algorithm is ill-conditioning. At the end of each outer Jacobi-Davidson iteration, as the Ritz values approach a multiple eigenvalue or a cluster of eigenvalues, the inner system (11.79) becomes poorly conditioned. This makes it difficult for an iterative solver to compute even a crude approximation of the solution of the inner system.

All these problems can be partially solved by the techniques developed in the trace minimization method, i.e. the multiple dynamic shifting strategy, the implicit deflation technique ($d_{k,i}$ is required to be B-orthogonal to all the Ritz vectors obtained in the previous iteration step), and the dynamic stopping strategy. We call the modified algorithm a *Davidson-type trace minimization algorithm* [38].

The Davidson-Type Trace Minimization Algorithm

Let $s \geqslant p$ be the block size, $m \geqslant s$ be a given integer that limits the dimension of the subspaces. The Davidson-type trace minimization algorithm is given by Algorithm 11.15. The orthogonality requirement $d_i^{(k)} \perp_B X_k$ is essential in the original

---

Algorithm 11.15 The Davidson-type trace minimization algorithm.

Choose a block size $s \geqslant p$ and an $n \times s$ matrix V_1 such that $V_1^\top B V_1 = I_s$.

For $k = 1, 2, \ldots$ until convergence, do

1: Compute $W_k = A V_k$ and the interaction matrix $H_k = V_k^\top W_k$.
2: Compute the s eigenpairs (Y_k, Θ_k) of H_k. The eigenvalues are arranged in ascending order and the eigenvectors are chosen to be orthogonal.
3: Compute the corresponding Ritz vectors $X_k = V_k Y_k$.
4: Compute the residuals $R_k = W_k Y_k - B X_k \Theta_k$.
5: Test for convergence.
6: for $1 \leqslant i \leqslant s$, solve the indefinite system

$$[P(A - \sigma_{k,i} B) P] d_{k,i} = P r_{k,i}, \qquad X_k^\top B d_{k,i} = 0, \tag{11.80}$$

to a certain accuracy determined by the stopping criterion described in Sect. 11.5.2. The shift parameters $\sigma_{k,i}$, $1 \leqslant i \leqslant s$, are determined according to the dynamic shifting strategy described in Sect. 11.5.3.
7: If $\dim(V_k) \leqslant m - s$, then

$$V_{k+1} = \text{Mod}GS_B(V_k, \Delta_k),$$

else

$$V_{k+1} = \text{Mod}GS_B(X_k, \Delta_k).$$

End for

---

trace minimization algorithm for maintaining the trace reduction property (11.64). In the current algorithm, it appears primarily as an implicit deflation technique. A more efficient approach is to require $d_i^{(k)}$ to be B-orthogonal only to "good" Ritz vectors. The number of outer iterations realized by this scheme is decreased compared to the trace minimization algorithm in Sect. 11.5.3, and compared to the block Jacobi-Davidson algorithm. In the block Jacobi-Davidson algorithm, the number of outer iterations cannot be reduced further when the number of iterations for the inner systems is increased. However, in the Davidson-type trace minimization algorithm, the number of outer iterations decreases steadily with increasing the number of iterations for the inner systems. Note that reducing the number of outer iterations enhances the efficiency of implementation on parallel architectures.

11.5.5 Implementations of TRACEMIN

In this section, we outline two possible implementation schemes for algorithm TRACEMIN on a cluster of multicore nodes interconnected via a high performance network.

One version is aimed at seeking a few of the smallest eigenpairs, while the other is more suitable if we seek a larger number of eigenvalues in the interior of the spectrum, together with the corresponding eigenvectors. We refer to these versions as TRACEMIN_1 and TRACEMIN_2, respectively.

TraceMIN_1

In TRACEMIN_1, i.e. Algorithm 11.13, we assume that the matrices A and B of the generalized eigenvalue problem are distributed across multiple nodes. We need the following kernels and algorithms on a distributed memory architecture, in which multithreading is used within each node.

- Multiplication of a sparse matrix by a tall narrow dense matrix (multivectors)
- Global reduction
- B-orthonormalization
- Solving sparse saddle-point problems via direct or iterative schemes

All other operations in TRACEMIN_1 do not require internode communications, and can be expressed as simple multithreaded BLAS or LAPACK calls. More specifically, the computation of all the eigenpairs of H_k can be realized on a single node using a multithreaded LAPACK eigensolver since H_k is of a relatively small size $s \ll n$.

In order to minimize the number of costly global communications, namely the all-reduce operations, one should group communications in order to avoid low parallel efficiency.

Sparse Matrix-multivector Multiplication

A sparse matrix multiplication $Y = AX$ requires communication of some elements of the tall-dense matrix X (or a single vector x). The amount of communication required depends on the sparsity pattern of the matrix. For example, a narrow banded matrix requires only nearest neighbor communication, but a general sparse matrix requires more internode communications. Once all the relevant elements of X are obtained on a given node, one may use a multithreaded sparse matrix multiplication scheme.

Global Reduction

There are two types of global reduction in our algorithm: multiplying two dense matrices, and concurrently computing the 2-norms of many vectors. In the first case, one may call a multithreaded dense matrix multiplication routine on each node, and perform all-reduce to obtain the result. In the second, each node repeatedly calls a dot product routine, followed by one all-reduce to obtain all the inner products.

B-orthonormalization

This may be achieved via the eigendecomposition of matrices of the form $W^\top B W$ or $W^\top A W$ to obtain a section of the generalized eigenvalue problem, i.e. to obtain a matrix V (with s columns) for which $V^\top A V$ is diagonal, and $V^\top B V$ is the identity matrix of order s. This algorithm requires one call to a sparse matrix-multivector multiplication kernel, one global reduction operation to compute H, one call to a multithreaded dense eigensolver, one call to a multithreaded dense matrix-matrix multiplication routine, and s calls to a multithreaded vector scaling routine. Note that only two internode communication operations are necessary. In addition, one can take full advantage of the multicore architecture of each node.

Linear System Solver

TRACEMINdoes not require a very small relative residual in solving the saddle point problem, hence we may use a modest stopping criterion when solving the linear system (11.64). Further, one can simultaneously solve the s independent linear systems via a single call to the CG scheme. The main advantage of such a procedure is that one can group the internode communications, thereby reducing the associated cost of the CG algorithm as compared with solving the systems one at a time. For instance, in the sparse matrix-vector multiplication and the global reduction schemes outlined above, grouping the communications results in fewer MPI communication operations.

We now turn our attention to TRACEMIN_2, which efficiently computes a larger number of eigenpairs corresponding to any interval in the spectrum.

TRACEMIN_2: **Trace Minimization with Multisectioning**

In this implementation of TRACEMIN, similar to our implementation of TREPS (see Sect. 11.1.2), we use multisectioning to divide the interval under consideration into a number of smaller subintervals. Our hybrid MPI+OpenMP approach assigns distinct subsets of the subintervals to different nodes. Unlike the implementation of TRACEMIN_1, we assume that each node has a local memory capable of storing all elements of A and B. For each subinterval $[a_i, a_{i+1}]$, we compute the $LDL^\top$—factorization of $A - a_i B$ and $A - a_{i+1} B$ to determine the inertia at each endpoint of the interval. This yields the number of eigenvalues of $Ax = \lambda Bx$ that lie in the subinterval $[a_i, a_{i+1}]$. TRACEMIN_1 is then used on each node, taking full advantage of as many cores as possible, to compute the lowest eigenpairs of $(A - \sigma B)x = (\lambda - \sigma)Bx$, in which σ is the mid-point of each corresponding subinterval. Thus, TRACEMIN_2 has two levels of parallelism: multiple nodes working concurrently on different sets of subintervals, with each node using multiple cores for implementing the TRACEMIN_1 iterations on a shared memory architecture. Note that this algorithm requires no internode communication after the intervals are selected, which means it scales extremely well. We employ a recursive scheme to divide the interval into subintervals. We select a variable n_e, denoting the maximum number of eigenvalues allowed to belong to any subinterval. Consequently, any interval containing more than n_e eigenvalues is divided in half. This process is repeated until we have many subintervals, each of which containing less than or equal to n_e eigenvalues. We then assign the subintervals amongst the nodes so that each node is in charge of roughly the same number of subintervals. Note that TRACEMIN_2 requires only one internode communication since the intervals can be subdivided in parallel.

11.6 The Sparse Singular-Value Problem

11.6.1 Basics

Algorithms for computing few of the singular triplets of large sparse matrices constitute a powerful computational tool that has a significant impact on numerous science and engineering applications. The singular-value decomposition (partial or complete) is used in data analysis, model reduction, matrix rank estimation, canonical correlation analysis, information retrieval, seismic reflection tomography, and real-time signal processing.

In what follows, similar to the basic material we introduced in Chap. 8 about the singular-value decomposition, we introduce the notations we use in this chapter, together with additional basic facts related to the singular-value problem.

The numerical accuracy of the ith approximate singular triplet $(\tilde{u}_i, \tilde{\sigma}_i, \tilde{v}_i)$ of the real matrix A can be determined via the eigensystem of the $2 - cyclic$ matrix

$$A_{aug} = \begin{pmatrix} 0 & A \\ A^\top & 0 \end{pmatrix}, \tag{11.81}$$

see Theorem 8.1, and is measured by the norm of the residual vector r_i given by

$$\| r_i \|_2 = \left[\| A_{aug} \begin{pmatrix} \tilde{u}_i \\ \tilde{v}_i \end{pmatrix} - \tilde{\sigma}_i \begin{pmatrix} \tilde{u}_i \\ \tilde{v}_i \end{pmatrix} \|_2 \right] / \left[\| \tilde{u}_i \|_2^2 + \| \tilde{v}_i \|_2^2 \right]^{\frac{1}{2}},$$

which can also be written as

$$\| r_i \|_2^2 = \left[\| A\tilde{v}_i - \tilde{\sigma}_i\tilde{u}_i \|_2^2 + \| A^\top\tilde{u}_i - \tilde{\sigma}_i\tilde{v}_i \|_2^2 \right] / \left[\| \tilde{u}_i \|_2^2 + \| \tilde{v}_i \|_2^2 \right], \tag{11.82}$$

with the backward error [74], given by,

$$\eta_i = \max \left\{ \| A\tilde{v}_i - \tilde{\sigma}_i\tilde{u}_i \|_2 \, , \, \| A^\top\tilde{u}_i - \tilde{\sigma}_i\tilde{v}_i \|_2^2 \right\}$$

could be used as a measure of absolute accuracy. Introducing a normalizing factor, the above expression could be used for assessing the relative error.

Alternatively, as outlined in Chap. 8, we may compute the SVD of A indirectly via the eigenpairs of either the $n \times n$ matrix $A^\top A$ or the $m \times m$ matrix $AA^\top$. Thus, if

$$V^\top (A^\top A) V = \mathrm{diag}(\sigma_1^2, \sigma_2^2, \ldots, \sigma_r^2, \underbrace{0, \ldots, 0}_{n-r}),$$

where $r = \min(m, n)$, and σ_i is the ith nonzero singular value of A corresponding to the right singular vector v_i, then the corresponding left singular vector, u_i, is obtained as $u_i = \frac{1}{\sigma_i} A v_i$. Similarly, if

$$U^\top (A A^\top) U = \mathrm{diag}(\sigma_1^2, \sigma_2^2, \ldots, \sigma_r^2, \underbrace{0, \ldots 0}_{m-r}),$$

where σ_i is the ith nonzero singular value of A corresponding to the left singular vector u_i, then the corresponding right singular vector, v_i, is obtained as $v_i = \frac{1}{\sigma_i} A^\top u_i$.

As stated in Chap. 8, computing the SVD of A via the eigensystems of either $A^\top A$ or $A A^\top$ may be adequate for determining the largest singular triplets of A, but some loss of accuracy is observed for the smallest singular triplets, e.g. see also [23]. In fact, if the smallest singular value of A is smaller than $(\sqrt{\mathbf{u}} \|A\|)$, then it will be computed as the zero eigenvalue of $A^\top A$ (or $A A^\top$). Thus, it is preferable to compute the smallest singular values of A via the eigendecomposition of the augmented matrix

$$A_{aug} = \begin{pmatrix} 0 & A \\ A^\top & 0 \end{pmatrix}.$$

Note that, whereas the square of the smallest and largest singular values of A are the lower and upper bounds of the spectrum of $A^\top A$ or $A A^\top$, the smallest singular values of A lie in the middle of the spectrum of A_{aug} in (11.81). Further, similar to (11.82), the norms of the residuals corresponding to the ith eigenpairs of $A^\top A$ and $A A^\top$, are given by

$$\| r_i \|_2 = \| A^\top A \tilde{v}_i - \tilde{\sigma}_i^2 \tilde{v}_i \|_2 \, / \, \| \tilde{v}_i \|_2 \quad \text{and} \quad \| r_i \|_2 = \| A A^\top \tilde{u}_i - \tilde{\sigma}_i^2 \tilde{u}_i \|_2 \, / \, \| \tilde{u}_i \|_2,$$

respectively.

When A is a square nonsingular matrix, it may be advantageous (in certain cases) to compute the needed few largest singular triplets of A^{-1} which are $\frac{1}{\sigma_n} \geq \cdots \geq \frac{1}{\sigma_1}$. This approach has the drawback of needing to solve linear systems involving the matrix A. If a suitable parallel direct sparse solver is available, this strategy provides a robust algorithm. The resulting computational scheme becomes of value in enhancing parallelism for the subspace method described below. This approach can also be extended to rectangular matrices of full column rank if the upper triangular matrix R in the orthogonal factorization $A = QR$ is available:

Proposition 11.3 *Let $A \in \mathbb{R}^{m \times n}$ ($m \geq n$) be of rank n. Let*

$$A = QR, \text{ where } Q \in \mathbb{R}^{m \times n} \text{ and } R \in \mathbb{R}^{n \times n},$$
$$\text{such that } Q^\top Q = I_n \text{ and } R \text{ upper triangular.}$$

The singular values of R are the same as those of A.

Consequently, the smallest singular values of A can be computed from the largest eigenvalues of $(R^{-1}R^{-\top})$ or of $\begin{pmatrix} 0 & R^{-1} \\ R^{-\top} & 0 \end{pmatrix}$.

Sensitivity of the Smallest Singular Value

In order to compute the smallest singular value in a reliable way, one must investigate the sensitivity of the singular values with respect to perturbations of the matrix under consideration.

Theorem 11.13 *Let $A \in \mathbb{R}^{m \times n}$ ($m \geq n$) be of rank n and $\Delta \in \mathbb{R}^{m \times n}$. If the singular values of A and $A + \Delta$ are respectively denoted by,*

$$\sigma_1 \geq \sigma_2 \geq \cdots \geq \sigma_n, \text{ and}$$
$$\tilde{\sigma}_1 \geq \tilde{\sigma}_2 \geq \cdots \geq \tilde{\sigma}_n.$$

then,

$$|\sigma_i - \tilde{\sigma}_i| \leq \|\Delta\|_2, \text{ for } i = 1, \ldots, n.$$

Proof see [75].

When applied to the smallest singular value, this result yields the following.

Proposition 11.4 *The relative condition number of the smallest singular value of a nonsingular matrix A is equal to $\kappa_2(A) = \frac{\sigma_1}{\sigma_n}$.*

This means that the smallest singular value of an ill-conditioned matrix cannot be computed with high accuracy even with a backward-stable algorithm.

In [76] it is shown that for some special class of matrices, e.g. tall and narrow sparse matrices, an accurate computation of the smallest singular value may be obtained via a combination of a parallel orthogonal factorization scheme of A, with column pivoting, and a one-sided Jacobi algorithm (see Chap. 8).

Since the nonzero singular values are roots of a polynomial (e.g. roots of the characteristic polynomial of the augmented matrix), then when simple, they are differentiable with respect to the entries of the matrix. More precisely, one can state that:

Theorem 11.14 *Let σ be a nonzero simple singular value of the matrix $A = (\alpha_{ij})$ with $u = (\mu_i)$ and $v = (v_i)$ being the corresponding normalized left and right singular vectors. Then, the singular value is differentiable with respect to the matrix A, or*

$$\frac{\partial \sigma}{\partial \alpha_{ij}} = \mu_i v_j, \ \forall i, \ j = 1, \ldots, n.$$

Proof See [77].

The effect of a perturbation of the matrix on the singular vectors can be more significant than that on the singular values. The sensitivity of the singular vectors depends upon the distribution of the singular values. When a simple singular value is not well separated from the rest, the corresponding left and right singular vectors are poorly determined. This is made precise by the theorem below, see [75], which we state here without proof. Let $A \in \mathbb{R}^{n \times m}$ ($n \geq m$) have the singular value decomposition

$$U^{\top} A V = \begin{pmatrix} \Sigma \\ 0 \end{pmatrix}.$$

Let U and V be partitioned as $U = (u_1 \ U_2 \ U_3)$ and $V = (v_1 \ V_2)$ where $u_1 \in \mathbb{R}^n$, $U_2 \in \mathbb{R}^{n \times (m-1)}$, $U_3 \in \mathbb{R}^{n \times (n-m)}$, $v_1 \in \mathbb{R}^m$ and $U_2 \in \mathbb{R}^{m \times (m-1)}$. Consequently,

$$U^{\top} A V = \begin{pmatrix} \sigma_1 & 0 \\ 0 & \Sigma_2 \\ 0 & 0 \end{pmatrix}.$$

Let A be perturbed by the matrix E, where

$$U^{\top} E V = \begin{pmatrix} \gamma_{11} & g_{12}^{\top} \\ g_{21} & G_{22} \\ g_{31} & G_{32} \end{pmatrix}.$$

Theorem 11.15 *Let $h = \sigma_1 g_{12} + \Sigma_2 g_{21}$, and $\tilde{A} = A + E$. If $(\sigma_1 I - \Sigma_2)$ is nonsingular (i.e. if σ_1 is a simple singular value of A), then the matrix*

$$U^{\top} \tilde{A} V = \begin{pmatrix} \sigma_1 + \gamma_{11} & g_{12}^{\top} \\ g_{21} & \Sigma_2 + G_{22} \\ g_{31} & G_{32} \end{pmatrix}$$

has a right singular vector of the form

$$\begin{pmatrix} 1 \\ (\sigma_1^2 I - \Sigma_2^2)^{-1} h \end{pmatrix} + O(\|E\|^2).$$

Next, we present a selection of parallel algorithms for computing the extreme singular values and the corresponding vectors of a large sparse matrix $A \in \mathbb{R}^{m \times n}$. In particular, we present the simultaneous iteration method and two Lanczos schemes for computing few of the largest singular eigenvalues and the corresponding eigenvectors of A. Our parallel algorithms of choice for computing a few of the smallest singular triplets, however, are the trace minimization and the Davidson schemes.

11.6.2 Subspace Iteration for Computing the Largest Singular Triplets

Subspace iteration, presented in Sect. 11.1, can be used to obtain the largest singular triplets via obtaining the dominant eigenpairs of the symmetric matrix $G = \tilde{A}_{aug}$, where

$$\tilde{A}_{aug} = \begin{pmatrix} \gamma I_m & A \\ A^\top & \gamma I_n \end{pmatrix}, \tag{11.83}$$

in which the shift parameter γ is chosen to assure that G is positive definite. This method generates the sequence

$$Z_k = G^k Z_0,$$

where the initial iterate is the $n \times s$ matrix $Z_0 = (z_1, z_2, \ldots, z_s)$, in which $s = 2p$ with p being the number of desired largest singular values. Earlier, we presented the basic form of simultaneous iteration without Rutishauser's classic Chebyshev acceleration in RITZIT (see [4]). This particular algorithm incorporates both a Rayleigh-Ritz procedure and an acceleration scheme via Chebyshev polynomials. The iteration which embodies the RITZIT procedure is given in Algorithm 11.16. The Rayleigh Quotient matrix, H_k, in step 3 is essentially the projection of G^2 onto the subspace spanned by the columns of Z_{k-1}. The three-term recurrence in step 7 follows from the mapping of the Chebyshev polynomial of the first kind, of degree q, $\mathcal{T}_q(x)$, onto the interval $[-e, e]$, where e is chosen to be slightly smaller than the smallest eigenvalue of the SPD matrix H_k. This use of Chebyshev polynomials has the desired effect of damping out the unwanted eigenvalues of G and producing the improved rate of convergence: $O(\mathcal{T}_k(\theta_s)/\mathcal{T}_k(\theta_1))$ which is considerably higher than the rate of convergence without the Chebyshev acceleration: $O(\theta_s/\theta_1)$, where $(\theta_1 \geq \theta_2 \geq \cdots \geq \theta_s)$ are the eigenvalues of H_k, or the square root of the diagonal matrix Δ_k^2 in step 4 of Algorithm 11.16.

Algorithm 11.16 SISVD: inner iteration of the subspace iteration as implemented in Rutishauser's RITZIT.

1: Compute $C_k = GZ_{k-1}$;   //without assembling G.
2: Factor $C_k = Q_k R_k$;
3: Form $H_k = R_k R_k^\top$;
4: Factor $H_k = P_k \Delta_k^2 P_k^\top$;
5: Form $Z_k = Q_k P_k$;
6: **do** j=2:q,
7:     $Z_{k+j} = \frac{2}{e} G Z_{k+j-1} - Z_{k+j-2}$;
8: **end**

The orthogonal factorization in step 2 of Algorithm 11.16 may be computed by a modified Gram-Schmidt procedure or by Householder transformations provided that the orthogonal matrix Q_k is explicitly available for the computation of Z_k in step 5. On parallel architectures, especially those having hierarchical memories, one may achieve higher performance by using either a block Gram-Schmidt (Algorithm B2GS) or the block Householder orthogonalization method in step 2. Further, we could use the parallel two- or one-sided Jacobi method, described earlier, to obtain the spectral decomposition in step 4 of RITZIT (as originally suggested in [47]). In fact, when appropriately adapted for symmetric positive definite matrices, the one-sided Jacobi scheme can realize high performance on parallel architectures (e.g. see [78]), provided that the dimension of the current subspace, s, is not too large. For larger subspace size s, an optimized implementation of the parallel Cuppen's algorithm [79] may be used instead in step 4.

The success of Rutishauser's subspace iteration method using Chebyshev acceleration relies upon the following strategy for limiting the degree of the Chebyshev polynomial, $\mathscr{T}_q\left(\frac{x}{e}\right)$, on the interval $[-e, e]$, where $e = \theta_s$ (assuming we use a block of s vectors, and $q = 1$ initially):

$$q_{new} = \min\{2q_{old}, \hat{q}\}, \quad \text{where}$$

$$\hat{q} = \begin{cases} 1, & \text{if } \theta_s < \xi_1\theta_1 \\ 2 \times \max\left(\dfrac{\xi_2}{\text{arccosh}\left(\frac{\theta_1}{\theta_s}\right)}, 1\right) & \text{otherwise.} \end{cases} \tag{11.84}$$

Here, $\xi_1 = 0.04$ and $\xi_2 = 4$. The polynomial degree of the current iteration is then taken to be $q = q_{new}$. It can easily be shown that the strategy in (11.84) insures that

$$\left\|\mathscr{T}_q\left(\frac{\theta_1}{\theta_s}\right)\right\|_2 = \cosh\left[q \,\text{arccosh}\left(\frac{\theta_1}{\theta_s}\right)\right] \leq \cosh(8) < 1500.$$

Although this has been successful for RITZIT, we can generate several variations of polynomial-accelerated subspace iteration schemes SISVD using a more flexible bound. Specifically, we can consider an *adaptive* strategy for selecting the degree q in which ξ_1 and ξ_2 are treated as control parameters for determining the *frequency* and the *degree* of polynomial acceleration, respectively. In other words, large (small) values of ξ_1, inhibit (invoke) polynomial acceleration, and large (small) values of ξ_2 yield larger (smaller) polynomial degrees when acceleration is selected. Correspondingly, the number of matrix-vector multiplications will increase with ξ_2 and the total number of iterations may well increase with ξ_1. Controlling the parameters, ξ_1 and ξ_2, allows us to monitor the method's complexity so as to maintain an optimal balance between the dominating kernels (specifically, sparse matrix—vector multiplication, orthogonalization, and spectral decomposition). We demonstrate the use of such controls in the polynomial acceleration-based trace minimization method for computing a few of the smallest singular triplets in Sect. 11.6.4.

11.6.3 The Lanczos Method for Computing a Few of the Largest Singular Triplets

The single-vector Lanczos tridiagonalization procedures (with and without reorthogonalization) and the corresponding Lanczos eigensolver as well as its block version, discussed above in Sect. 11.2, may be used to obtain an approximation of the extreme singular values and the associated singular vectors by considering the standard eigenvalue problem for the $(m + n) \times (m + n)$ symmetric indefinite matrix A_{aug}:

$$A_{aug} = \begin{pmatrix} 0 & A \\ A^\top & 0 \end{pmatrix}. \tag{11.85}$$

Note that, the largest singular triplets of A will be obtained with much higher accuracy than their smallest counterparts using the Lanczos method.

The Single-Vector Lanczos Method

Using Algorithms 11.5 or 11.6 for the tridiagonalization of A_{aug}, which is referenced only through matrix-vector multiplications, we generate elements of the corresponding tridiagonal matrices to be used by the associated Lanczos eigensolvers: Algorithms 11.8 or 11.9, respectively. We denote this method by LASVD.

The Block Lanczos Method

As mentioned earlier, one could use the block version of the Lanczos method as an alternate to the single-vector scheme LASVD. The resulting block version BLSVD uses block three-term recurrence relations which require sparse matrix-tall dense matrix multiplications, dense matrix multiplications, and dense matrix orthogonal factorizations. These are primitives that achieve higher performance on parallel architectures. In addition, this block version of the Lanczos algorithm is more robust for eigenvalue problems with multiple or clustered eigenvalues. Again, we consider the standard eigenvalue problem involving the 2-cyclic matrix A_{aug}. Exploiting the structure of the matrix A_{aug}, we can obtain an alternative form for the Lanczos recursion (11.9). If we apply the Lanczos recursion given by (11.9) to A_{aug} with a starting vector $\tilde{u} = (u, 0)^\top$ such that $\|\tilde{u}\|_2 = 1$, then all the diagonal entries of the real symmetric tridiagonal Lanczos matrices generated are identically zero. In fact, this Lanczos recursion, for $i = 1, 2, \ldots, k$, reduces to

$$\begin{aligned} \beta_{2i} v_i &= A^\top u_i - \beta_{2i-1} v_{i-1}, \\ \beta_{2i+1} u_{i+1} &= A v_i - \beta_{2i} u_i. \end{aligned} \tag{11.86}$$

where $u_1 \equiv u$, $v_0 \equiv 0$, and $\beta_1 \equiv 0$. Unfortunately, (11.86) can only compute the distinct singular values of A but not their multiplicities. Following the block Lanczos recursion as done in Algorithm 11.7, (11.86) can be represented in matrix form as

$$\begin{aligned} A^\top \hat{U}_k &= \hat{V}_k J_k^\top + Z_k, \\ A \hat{V}_k &= \hat{U}_k J_k + \hat{Z}_k, \end{aligned} \tag{11.87}$$

where $\hat{U}_k = (u_1, \ldots, u_k)$, $\hat{V}_k = (v_1, \ldots, v_k)$, with J_k being a $k \times k$ bidiagonal matrix in which $(J_k)_{j\,j} = \beta_{2j}$ and $(J_k)_{j\,j+1} = \beta_{2j+1}$. In addition, Z_k, $\tilde{Z}_k$ contain remainder terms. It is easy to show that the nonzero singular values of J_k are the same as the positive eigenvalues of

$$K_k \equiv \begin{pmatrix} O & J_k \\ J_k^\top & O \end{pmatrix}. \tag{11.88}$$

For the block analogue of (11.87), we make the simple substitutions

$$u_i \leftrightarrow U_i, \quad v_i \leftrightarrow V_i,$$

where U_i and V_i are matrices of order $m \times b$ and $n \times b$, respectively, with b being the current block size. The matrix J_k is now a block-upper bidiagonal matrix of order bk

$$J_k \equiv \begin{pmatrix} S_1 & R_1^\top & & & \\ & S_2 & R_2^\top & & \\ & & \ddots & \ddots & \\ & & & \ddots & R_{k-1}^\top \\ & & & & S_k \end{pmatrix}, \tag{11.89}$$

in which S_i and R_i are $b \times b$ upper-triangular matrices. If U_i and V_i form mutually orthogonal sets of bk vectors so that $\hat{U}_k$ and $\hat{V}_k$ are orthonormal matrices, then the singular values of the matrix J_k will be identical to those of the original matrix A. Given the upper block-bidiagonal matrix J_k, we approximate the singular triplets of A by first computing the singular triplets of J_k. To determine the left and right singular vectors of A from those of J_k, we must retain the Lanczos vectors of $\hat{U}_k$ and $\hat{V}_k$. Specifically, if $\{\sigma_i^{(k)}, y_i^{(k)}, z_i^{(k)}\}$ is the ith singular triplet of J_k, then the approximation to the ith singular triplet of A is given by $\{\sigma_i^{(k)}, \hat{U}_k y_i^{(k)}, \hat{V}_k z_i^{(k)}\}$, where $\hat{U}_k y_i^{(k)}$, $\hat{V}_k z_i^{(k)}$ are the left and right approximate singular vectors, respectively. The computation of singular triplets for J_k requires two phases. The first phase reduces J_k to an upper-bidiagonal matrix C_k having diagonal elements $\{\alpha_1, \alpha_2, \ldots, \alpha_{bk}\}$ and superdiagonal elements $\{\beta_1, \beta_2, \ldots, \beta_{bk-1}\}$ via a finite sequence of orthogonal transformations (thus preserving the singular values of J_k). The second phase reduces C_k to the diagonal form by a modified QR algorithm. This diagonalization procedure is discussed in detail in [80]. The resulting diagonalized C_k will yield the approximate singular values of A, while the corresponding left and right singular vectors are determined through multiplications by all the left and right transformations used in both phases of the SVD of J_k.

There are a few options for the reduction of J_k to the bidiagonal matrix, C_k. The use of either Householder or Givens reductions that zero out or *chase off* elements generated above the first super-diagonal of J_k was advocated in [81]. This bidiagonalization, and the subsequent diagonalization processes offer only poor data locality and limited parallelism. For this reason, one should adopt instead the single vector Lanczos bidiagonalization recursion given by (11.86) and (11.87) as the strategy of choice for reducing the upper block-bidiagonal matrix J_k to the bidiagonal form (C_k), i.e.

$$\begin{aligned}
J_k^\top \hat{Q} &= \hat{P} C_k^\top, \\
J_k \hat{P} &= \hat{Q} C_k,
\end{aligned} \tag{11.90}$$

or

$$\begin{aligned}
J_k p_j &= \alpha_j q_j + \beta_{j-1} q_{j-1}, \\
J_k^\top q_j &= \alpha_j p_j + \beta_j p_{j+1},
\end{aligned} \tag{11.91}$$

where $\hat{P} = (p_1, p_2, \ldots, p_{bk})$ and $\hat{Q} = (q_1, q_2, \ldots, q_{bk})$ are orthonormal matrices of order $bk \times bk$. Note that the recursions in (11.91) which require banded matrix-vector multiplications can be realized via optimized level-2 BLAS routines. For orthogonalization of the outermost Lanczos vectors, $\{U_i\}$ and $\{V_i\}$, as well as the innermost Lanczos vectors, $\{p_i\}$ and $\{q_i\}$, one can apply a complete reorthogonalization [10] strategy to insure robust approximation of the singular triplets of the matrix A. Such a hybrid Lanczos approach which incorporates in each outer iteration of the block Lanczos SVD recursion, inner iterations of the single-vector Lanczos bidiagonalization procedure has been introduced in [82]. As an alternative to the outer recursion given by (11.87), which is derived from the 2-cyclic matrix A_{aug}, Algorithm 11.17 depicts the simplified outer block Lanczos recursion for approximating primarily the the largest eigenpairs of $A^\top A$. Combining the equations in (11.87), we obtain

$$A^\top A \hat{V}_k = \hat{V}_k H_k,$$

where $H_k = J_k^\top J_k$ is the symmetric block-tridiagonal matrix

$$H_k \equiv \begin{pmatrix}
S_1 & R_1^\top & & & & \\
R_1 & S_2 & R_2^\top & & & \\
 & R_2 & \cdot & \cdot & & \\
 & & \cdot & \cdot & \cdot & \\
 & & & \cdot & \cdot & R_{k-1}^\top \\
 & & & & R_{k-1} & S_k
\end{pmatrix}, \tag{11.92}$$

Applying the block Lanczos recursion [10] in Algorithm 11.17 for computing the eigenpairs of the $n \times n$ symmetric positive definite matrix $A^\top A$, the tridiagonalization of H_k via an inner Lanczos recursion follows from a simple application of (11.9). Analogous to the reduction of J_k in (11.89), computation of the eigenpairs of the resulting tridiagonal matrix can be performed via a Jacobi or a QR-based symmetric eigensolver.

Algorithm 11.17 BLSVD: hybrid Lanczos outer iteration (Formation of symmetric block-tridiagonal matrix H_k).

1: Choose $V_1 \in \mathbb{R}^{n \times b}$ orthonormal and $c = \max(b, k)$.
2: Compute $S_1 = V_1^\top A^\top A V_1$, $(V_0, R_0^\top = 0$ initially$)$.
3: **do** $i = 1 : k$, (where $k = \lfloor c/b \rfloor$)
4:     Compute $Y_{i-1} = A^\top A V_{i-1} - V_{i-1} S_{i-1} - V_{i-1} R_{i-2}^\top$;
5:     Orthogonalize Y_{i-1} against $\{V_\ell\}_{\ell=0}^{i-1}$;
6:     Factor $Y_{i-1} = V_i R_{i-1}$;
7:     Compute $S_i = V_i^\top A^\top A V_i$;
8: **end**

Clearly, $A^\top A$ should never be formed explicitly. The above algorithms require only sparse matrix-vector multiplications involving A and $A^\top$. Higher parallel scalability is realized in the outer (block) Lanczos iterations by the multiplication of A and $A^\top$ by b vectors rather than multiplication by a single vector. A stable variant of the block Gram-Schmidt orthogonalization [4], which requires efficient dense matrix-vector multiplication (level-2 BLAS) routines or efficient dense matrix-matrix multiplication (level-3 BLAS) routines as Algorithm B2GS, is used to produce the orthogonal projections of Y_i (i.e. R_{i-1}) and W_i (i.e. S_i) onto $\tilde{V}^\perp$ and $\tilde{U}^\perp$, respectively, where U_0 and V_0 contain the converged left and right singular vectors, respectively, and

$$\tilde{V} = (V_0, V_1, \ldots, V_{i-1}) \text{ and } \tilde{U} = (U_0, U_1, \ldots, U_{i-1}).$$

11.6.4 The Trace Minimization Method for Computing the Smallest Singular Triplets

Another candidate subspace method for the SVD of sparse matrices is based upon the trace minimization algorithm, discussed earlier in this chapter, for the generalized eigenvalue problem

$$Hx = \lambda Gx, \tag{11.93}$$

where H and G are symmetric, with G being positive definite. In order to compute the t smallest singular triplets of an $m \times n$ matrix A, we could set $H = A^\top A$, and $G = I_n$. If $\mathcal{Y}$ is defined as the set of all $n \times p$ matrices Y for which $Y^\top Y = I_p$, where $p = 2t$, then as illustrated before, we have

$$\min_{Y \in \mathcal{Y}} \text{ trace}(Y^\top H Y) = \sum_{i=1}^{p} \tilde{\sigma}_{n-i+1}^2, \tag{11.94}$$

where $\tilde{\sigma}_i$ is a singular value of A, $\lambda_i = \tilde{\sigma}_i^2$ is an eigenvalue of H, and $\tilde{\sigma}_1 \geq \tilde{\sigma}_2 \geq \cdots \geq \tilde{\sigma}_n$. In other words, given an $n \times p$ matrix Y which forms a *section* of the eigenvalue problem

$$Hz = \lambda z, \tag{11.95}$$

i.e.

$$Y^\top HY = \tilde{\Sigma}, \quad Y^\top Y = I_p, \tag{11.96}$$

$$\tilde{\Sigma} = \mathrm{diag}(\tilde{\sigma}_n^2, \tilde{\sigma}_{n-1}^2, \ldots, \tilde{\sigma}_{n-p+1}^2),$$

our trace minimization scheme TRSVD generates the sequence Y_k, whose first t columns converge to the t left singular vectors corresponding to the t-smallest singular values of A.

Polynomial Acceleration Techniques for TRSVD

Convergence of the trace minimization algorithm TRSVD can be accelerated via either a shifting strategy as discussed earlier in Sect. 11.5.3, via a Chebyshev acceleration strategy as illustrated for subspace iterations SISVD, or via a combination of both strategies. For the time being we focus first on Chebyshev acceleration. Hence, in order to dampen the unwanted (i.e. largest) singular values of A, we need to consider applying the trace minimization scheme to the generalized eigenvalue problem

$$x = \frac{1}{P_q(\lambda)} P_q(H)x, \tag{11.97}$$

where $P_q(x) = T_q(x) + \varepsilon$, and $T_q(x)$ is the Chebyshev polynomial of degree q with ε chosen so that $P_q(H)$ is (symmetric) positive definite. The appropriate quadratic minimization problem here can be expressed as

$$\text{minimize } ((y_j^{(k)} - d_j^{(k)})^\top (y_j^{(k)} - d_j^{(k)})) \tag{11.98}$$

subject to the constraints

$$Y^\top P_q(H)d_j^{(k)} = 0, \quad j = 1, 2, \ldots, p.$$

In effect, we approximate the smallest eigenvalues of H as the *largest* eigenvalues of the matrix $P_q(H)$ in which the gaps between its eigenvalues are considerably larger than those between the eigenvalues of H.

Although the additional number of sparse matrix-vector multiplications involving $P_q(H)$ could be significantly higher than before, especially for large degrees q, the saddle-point problems that need to be solved in each outer trace minimization iteration become of the form,

$$\begin{pmatrix} I & P_q(H)Y_k \\ Y_k^\top P_q(H) & 0 \end{pmatrix} \begin{pmatrix} d_j^{(k)} \\ l \end{pmatrix} = \begin{pmatrix} y_j^{(k)} \\ 0 \end{pmatrix}, \quad j = 1, 2, \ldots, p. \tag{11.99}$$

which are much easier to solve. It can be shown that the updated eigenvector approximation, $y_j^{(k+1)}$, is determined by

$$y_j^{(k+1)} = y_j^{(k)} - d_j^{(k)} = P_q(H)Y_k \left[Y_k^\top P_q^2(H)Y_k \right]^{-1} Y_k^\top P_q(H) y_j^{(k)}.$$

Thus, we may not need to use an iterative solver for determining Y_{k+1} since the matrix $[Y_k^\top P_q^2(H)Y_k]^{-1}$ is of relatively small order p. Using the orthogonal factorization

$$P_q(H)Y_k = \hat{Q}\hat{R},$$

we have

$$[Y_k^\top P_q^2(H)Y_k]^{-1} = \hat{R}^{-\top}\hat{R}^{-1},$$

where the polynomial degree, q, is determined by the strategy adopted in SISVD.

Adding a Shifting Strategy for TRSVD

As discussed in [37], we can also accelerate the convergence of the Y_k's to eigenvectors of H by incorporating Ritz shifts (see [1]) into TRSVD. Specifically, we modify the symmetric eigenvalue problem as follows,

$$(H - v_j^{(k)}I)z_j = (\lambda_j - v_j^{(k)})z_j, \quad j = 1, 2, \ldots, s, \tag{11.100}$$

where $v_j^{(k)} = (\tilde{\sigma}_{n-j+1}^{(k)})^2$ is the j-th approximate eigenvalue at the k-th iteration of TRSVD, with λ_j, z_j being an exact eigenpair of H. In other words, we simply use our most recent approximations to the eigenvalues of H from our k-th section within TRSVD as Ritz shifts. As was shown by Wilkinson in [83], the Rayleigh quotient iteration associated with (11.100) will ultimately achieve cubic convergence to the square of an exact singular value of A, σ_{n-j+1}^2, provided $v_j^{(k)}$ is sufficiently close to σ_{n-j+1}^2. However, since we have $v_j^{(k+1)} < v_j^{(k)}$ for all k, i.e. we approximate the eigenvalues of H from above resulting in $H - v_j^{(k)}I$ not being positive definite and hence requiring adopting an appropriate linear system solver. Algorithm 11.18 outlines the basic steps of TRSVD that appropriately utilize polynomial (Chebyshev) acceleration prior to using Ritz shifts. It is important to note that once a shifting has been invoked (Steps 8–15) we abandon the use of Chebyshev polynomials $P_q(H)$ and solve the resulting saddle-point problems using appropriate solvers that take into account that the $(1, 1)$ block could be indefinite. The *context switch* from either non-accelerated (or polynomial-accelerated) trace minimization iterations to trace minimization iterations with Ritz shifts, is accomplished by monitoring the reduction of the residuals in (11.82) for isolated eigenvalues $(r_j^{(k)})$ or clusters of eigenvalues $(R_j^{(k)})$.

Algorithm 11.18 TRSVD: trace minimization with Chebyshev acceleration and Ritz shifts.

1: Choose an initial $n \times p$ subspace iterate $Y_0 = [y_1^{(0)}, y_2^{(0)}, \ldots, y_p^{(0)}]$;
2: Form a section, i.e. determine Y_0 such that $Y_0^\top P_q(H)Y_0 = I_p$ and $Y_0^\top Y_0 = \Gamma_0$ where Γ_0 is diagonal;
3: **do** $k = 0, 1, 2, \ldots$ until convergence,
4:     Determine the approximate singular values $\Sigma_k = \mathrm{diag}(\tilde{\sigma}_n^{(k)}, \cdots, \tilde{\sigma}_{n-p+1}^{(k)})$ from the Ritz values of H corresponding to the columns of Y_k;
5:     $R_k = HY_k - Y_k \Sigma_k^2$;   *//Compute residuals*
6:     Analyze the current approximate spectrum (Gerschgorin disks determine n_c groups $\mathscr{G}_j$ of eigenvalues)
7:     *//Invoke Ritz shifting strategy ([37])*
8:     **do** $\ell = 1 : n_c$,
9:        **if** $\mathscr{G}_\ell = \{\tilde{\sigma}_{n-j+1}^{(0)}\}$ includes a unique eigenvalue, **then**
10:           Shift is selected if $\|r_j^{(k)}\|_2 \leq \eta \|r_j^{(k_0)}\|_2$, where $\eta \in [10^{-3}, 10^0]$ and $k_0 < k$;
11:        **else**
12:           *//$\mathscr{G}_\ell$ is a cluster of c eigenvalues*
              Shift is selected if $\|R_\ell^{(k)}\|_F \leq \eta \|R_\ell^{(k_0)}\|_F$, where $R_\ell^{(k)} \equiv \{r_j^{(k)}, \ldots, r_{j+c-1}^{(k)}\}$ and $k_0 < k$;
13:        **end if**
14:        Disable polynomial acceleration if shifting is selected;
15:     **end**
16:     *//Deflation* Reduce subspace dimension, p, by number of the H-eigenpairs accepted;
17:     Adjust the polynomial degree q for $P_q(H)$ in iteration $k + 1$ (if needed).
18:     Update subspace iterate $Y_{k+1} = Y_k - \Delta_k$ as in (11.99) or for the shifted problem;
19: **end**

11.6.5 Davidson Methods for the Computation of the Smallest Singular Values

The smallest singular value of a matrix A may be obtained by applying one of the various versions of the Davidson methods to obtain the smallest eigenvalue of the matrix $C = A^\top A$, or to obtain the innermost positive eigenvalue of the 2-cyclic augmented matrix in (11.81). We assume that one has the basic kernels for matrix-vector multiplications using either A or $A^\top$. Multiplying $A^\top$ by a vector is often considered as a drawback. Thus, whenever possible, the so-called "transpose-free" methods should be used. Even though one can avoid such a drawback when dealing with the interaction matrix $H_k = V_k^\top C V_k = (AV_k)^\top AV_k$, we still have to compute the residuals corresponding to the Ritz pairs which do involve multiplication of the transpose of a matrix by a vector.

For the regular single-vector Davidson method, the correction vector is obtained by approximately solving the system

$$A^\top A t_k = r_k. \tag{11.101}$$

Obtaining an exact solution of (11.101) would yield the Lanczos algorithm applied to C^{-1}. Once the Ritz value approaches the square of the sought after smallest singular value, it is recommended that we solve (11.101) without any shifts; the benefit is that we deal with a fixed symmetric positive definite system matrix.

The approximate solution of (11.101) can be obtained by performing a fixed number of iterations of the Conjugate Gradient scheme, or by solving an approximate linear system $M t_k = r_k$ with a direct method, where M is obtained from an approximate factorization of A or $A^\top A$ [84]:

Incomplete LU factorization of A: Here, $M = U^{-1} L^{-1} L^{-\top} U^{-\top}$, where L and U are the products of an incomplete LU factorization of A where one drops entries of the reduced matrix A which are below a given threshold. This version of the Davidson method is called DAVIDLU.

Incomplete QR factorization of A: Here, $M = R^{-1} R^{-\top}$, where R is the upper triangular factor of an incomplete QR factorization of A. This version of the Davidson method is called DAVIDQR.

Incomplete Cholesky of $A^T A$: Here, $M = L^{-\top} L^{-1}$ where L is the lower triangular factor of an incomplete Cholesky factorization of the normal equations. This version of the Davidson method is called DAVIDIC.

Even though the construction of any of the above approximate factorizations may fail, experiments presented in [84] show the effectiveness of the above three preconditioners whenever they exist.

The corresponding method is expressed by algorithm 11.19. At step 11, the preconditioner C is defined by one of the methods DAVIDLU, DAVIDIC, or DAVIDQR. The steps 3, 4, and 13 must be implemented in such a way such that redundant computations are skipped.

Similar to trace minimization, the Jacobi-Davidson method can be used directly on the matrix $A^\top A$ to compute the smallest eigenvalue and the corresponding eigenvector. In [85] the Jacobi-Davidson method has been adapted for obtaining the singular values of A by considering the eigenvalue problem corresponding to the 2-cyclic augmented matrix.

11.6.6 Refinement of Left Singular Vectors

Having determined approximate singular values, $\tilde{\sigma}_i$ and their corresponding right singular vectors, $\tilde{v}_i$, to a user-specified tolerance for the residual

$$\hat{r}_i = A^\top A \tilde{v}_i - \tilde{\sigma}_i^2 \tilde{v}_i, \tag{11.102}$$

Algorithm 11.19 Computing smallest singular values by the block Davidson method.

Input: $A \in \mathbb{R}^{m \times n}$, $p \geq 1$, $V \in \mathbb{R}^{n \times p}$, and $2p \leq \ell \leq n \leq m$.
Output: The p smallest singular values $\{\sigma_1, \cdots, \sigma_p\} \subset \Lambda(A)$ and their corresponding right singular vectors $X = [x_1, \cdots, x_p] \in \mathbb{R}^{m \times p}$.
1: $V_1 = \mathrm{MGS}(V)$; $k = 1$;
2: **repeat**
3:    compute $U_k = AV_k$ and $W_k = A^\top U_k$;
4:    compute the interaction matrix $H_k = V_k{}^\top W_k$;
5:    compute the p smallest eigenpairs $(\sigma_{k,i}^2, y_{k,i})$ of H_k, for $i = 1, \cdots, p$;
6:    compute the p corresponding Ritz vectors $X_k = V_k Y_k$ where $Y_k = [y_{k,1}, \cdots y_{k,p}]$;
7:    compute the residuals $R_k = [r_{k,1}, \cdots, r_{k,p}] = W_k Y_k - X_k \,\mathrm{diag}(\sigma_{k,1}^2, \cdots, \sigma_{k,p}^2)$;
8:    **if** convergence **then**
9:       Exit;
10:   **end if**
11:   compute the new block $T_k = [t_{k,1}, \ldots, t_{k,p}]$ where $t_{k,i} = Cr_{k,i}$;
12:   **if** $\dim(V_k) \leq \ell - p$, **then**
13:      $V_{k+1} = \mathrm{MGS}([V_k, T_k])$;
14:   **else**
15:      $V_{k+1} = \mathrm{MGS}([X_k, T_k])$;
16:   **end if**
17:   $k = k + 1$;
18: **until** Convergence
19: $X = X_k$; $\sigma_i = \sigma_{k,i}$ for $i = 1, \cdots, p$.

---

we must then obtain an approximation to the corresponding left singular vector, u_i, via

$$u_i = \frac{1}{\tilde{\sigma}_i} A v_i, \tag{11.103}$$

As mentioned in Sect. 11.6.1, however, it is quite possible that square roots of the approximate eigenvalues of $A^\top A$ will be poor approximations to those singular values of A which are extremely small. This phenomenon, of course, will lead to poor approximations to the left singular vectors. Even if $\tilde{\sigma}_i$ is an acceptable singular value approximation, the residual corresponding to the singular triplet $\{\tilde{\sigma}_i, \tilde{u}_i, \tilde{v}_i\}$, defined by (11.82), has a 2-norm bounded from above by

$$\|r_i\|_2 \leq \|\hat{r}_i\|_2 / \left[\tilde{\sigma}_i \left(\left[\|\tilde{u}_i\|_2^2 + \|\tilde{v}_i\|_2^2 \right] \right)^{\frac{1}{2}} \right], \tag{11.104}$$

where $\hat{r}_i$ is the residual given in (11.102) for the symmetric eigenvalue problem for $A^\top A$ or $(\gamma^2 I_n - A^\top A)$. Scaling by $\tilde{\sigma}_i$ can easily lead to significant loss of accuracy in estimating the singular triplet residual norm, $\|r_i\|_2$, especially when $\tilde{\sigma}_i$ approaches the machine unit roundoff.

One remedy is to refine the initial approximation of the left singular vectors, corresponding to the few computed singular values and right singular vectors, via inverse iteration. To achieve this, consider the following equivalent eigensystem,

$$\begin{pmatrix} \gamma I_n & A^\top \\ A & \gamma I_m \end{pmatrix} \begin{pmatrix} v_i \\ u_i \end{pmatrix} = (\gamma + \sigma_i) \begin{pmatrix} v_i \\ u_i \end{pmatrix}, \tag{11.105}$$

where $\{\sigma_i, u_i, v_i\}$ is the ith singular triplet of A, and

$$\gamma = \min(1, \max\{\sigma_i\}). \tag{11.106}$$

One possible refinement recursion (via inverse iteration) is thus given by

$$\begin{pmatrix} \gamma I_n & A^\top \\ A & \gamma I_m \end{pmatrix} \begin{pmatrix} \tilde{v}_i^{(k+1)} \\ \tilde{u}_i^{(k+1)} \end{pmatrix} = (\gamma + \tilde{\sigma}_i) \begin{pmatrix} \tilde{v}_i^{(k)} \\ \tilde{u}_i^{(k)} \end{pmatrix}, \tag{11.107}$$

where $\{\tilde{\sigma}_i, \tilde{u}_i, \tilde{v}_i\}$ is the ith computed smallest singular triplet. By applying block
Gaussian elimination to (11.107) we obtain a more optimal form (reduced system)
of the recursion

$$\begin{pmatrix} \gamma I_n & A^\top \\ 0 & \gamma I_m - \frac{1}{\gamma} A A^\top \end{pmatrix} \begin{pmatrix} \tilde{v}_i^{(k+1)} \\ \tilde{u}_i^{(k+1)} \end{pmatrix} = (\gamma + \tilde{\sigma}_i) \begin{pmatrix} \tilde{v}_i^{(k)} \\ \tilde{u}_i^{(k)} - \frac{1}{\gamma} A \tilde{v}_i^{(k)} \end{pmatrix}. \tag{11.108}$$

Our iterative refinement strategy for an approximate singular triplet of A, $\{\tilde{\sigma}_i, \tilde{u}_i, \tilde{v}_i\}$,
is then defined by the last m equations of (11.108), i.e.

$$\left(\gamma I_m - \frac{1}{\gamma} A A^\top \right) \tilde{u}_i^{(k+1)} = (\gamma + \tilde{\sigma}_i) \left(\tilde{u}_i^{(k)} - \frac{1}{\gamma} A \tilde{v}_i \right), \tag{11.109}$$

where the superscript k is dropped from $\tilde{v}_i$ since we refine only our left singular
vector approximation, $\tilde{u}_i$. If $\tilde{u}_i^{(0)} \equiv u_i$ from (11.105), then (11.109) can be rewritten
as

$$\left(\gamma I_m - \frac{1}{\gamma} A A^\top \right) \tilde{u}_i^{(k+1)} = (\gamma - \tilde{\sigma}_i^2 / \gamma) \tilde{u}_i^{(k)}, \tag{11.110}$$

followed by the normalization

$$\tilde{u}_i^{(k+1)} = \tilde{u}_i^{(k+1)} / \|\tilde{u}_i^{(k+1)}\|_2.$$

It is easy to show that the left-hand-side matrix in (11.109) is symmetric positive
definite provided (11.106) holds. Accordingly, we may use parallel conjugate gradient iterations to refine each singular triplet approximation. Hence, the refinement
procedure outlined in may be considered as a *black box* procedure to follow the
eigensolution of $A^\top A$ or $\gamma^2 I_n - A^\top A$ by any one of the above candidate methods
for the sparse singular-value problem. The iterations in steps 2–9 of the refinement

scheme in Algorithm 11.20 terminate once the norms of the residuals of all p approximate singular triplets ($\|r_i\|_2$) fall below a user-specified tolerance or after k_{max} iterations.

Algorithm 11.20 Refinement procedure for the left singular vector approximations obtained via scaling.

Input: $A \in \mathbb{R}^{m \times n}$, p approximate singular values $\Sigma = \text{diag}(\tilde{\sigma}_1, \cdots, \tilde{\sigma}_p)$ and their corresponding approximate right singular vectors $V = [\tilde{v}_1, \cdots, \tilde{v}_p]$.

1: $U_0 = AV\Sigma^{-1}$;   //By definition: $U_k = [\tilde{u}_1^{(k)}, \cdots, \tilde{u}_p^{(k)}]$

2: **do** $j = 1 : p$,

3:    $k = 0$;

4:    **while** $\|A\tilde{v}_j - \tilde{\sigma}_j \tilde{u}_j^{(k)}\| > \tau$,

5:       $k := k + 1$;

6:       Solve $\left(\gamma I_m - \frac{1}{\gamma} AA^\top\right)\tilde{u}_i^{(k+1)} = (\gamma - \tilde{\sigma}_i^2/\gamma)\tilde{u}_i^{(k)}$;   //See (11.110)

7:       Set $\tilde{u}_i^{(k+1)} = \tilde{u}_i^{(k+1)}/\|\tilde{u}_i^{(k+1)}\|_2$;

8:    **end while**

9: **end**

References

1. Parlett, B.N.: The Symmetric Eigenvalue Problem. Prentice Hall, Englewood Cliffs (1980)
2. Bauer, F.: Das verfahren der treppeniteration und verwandte verfahren zur losung algebraischer eigenwertprobleme. ZAMP **8**, 214–235 (1957)
3. Wilkinson, J.H.: The Algebraic Eigenvalue Problem. Oxford University Press, New York (1965)
4. Rutishauser, H.: Simultaneous iteration method for symmetric matrices. Numer. Math. **16**, 205–223 (1970)
5. Stewart, G.W.: Simultaneous iterations for computing invariant subspaces of non-Hermitian matrices. Numer. Math. **25**, 123–136 (1976)
6. Stewart, W.J., Jennings, A.: Algorithm 570: LOPSI: a simultaneous iteration method for real matrices [F2]. ACM Trans. Math. Softw. **7**(2), 230–232 (1981). doi:10.1145/355945.355952
7. Saad, Y.: Numerical Methods for Large Eigenvalue Problems. Halstead Press, New York (1992)
8. Sameh, H., Lermit, J., Noh, K.: On the intermediate eigenvalues of symmetric sparse matrices. BIT 185–191 (1975)
9. Bunch, J., Kaufman, K.: Some stable methods for calculating inertia and solving symmetric linear systems. Math. Comput. **31**, 162–179 (1977)
10. Golub, G., Van Loan, C.: Matrix Computations, 4th edn. Johns Hopkins University Press, Baltimore (2013)
11. Duff, I., Gould, N.I.M., Reid, J.K., Scott, J.A., Turner, K.: The factorization of sparse symmetric indefinite matrices. IMA J. Numer. Anal. **11**, 181–204 (1991)
12. Duff, I.: MA57-a code for the solution of sparse symmetric definite and indefinite systems. ACM TOMS 118–144 (2004)
13. Kalamboukis, T.: Tridiagonalization of band symmetric matrices for vector computers. Comput. Math. Appl. **19**, 29–34 (1990)
14. Lang, B.: A parallel algorithm for reducing symmetric banded matrices to tridiagonal form. SIAM J. Sci. Comput. **14**(6), 1320–1338 (1993). doi:10.1137/0914078

15. Philippe, B., Vital, B.: Parallel implementations for solving generalized eigenvalue problems with symmetric sparse matrices. Appl. Numer. Math. **12**, 391–402 (1993)
16. Carey, C., Chen, H.C., Golub, G., Sameh, A.: A new approach for solving symmetric eigenvalue problems. Comput. Sys. Eng. **3**(6), 671–679 (1992)
17. Golub, G., Underwood, R.: The block Lanczos method for computing eigenvalues. In: Rice, J. (ed.) Mathematical Software III, pp. 364–377. Academic Press, New York (1977)
18. Underwood, R.: An iterative block Lanczos method for the solution of large sparse symmetric eigenproblems. Technical Report STAN-CS-75-496, Computer Science, Stanford University, Stanford (1975)
19. Kaniel, S.: Estimates for some computational techniques in linear algebra. Math. Comput. **20**, 369–378 (1966)
20. Paige, C.: The computation of eigenvalues and eigenvectors of very large sparse matrices. Ph.D. thesis, London University, London (1971)
21. Meurant, G.: The Lanczos and Conjugate Gradient Algorithms: from Theory to Finite Precision Computations (Software, Environments, and Tools). SIAM, Philadelphia (2006)
22. Paige, C.C.: Accuracy and effectiveness of the Lanczos algorithm for the symmetric eigenproblem. Linear Algebra Appl. **34**, 235–258 (1980)
23. Cullum, J.K., Willoughby, R.A.: Lanczos Algorithms for Large Symmetric Eigenvalue Computations. SIAM, Philadelphia (2002)
24. Lehoucq, R., Sorensen, D.: Deflation techniques for an implicitly restarted Arnoldi iteration. SIAM J. Matrix Anal. Appl. **17**, 789–821 (1996)
25. Lehoucq, R., Sorensen, D., Yang, C.: ARPACK User's Guide: Solution of Large-Scale Eigenvalue Problems With Implicitly Restarted Arnoldi Methods. SIAM, Philadelphia (1998)
26. Calvetti, D., Reichel, L., Sorensen, D.C.: An implicitly restarted Lanczos method for large symmetric eigenvalue problems. Electron. Trans. Numer. Anal. **2**, 1–21 (1994)
27. Sorensen, D.: Implicit application of polynomial filters in a k-step Arnoldi method. SIAM J. Matrix Anal. Appl. **13**, 357–385 (1992)
28. Lewis, J.G.: Algorithms for sparse matrix eigenvalue problems. Technical Report STAN-CS-77-595, Department of Computer Science, Stanford University, Palo Alto (1977)
29. Ruhe, A.: Implementation aspects of band Lanczos algorithms for computation of eigenvalues of large sparse symmetric matrices. Math. Comput. **33**, 680–687 (1979)
30. Scott, D.: Block lanczos software for symmetric eigenvalue problems. Technical Report ORNL/CSD-48, Oak Ridge National Laboratory, Oak Ridge (1979)
31. Baglama, J., Calvetti, D., Reichel, L.: IRBL: an implicitly restarted block Lanczos method for large-scale Hermitian eigenproblems. SIAM J. Sci. Comput. **24**(5), 1650–1677 (2003)
32. Chen, H.C., Sameh, A.: Numerical linear algebra algorithms on the cedar system. In: Noor, A. (ed.) Parallel Computations and Their Impact on Mechanics, Applied Mechanics Division, vol. 86, pp. 101–125. American Society of Mechanical Engineers (1987)
33. Chen, H.C.: The sas domain decomposition method. Ph.D. thesis, University of Illinois at Urbana-Champaign (1988)
34. Davidson, E.: The iterative calculation of a few of the lowest eigenvalues and corresponding eigenvectors of large real-symmetric matrices. J. Comput. Phys. **17**, 817–825 (1975)
35. Morgan, R., Scott, D.: Generalizations of Davidson's method for computing eigenvalues of sparse symmetric matrices. SIAM J. Sci. Stat. Comput. **7**, 817–825 (1986)
36. Crouzeix, M., Philippe, B., Sadkane, M.: The Davidson method. SIAM J. Sci. Comput. **15**, 62–76 (1994)
37. Sameh, A.H., Wisniewski, J.A.: A trace minimization algorithm for the generalized eigenvalue problem. SIAM J. Numer. Anal. **19**(6), 1243–1259 (1982)
38. Sameh, A., Tong, Z.: The trace minimization method for the symmetric generalized eigenvalue problem. J. Comput. Appl. Math. **123**, 155–170 (2000)
39. Sleijpen, G., van der Vorst, H.: A Jacobi-Davidson iteration method for linear eigenvalue problems. SIAM J. Matrix Anal. Appl. **17**, 401–425 (1996)
40. Simoncini, V., Eldén, L.: Inexact Rayleigh quotient-type methods for eigenvalue computations. BIT Numer. Math. **42**(1), 159–182 (2002). doi:10.1023/A:1021930421106

41. Bathe, K., Wilson, E.: Large eigenvalue problems in dynamic analysis. ASCE J. Eng. Mech. Div. **98**, 1471–1485 (1972)
42. Bathe, K., Wilson, E.: Solution methods for eigenvalue problems in structural mechanics. Int. J. Numer. Methods Eng. **6**, 213–226 (1973)
43. Grimm, R., Greene, J., Johnson, J.: Computation of the magnetohydrodynamic spectrum in axissymmetric toroidal confinement systems. Method Comput. Phys. **16** (1976)
44. Gruber, R.: Finite hybrid elements to compute the ideal magnetohydrodynamic spectrum of an axisymmetric plasma. J. Comput. Phys. **26**, 379–389 (1978)
45. Stewart, G.: A bibliographical tour of the large, sparse generalized eigenvalue problems. In: Banch, J., Rose, D. (eds.) Sparse Matrix Computations, pp. 113–130. Academic Press, New York (1976)
46. van der Vorst, H., Golub, G.: One hundred and fifty years old and still alive: eigenproblems. In: Duff, I., Watson, G. (eds.) The State of the Art in Numerical Analysis, pp. 93–119. Clarendon Press, Oxford (1997)
47. Rutishauser, H.: Computational aspects of f. l. bauer's simultaneous iteration method. Numerische Mathematik **13**(1), 4–13 (1969). doi:10.1007/BF02165269
48. Clint, M., Jennings, A.: The evaluation of eigenvalues and eigenvectors of real symmetric matrices by simultaneous iteration. Computers **13**, 76–80 (1970)
49. Levin, A.: On a method for the solution of a partial eigenvalue problem. J. Comput. Math. Math. Phys. **5**, 206–212 (1965)
50. Stewart, G.: Accelerating the orthogonal iteration for the eigenvalues of a Hermitian matrix. Numer. Math. **13**, 362–376 (1969)
51. Sakurai, T., Sugiura, H.: A projection method for generalized eigenvalue problems using numerical integration. J. Comput. Appl. Math. **159**, 119–128 (2003)
52. Tang, P., Polizzi, E.: FEAST as a subspace iteration eigensolver accelerated by approximate spectral projection. SIAM J. Matrix Anal. Appl. **35**(2), 354–390 (2014)
53. Lanczos, C.: An iteration method for the solution of the eigenvalue problem of linear differential and integral operators. J. Res. Natl. Bur. Stand. **45**, 225–280 (1950)
54. Fokkema, D.R., Sleijpen, G.A.G., van der Vorst, H.A.: Jacobi-Davidson style QR and QZ algorithms for the reduction of matrix pencils. SIAM J. Sci. Comput. **20**(1), 94–125 (1998)
55. Sleijpen, G., Booten, A., Fokkema, D., van der Vorst, H.: Jacobi-Davidson type methods for generalized eigenproblems and polynomial eigenproblems. BIT **36**, 595–633 (1996)
56. Cullum, J., Willoughby, R.: Lanczos and the computation in specified intervals of the spectrum of large, sparse, real symmetric matrices. In: Duff, I., Stewart, G. (eds.) Proceedings of the Sparse Matrix 1978. SIAM (1979)
57. Parlett, B., Scott, D.: The Lanczos algorithm with selective orthogonalization. Math. Comput. **33**, 217–238 (1979)
58. Simon, H.: The Lanczos algorithm with partial reorthogonalization. Math. Comput. **42**, 115–142 (1984)
59. Cullum, J., Willoughby, R.: Computing eigenvalues of very large symmetric matrices—an implementation of a Lanczos algorithm with no reorthogonalization. J. Comput. Phys. **44**, 329–358 (1984)
60. Ericsson, T., Ruhe, A.: The spectral transformation Lanczos method for the solution of large sparse generalized symmetric eigenvalue problems. Math. Comput. **35**, 1251–1268 (1980)
61. Grimes, R., Lewis, J., Simon, H.: A shifted block Lanczos algorithm for solving sparse symmetric generalized Eigenproblems. SIAM J. Matrix Anal. Appl. **15**, 228–272 (1994)
62. Kalamboukis, T.: A Lanczos-type algorithm for the generalized eigenvalue problem $ax = \lambda bx$. J. Comput. Phys. **53**, 82–89 (1984)
63. Liu, B.: The simultaneous expansion for the solution of several of the lowest eigenvalues and corresponding eigenvectors of large real-symmetric matrices. In: Moler, C., Shavitt, I. (eds.) Numerical Algorithms in Chemistry: Algebraic Method, pp. 49–53. University of California, Lawrence Berkeley Laboratory (1978)
64. Stathopoulos, A., Saad, Y., Fischer, C.: Robust preconditioning of large, sparse, symmetric eigenvalue problems. J. Comput. Appl. Math. 197–215 (1995)

65. Wu, K.: Preconditioning techniques for large eigenvalue problems. Ph.D. thesis, University of Minnesota (1997)
66. Jacobi, C.: Über ein leichtes verfahren die in der theorie der säculärstörungen vorkom menden gleichungen numerisch aufzulösen. Crelle's J. für reine und angewandte Mathematik **30**, 51–94 (1846)
67. Beckenbach, E., Bellman, R.: Inequalities. Springer, New York (1965)
68. Kantorovič, L.: Functional analysis and applied mathematics (Russian). Uspekhi Mat. Nauk. **3**, 9–185 (1948)
69. Newman, M.: Kantorovich's inequality. J. Res. Natl. Bur. Stand. B. Math. Math. Phys. **64B**(1), 33–34 (1959). http://nvlpubs.nist.gov/nistpubs/jres/64B/jresv64Bn1p33_A1b.pdf
70. Benzi, M., Golub, G., Liesen, J.: Numerical solution of Saddle-point problems. Acta Numerica pp. 1–137 (2005)
71. Elman, H., Silvester, D., Wathen, A.: Performance and analysis of Saddle-Point preconditioners for the discrete steady-state Navier-Stokes equations. Numer. Math. **90**, 641–664 (2002)
72. Paige, C.C., Saunders, M.A.: Solution of sparse indefinite systems of linear equations. SIAM J. Numer. Anal. **12**(4), 617–629 (1975)
73. Daniel, J., Gragg, W., Kaufman, L., Stewart, G.: Reorthogonalization and stable algorithms for updating the Gram-Schmidt QR factorization. Math. Comput. **136**, 772–795 (1976)
74. Sun, J.G.: Condition number and backward error for the generalized singular value decomposition. SIAM J. Matrix Anal. Appl. **22**(2), 323–341 (2000)
75. Stewart, G.G., Sun, J.: Matrix Perturbation Theory. Academic Press, Boston (1990)
76. Demmel, J., Gu, M., Eisenstat, S.I., Slapničar, K., Veselić, Z.D.: Computing the singular value decomposition with high accuracy. Linear Algebra Appl. **299**(1–3), 21–80 (1999)
77. Sun, J.: A note on simple non-zero singular values. J. Comput. Math. **6**(3), 258–266 (1988)
78. Berry, M., Sameh, A.: An overview of parallel algorithms for the singular value and symmetric eigenvalue problems. J. Comput. Appl. Math. **27**, 191–213 (1989)
79. Dongarra, J., Sorensen, D.C.: A fully parallel algorithm for the symmetric eigenvalue problem. SIAM J. Sci. Stat. Comput. **8**(2), s139–s154 (1987)
80. Golub, G., Reinsch, C.: Singular Value Decomposition and Least Squares Solutions. Springer (1971)
81. Golub, G., Luk, F., Overton, M.: A block Lanczos method for computing the singular values and corresponding singular vectors of a matrix. ACM Trans. Math. Softw. **7**, 149–169 (1981)
82. Berry, M.: Large scale singular value decomposition. Int. J. Supercomput. Appl. **6**, 13–49 (1992)
83. Wilkinson, J.: Inverse Iteration in Theory and in Practice. Academic Press (1972)
84. Philippe, B., Sadkane, M.: Computation of the fundamental singular subspace of a large matrix. Linear Algebra Appl. **257**, 77–104 (1997)
85. Hochstenbach, M.: A Jacobi-Davidson type SVD method. SIAM J. Sci. Comput. **23**(2), 606–628 (2001)

Part IV

Matrix Functions and Characteristics

Chapter 12
Matrix Functions and the Determinant

Many applications require the numerical evaluation of matrix functions, such as polynomials and rational or transcendental functions with matrix arguments (e.g. the exponential, the logarithm or trigonometric), or scalar functions of the matrix elements, such as the determinant. When the underlying matrices are large, it is important to have methods that lend themselves to parallel implementation.

As with iterative methods, two important applications are the numerical solution of ODEs and the computation of network metrics. Regarding the former, it has long been known that many classical schemes for solving ODEs amount to approximating the action of the matrix exponential on a vector [1]. Exponential integrators have become a powerful tool for solving large scale ODEs (e.g. see [2] and references therein) and software has been developed to help in this task on uniprocessors and (to a much lesser extent) parallel architectures (e.g. see [3–6]). Recall also that in our discussion of rapid elliptic solvers in Sect. 6.4 we have already encountered computations with matrix rational functions.

Matrix function-based metrics for networks have a long history. It was shown as early as 1949 in [7] that if A is an adjacency matrix for a digraph, then the elements of A^k show the number of paths of length k that connect any two nodes. A few years later, a status score was proposed for each node of a social network based on the underlying link structure [8]. Specifically, the score used the values of the product of the resolvent $(\zeta I - A)^{-1}$ with a suitable vector. These ideas have resurfaced with ranking algorithms such as Google's PageRank and HITS [9–11] and interest in this topic has peaked in the last decade due in part to the enormous expansion of the Web and social networks. For example, researchers are studying how to compute such values fast for very large networks; cf. [12].

Some network metrics use the matrix exponential [13, 14], the matrix hyperbolic sine and cosine functions [15–17] or special series and polynomials [18–20]. In other areas, such as multivariate statistics, economics, physics, and uncertainty quantification, one is interested in the determinant, the diagonal of the matrix inverse and its trace; cf. [21–23].

In this chapter we consider a few parallel algorithms for evaluating matrix rational functions, the matrix exponential and the determinant. The monographs [24, 25] and

© Springer Science+Business Media Dordrecht 2016
E. Gallopoulos et al., *Parallelism in Matrix Computations*,
Scientific Computation, DOI 10.1007/978-94-017-7188-7_12

[26], are important references regarding the theoretical background and the design of high quality algorithms for problems of this type on uniprocessors. Reference [27] is a useful survey of software for matrix functions. Finally, it is worth noting that the numerical evaluation of matrix functions is an extremely active area of research, and many interesting developments are currently under consideration or yet to come.

12.1 Matrix Functions

It is well known from theory that if A is diagonalizable with eigenvalues $\{\lambda_i\}_{i=1}^n$ and if the scalar function $f(\zeta)$ is such that $f(\lambda_i)$ exists for $i = 1, \ldots, n$ and $Q^{-1}AQ = \Lambda$ is the matrix of eigenvalues, then the matrix function is defined as $f(A) = Qf(\Lambda)Q^{-1}$. If A is non-diagonalizable, and $Q^{-1}AQ = \mathrm{diag}(J_1, \ldots, J_p)$ is its Jordan canonical form, then $f(A)$ is defined as $f(A) = Q\mathrm{diag}(f(J_1), \ldots, f(J_p))Q^{-1}$, assuming that for each eigenvalue λ_i, the derivatives $\{f(\lambda_i), f^{(1)}(\lambda_i), \ldots, \frac{f^{(n_i-1)}(\lambda_i)}{(n_i-1)!}\}$ exist where n_i is the size of the largest Jordan block containing λ_i and $f(J_i)$ is the Toeplitz upper triangular matrix with first row the vector

$$(f(\lambda_i), f^{(1)}(\lambda_i), \ldots, \frac{f^{(m_i-1)}(\lambda_i)}{(m_i - 1)!}).$$

A matrix function can also be defined using the Cauchy integral

$$f(A) = \frac{1}{2\pi\iota} \int_\Gamma f(\zeta)(\zeta I - A)^{-1}d\zeta, \tag{12.1}$$

where Γ is a closed contour of winding number one enclosing the eigenvalues of A.

A general template for computations involving matrix functions is

$$D + Cf(A)B \tag{12.2}$$

where f is the function under consideration defined on the spectrum of the square matrix A and B, C, D are of compatible shapes. A large variety of matrix computations, including most of the BLAS, matrix powers and inversion, linear systems with multiple right-hand sides and bilinear forms can be cast as in (12.2). We focus on *rational functions* (that is ratios of polynomials) and the *matrix exponential*. Observe that the case of functions with a linear denominator amounts to matrix inversion or solving linear systems, which have been addressed earlier in this book.

Polynomials and rational functions are primary tools in the practical approximation of scalar functions. Moreover, as is well known from theory, matrix functions can also be defined by means of a unique polynomial (the Hermite interpolating polynomial) that depends on the underlying matrix and its degree is at most that of the minimal polynomial. Even though, in general, it is impractical to use it, its existence provides a strong motivation for using polynomials. Finally, rational functions are sometimes even more effective than polynomials in approximating scalar functions. Since their manipulation in a matrix setting involves the solution of linear systems, an interesting question, that we also discuss, is how to manipulate matrix rational functions efficiently in a parallel environment.

We have already encountered matrix rational functions in this book. Recall our discussion of algorithms BCR and EES in Sect. 6.4, where we used ratios of Chebyshev polynomials, and described the advantage of using partial fractions in a parallel setting. In this section we extend this approach to more general rational functions. Specifically, we consider the case where f in (12.2) is a rational function denoted by

$$r(\zeta) = \frac{q(\zeta)}{p(\zeta)} \tag{12.3}$$

where the polynomials q and p are such that $r(A)$ exists.

Remark 12.1 Unless noted otherwise, we make the following assumptions for the rational function in (12.3): (i) the degree of q (denoted by deg q) is no larger than the degree of p. (ii) The polynomials p and q have no common roots. (iii) The roots of p (the poles of r) are mutually distinct. (iv) The degrees of p and q are much smaller than the matrix size. (v) We also assume that the coefficients have been normalized so that p is monic. Because these assumptions hold for a large number of computations of interest, we will be referring to them as *standard assumptions* for the rational function.

Since $r(A)$ exists, the set of poles of r and the set of eigenvalues of A have empty intersection. Assumption (iv) is directly related to our focus on large matrices and the fact that it is impractical and in the context of most problems, not necessary anyway, to compute with matrix polynomials and rational functions of very high degree.

In the sequel, we assume that the roots of scalar polynomials as well as partial fraction coefficients are readily available, as the cost of their computation is insignificant relative to the matrix operations. The product form representation of p is

$$p(\zeta) = \prod_{j=1}^{d} (\zeta - \tau_j), \text{ where } d = \deg p.$$

Two cases of the template (12.2) amount to computing one of the following:

$$\text{either } r(A) = (p(A))^{-1}q(A), \text{ or } x = (p(A))^{-1}q(A)b. \tag{12.4}$$

The vector x can be computed by first evaluating $r(A)$ and then multiplying by b. As with matrix inversion, we can approximate x without first computing $r(A)$ explicitly.

12.1.1 Methods Based on the Product Form
of the Denominator

Algorithm 12.1 is one way of evaluating $x = (p(A))^{-1}b$ when $q \equiv 1$. A parallel implementation can only exploit the parallelism available in the solution of each linear system. Algorithm 12.2, on the other hand, contains a parallel step at each one of the d outer iterations. In both cases, because the matrices are shifts of the same A, it might be preferable to first reduce A to an orthogonally similar Hessenberg form. We comment more extensively on this reduction in the next subsection and in Sect. 13.1.2.

Algorithm 12.1 Computing $x = (p(A))^{-1}b$ when $p(\zeta) = \prod_{j=1}^{d}(\zeta - \tau_j)$ with τ_j mutually distinct.

Input: $A \in \mathbb{R}^{n \times n}$, $b \in \mathbb{R}^n$ and values τ_j distinct from the eigenvalues of A.
Output: Solution $(p(A))^{-1}b$.
1: $x_0 = b$
2: **do** $j = 1 : d$
3:     solve $(A - \tau_j I)x_j = x_{j-1}$
4: **end**
5: return $x = x_d$

Algorithm 12.2 Computing $X = (p(A))^{-1}$ when $p(\zeta) = \prod_{j=1}^{d}(\zeta - \tau_j)$ with τ_j mutually distinct

---

Input: $A \in \mathbb{R}^{n \times n}$, $b \in \mathbb{R}^n$ and values τ_j distinct from the eigenvalues of A.
Output: $(p(A))^{-1}$.
1: $x_i^{(0)} = e_i$ for $i = 1, \ldots, n$ and $X = (x_1^{(0)}, \ldots, x_n^{(0)})$.
2: **do** $j = 1 : d$
3:     **doall** $i = 1 : n$
4:         solve $(A - \tau_j I)x_i^{(j)} = x_i^{(j-1)}$
5:     **end**
6: **end**
7: return $X = (x_1^{(d)}, \ldots, x_n^{(d)})$

We next consider evaluating r in the case that the numerator, q, is non-trivial and/or its degree is larger than the denominator (violating for the purposes of this discussion standard assumption (i)). If q is in power form, then Horner's rule can be used. The parallel evaluation of scalar polynomials has been discussed in the literature and in this book (cf. the parallel Horner scheme in Sect. 3.3). See also [28–34]. The presence of matrix arguments introduces some challenges as the relative costs of the operations are very different. In particular, the costs of multiplying by a scalar coefficient (in the power form), subtracting a scalar from the diagonal (in the product form) or adding two matrices are much smaller than the cost of matrix multiplication.

A method that takes into account these differences and uses approximately $2\sqrt{d}$ matrix multiplications was presented in [24] based on an original algorithm (sometimes called Paterson-Stockmeyer) from Ref. [35]. The basic idea is to express $p(\zeta)$ as a polynomial in the variable ζ^s for suitable s, with coefficients that are also polynomials, of degree d/s in ζ. The parallelism in the above approach stems first from the matrix multiplications; cf. Sect. 2.2 for related algorithms for dense matrices. Further opportunities are also available. For example, when $d = 6$ and $s = 3$, the expression used in [24] is

$$p(A) = \pi_6 I (A^3)^2 + (\pi_5 A^2 + \pi_4 A + \pi_3 I)A^3 + (\pi_2 A^2 + \pi_1 A + \pi_0 I),$$

Evidently, the matrix terms in parentheses can be evaluated in parallel. A parallel implementation on GPUs was considered in [36].

We next sketch a slightly different approach. Consider, for example, a polynomial of degree 7,

$$p(\zeta) = \sum_{j=0}^{7} \pi_j \zeta^j.$$

This can be written as a cubic polynomial in ζ^2, with coefficients that are linear in ζ,

$$\sum_{j=0}^{3} (\pi_{2j} + \pi_{2j+1}\zeta)\zeta^{2j}$$

or as a linear polynomial in ζ^4 with coefficients that are cubic polynomials in ζ,

$$p(\zeta) = \sum_{j=0}^{1} (\pi_{4j} + \pi_{4j+1}\zeta + \pi_{4j+2}\zeta^2 + \pi_{4j+3}\zeta^3)\zeta^{4j}.$$

These can be viewed as modified instances of a recursive approach described in [28] that consists of expressing p with "polynomial multiply-and-add" operations, where the one of the two summands consists of the terms of even degree and the other is the product of ζ with another even polynomial. Assuming for simplicity that $d = 2^k - 1$, we can write p as a polynomial multiply-and-add

$$p_{2^k-1}(\zeta) = p^{(1)}_{2^{k-1}-1}(\zeta) + \zeta^{2^{k-1}} \hat{p}^{(1)}_{2^{k-1}-1}(\zeta) \tag{12.5}$$

where we use the subscript to explicitly denote the maximum degree of the respective polynomials. Note that the process can be applied recursively on each term $p^{(1)}_j$ and $\hat{p}^{(1)}_j$. We can express this entirely in matrix form as follows. Define the matrices of order $2n$

$$M_j = \begin{pmatrix} A & 0 \\ \pi_j I & I \end{pmatrix}, \quad j = 0, 1, \ldots, 2^k - 1. \tag{12.6}$$

Then it follows that

$$M_{2^k-1} M_{2^k-2} \cdots M_0 = \begin{pmatrix} A^{2^k} & 0 \\ p_{2^k-1}(A) & I \end{pmatrix}, \tag{12.7}$$

so the sought polynomial is in block position $(2, 1)$ of the product. Moreover,

$$M_{2^k-1} M_{2^k-2} \cdots M_{2^{k-1}} = \begin{pmatrix} A^{2^{k-1}} & 0 \\ \hat{p}^{(1)}_{2^{k-1}-1}(A) & I \end{pmatrix}$$

and

$$M_{2^{k-1}-1} M_{2^{k-1}-2} \cdots M_0 = \begin{pmatrix} A^{2^{k-1}} & 0 \\ p^{(1)}_{2^{k-1}-1}(A) & I \end{pmatrix}$$

where the terms $p^{(1)}_{2^{k-1}-1}, \hat{p}^{(1)}_{2^{k-1}-1}$ are as in decomposition (12.5). The connection of the polynomial multiply-and-add approach (12.5) with the product form (12.7) using the terms M_j defined in (12.6), motivates the design of algorithms for the parallel computation of the matrix polynomial based on a parallel fan-in approach, e.g. as that proposed to solve triangular systems in Sect. 3.2.1 (Algorithm 3.2). With no limit on the number of processors, such a scheme would take $\log d$ stages each consisting of a matrix multiply (squaring) and a matrix multiply-and-add. See also [37] for interesting connections between these approaches. For a limited number of processors, the Paterson-Stockmeyer algorithm can be applied in each processor to evaluate a low degree polynomial corresponding to the multiplication of some of the terms M_j. Subsequently, the fan-in approach can be applied to combine these intermediate results.

We also need to note that the aforementioned methods require more storage compared to the standard Horner scheme.

12.1.2 Methods Based on Partial Fractions

An alternative approach to the methods of the previous subsection is based on the partial fraction representation. This not only highlights the linear aspects of rational functions, as noted in [26], but it simplifies their computation in a manner that engenders parallel processing. The partial fraction representation also emerges in the approximation of matrix functions via numerical quadrature based on the Cauchy integral form (12.1); cf. [5, 38].

In the simplest case of linear p and q, that is solving

$$(A - \tau I)x = (A - \sigma I)b, \tag{12.8}$$

the Laurent expansion for the rational function around the pole τ can be written as

$$\frac{\zeta - \sigma}{\zeta - \tau} = (\tau - \sigma)\frac{1}{\zeta - \tau} + 1,$$

and so the matrix-vector multiplications can be completely avoided by replacing (12.8) with

$$x = b + (\tau - \sigma)(A - \tau I)^{-1}b;$$

cf. [39]. Recall that we have already made use of partial fractions to enable the parallel evaluation of some special rational functions in the rapid elliptic solvers of Sect. 6.4. The following result is well known; see e.g. [40].

Theorem 12.1 *Let p, q be polynomials such that* deg $q \le$ deg p, *and let $\tau_1, \ldots, \tau_d$ be the distinct zeros of p, with multiplicities of $\mu_1, \ldots, \mu_d$, respectively, and distinct from the roots of q. Then at each τ_j, the rational function $r(\zeta) = \frac{q(\zeta)}{p(\zeta)}$ has a pole of order at most μ_j, and the partial fraction representation of r is given by,*

$$r(\zeta) = \rho_0 + \sum_{i=1}^{d} \pi_i(\zeta), \ \text{with } \pi_i(\zeta) = \sum_{j=1}^{\mu_j} \gamma_{i,j}(\zeta - \tau_j)^{-j}$$

where $\pi_i(\zeta)$ denotes the principal part of r at the pole τ_j, $\rho_0 = \lim_{|z| \to \infty} r(\zeta)$ and the constants $\gamma_{i,j}$ are the partial fraction expansion coefficients of r.

When the standard assumptions hold for the rational function, the partial fraction expansion is

$$r(A) = \rho_0 I + \sum_{j=1}^{d} \rho_j (A - \tau_j I)^{-1} \ \text{with } \rho_j = \frac{q(\tau_j)}{p'(\tau_j)}, \tag{12.9}$$

which exhibits ample opportunities for parallel evaluation. One way to compute $r(A)$ is to first evaluate the terms $(A - \tau_j I)^{-1}$ for $j = 1, \ldots, d$ simultaneously, followed by the weighted summation. To obtain the vector $r(A)b$, we can first solve the d systems $(A - \tau_j I)x_j = b$ simultaneously, followed by computing $x = b + \sum_{j=1}^{d} \rho_j x_j$. The latter can also be written as an MV operation,

$$x = b + Xr, \ \text{where } X = (x_1, \ldots, x_d), \ \text{and } r = (\rho_1, \ldots, \rho_d)^{\top}. \tag{12.10}$$

Algorithm 12.3 uses partial fractions to compute $r(A)b$.

Algorithm 12.3 Computing $x = (p(A))^{-1}q(A)b$ when $p(\zeta) = \prod_{j=1}^{d}(\zeta - \tau_j)$ and the roots τ_j are mutually distinct

---

Input: $A \in \mathbb{R}^{n \times n}$, $b \in \mathbb{R}^n$, values $\{\tau_1, \ldots, \tau_d\}$ (roots of p).
Output: Solution $x = (p(A))^{-1}q(A)b$.
1: **doall** $j = 1 : d$
2:     compute coefficient $\rho_j = \dfrac{q(\tau_j)}{p'(\tau_j)}$
3:     solve $(A - \tau_j I)x_j = b$
4: **end**
5: set $\rho_0 = \lim_{|z| \to \infty} \frac{q(\zeta)}{p(\zeta)}$, $r = (\rho_1, \ldots, \rho_d)^\top$, $X = (x_1, \ldots, x_d)$
6: compute $x = \rho_0 b + Xr$ //it is important to compute this step in parallel

---

The partial fraction coefficients are computed only once for any given function, thus under the last of the standard assumptions listed in Remark 12.1, the cost of line 2 is not taken into account; cf. [40] for a variety of methods. On d processors, each can undertake the solution of one linear system (line 3 of Algorithm 12.3). Except for line 6, where one needs to compute the linear combination of the d solution vectors, the computations are totally independent. Thus, the cost of this algorithm on d processors is equal to a (possibly complex) linear system solve per processor and a dense MV. For general dense matrices of size n, the cost of solving the linear systems dominates. For structured matrices, the cost of each system solve could be comparable or even less than that of a single dense sequential MV. For tridiagonals of order n, for example, the cost of a single sequential MV, in line 6, dominates the $O(n)$ cost of the linear system solve in line 3. Thus, it is important to consider using a parallel algorithm for the dense MV. If the number of processors is smaller than d then more systems can be assigned to each processor, whereas if it is larger, more processors can participate in the solution of each system using some parallel algorithm.

It is straightforward to modify Algorithm 12.3 to compute $r(A)$ directly. Line 3 becomes an inversion while line 6 a sum of d inverses, costing $O(n^2 d)$. For structured matrices for which the inversion can be accomplished at a cost smaller than $O(n^3)$, the cost of the final step can become comparable or even larger, assuming that the inverses have no special structure.

Next, we consider in greater detail the solution of the linear systems in Algorithm 12.3. Notice that all matrices involved are simple diagonal shifts of the form $(A - \tau_j I)$. One possible preprocessing step is to transform A to an orthogonally similar upper Hessenberg or (if symmetric) tridiagonal form; cf. [41] for an early use of this observation, [42] for its application in computing the matrix exponential and [43] for the case of reduction to tridiagonal form when the matrix is symmetric.

This preprocessing, incorporated in Algorithm 12.4, leads to considerable savings since the reduction step is performed only once while the Hessenberg linear systems in line 5 require only $O(n^2)$ operations. The cost drops to $O(n)$ when the matrix is symmetric since the reduced system becomes tridiagonal. In both cases, the dominant cost is the reduction in line 1; e.g. see [44, 45] for parallel solvers of such linear systems. In Chap. 13 we will make use of the same preprocessing step for computing the matrix pseudospectrum.

Algorithm 12.4 Compute $x = (p(A))^{-1}q(A)b$ when $p(\zeta) = \prod_{j=1}^{d}(\zeta - \tau_j)$ and the roots τ_j are mutually distinct

---

Input: $A \in \mathbb{R}^{n \times n}$, $b \in \mathbb{R}^n$, values $\{\tau_1, \ldots, \tau_d\}$ (roots of p).
Output: Solution $x = (p(A))^{-1}q(A)b$.
1: $[Q, H] = \texttt{hess}(A)$ //Q obtained as product of Householder transformations so that $Q^\top A Q = H$ is upper Hessenberg
2: compute $\hat{b} = Q^\top b$ //exploit the product form
3: **doall** $j = 1 : d$
4:   compute coefficient $\rho_j = \dfrac{q(\tau_j)}{p'(\tau_j)}$
5:   solve $(H - \tau_j I)\hat{x}_j = \hat{b}$
6:   compute $x_j = Q\hat{x}_j$ //exploit the product form
7: **end**
8: set $\rho_0 = \lim_{|z|\to\infty}\frac{q(\zeta)}{p(\zeta)}$, $r = (\rho_1, \ldots, \rho_d)^\top$, $X = Q(\hat{x}_1, \ldots, \hat{x}_d)$
9: compute and return $x = \rho_0 b + Xr$ //it is important to compute this step in parallel

---

Note that when the rational function is real, any complex poles appear in conjugate pairs. Then the partial fraction coefficients also appear in conjugate pairs which makes possible significant savings. For example, if (ρ, τ) and $(\bar{\rho}, \bar{\tau})$ are two such conjugate pairs, then

$$\rho(A - \tau I)^{-1} + \bar{\rho}(A - \bar{\tau}I)^{-1} = 2\Re\left(\rho(A - \tau I)^{-1}\right). \tag{12.11}$$

Therefore, pairs of linear system solves can be combined into one.

Historical Remark

The use of partial fraction decomposition to enable parallel processing can be traced back to an ingenious idea proposed in [46] for computing ζ^d, where ζ is a scalar and d a positive integer, using $O(\log d)$ parallel additions and a few parallel operations. Using this as a basis, methods for computing expressions such as $\{\zeta^2, \zeta^3, \ldots, \zeta^d\}$, $\prod_1^d(\zeta + \alpha_i)$, $\sum_0^d \alpha_i \zeta^i$ have also been developed. We briefly outline the idea since it is at the core of the technique we described above. Let $p(\omega) = \omega^d - 1$ which in factored form is $p(\omega) = \prod_{k=1}^{d}(\omega - \omega_k)$, where $\omega_k = e^{2\pi\iota\frac{k-1}{d}}$ $(k = 1, \ldots, d)$ are the dth roots of unity. Then, we can compute the logarithmic derivative p'/p explicitly as well as from the product form,

$$(p(\omega))' = \sum_{j=1}^{d}\prod_{\substack{k=1\\k\neq j}}^{d}(\omega - \omega_k).$$

$$\frac{p'}{p} = \frac{\zeta^{d-1}d}{\zeta^d - 1} = \sum_{j=1}^{d} \frac{1}{\zeta - \omega_j}$$

and therefore

$$\zeta^d = \frac{1}{1 - \frac{d}{\zeta y}} \qquad \text{where } y = \sum_{j=1}^{d} \frac{1}{\zeta - \omega_j}.$$

Therefore, ζ^d can be computed by means of Algorithm 12.5.

Algorithm 12.5 Computing ζ^d using partial fractions.

Input: $\zeta \in \mathbb{C}$, positive integer d, and values $\omega_j = e^{2\pi \iota \frac{j-1}{d}}$ $(j = 1, ..., d)$.
Output: $\tau = \zeta^d$.
1: **doall** $j = 1 : d$
2:    compute $\psi_j = (\zeta - \omega_j)^{-1}$
3: **end**
4: compute the sum $y = \sum_{j=1}^{d} \psi_j$
5: compute $\tau = \frac{1}{1 - \frac{d}{\zeta y}}$

On d processors, the total cost of the algorithm is $\log d$ parallel additions for the reduction, one parallel division and a small number of additional operations. At first sight, the result is somewhat surprising; why should one go through this process to compute powers in $O(\log d)$ parallel additions rather than apply repeated squaring to achieve the same in $O(\log d)$ (serial) multiplications? The reason is that the computational model in [46] was of a time when the cost of (scalar) multiplication was higher than addition. Today, this assumption holds if we consider matrix operations. In fact, [46] briefly mentions the possibility of using partial fraction expansions with matrix arguments. Note that similar tradeoffs drive other fast algorithms, such as the $3M$ method for complex matrix multiplication, and Strassen's method; cf. Sect. 2.2.2.

12.1.3 Partial Fractions in Finite Precision

Partial fraction expansions are very convenient for introducing parallelism but one must be alert to roundoff effects in floating-point arithmetic. In particular, there are cases where the partial fraction expansion contains terms that are large and of mixed sign. If the computed result is small relative to the summands, it could be polluted by the effect of catastrophic cancellation. For instance, building on an example from [47], the expansion of $p(\zeta) = (\zeta - \alpha)(\zeta - \delta)(\zeta + \delta)$ is

$$\frac{1}{p(\zeta)} = \frac{(\alpha^2 - \delta^2)^{-1}}{\zeta - \alpha} + \frac{(2\delta(\delta - \alpha))^{-1}}{\zeta - \delta} + \frac{(2\delta(\delta + \alpha))^{-1}}{\zeta + \delta}. \tag{12.12}$$

When $|\delta|$ is very small and $|\alpha| \gg |\delta|$ the $O(\frac{1}{\delta})$ factors in the last two terms lead to catastrophic cancellation.

In this example, the danger of catastrophic cancellation is evident from the presence of two nearby poles that trigger the generation of large partial fraction coefficients of different sign. As shown in [47], however, cancellation can also be caused by the distribution of the poles. For instance, if $p(\zeta) = \prod_{j=1}^{d}(\zeta - \frac{j}{d})$ then

$$\frac{1}{p(\zeta)} = \sum_{j=1}^{d} \frac{\rho_j}{\zeta - \frac{j}{d}}, \quad \rho_j = \frac{d^{d-1}}{\prod_{\substack{k=1 \\ k \neq j}}^{d}(j - k)}. \tag{12.13}$$

When $d = 20$, then $\rho_{20} = -\rho_1 = 20^{19}/19! \approx 4.3 \times 10^{7}$. Finally note that even if no catastrophic cancellation takes place, multiplications with large partial fraction coefficients will magnify any errors that are already present in each partial fraction term.

In order to prevent large coefficients that can lead to cancellation effects when evaluating partial fraction expansions, we follow [47] and consider hybrid representations for the rational polynomial, standing between partial fractions and the product form. Specifically, we use a modification of the *incomplete partial fraction decomposition (IPF* for short) that was devised in [48] to facilitate computing partial fraction coefficients corresponding to denominators with quadratic factors which also helps one avoid complex arithmetic; see also [40]. For example, if $p(\zeta) = (\zeta - \alpha)(\zeta^2 + \delta^2)$ is real with one real root α and two purely imaginary roots, $\pm \iota \delta$, then an *IPF* that avoids complex coefficients is

$$\frac{1}{p(\zeta)} = \frac{1}{(\alpha^2 + \delta^2)} \left(\frac{1}{\zeta - \alpha} - \frac{\zeta + \alpha}{\zeta^2 + \delta^2} \right).$$

The coefficients appearing in this *IPF* are $\pm(\alpha^2 + \delta^2)^{-1}$ and $-\alpha(\alpha^2 + \delta^2)^{-1}$. They are all real and, fortuitously, bounded for very small values of δ.

In general, assuming that the rational function $r = q/p$ satisfies the standard assumptions and that we have a factorization of the denominator $p(\zeta) = \prod_{l=1}^{\mu} p_l(\zeta)$ into non-trivial polynomial factors $p_1, \ldots, p_\mu$, then an *IPF* based on these factors is

$$r(\zeta) = \sum_{l=1}^{\mu} \frac{h_l(\zeta)}{p_l(\zeta)}. \tag{12.14}$$

where the polynomials h_l are such that $\deg h_l \leq \deg p_l$ and their coefficients can be computed using algorithms from [40]. It must be noted that the characterization "incomplete" does not mean that it is approximate but rather that it is not fully developed as a sum of d terms with linear denominators or powers thereof and thus does not fully reveal some of the distinct poles.

Instead of this *IPF*, that we refer to as *additive* because it is written as a sum of terms, we can use a *multiplicative IPF* of the form

$$r(\zeta) = \frac{q(\zeta)}{p(\zeta)} = \prod_{l=1}^{\mu} \frac{g_l(\zeta)}{g_l(\zeta)}, \tag{12.15}$$

where $q_1, \ldots, q_\mu$ are suitable polynomials with deg $g_l \le$ deg p_l and each term q_l/p_l is expanded into its partial fractions.

The goal is to construct this decomposition and in turn the partial fraction expansions so that none of the partial fraction coefficients exceeds in absolute value some selected threshold, τ, and the number of terms, μ, is as small as possible. As in [47] we expand each term above and obtain the partial fraction representation

$$r(\zeta) = \prod_{l=1}^{\mu} \left(\rho_0^{(l)} + \sum_{j=1}^{k_l} \frac{\rho_j^{(l)}}{\zeta - \tau_{j,l}} \right), \quad \sum_{l=1}^{\mu} k_l = \deg p,$$

where the poles have been re-indexed from $\{\tau_1, \ldots, \tau_d\}$ to $\{\tau_{1,1}, \ldots, \tau_{k_1,1}, \tau_{1,2}, \ldots, \tau_{k_\mu,\mu}\}$. One idea is to perform a brute force search for an *IPF* that returns coefficients that are not too large. For instance, for the rational function in Eq. (12.13) above, the partial fraction coefficients when $d = 6$ are

$$\rho_1 = -\rho_6 = 64.8, \rho_2 = -\rho_5 = -324, \rho_3 = -\rho_4 = 648 \tag{12.16}$$

If we set $\mu = 2$ with $k_1 = 5$ and $k_2 = 1$ then we obtain the possibilities shown in Table 12.1. Because of symmetries, some values are repeated. Based on these values, we conclude that the smallest of the maximal coefficients are obtained when we select p_1 to be one of

$$\prod_{\substack{j=1 \\ j \ne 3}}^{6} \left(\zeta - \frac{j}{6} \right) \quad \text{or} \quad \prod_{\substack{j=1 \\ j \ne 4}}^{6} \left(\zeta - \frac{j}{6} \right)$$

Table 12.1 Partial fraction coefficients for $\frac{1}{p_{1,k}(\zeta)}$ when $p_{1,k}(\zeta) = \prod_{\substack{j=1 \\ j \ne k}}^{6} (\zeta - \frac{j}{6})$ for the incomplete partial fraction expansion of $\frac{1}{p_{1,k}(\zeta)} \frac{1}{(\zeta - \frac{k}{n})}$

$k = 1$	2	3	4	5	6
54.0	43.2	32.4	21.6	10.8	54.0
-216.0	-162.0	-108.0	-54.0	-108.0	-216.0
324.0	216.0	108.0	108.0	216.0	324.0
-216.0	-108.0	-54.0	-108.0	-162.0	-216.0
54.0	10.8	21.6	32.4	43.2	54.0

If we select $\mu = 2$, $k_1 = 4$ and $k_2 = 2$, the coefficients are as tabulated in Table 12.2 (omitting repetitions).

Using $\mu = 2$ as above but $k_1 = k_2 = 3$, the coefficients obtained are much smaller as the following expansion shows:

$$
\frac{1}{p_1(\zeta)}\frac{1}{p_2(\zeta)} = \left(\frac{1}{(\zeta - \frac{1}{6})(\zeta - \frac{3}{6})(\zeta - \frac{5}{6})}\right)\left(\frac{1}{(\zeta - \frac{2}{6})(\zeta - \frac{4}{6})(\zeta - 1)}\right)
$$
$$
= \left(\frac{9}{2(\zeta - \frac{1}{6})} - \frac{9}{\zeta - \frac{3}{6}} + \frac{9}{2(\zeta - \frac{5}{6})}\right)\left(\frac{9}{2(\zeta - \frac{2}{6})} - \frac{9}{\zeta - \frac{4}{6}} + \frac{9}{2(\zeta - 1)}\right).
$$

From the above we conclude that when $\mu = 2$, there exists an *IPF* with maximum coefficient equal to 9, which is a significant reduction from the maximum value of 648, found in Eq. (12.16) for the usual partial fractions ($\mu = 1$).

If the rational function is of the form $1/p$, an exhaustive search like the above can identify the *IPF* decomposition that will minimize the coefficients. For the parallel setting, one has to consider the tradeoff between the size of the coefficients, determined by τ, and the level of parallelism, determined by μ. For a rational function with numerator and denominator degrees $(0, d)$ the total number of cases that must be examined is equal to the total number of k-partitions of d elements, for $k = 1, \ldots, d$. For a given k, the number of partitions is equal to the Stirling number of the 2nd kind, $S(d, k)$, while their sum is Bell number, $\mathscr{B}(s)$. These grow rapidly, for instance the $\mathscr{B}(6) = 203$ whereas $\mathscr{B}(10) = 115{,}975$ [49]. Therefore, as d grows larger the procedure can become very costly if not prohibitive. Moreover, we also have to address the computation of appropriate factorizations that are practical when the numerator is not trivial. One option is to start the search and stop as soon as an *IPF* with coefficients smaller than some selected threshold, say τ, has been computed.

We next consider an effective approach for computing *IPF*, proposed in [47] and denoted by $IPF(\tau)$, where the value τ is an upper bound for the coefficients that is set by the user. When $\tau = 0$ no decomposition is applied and when $\tau = \infty$ (or sufficiently large) it is the usual decomposition with all d terms. We outline the application of the method for rational functions with numerator 1 and denominator degree n. The partial fraction coefficients in this case are

$$
\left(\frac{dp}{dz}(\tau_i)\right)^{-1} = \prod_{\substack{j=1 \\ j \neq i}}^{d}(\tau_i - \tau_j), \ i = 1, \ldots, d
$$

so their magnitude is determined by the magnitude of these products. To minimize their value, the Leja ordering of the poles is used. Recalling Definition 9.5, if the set of poles is $\mathscr{T} = \{\tau_j\}_{j=1}^{d}$ we let $\tau_1^{(1)}$ satisfy $\tau_{1,1} = \arg\min_{\theta \in \mathscr{T}} |\theta|$ and let $\tau_{k,1} \in \mathscr{T}$ be such that

$$
\prod_{j=1}^{k-1} |\tau_{k,1} - \tau_{j,1}| = \max_{\theta \in \mathscr{T}} \prod_{j=1}^{k-1} |\theta - \tau_{j,1}|, \ \tau_{k,1} \in \mathscr{T}, \ k = 1, 2, \ldots.
$$

Table 12.2 Partial fraction coefficients for $\frac{1}{p_{1,(k,i)}(\zeta)}$ when $p_{1,(k,i)}(\zeta) = \prod_{\substack{j=1 \\ j \neq k,i}}^{6} (\zeta - \frac{j}{6})$ for the incomplete partial fraction expansion of $\frac{1}{p_{1,(k,i)}(\zeta)} \frac{1}{(\zeta - \frac{k}{n})(\zeta - \frac{i}{n})}$

(1, 2)	(1, 3)	(1, 4)	(1, 5)	(1, 6)	(2, 3)	(2, 4)	(2, 5)	(2, 6)	(3, 4)	(3, 5)	(3, 6)	(4, 5)
36.0	27.0	18.0	9.0	36.0	21.6	14.4	7.2	27.0	10.8	5.4	18.0	3.6
−108.0	−72.0	−36.0	−54.0	−108.0	−54.0	−27.0	−36.0	−72.0	−18.0	−18.0	−36.0	−36.0
108.0	54.0	36.0	72.0	108.0	36.0	18.0	36.0	54.0	18.0	27.0	36.0	54.0
−36.0	−9.0	−18.0	−27.0	−36.0	−3.6	−5.4	−7.2	−9.0	−10.8	−14.4	−18.0	−21.6
6.0	3.0	2.0	1.5	1.2	6.0	3.0	2.0	1.5	6.0	3.0	2.0	6.0
−6.0	−3.0	−2.0	−1.5	−1.2	−6.0	−3.0	−2.0	−1.5	−6.0	−3.0	−2.0	−6.0

The pairs (k, i) in the header row indicate the poles left out to form the first factor. The partial fraction coefficients for the second factor $((\zeta - \frac{k}{n})(\zeta - \frac{i}{n}))^{-1}$ are listed in the bottom two rows. Since these are smaller than the coefficients for the first factor, we need not be concerned about their values

The idea is then to compute the coefficients of the partial fraction decomposition of $\left(\prod_{j=1}^{k}(\zeta - \tau_{j,1})\right)^{-1}$ for increasing values of k until one or more coefficients exceed the threshold τ. Assume this happens for some value of $k = k_1 + 1$. Then set as first factor of the sought $IPF(\tau)$ the term $(p_1(\zeta))^{-1}$ where

$$p_1(\zeta) = \prod_{j=1}^{k_1}(\zeta - \tau_{j,1}),$$

and denote the coefficients $\rho_j^{(l)}$, $j = 1, \ldots, k$. Next, remove the poles $\{\tau_{j,1}\}_{j=1}^{k_1}$ from $\mathcal{T}$, perform Leja ordering to the remaining ones and repeat the procedure to form the next factor. On termination, the $IPF(\tau)$ representation is

$$\prod_{j=1}^{d}(\zeta - \tau_j)^{-1} = \prod_{l=1}^{\mu}\left(\sum_{j=1}^{k_l} \frac{\rho_j^{(l)}}{z - \tau_{j,l}}\right)$$

This approach is implemented in Algorithm 12.6. The algorithm can be generalized to provide the $IPF(\tau)$ representation of rational functions $r = q/p$ with $\deg q \leq \deg p$; more details are provided in [47].

Generally, the smaller the value of τ, the larger the number of factors μ in this representation. If $\tau = 0$, then Algorithm 12.6 yields the product form representation of $1/p$ with the poles ordered so that their magnitude increases with their index. This ordering is appropriate unless it causes overflow; cf. [50] regarding the Leja ordering in product form representation.

There are no available bounds for the magnitude of the incomplete partial fraction coefficients when the Leja ordering is applied on general sets $\mathcal{T}$. We can, however, bound the product of the coefficients, $\rho_j^{(d)}$, of the simple partial fraction decomposition $(IPF(\infty))$ $\prod_{j=1}^{d} |\rho_j^{(d)}|$ with d, which motivates the use of the Leja ordering. As was shown in [47, Theorem 4.3], if

$$r_d(\zeta) = \frac{1}{p_d(\zeta)}, \quad \text{where } p_d(\zeta) = \prod_{j=1}^{d}(\zeta - \tau_j^{(d)}),$$

and for each d the poles $\tau_j^{(d)}$ are distinct and are the first d Leja points from a set $\mathcal{T}$ that is compact in $\mathbb{C}$ and its complement is connected and regular for the Dirichlet problem, then if we denote by $\hat{\rho}_j^{(d)}$ the partial fraction coefficients for any other arbitrary set of pairwise distinct points from $\mathcal{T}$, then

$$\prod_{j=1}^{d} |\rho_j^{(d)}| \leq \chi^{-d(d-1)}, \quad \lim_{d\to\infty} \prod_{j=1}^{d} |\rho_j^{(d)}|^{\frac{1}{d(d-1)}} = \chi^{-1} \leq \lim_{d\to\infty} |\hat{\rho}_j^{(d)}|^{\frac{1}{d(d-1)}},$$

where χ is the transfinite diameter of $\mathcal{T}$.

Algorithm 12.6 Computing the $IPF(\tau)$ representation of $(p(\zeta))^{-1}$ when $p(\zeta) = \prod_{j=1}^{d}(\zeta - \tau_j)$ and the roots τ_j are mutually distinct

---

Input: $\mathcal{T} = \{\tau_j\}_{j=1}^{d}, \tau \geq 0$;

Output: $\bigcup_{l=1}^{\mu}\{\tau_{j,l}\}_{j=1}^{k_l}, \bigcup_{l=1}^{\mu}\{\rho_j^{(l)}\}$, where $\sum_{l=1}^{\mu} k_l = d$;

1: $j = 1, \mu = 1, k = 0$;
2: **while** $j \leq d$
3:     $k = k + 1$; $//j - 1$ poles already selected and $k - 1$ poles in present factor of the $IPF(\tau)$
        representation
4:     **if** $k = 1$ **then**
5:         choose $\tau_{1,\mu} \in \mathcal{T}$ such that $|\tau_{1,\mu}| = \min_{t \in \mathcal{T}} |t|$;
6:         $\rho_1^{(\mu)} = 1; j = j + 1$;
7:     **else**
8:         choose $\tau_{k,\mu} \in \mathcal{T}$ such that $\prod_{l=1}^{k-1} |\tau_{k,\mu} - \tau_{l,\mu}| = \max_{t \in \mathcal{T}} \prod_{l=1}^{k-1} |t - \tau_{l,m}|$;
9:         **do** $l = 1 : k - 1$
10:             $\tilde{\rho}_l^{(\mu)} = \rho_l^{(\mu)}(\tau_{l,\mu} - \tau_{k,\mu})^{-1}$;
11:         **end**
12:         $\tilde{\rho}_l^{(\mu)} = \prod_{l=1}^{k-1}(\tau_{k,\mu} - \tau_{l,\mu})^{-1}$;
13:         **if** $\max_{1 \leq l \leq k} |\tilde{\rho}_l^{(\mu)}| \leq \tau$ **then**
14:             **do** $l = 1 : k$
15:                 $\rho_l^{(\mu)} = \tilde{\rho}_l^{(\mu)}$;
16:             **end**
17:             $j = j + 1$
18:         **else**
19:             $\mathcal{T} = \mathcal{T} \setminus \{\tau_{l,\mu}\}_{l=1}^{k-1}; k_\mu = k - 1; k = 0; \mu = \mu + 1$; $//$begin new factor
20:         **end if**
21:     **end if**
22: **end while**

12.1.4 Iterative Methods and the Matrix Exponential

We next extend our discussion to include cases that the underlying matrix is so large that iterative methods become necessary. It is then only feasible to compute $f(A)B$, where B consists of one or few vectors, rather than $f(A)$. A typical case is in solving large systems of differential equations, e.g. those that occur after spatial discretization of initial value problems of parabolic type. Then, the function in the template (12.2) is related to the exponential; a large number of *exponential integrators*, methods suitable for these problems, have been developed; cf. [2] for a survey. See also [51–55]. The discussion that follows focuses on the computation of $\exp(A)b$.

The two primary tools used for the effective approximation of $\exp(A)b$ when A is large, are the partial fraction representation of rational approximations to the exponential and Krylov projection methods. The early presentations [56–60] on the combination of these principles for computing the exponential also contained proposals for the parallel computation of the exponential. This was a direct consequence of the fact that partial fractions and the Arnoldi procedure provide opportunities for large and medium grain parallelism. These tools are also applicable for more general matrix functions. Moreover, they are of interest independently of parallel processing.

For this reason, the theoretical underpinnings of such methods for the exponential and other matrix functions have been the subject of extensive research; see e.g. [61–65].

One class of rational functions are the Padé rational approximants. Their numerator and denominator polynomials are known analytically; cf. [66, 67]. Padé, like Taylor series, are designed to provide good local approximations, e.g. near 0. Moreover, the roots of the numerator and denominator are simple and contain no common values. For better approximation, the identity $\exp(A) = (\exp(Ah))^{1/h}$ is used to cluster the eigenvalues close to the origin. The other possibility is to construct a rational Chebyshev approximation on some compact set in the complex plane. The Chebyshev rational approximation is primarily applicable to matrices that are symmetric negative definite and has to be computed numerically; a standard reference for the power form coefficients of the numerator and denominator polynomials with $\deg p = \deg q$ ("diagonal approximations") ranging from 1 to 30 is [68]. Reference [58] describes parallel algorithms based on the Padé and Chebyshev rational approximations, illustrating their effectiveness and the advantages of the Chebyshev approximation [68–70] for negative definite matrices. An alternative to the delicate computations involved in the Chebyshev rational approximation is to use the Caratheodory-Fejér (CF) method to obtain good rational approximations followed by their partial fraction representation; cf. [5, 26].

Recall that any complex poles appear in conjugate pairs. These can be grouped as suggested in Eq. (12.11) to halve the number of solves with $(A - \tau I)$, where τ is the one of the conjugate pairs. In the common case that A is hermitian and τ is complex, $(A - \tau I)$ is complex symmetric and there exist Lanczos and CG type iterative methods that take advantage of this special structure; cf. [71–73]. Another possibility is to combine terms corresponding to conjugate shifts and take advantage of the fact that

$$(A - \tau I)(A - \bar{\tau} I) = A^2 - 2\Re\tau A + |\tau|^2 I = A(A - 2\Re\tau I) + |\tau|^2 I.$$

is a real matrix. Then, MV operations in a Krylov method would only involve real arithmetic.

Consider now the solution of the linear systems corresponding to the partial fraction terms. As noted in the previous subsection, if one were able to use a direct method, it would be possible to save computations to lower redundancy by first reducing the matrix to Hessenberg form. When full reduction is prohibitive because of the size of A, one can apply a "partial" approach using the Arnoldi process to build an orthonormal basis, V_ν, for the Krylov subspace $\mathcal{K}_\nu(A, b)$, where $\nu \ll n$. The parallelism in Krylov methods was discussed in Sect. 9.3.2 of Chap. 9. Here, however, it also holds that

$$V_\nu^\top (A - \tau I) V_\nu = H_\nu - \tau I_\nu$$

for any τ and so the same basis reduces not only A but all shifted matrices to Hessenberg form [74, 75]. This is the well known shift invariance property of Krylov subspaces that can be expressed by $\mathcal{K}_\nu(A, b) = \mathcal{K}_\nu(A - \tau I, b)$; cf. [76]. This property implies that methods such as FOM and GMRES, can proceed by first computing the basis via Arnoldi followed by the solution of (different) small systems, followed by computing the solution to each partial fraction term using the same basis and then combining using the partial fraction coefficients. For example, if FOM can be used, then setting $\beta = \|b\|$, an approximation to $r(A)b$ from $\mathcal{K}_\nu(A, b)$ is

$$\tilde{x} = \beta V_\nu \sum_{j=1}^{d} \rho_j (H_\nu - \tau_j I)^{-1} e_1 = \beta V_\nu r(H_\nu) e_1, \tag{12.17}$$

on the condition that all matrices $(H_\nu - \tau_j I)$ are invertible. The above approach can also be extended to handle restarting; cf. [77] when solving the shifted systems. It is also possible to solve the multiply shifted systems by Lanczos based approaches including BICGSTAB, QMR and transpose free QMR; cf. [78, 79].

Consider now (12.17). By construction, if $r(H_\nu)$ exists, it must also provide a rational approximation to the exponential. In other words, we can write

$$\exp(A)b \approx \beta V_\nu \exp(H_\nu) e_1. \tag{12.18}$$

This is the Krylov approach for approximating the exponential, cf. [57, 58, 80, 81]. Note that the only difference between (12.17) and (12.18) is that $r(H_\nu)$ was replaced by $\exp(H_\nu)$. It can be shown that the term $\beta V_\nu \exp(H_\nu) e_1$ is a polynomial approximation of $\exp(A)b$ with a polynomial of degree $\nu - 1$ which interpolates the exponential, in the Hermite sense, on the set of eigenvalues of H_ν; [82]. In fact, for certain classes of matrices, this type of approximation is extremely accurate; cf. [61] and references therein. Moreover, Eq. (12.18) provides a a framework for approximating general matrix functions using Krylov subspaces. We refer to [83] for generalizations and their parallel implementations.

Consider, for example, using transpose free QMR. Then there is a multiply shifted version of TFQMR [78] where each iteration consists of computations with the common Krylov information to be shared between all systems and computations that are specific to each term, building or updating data special to each term. As described in [56], at each iteration of multiply shifted TFQMR, the dimension of the underlying Krylov subspace increases by 2. The necessary computations are of two types, one set that is used to advance the dimension of the Krylov subspace that are independent of the total number of systems and consist of 2 MVs, 4 dot products and 6 vector updates. The other set consists of computations specific to each term, namely 9 vector updates and a few scalar ones, that can be conducted in parallel. One possibility is to stop the iterations when an accurate approximation has been obtained for all systems. Finally, there needs to be an MV operation to combine the partial solutions as in (12.10). Because the roots τ_j are likely to be complex, we expect that this BLAS2 operation will contain complex elements.

An important characteristic of this approach is that it has both large grain parallelism because of the partial fraction decomposition as well as medium grain parallelism because of the shared computations with A and vectors of that size. Recall that the shared computations occur because of our desire to reduce redundancy by exploiting the shift invariance of Krylov subspaces.

The exploitation of the multiple shifts reduces redundancy but also curtails parallelism. Under some conditions (e.g. high communication costs) this might not be desirable. A more general approach that provides more flexibility over the amount of large and medium grain parallelism was proposed in [56]. The ideas is to organize the partial fraction terms into groups, and to express the rational function as a double sum, say

$$r(A) = \rho_0 I + \sum_{l=1}^{k} \sum_{j \in \mathscr{I}_j} \rho_j^{(l)} (A - \zeta_j I)^{-1}$$

where the index sets $\mathscr{I}_l, l = 1, \ldots, k$ are a partition of $\{1, 2, \ldots, \deg p\}$ and the sets of coefficients $\{\rho_j^{(1)}, \ldots, \rho_j^{(l)}\}$ are a partition of the set $\{\rho_1, \ldots, \rho_d\}$. Then we can build a hybrid scheme in which the inner sum is constructed using the multiply shifted approach, but the k components of the outer sum are treated completely independently. The extreme cases are $k = \deg p$, in which all systems are treated independently, and $k = 1$, in which all systems are solved with a single instance of the multiply shifted approach. This flexible hybrid approach was used on the Cedar vector multiprocessor but can be useful in the context of architectures with hierarchical parallelism, that are emerging as a dominant paradigm in high performance computing.

In the more general case where we need to compute $\exp(A)B$, where B is a block of vectors that replaces b in (12.9), there is a need to solve a set of shifted systems for each column of B. This involves ample parallelism, but also replication of work, given what we know about the Krylov invariance to shifting. As before, it is natural to seek techniques (e.g. [84–86]) that will reduce the redundancy. The final choice, of course, will depend on the problem and on the characteristics of the underlying computational platform.

We conclude by recalling from Sect. 12.1.3 that the use of partial fractions requires care to avoid catastrophic cancellations and loss of accuracy. Table 12.3 indicates how large the partial fraction coefficients of some Padé approximants become as the degrees of the numerator and denominator increase. To avoid problems, we can apply incomplete partial fractions as suggested in Sect. 12.1.3. For example, in Table 12.4 we present results with the $IPF(\tau)$ algorithm that extends (12.6) to the case of rational functions that are the quotient of polynomials of equal degree. The algorithm was applied to evaluate the partial fraction representations of diagonal Padé approximations for the matrix exponential applied on a vector, that is $r_{d,d}(-A\delta)b$ as approximations for $\exp(-A\delta)b$ with $A = \frac{1}{h^2}[-1, 2, -1]_N$ using $h = 1/(N+1)$ with $N = 998$. The right-hand side $b = \sum_{j=1}^{N} \frac{1}{j} v_j$, where v_j are the eigenvectors of A, ordered so that v_j corresponds to the eigenvalue $\lambda_j = \frac{4}{h^2} \sin^2 \frac{j\pi}{2(N+1)}$. Under "components" we show the groups that formed from the application of the IPF

Table 12.3 Base 10 logarithm of partial fraction coefficient of largest

$\deg q$

	2	4	8	14	20
2	1.1				
4	1.3	2.3			
8	1.8	2.7	4.6		
14	2.5	3.5	5.3	8.1	
20	3.3	4.2	6.1	8.8	11.5

Magnitude of the Padé approximation of e^z with numerator and denominator degrees $\deg q$, $\deg p$. Data are from Ref. [47]

Table 12.4 Components and base 10 logarithm of maximum relative

d	τ	Components	Errors	d	τ	Components	Errors
14	0	$\{1, \ldots, 1\}$	-5.5	24	0	$\{1, \ldots, 1\}$	-13.1
	10^4	$\{8, 4, 2\}$	-5.5		10^4	$\{9, 5, 5, 3, 2\}$	-12.7
	10^8	$\{13, 1\}$	-5.5		10^8	$\{16, 8\}$	-9.1
	∞	$\{14\}$	-5.5		∞	$\{24\}$	-2.6
18	0	$\{1, \ldots, 1\}$	-8.2	28	0	$\{1, \ldots, 1\}$	-13.7
	10^4	$\{8, 5, 4, 1\}$	-8.2		10^4	$\{8, 6, 5, 5, 3, 1\}$	-12.6
	10^8	$\{15, 3\}$	-8.2		10^8	$\{18, 9, 1\}$	-8.6
	∞	$\{18\}$	-6.1		∞	$\{28\}$	-1.7

Errors when $\exp(-A\delta)b$ is evaluated based on the diagonal Padé approximant of $\exp(-A\delta)$ and using $IPF(\tau)$ for $\delta = 1 \times 10^{-5}$. Data extracted from [47]

algorithm. For example, when $d = 24$, the component set for $\tau = 10^8$ indicates that to keep the partial fraction coefficients below that value, the rational function was written as a product of two terms, one consisting of a sum of 16 elements, the other of a sum of 8. The resulting relative error in the final solution is approximately 10^{-9}. If instead one were to use the full partial fraction as a sum of 28 terms, the error would be $10^{-2.6}$.

12.2 Determinants

The determinant of a matrix $A \in \mathbb{C}^{n \times n}$ is easily computed from its LU factorization: if $PAQ = LU$, where P, and Q are permutation matrices, L is a unit lower triangular matrix, and $U = (v_{ij})$ an upper-triangular matrix, then $\det A = \det P \times \det Q \times \prod_{i=1}^{n} v_{ii}$. Using Gaussian Elimination with partial pivoting, we obtain the factorization of the dense matrix A, $PA = LU$. This guarantees that the computed determinant of U is the exact determinant of a slightly perturbed A (see [87]). The main problem we face here is how to guarantee that no over- or underflow occurs when accumulating the multiplication of the diagonal entries of

U. The classical technique consists of *normalizing* this quantity by representing the determinant by the triplet (ρ, K, n) where

$$\det\, A = \rho K^n, \tag{12.19}$$

in which $|\rho| = 1$ and $\ln K = \frac{1}{n} \sum_{i=1}^{n} \ln |v_{ii}|$.

For a general dense matrix, the complexity of the computation is $O(n^3)$, and $O(n^2)$ for a Hessenberg matrix. For the latter, a variation of the above procedure is the use of Hyman's method (see [87] and references herein). In the present section, we consider techniques that are more suitable for large structured sparse matrices.

12.2.1 Determinant of a Block-Tridiagonal Matrix

The computation of the determinant of a nonsingular large sparse matrix can be obtained by using any of the parallel sparse LU factorization schemes used in MUMPS [88], or SuperLU [89], for example. On distributed memory parallel architectures, such schemes realize very modest parallel efficiencies at best. If the sparse matrices can be reordered into the block-tridiagonal form, however, we can apply the Spike factorization scheme (see Sect. 5.2) to improve the parallel scalability of computing matrix determinants. This approach is described in [90].

Let the reordered sparse matrix A be of the form,

$$\begin{pmatrix} A_1 & A_{1,2} & 0 & \dots & & 0 \\ A_{2,1} & A_2 & \ddots & & \ddots & \vdots \\ 0 & \ddots & \ddots & & \ddots & 0 \\ \vdots & & \ddots & \ddots & \ddots & A_{q-1,q} \\ 0 & \dots & & 0 & A_{q,q-1} & A_q \end{pmatrix}, \tag{12.20}$$

where for $i = 1, \dots, q - 1$ the blocks $A_{i+1,i}$ and $A_{i,i+1}$ are coupling matrices defined by:

$$A_{i,i+1} = \begin{pmatrix} 0 & 0 \\ B_i & 0 \end{pmatrix},$$

$$A_{i+1,i} = \begin{pmatrix} 0 & C_{i+1} \\ 0 & 0 \end{pmatrix};$$

with $A_i \in \mathbb{C}^{n_i \times n_i}$, $B_i \in \mathbb{C}^{b_i \times l_i}$, $C_{i+1} \in \mathbb{C}^{c_{i+1} \times r_{i+1}}$. We assume throughout that $c_i \leq n_i, b_i \leq n_i$ and $l_{i-1} + r_{i+1} \leq n_i$.

Consider the Spike factorization scheme in Sect. 5.2.1, where, without loss of generality, we assume that $D = \mathrm{diag}(A_1, A_2, \ldots, A_q)$ is nonsingular, and for $i = 1, \ldots, q$, we have the factorizations $P_i A_i Q_i = L_i U_i$ of the sparse diagonal blocks A_i. Let $P = \mathrm{diag}(P_1, P_2, \ldots, P_q)$, $Q = \mathrm{diag}(Q_1, Q_2, \ldots, Q_q)$ and

$$S = D^{-1}A = I_n + T,$$

where

$$I_n = \mathrm{diag}(I_{s_1}, I_{r_2}, I_{l_1}, I_{s_2}, I_{r_3}, I_{l_2} \cdots, I_{r_q}, I_{l_{q-1}}, I_{s_q}),$$

and

$$T = \begin{pmatrix}
0_{s_1} & 0 & V_1' & 0 & 0 \\
0 & 0_{r_2} & V_1^b & 0 & 0 \\
0 & W_2^t & 0_{l_1} & 0 & 0 & V_2^t \\
0 & W_2' & 0 & 0_{s_2} & 0 & V_2' \\
0 & W_2^b & 0 & 0 & 0_{r_3} & V_2^b \\
& & & W_3^t & 0_{l_2} & 0 & 0 & V_3^t \\
& & & W_3' & 0 & 0_{s_3} & 0 & V_3' \\
& & & W_3^b & 0 & 0 & 0_{r_4} & V_3^b \\
& & & & & \ddots & & & 0 & V_{q-1}^t \\
& & & & & & & & 0 & V_{q-1}' \\
& & & & & & & 0_{r_q} & 0 & V_{q-1}^b \\
& & & & & & 0 & W_q^t & 0_{l_{q-1}} & 0 \\
& & & & & & 0 & W_q' & 0 & 0_{s_q}
\end{pmatrix},$$

and where the right and left spikes, respectively, are given by [91]:

$$V_1 \in \mathbb{C}^{n_1 \times l_1} \equiv \begin{vmatrix} V_1' \\ V_1^b \end{vmatrix} \begin{matrix} s_1 \\ r_2, \end{matrix} \tag{12.21}$$

$$= (A_1)^{-1}\begin{pmatrix} 0 \\ B_1 \end{pmatrix} = Q_1 U_1^{-1} L_1^{-1} P_1 \begin{pmatrix} 0 \\ B_1 \end{pmatrix}, \tag{12.22}$$

and for $i = 2, \ldots, q - 1$,

$$V_i \in \mathbb{C}^{n_i \times l_i} \equiv \begin{vmatrix} V_i^t \\ V_i' \\ V_i^b \end{vmatrix} \begin{matrix} l_{i-1} \\ s_i \\ r_{i+1}, \end{matrix} \tag{12.23}$$

$$= (A_i)^{-1}\begin{pmatrix} 0 \\ B_i \end{pmatrix} = Q_i U_i^{-1} L_i^{-1} P_i \begin{pmatrix} 0 \\ B_i \end{pmatrix}, \tag{12.24}$$

$$W_i \in \mathbb{C}^{n_i \times r_i} \equiv \begin{vmatrix} W_i^t & l_{i-1} \\ W_i' & s_i \\ W_i^b & r_{i+1} \end{vmatrix} \tag{12.25}$$

$$= (A_i)^{-1} \begin{pmatrix} C_i \\ 0 \end{pmatrix} = Q_i U_i^{-1} L_i^{-1} P_i \begin{pmatrix} C_i \\ 0 \end{pmatrix}, \tag{12.26}$$

and

$$W_q \in \mathbb{C}^{n_q \times r_q} \equiv \begin{vmatrix} W_q^t & l_{q-1} \\ W_q' & s_q \end{vmatrix} \tag{12.27}$$

$$= (A_q)^{-1} \begin{pmatrix} C_q \\ 0 \end{pmatrix} = Q_q U_q^{-1} L_q^{-1} P_q \begin{pmatrix} C_q \\ 0 \end{pmatrix}. \tag{12.28}$$

With this partition, it is clear that:

$$\det\ A = \left(\prod_{i=1}^{q} \mathrm{sign}\ P_i \times \mathrm{sign}\ Q_i \times \det\ U_i \right) \det\ S, \tag{12.29}$$

where sign P stands for the signature of permutation P.

However, as we have seen in Sect.5.2.1, the Spike matrix S can be reordered to result in a 2×2 block-upper triangular matrix with one of the diagonal blocks the identity matrix and the other, $\hat{S}$, being the coefficient matrix of the nonsingular reduced system, see (5.10) and (5.11), which is of order $n - \sum_{k=1}^{q} s_k$. In other words,

$$\det\ S = \det\ \hat{S}.$$

Thus, the computation of the determinant of $\tilde{S}$ requires an LU-factorization of $\hat{S}$, which can be realized via one of the parallel sparse direct solvers MUMPS, SuperLU, or IBM's WSMP.

12.2.2 Counting Eigenvalues with Determinants

The localization of eigenvalues of a given matrix A is of interest in many scientific applications. When the matrix is real symmetric or complex Hermitian, a procedure based on the computation of Sturm sequences allows the safe application of bisections on real intervals to localize the eigenvalues as shown in Sects. 8.2.3 and 11.1.2. The problem is much harder for real nonsymmetric or complex non-Hermitian matrices and especially for non-normal matrices.

Let us assume that some Jordan curve Γ is given in the complex plane, and that one wishes to count the number of eigenvalues of the matrix A that are surrounded by Γ. Several procedures have been proposed for such a task in [92] and, more recently, a complete algorithm has been proposed in [93] which we present below.

The number of surrounded eigenvalues is determined by evaluating the integral from the Cauchy formula (see e.g. [94, 95]):

$$N_\Gamma = \frac{1}{2i\pi} \int_\Gamma \frac{f'(z)}{f(z)} dz, \tag{12.30}$$

where $f(z) = \det(zI - A)$ is the characteristic polynomial of A. This integral is also considered in work concerning nonlinear eigenvalue problems [96, 97].

Let us assume that $\Gamma = \bigcup_{i=0}^{N-1} [z_i, z_{i+1}]$ is a polygonal curve where $[z_i, z_{i+1}] \subset \mathbb{C}$ denotes the line segment with end points z_i and z_{i+1}. This is a user-defined curve which approximates the initial Jordan curve. Hence,

$$f(z + h) = f(z) \ \det(I + hR(z)). \tag{12.31}$$

where $R(z) = (zI - A)^{-1}$ is the resolvent. Also, let $\Phi_z(h) = \det(I + hR(z))$, then

$$\int_z^{z+h} \frac{f'(z)}{f(z)} dz = \ln(\Phi_z(h))$$
$$= \ln|\Phi_z(h)| + i \ \arg(\Phi_z(h)).$$

The following lemma determines the stepsize control which guarantees proper integration. The branch (i.e. a determination $\arg_0$ of the argument), which is to be followed along the integration process, is fixed by selecting an origin $z_0 \in \Gamma$ and by insuring that

$$\arg_0(f(z_0)) = \text{Arg}(f(z_0)). \tag{12.32}$$

Lemma 12.1 (Condition (A)) *Let z and h be such that $[z, z + h] \subset \Gamma$. If*

$$|\text{Arg}(\Phi_z(s))| < \pi, \ \ \forall s \in [0, h], \tag{12.33}$$

then,

$$\arg_0(f(z + h)) = \arg_0(f(z)) + \text{Arg}(\Phi_z(h)), \tag{12.34}$$

where $\arg_0$ is determined as in (12.32) by a given $z_0 \in \Gamma$.

Proof 1 See [93].

Condition (A) is equivalent to

$$\Phi_z(s) \notin (-\infty, 0], \forall s \in [0, h].$$

It can be replaced by a more strict condition. Since

$$\Phi_z(s) = 1 + \delta, \ \text{with} \ \delta = \rho e^{i\theta},$$

a sufficient condition for (12.33) to be satisfied, which we denote as **Condition (B)**, is $\rho < 1$, i.e.

$$|\Phi_z(s) - 1| < 1, \quad \forall s \in [0, h]. \tag{12.35}$$

This condition can be approximated by considering the tangent of $\Psi_z(s) = 1 + s\Phi_z'(0)$, and, substituting it in (12.35) to obtain **Condition (C)**:

$$|h| < \frac{1}{|\Phi_z'(0)|}. \tag{12.36}$$

The derivative $\Phi_z'(0)$ is given by :

$$\Phi_z'(0) = \text{trace}(R(z)). \tag{12.37}$$

e.g. see [21, 97–100]. The most straightforward procedure, but not the most efficient, consists of approximating the derivative with the ratio

$$\Phi_z'(0) \approx \frac{\Phi_z(s) - 1}{s},$$

where $s = \alpha h$ with an appropriately small α. Therefore, the computation imposes an additional LU factorization for evaluating the quantity $\Phi_z(s)$. This approach doubles the computational effort when compared to computing only one determinant per vertex as needed for the integration.

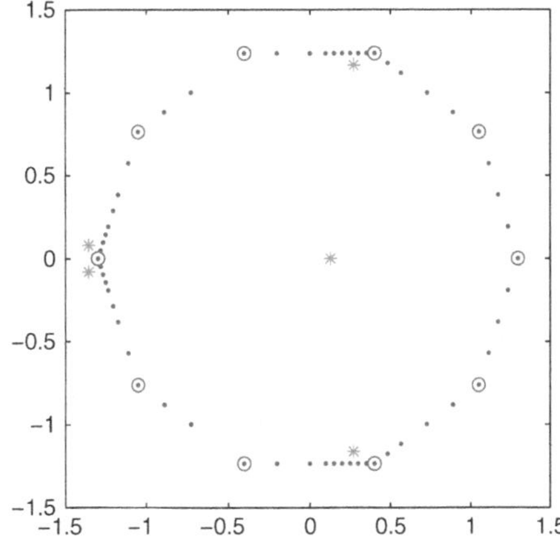

Fig. 12.1 Application of EIGENCNTon a random matrix of order 5. The eigenvalues are indicated by the stars. The *polygonal line* is defined by the *10 points with circles*; the other *points* of the line are automatically introduced to insure the conditions as specified in [93]

A short description of the procedure EIGENCNT is given in Algorithm 12.7. For a detailed description, see [93]. The algorithm exhibits a parallel loop in which the only critical section is related to the management of the ordered list of vertices $\mathscr{L}$. For a real matrix A and when the line Γ is symmetric with respect to the real axis, the computational cost is halved by only integrating in the half complex plane of nonnegative imaginary values.

Algorithm 12.7 EIGENCNT: counting eigenvalues surrounded by a curve.

Input: $A \in \mathbb{R}^{n \times n}$ and a polygonal line Γ.
Output: The number *neg* of algebraic eigenvalues of A which are surrounded by Γ.
1: $\mathscr{L}$= ordered list of the vertices of Γ; $Status(\mathscr{L})$ = "no information";   //The last element of the list is equal to the first one.
2: **while** $\mathscr{L} \neq$ "ready",
3:    **doall** $z \in \mathscr{L}$,
4:      **if** $Status(z) \neq$ "ready", **then**
5:        **if** $Status(z) =$ "no information", **then**
6:          Compute Φ_z; Store the result in $Integral(I(z))$;   //$I(z)$ is the rank of z in the list $\mathscr{L}$.
7:        **end if**
8:        Compute $\Phi_z'(0)$;
9:        Control the stepsize and either insert points in $\mathscr{L}$ with status "noinformation", or set $Status(z) =$ "ready".
10:      **end if**
11:    **end**
12: **end while**
13: $Integral(1:N) = Integral(2:N+1)/Integral(1:N)$;   //$N$: number of vertices in $\mathscr{L}$;
14: $neg = \mathrm{round}(\frac{\sum_{k=1}^{N} \arg\, Integral(k)}{2\pi})$;

We illustrate the consequence of the stepsize control on a random matrix of order 5. The polygonal line Γ is determined by 10 points regularly spaced on a circle with center at 0 and radius 1.3. Figure 12.1 depicts the eigenvalues of A, the line Γ, and the points that are automatically inserted by the procedure. The figure illustrates that, when the line gets closer to an eigenvalue, the segment length becomes smaller.

References

1. Varga, R.: Matrix Iterative Analysis. Springer Series in Computational Mathematics, 2nd edn. Springer, Berlin (2000)
2. Hochbruck, M., Ostermann, A.: Exponential integrators. Acta Numer. **19**, 28–209 (2012). doi:10.1017/S0962492910000048
3. Sidje, R.: EXPOKIT: software package for computing matrix exponentials. ACM Trans. Math. Softw. **24**(1), 130–156 (1998)
4. Berland, H., Skaflestad, B., Wright, W.M.: EXPINT—A MATLAB package for exponential integrators. ACM Trans. Math. Softw. **33**(1) (2007). doi:10.1145/1206040.1206044. http://doi.acm.org/10.1145/1206040.1206044
5. Schmelzer, T., Trefethen, L.N.: Evaluating matrix functions for exponential integrators via Carathéodory-Fejér approximation and contour integrals. ETNA **29**, 1–18 (2007)

6. Skaflestad, B., Wright, W.: The scaling and modified squaring method for matrix functions related to the exponential. Appl. Numer. Math. **59**, 783–799 (2009)
7. Festinger, L.: The analysis of sociograms using matrix algebra. Hum. Relat. **2**, 153–158 (1949)
8. Katz, L.: A new status index derived from sociometric index. Psychometrika **18**(1), 39–43 (1953)
9. Brin, S., Page, L.: The anatomy of a large-scale hypertextual web search engine. In: Proceedings of 7th International Conference World Wide Web, pp. 107–117. Elsevier Science Publishers B.V., Brisbane (1998)
10. Kleinberg, J.: Authoritative sources in a hyperlinked environment. J. ACM **46**, 604–632 (1999)
11. Langville, A., Meyer, C.: Google's PageRank and Beyond: The Science of Search Engine Rankings. Princeton University Press, Princeton (2006)
12. Bonchi, F., Esfandiar, P., Gleich, D., Greif, C., Lakshmanan, L.: Fast matrix computations for pairwise and columnwise commute times and Katz scores. Internet Math. **8**(2011), 73–112 (2011). http://projecteuclid.org/euclid.im/1338512314
13. Estrada, E., Higham, D.: Network properties revealed through matrix functions. SIAM Rev. **52**, 696–714. Published online November 08, 2010
14. Fenu, C., Martin, D., Reichel, L., Rodriguez, G.: Network analysis via partial spectral factorization and Gauss quadrature. SIAM J. Sci. Comput. **35**, A2046–A2068 (2013)
15. Estrada, E., Rodríguez-Velázquez, J.: Subgraph centrality and clustering in complex hypernetworks. Phys. A: Stat. Mech. Appl. **364**, 581–594 (2006). http://www.sciencedirect.com/science/article/B6TVG-4HYD5P6-3/1/69cba5c107b2310f15a391fe982df305
16. Benzi, D., Ernesto, E., Klymko, C.: Ranking hubs and authorities using matrix functions. Linear Algebra Appl. **438**, 2447–2474 (2013)
17. Estrada, E., Hatano, N.: Communicability Graph and Community Structures in Complex Networks. CoRR arXiv:abs/0905.4103 (2009)
18. Baeza-Yates, R., Boldi, P., Castillo, C.: Generic damping functions for propagating importance in link-based ranking. J. Internet Math. **3**(4), 445–478 (2006)
19. Kollias, G., Gallopoulos, E., Grama, A.: Surfing the network for ranking by multidamping. IEEE TKDE (2013). http://www.computer.org/csdl/trans/tk/preprint/06412669-abs.html
20. Kollias, G., Gallopoulos, E.: Multidamping simulation framework for link-based ranking. In: A. Frommer, M. Mahoney, D. Szyld (eds.) Web Information Retrieval and Linear Algebra Algorithms, no. 07071 in Dagstuhl Seminar Proceedings. Internationales Begegnungs- und Forschungszentrum für Informatik (IBFI), Schloss Dagstuhl, Germany (2007). http://drops.dagstuhl.de/opus/volltexte/2007/1060
21. Bai, Z., Fahey, G., Golub, G.: Some large-scale matrix computation problems. J. Comput. Appl. Math. **74**(1–2), 71–89 (1996). doi:10.1016/0377-0427(96)00018-0. http://www.sciencedirect.com/science/article/pii/0377042796000180
22. Bekas, C., Curioni, A., Fedulova, I.: Low-cost data uncertainty quantification. Concurr. Comput.: Pract. Exp. (2011). doi:10.1002/cpe.1770. http://dx.doi.org/10.1002/cpe.1770
23. Stathopoulos, A., Laeuchli, J., Orginos, K.: Hierarchical probing for estimating the trace of the matrix inverse on toroidal lattices. SIAM J. Sci. Comput. **35**, S299–S322 (2013). http://epubs.siam.org/doi/abs/10.1137/S089547980036869X
24. Higham, N.: Functions of Matrices: Theory and Computation. SIAM, Philadelphia (2008)
25. Golub, G., Meurant, G.: Matrices. Moments and Quadrature with Applications. Princeton University Press, Princeton (2010)
26. Trefethen, L.: Approximation Theory and Approximation Practice. SIAM, Philadelphia (2013)
27. Higham, N., Deadman, E.: A catalogue of software for matrix functions. version 1.0. Technical Report 2014.8, Manchester Institute for Mathematical Sciences, School of Mathematics, The University of Manchester (2014)
28. Estrin, G.: Organization of computer systems: the fixed plus variable structure computer. In: Proceedings Western Joint IRE-AIEE-ACM Computer Conference, pp. 33–40. ACM, New York (1960)

29. Lakshmivarahan, S., Dhall, S.K.: Analysis and Design of Parallel Algorithms: Arithmetic and Matrix Problems. McGraw-Hill Publishing, New York (1990)

30. Maruyama, K.: On the parallel evaluation of polynomials. IEEE Trans. Comput. **C-22**(1) (1973)

31. Munro, I., Paterson, M.: Optimal algorithms for parallel polynomial evaluation. J. Comput. Syst. Sci. **7**, 189–198 (1973)

32. Pan, V.: Complexity of computations with matrices and polynomials. SIAM Rev. **34**(2), 255–262 (1992)

33. Reynolds, G.: Investigation of different methods of fast polynomial evaluation. Master's thesis, EPCC, The University of Edinburgh (2010)

34. Eberly, W.: Very fast parallel polynomial arithmetic. SIAM J. Comput. **18**(5), 955–976 (1989)

35. Paterson, M., Stockmeyer, L.: On the number of nonscalar multiplications necessary to evaluate polynomials. SIAM J. Comput. **2**, 60–66 (1973)

36. Alonso, P., Boratto, M., Peinado, J., Ibáñez, J., Sastre, J.: On the evaluation of matrix polynomials using several GPGPUs. Technical Report, Department of Information Systems and Computation, Universitat Politécnica de Valéncia (2014)

37. Bernstein, D.: Fast multiplication and its applications. In: Buhler, J., Stevenhagen, P. (eds.) Algorithmic number theory: lattices, number fields, curves and cryptography, Mathematical Sciences Research Institute Publications (Book 44), pp. 325–384. Cambridge University Press (2008)

38. Trefethen, L., Weideman, J., Schmelzer, T.: Talbot quadratures and rational approximations. BIT Numer. Math. **46**, 653–670 (2006)

39. Swarztrauber, P.N.: A direct method for the discrete solution of separable elliptic equations. SIAM J. Numer. Anal. **11**(6), 1136–1150 (1974)

40. Henrici, P.: Applied and Computational Complex Analysis. Wiley, New York (1974)

41. Enright, W.H.: Improving the efficiency of matrix operations in the numerical solution of stiff differential equations. ACM TOMS **4**(2), 127–136 (1978)

42. Choi, C.H., Laub, A.J.: Improving the efficiency of matrix operations in the numerical solution of large implicit systems of linear differential equations. Int. J. Control **46**(3), 991–1008 (1987)

43. Lu, Y.: Computing a matrix function for exponential integrators. J. Comput. Appl. Math. **161**(1), 203–216 (2003)

44. Ltaief, H., Kurzak, J., Dongarra., J.: Parallel block Hessenberg reduction using algorithms-by-tiles for multicore architectures revisited. Technical Report. Innovative Computing Laboratory, University of Tennessee (2008)

45. Quintana-Ortí, G., van de Geijn, R.: Improving the performance of reduction to Hessenberg form. ACM Trans. Math. Softw. **32**, 180–194 (2006)

46. Kung, H.: New algorithms and lower bounds for the parallel evaluation of certain rational expressions and recurrences. J. Assoc. Comput. Mach. **23**(2), 252–261 (1976)

47. Calvetti, D., Gallopoulos, E., Reichel, L.: Incomplete partial fractions for parallel evaluation of rational matrix functions. J. Comput. Appl. Math. **59**, 349–380 (1995)

48. Henrici, P.: An algorithm for the incomplete partial fraction decomposition of a rational function into partial fractions. Z. Angew. Math. Phys. **22**, 751–755 (1971)

49. Graham, R., Knuth, D., Patashnik, O.: Concrete Mathematics. Addison-Wesley, Reading (1989)

50. Reichel, L.: The ordering of tridiagonal matrices in the cyclic reduction method for Poisson's equation. Numer. Math. **56**(2/3), 215–228 (1989)

51. Butcher, J., Chartier, P.: Parallel general linear methods for stiff ordinary differential and differential algebraic equations. Appl. Numer. Math. **17**(3), 213–222 (1995). doi:10.1016/0168-9274(95)00029-T. http://www.sciencedirect.com/science/article/pii/016892749500029T. Special Issue on Numerical Methods for Ordinary Differential Equations

52. Chartier, P., Philippe, B.: A parallel shooting technique for solving dissipative ODE's. Computing **51**(3–4), 209–236 (1993). doi:10.1007/BF02238534

53. Chartier, P.: L-stable parallel one-block methods for ordinary differential equations. SIAM J. Numer. Anal. **31**(2), 552–571 (1994). doi:10.1137/0731030

54. Gander, M., Vandewalle, S.: Analysis of the parareal time-parallel time-integration method. SIAM J. Sci. Comput. **29**(2), 556–578 (2007)

55. Maday, Y., Turinici, G.: A parareal in time procedure for the control of partial differential equation. C. R. Math. Acad. Sci. Paris **335**(4), 387–392 (2002)

56. Baldwin, C., Freund, R., Gallopoulos, E.: A parallel iterative method for exponential propagation. In: D. Bailey, R. Schreiber, J. Gilbert, M. Mascagni, H. Simon, V. Torczon, L. Watson (eds.) Proceedings of Seventh SIAM Conference on Parallel Processing for Scientific Computing, pp. 534–539. SIAM, Philadelphia (1995). Also CSRD Report No. 1380

57. Gallopoulos, E., Saad, Y.: Efficient solution of parabolic equations by Krylov approximation methods. SIAM J. Sci. Stat. Comput. 1236–1264 (1992)

58. Gallopoulos, E., Saad, Y.: On the parallel solution of parabolic equations. In: Proceedings of the 1989 International Conference on Supercomputing, pp. 17–28. Herakleion, Greece (1989)

59. Gallopoulos, E., Saad, Y.: Efficient parallel solution of parabolic equations: implicit methods on the Cedar multicluster. In: J. Dongarra, P. Messina, D.C. Sorensen, R.G. Voigt (eds.) Proceedings of Fourth SIAM Conference Parallel Processing for Scientific Computing, pp. 251–256. SIAM, (1990) Chicago, December 1989

60. Sidje, R.: Algorithmes parallèles pour le calcul des exponentielles de matrices de grandes tailles. Ph.D. thesis, Université de Rennes I (1994)

61. Lopez, L., Simoncini, V.: Analysis of projection methods for rational function approximation to the matrix exponential. SIAM J. Numer. Anal. **44**(2), 613–635 (2006)

62. Popolizio, M., Simoncini, V.: Acceleration techniques for approximating the matrix exponential operator. SIAM J. Matrix Anal. Appl. **30**, 657–683 (2008)

63. Frommer, A., Simoncini, V.: Matrix functions. In: Schilders, W., van der Vorst, H.A., Rommes, J. (eds.) Model Order Reduction: Theory, Research Aspects and Applications, pp. 275–303. Springer, Berlin (2008)

64. van den Eshof, J., Hochbruck, M.: Preconditioning Lanczos approximations to the matrix exponential. SIAM J. Sci. Comput. **27**, 1438–1457 (2006)

65. Gu, C., Zheng, L.: Computation of matrix functions with deflated restarting. J. Comput. Appl. Math. **237**(1), 223–233 (2013). doi:10.1016/j.cam.2012.07.020. http://www.sciencedirect.com/science/article/pii/S037704271200310X

66. Varga, R.S.: On higher order stable implicit methods for solving parabolic partial differential equations. J. Math. Phys. **40**, 220–231 (1961)

67. Baker Jr, G., Graves-Morris, P.: Padé Approximants. Part I: Basic Theory. Addison Wesley, Reading (1991)

68. Carpenter, A.J., Ruttan, A., Varga, R.S.: Extended numerical computations on the 1/9 conjecture in rational approximation theory. In: Graves-Morris, P.R., Saff, E.B., Varga, R.S. (eds.) Rational Approximation and Interpolation. Lecture Notes in Mathematics, vol. 1105, pp. 383–411. Springer, Berlin (1984)

69. Cody, W.J., Meinardus, G., Varga, R.S.: Chebyshev rational approximations to e^{-x} in $[0, +\infty)$ and applications to heat-conduction problems. J. Approx. Theory **2**(1), 50–65 (1969)

70. Cavendish, J.C., Culham, W.E., Varga, R.S.: A comparison of Crank-Nicolson and Chebyshev rational methods for numerically solving linear parabolic equations. J. Comput. Phys. **10**, 354–368 (1972)

71. Freund, R.W.: Conjugate gradient-type methods for linear systems with complex symmetric coefficient matrices. SIAM J. Sci. Stat. Comput. **13**(1), 425–448 (1992)

72. Axelsson, O., Kucherov, A.: Real valued iterative methods for solving complex symmetric linear systems. Numer. Linear Algebra Appl. **7**(4), 197–218 (2000).doi:10.1002/1099-1506(200005)http://dx.doi.org/10.1002/1099-1506(200005)7:4<197::AID-NLA194>3.0.CO;2-S

73. Howle, V., Vavasis, S.: An iterative method for solving complex-symmetric systems arising in electrical power modeling. SIAM J. Matrix Anal. Appl. **26**, 1150–1178 (2005)

74. Datta, B.N., Saad, Y.: Arnoldi methods for large Sylvester-like observer matrix equations, and an associated algorithm for partial spectrum assignment. Linear Algebra Appl. **154–156**, 225–244 (1991)

75. Gear, C.W., Saad, Y.: Iterative solution of linear equations in ODE codes. SIAM J. Sci. Stat. Comput. **4**, 583–601 (1983)

76. Parlett, B.N.: The Symmetric Eigenvalue Problem. Prentice Hall, Englewood Cliffs (1980)

77. Simoncini, V.: Restarted full orthogonalization method for shifted linear systems. BIT Numer. Math. **43**(2), 459–466 (2003)

78. Freund, R.: Solution of shifted linear systems by quasi-minimal residual iterations. In: Reichel, L., Ruttan, A., Varga, R. (eds.) Numerical Linear Algebra, pp. 101–121. W. de Gruyter, Berlin (1993)

79. Frommer, A.: BiCGStab(ℓ) for families of shifted linear systems. Computing **7**(2), 87–109 (2003)

80. Druskin, V., Knizhnerman, L.: Two polynomial methods of calculating matrix functions of symmetric matrices. U.S.S.R. Comput. Math. Math. Phys. **29**, 112–121 (1989)

81. Friesner, R.A., Tuckerman, L.S., Dornblaser, B.C., Russo, T.V.: A method for exponential propagation of large systems of stiff nonlinear differential equations. J. Sci. Comput. **4**(4), 327–354 (1989)

82. Saad, Y.: Analysis of some Krylov subspace approximations to the matrix exponential operator. SIAM J. Numer. Anal. **29**, 209–228 (1992)

83. Güttel, S.: Rational Krylov methods for operator functions. Ph.D. thesis, Technischen University Bergakademie Freiberg (2010)

84. Simoncini, V., Gallopoulos, E.: A hybrid block GMRES method for nonsymmetric systems with multiple right-hand sides. J. Comput. Appl. Math. **66**, 457–469 (1996)

85. Darnell, D., Morgan, R.B., Wilcox, W.: Deflated GMRES for systems with multiple shifts and multiple right-hand sides. Linear Algebra Appl. **429**, 2415–2434 (2008)

86. Soodhalter, K., Szyld, D., Xue, F.: Krylov subspace recycling for sequences of shifted linear systems. Appl. Numer. Math. **81**, 105–118 (2014)

87. Higham, N.: Accuracy and Stability of Numerical Algorithms, 2nd edn. SIAM, Philadelphia (2002)

88. MUMPS: A parallel sparse direct solver. http://graal.ens-lyon.fr/MUMPS/

89. SuperLU (Supernodal LU). http://crd-legacy.lbl.gov/~xiaoye/SuperLU/

90. Kamgnia, E., Nguenang, L.B.: Some efficient methods for computing the determinant of large sparse matrices. ARIMA J. **17**, 73–92 (2014). http://www.inria.fr/arima/

91. Polizzi, E., Sameh, A.: A parallel hybrid banded system solver: the SPIKE algorithm. Parallel Comput. **32**, 177–194 (2006)

92. Bertrand, O., Philippe, B.: Counting the eigenvalues surrounded by a closed curve. Sib. J. Ind. Math. **4**, 73–94 (2001)

93. Kamgnia, E., Philippe, B.: Counting eigenvalues in domains of the complex field. Electron. Trans. Numer. Anal. **40**, 1–16 (2013)

94. Rudin, W.: Real and Complex Analysis. McGraw Hill, New York (1970)

95. Silverman, R.A.: Introductory Complex Analysis. Dover Publications, Inc., New York (1972)

96. Bindel, D.: Bounds and error estimates for nonlinear eigenvalue problems. Berkeley Applied Math Seminar (2008). http://www.cims.nyu.edu/~dbindel/present/berkeley-oct08.pdf

97. Maeda, Y., Futamura, Y., Sakurai, T.: Stochastic estimation method of eigenvalue density for nonlinear eigenvalue problem on the complex plane. J. SIAM Lett. **3**, 61–64 (2011)

98. Bai, Z., Golub, G.H.: Bounds for the trace of the inverse and the determinant of symmetric positive definite matrices. Ann. Numer. Math. **4**, 29–38 (1997)

99. Duff, I., Erisman, A., Reid, J.: Direct Methods for Sparse Matrices. Oxford University Press Inc., New York (1989)

100. Golub, G.H., Meurant, G.: Matrices, Moments and Quadrature with Applications. Princeton University Press, Princeton (2009)

Chapter 13
Computing the Matrix Pseudospectrum

The ε-pseudospectrum, $\Lambda_\varepsilon(A)$, of a square matrix (pseudospectrum for short) is the locus of eigenvalues of $A + E$ for all possible E such that $\|E\| \leq \varepsilon$ for some matrix norm and given $\varepsilon > 0$; that is

$$\Lambda_\varepsilon(A) = \{z \in \mathbb{C} : z \in \Lambda(A + E), \quad \|E\| \leq \varepsilon\}, \tag{13.1}$$

where $\Lambda(A)$ denotes the set of eigenvalues of A. Unless mentioned otherwise we assume that the 2-norm is used. According to relation (13.1), the pseudospectrum is characterized by means of the eigenvalues of bounded perturbations of A. Pseudospectra have many interesting properties and provide information regarding the matrix that is sometimes more useful than the eigenvalues; the seminal monograph on the subject is [1].

The pseudospectral regions of a matrix are nested as ε varies, i.e. if $\varepsilon_1 > \varepsilon_2$ then necessarily $\Lambda_{\varepsilon_1}(A) \supset \Lambda_{\varepsilon_2}(A)$. Trivially then, $\Lambda_0(A) = \Lambda(A)$. Therefore, to describe the pseudospectrum for some ε it is sufficient to plot the boundary curve(s) $\partial \Lambda_\varepsilon(A)$ (there are more than one boundaries when $\Lambda_\varepsilon(A)$ is not simply connected).

To make the pseudospectrum a practical gauge we need fast methods for computing it. As we will see shortly, however, the computation of pseudospectra for large matrices is expensive, more so than other matrix characteristics (eigenvalues, singular values, condition number). On the other hand, as the success of the EIGTOOL package indicates (cf. Sect. 13.4), there is interest in using pseudospectra. Parallel processing becomes essential in this task. In this chapter we consider parallelism in algorithms for computing pseudospectra. We will see that there exist several opportunities for parallelization that go beyond the use of parallel kernels for BLAS type matrix operations. On the other hand, one must be careful not to create algorithms with abundantly redundant parallelism.

There exist two more characterizations of the pseudospectrum, that turn out to be more useful in practice than perturbation based relation (13.1). The second characterization is based on the smallest singular value of the shifted matrix $A - zI$:

$$\Lambda_\varepsilon(A) = \{z \in \mathbb{C} : \sigma_{\min}(A - zI) \leq \varepsilon\}, \tag{13.2}$$

© Springer Science+Business Media Dordrecht 2016

E. Gallopoulos et al., *Parallelism in Matrix Computations*,
Scientific Computation, DOI 10.1007/978-94-017-7188-7_13

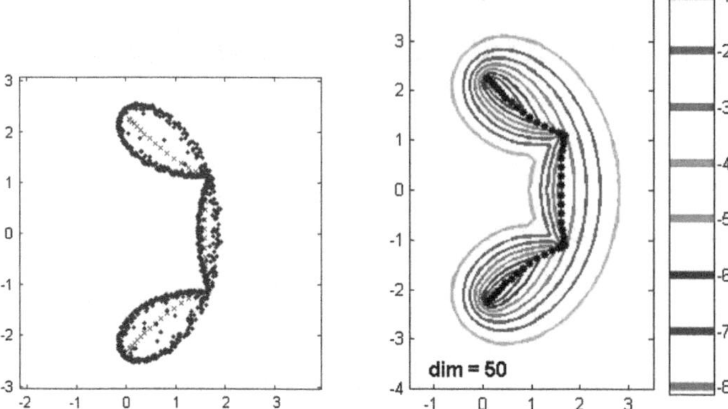

Fig. 13.1 Illustrations of pseudospectra for matrix `grcar` of order $n = 50$. The *left* frame was computed using function `ps` from the Matrix Computation Toolbox that is based on relation 13.1 and shows the eigenvalues of matrices $A + E_j$ for random perturbations $E_j \in \mathbb{C}^{50 \times 50}$, $j = 1, \ldots, 10$ where $\|E_j\| \leq 10^{-3}$. The frame on the *right* was computed using the EigTool package and is based on relation 13.2; it shows the level curves defined by $\{z : s(z) \leq \varepsilon\}$ for $\varepsilon = 10^{-1}$ *down* to 10^{-10}

For brevity, we also write $s(z)$ to denote $\sigma_{\min}(A - zI)$. Based on this characterization, for any $\varepsilon > 0$, we can classify any point z to be interior to $\Lambda_\varepsilon(A)$ if $s(z) < \varepsilon$ or to be exterior when $s(z) > \varepsilon$. By convention, a point of $\partial \Lambda_\varepsilon(A)$ is assimilated to an interior point.

The third characterization is based on the resolvent of A:

$$\Lambda_\varepsilon(A) = \{z \in \mathbb{C} : \|(A - zI)^{-1}\| \geq \varepsilon^{-1}\}. \tag{13.3}$$

We denote the resolvent matrix by $R(z) = (A - zI)^{-1}$. A basic fact is the following:

Proposition 13.1 *The sets defined by (13.1)–(13.3) are identical.*

Figure 13.1 illustrates the pseudospectrum of matrix `grcar` of order 50 computed using algorithms based on the first two definitions. Matrix `grcar` is pentadiagonal Toeplitz corresponding to the symbol $t(z) = z^{-2} + z^{-1} + 1 - z$ and is frequently used to benchmark pseudospectrum codes.

13.1 Grid Based Methods

13.1.1 Limitations of the Basic Approach

Characterization (13.1) appears to necessitate the computation of all eigenvalues of matrices of the form $A + E$ where E is a bounded perturbation. This is an expensive task that becomes impractical for large matrices. Not only that, but it is not clear how to select the perturbations so as to achieve a satisfactory approximation. In particular, there is no guarantee that any of the randomly selected perturbations will cause the extreme dislocations that are possible in the eigenvalues of A if any ε-bounded perturbation is allowed.

A straightforward algorithm for computing pseudospectra is based on characterization (13.2). We list it as Algorithm 13.1, and call it GRID.

Algorithm 13.1 GRID: Computing $\Lambda_\varepsilon(A)$ based on Def. (13.2).

Input: $A \in \mathbb{R}^{n \times n}, l \geq 1$ positive values in decreasing order $\mathscr{E} = \{\varepsilon_1, \varepsilon_2, ..., \varepsilon_l\}$
Output: plot of $\partial\Lambda_\varepsilon(A)$ for all $\varepsilon \in \mathscr{E}$
1: construct $\Omega \supseteq \Lambda_\varepsilon(A)$ for the largest ε of interest and a grid Ω_h discretizing Ω with gridpoints z_k.
2: **doall** $z_k \in \Omega_h$
3:     compute $s(z_k) = \sigma_{\min}(z_k I - A)$
4: **end**
5: Plot the l contours of $s(z_k)$ for all values of $\mathscr{E}$.

GRID is simple to implement and offers parallelism at several levels: large grain, since the singular value computations at each point are independent, as well as medium and fine grain grain parallelism when computing each $s(z_k)$. The sequential cost is typically modeled by

$$\mathbf{T}_1 = |\Omega_h| \times C_{\sigma_{\min}} \tag{13.4}$$

where $|\Omega_h|$ denotes the number of nodes of Ω_h and $C_{\sigma_{\min}}$ is the average cost for for computing $s(z)$. The total sequential cost becomes rapidly prohibitive as the size of A and/or the number of nodes increase. Given that the cost of computing $s(z)$ is at least $O(n^2)$ and even $O(n^3)$ for dense matrix methods, and that a typical mesh could easily contain $O(10^4)$ points, the cost can be very high even for matrices of moderate size. For large enough $|\Omega_h|$ relative to the number of processors p, GRID can be implemented with almost perfect speedup by simple static assignment of $|\Omega_h|/p$ computations of $s(z)$ per processor.

Load balancing also becomes an issue when the cost of computing $s(z)$ varies a lot across the domain. This could happen when iterative methods, such as those described in Chap. 11, Sects. 11.6.4 and 11.6.5, are used to compute $s(z)$.

Another obstacle is that the smallest singular value of a given matrix is usually the hardest to compute (much more so than computing the largest); moreover, this computation has to be repeated for as many points as necessary, depending on the resolution of Ω. Not surprisingly, therefore, as the size of the matrix and/or the number of mesh points increase the cost of the straightforward algorithm above also becomes prohibitive. Advances over GRID for computing pseudospectra are based on some type of dimensionality reduction on the domain or on the matrix, that lower the cost of the factors in the cost model (13.4). In order to handle large matrices, it is necessary to construct algorithms that combine or blend the two approaches.

13.1.2 Dense Matrix Reduction

Despite the potential for near perfect speedup on a parallel system, Algorithm 13.1 (GRID) entails much redundant computation. Specifically, $\sigma_{\min}$ is computed over different matrices, but all of them are shifts of one and the same matrix. In a sequential algorithm, it is preferable to first apply an unitary similarity transformation, say $Q^*AQ = T$, that reduces the matrix to complex upper triangular (via Schur factorization) or upper Hessenberg form. In both cases, the singular values remain intact, thus $\sigma_{\min}(A - zI) = \sigma_{\min}(T - zI)$. The gains from this preprocessing are substantial in the sequential case: the total cost lowers from $|\Omega_h|O(n^3)$ to that of the initial reduction, which is $O(n^3)$, plus $|\Omega_h|O(n^2)$. The cost of the reduction is amortized as the number of gridpoints increases.

We illustrate these ideas in Algorithm 13.2 (`GRID_fact`) which reduces A to Hessenberg or triangular form and then approximates each $s(z_k)$ from the (square root of the) smallest eigenvalue of $(z_k I - T)^*(z_k I - T)$. This is done using inverse Lanczos iteration, exploiting the fact that the cost at each iteration is quadratic since the systems to be solved at each step are Hessenberg or triangular.

For a parallel implementation, we need to decide upon the following:

1. How to evaluate the preprocessing steps (lines 2–5 of Algorithm 13.2). On a parallel system, the reduction step can either be replicated on one or more processors and the reduced matrix distributed to all processors participating in computing the values $\sigma_{\min}(T - z_j I)$. The overall cost will be that for the preprocessing plus $\frac{|\Omega_h|}{p} O(n^2)$ when $p \leq |\Omega_h|$. The cost of the reduction is amortized as the number of gridpoints allocated per processor increases.
2. How to partition the work between processors so as to achieve load balance: A system-level approach is to use queues. Specifically, one or more points in Ω_h can be assigned to a task that is added to a central queue. The first task in the queue is dispatched to the next available processor, until no more tasks remain. Every processor computes its corresponding value(s) of $s(z)$ and once finished, is assigned another task until none remains. A distributed queue can also be used for processors with multithreading capabilities, in which case the groups of points assigned to each task are added to a local queue and a thread is assigned to each point in turn. We also note that static load balancing strategies can also be used if the workload differential between gridpoints can be estimated beforehand.

Algorithm 13.2 GRID_FACT: computes $\Lambda_\varepsilon(A)$ using Def. 13.2 and factorization.

Input: $A \in \mathbb{R}^{n \times n}$, $l \geq 1$ positive values in decreasing order $\mathscr{E} = \{\varepsilon_1, \varepsilon_2, ..., \varepsilon_l\}$, logical variable
  `schur` set to 1 to apply Schur factorization.
Output: plot of $\partial\Lambda_\varepsilon(A)$ for all $\varepsilon \in \mathscr{E}$
1: Ω_h as in line 1 of Algorithm 13.1.
2: $T = \mathrm{hess}(A)$ //reduce A to Hessenberg form
3: **if** `schur` $== 1$ **then**
4:    $T = \mathrm{schur}(T, \,'\mathtt{complex}')$ //compute complex Schur factor T
5: **end if**
6: **doall** $z_k \in \Omega_h$
7:   compute $s(z_k) = \sqrt{\lambda_{\min}(z_k I - T)^*(z_k I - T)}$ //using inverse Lanczos iteration and
    exploiting the structure of T
8: **end**
9: Plot the l contours of $s(z_k)$ for all values of $\mathscr{E}$.

13.1.3 Thinning the Grid: The Modified GRID *Method*

We next consider an approach whose goal is to reduce the number of gridpoints where $s(z)$ needs to be evaluated. It turns out that it is possible to rapidly "thin the grid" by stripping disk shaped regions from Ω. Recall that for any given ε and gridpoint z, algorithm GRID uses the value of $s(z)$ to classify z as lying outside or inside $\Lambda_\varepsilon(A)$. In that sense, GRID makes pointwise use of the information it computes at each z. A much more effective idea is based on the fact that at any point where the minimum singular value $s(z)$ is evaluated, given any value of ε, either the point lies in the pseudospectrum or its boundary or a disk is constructed whose points are guaranteed to be exterior to $\Lambda_\varepsilon(A)$. The following result is key.

Proposition 13.2 ([2]) *If $s(z) = r > \varepsilon$ then*

$$D^\circ(z, r - \varepsilon) \cap \Lambda_\varepsilon(A) = \emptyset,$$

where $D^\circ(z, r - \varepsilon)$ is the open disk centered at z with radius $r - \varepsilon$.

This provides a mechanism for computing pseudospectra based on "inclusion-exclusion". At any point $s(z)$ is computed (as in GRID_FACT this is better done after initially reducing A to Hessenberg or Schur form), it is readily confirmed that the point is included in the pseudospectrum (as in GRID), otherwise it defines an "exclusion disk" centered at that point. Algorithm 13.3 (MOG) deploys this strategy and approximates the pseudospectrum by repeatedly pruning the initial domain that is assumed to completely enclose the pseudospectrum until no more exclusions can be applied. This is as robust as GRID but can be much faster. It can also be shown that as we move away from the sought pseudospectrum boundary, the exclusion disks become larger which makes the method rather insensitive to the choice of Ω compared to GRID. We mention in passing that a suitable enclosing region for the pseudospectrum that can be used as the initial Ω for both GRID and MOG is based on the field-of-values [3].

Algorithm 13.3 MoG: Computing $\Lambda_\varepsilon(A)$ based on inclusion-exclusion [2]

Input: $A \in \mathbb{R}^{n \times n}, l \geq 1$ and ε
Output: plot of $\Lambda_\varepsilon(A)$
1: construct $\Omega \supseteq \Lambda_\varepsilon(A)$ and a grid Ω_h discretizing Ω with gridpoints z_k
2: reduce A to its Hessenberg or Schur form, if feasible
3: **doall** $z_k \in \Omega_h$
4:    compute exclusion disk Δ.
5:    Set $\Omega_h = \Omega_h \setminus \Delta$ //Remove from Ω_h all gridpoints in Δ.
6: **end**

Figure 13.2 (from [4]) illustrates the application of MOG for matrix `triangle` (a Toeplitz matrix with symbol $z^{-1} + \frac{1}{4}z^2$ from [5]) of order 32. MOG required only 676 computations of $s(z)$ to approximate $\Lambda_\varepsilon(A)$ for $\varepsilon = 10^{-1}$ compared to the 2500 needed for GRID.

Unlike GRID, the preferred sequence of exclusions in MOG cannot be determined beforehand but depends on the order in which the gridpoints are swept since Ω_h is modified in the course of the computation. This renders a parallel implementation more challenging. Consider for instance a static allocation policy of gripoints, e.g. block partitioning. Then a single computation within one processor might exclude many points allocated to this and possibly other processors. In other words, if $s(z)$ is computed, then it is likely that values $s(z + \Delta z)$ at nearby points will not be needed. This is an action reverse from the spatial locality of reference principle in computer science and can cause load imbalance.

To handle this problem, it was proposed in [4] to create a structure accessible to all processors that contains information about the status of each gridpoint. Points are classified in three categories: *Active*, if $s(z)$ still needs to be computed, *inactive*, if they have been excluded, and *fixed*, if $s(z) \leq \varepsilon$. This structure must be updated as the computation proceeds. The pseudospectrum is available when there are no more *active* points.

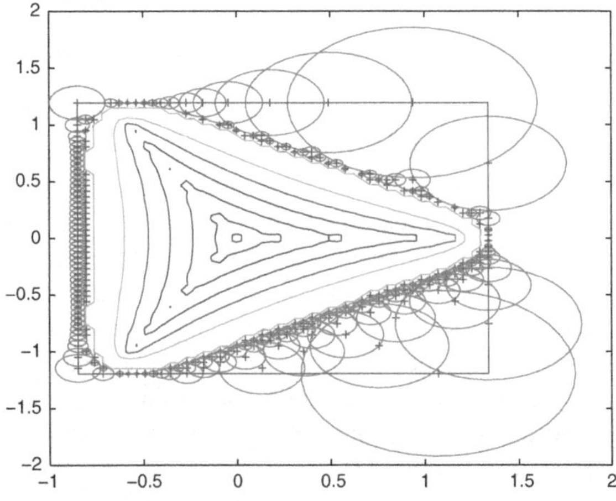

Fig. 13.2 Using MOG to compute the pseudospectrum of `triangle(32)` for $\varepsilon = 1e - 1$ (from [4])

Based on this organization, we construct a parallel MOG algorithm as follows. We first assume that a discretization Ω_h of a rectangular region Ω enclosing the pseudospectrum is available. Initially, the set of active nodes $\mathcal{N}$ is defined to be the set of all gridpoints. As the algorithm proceeds, this set is updated to contain only those gridpoints that have not been excluded in previous steps. The algorithm consists of three stages. The first stage aims to find the smallest bounding box $\mathcal{B}$ of $\partial \Lambda_\varepsilon(A)$, the perimeter of which does not contain any active points. The second stage performs a collapse of the perimeter of $\mathcal{B}$ towards $\partial \Lambda_\varepsilon(A)$. At the end of this stage no further exclusion are performed. The set of active nodes $\mathcal{N}$ contains only mesh nodes $z_{in} \in \Lambda_\varepsilon(A)$. In the third stage all remaining $s(z)$ for $z \in \mathcal{N}$ are computed.

The motivation behind the first stage is to attempt large exclusions that have small overlap. Since the exclusion disks are not known a priori, the heuristic is that gridpoints lying on the outer convex hull enclosing the set of active points offer a better chance for larger exclusions. The second stage is designed to separate those points that should remain active from those that should be marked inactive but have not been rendered so in stage 1 because of their special location; e.g. stage 1 does not exclude points that lie deep inside some concave region of the pseudospectrum. To see this, consider in Fig. 13.3 the case of matrix `grcar` (1000), where dots ('·') label gridpoints that remain active after the end of stage 1 and circles ('o') indicate gridpoints that remain active after stage 2 is completed.

A master node could take the responsibility of gathering the singular values and distributing the points, on the bounding box $\mathcal{B}$ (stage 1) and on the hull $\mathcal{H}$ (stage 2). The administrative task assigned to the master node (exclusions and construction of the hull) is light; it is thus feasible to also use the master node to compute singular triplets. In conclusion, stages 1 and 2 adopt a synchronized queue model: The master waits for all processes, including itself, to finish computing $s(z)$ for their currently allocated point and only then proceeds to dispatch the next set. Stage 3 is similar to the parallel implementation of GRID, the only difference being that the points where we need to compute $s(z)$ are known to satisfy $s(z) \leq \varepsilon$.

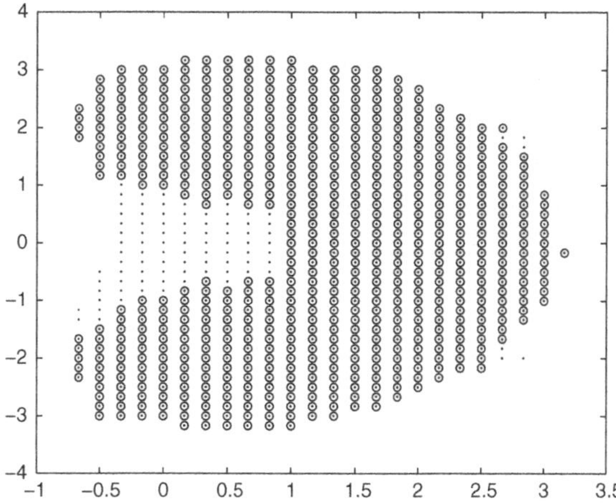

Fig. 13.3 Method MOG: outcome from stages 1 (*dots* '·') and 2 (*circle* 'o') for matrix `grcar(1000)` on 50×50 grid (from [4])

If the objective is to compute the $\Lambda_\varepsilon(A)$ for $\varepsilon_1 > \cdots > \varepsilon_s$, we can apply MOG to estimate the enclosing $\Lambda_{\varepsilon_1}(A)$ followed by GRID for all remaining points. An alternative is to take advantage of the fact that

$$\sigma_{\min}(A) - \varepsilon_1 < \sigma_{\min}(A) - \varepsilon_2 < \cdots < \sigma_{\min}(A) - \varepsilon_s.$$

Discounting negative and zero values, these define a sequence of concentric disks that do not contain any point of the pseudospectra for the corresponding values of ε. For example, the disk centered at z and radius $\sigma_{\min}(A) - \varepsilon_j$ has no intersection with $\Lambda_{\varepsilon_j}(A)$. This requires careful bookkeeping but its parallel implementation, would greatly reduce the need to use GRID.

13.2 Dimensionality Reduction on the Domain: Methods Based on Path Following

One approach that has the potential for dramatic dimensionality reduction on the domain is to trace the individual pseudospectral boundaries using some type of path following. We first review the original path following idea for pseudospectra and then examine three methods that enable parallelization. Interestingly, parallelization also helps to improve the numerical robustness of path following.

13.2.1 Path Following by Tangents

To trace the curve, we can use predictor-corrector techniques that require differential information from the curve, specifically tangents and normals. The following important result shows that such differential information at any point z is available if together with $\sigma_{\min}$ one computes the corresponding singular vectors (that is the minimum singular triplet $\{\sigma_{\min}, u_{\min}, v_{\min}\}$. It is useful here to note that $\Lambda_\varepsilon(A)$ can be defined implicitly by the equation

$$g(x, y) = \varepsilon, \quad \text{where } g(x, y) = \sigma_{\min}((x + iy)I - A), \tag{13.5}$$

(here we identify the complex plane $\mathbb{C}$ with $\mathbb{R}^2$). The key result is the following that is a generalization of Theorem 11.14 to the complex plane [6]:

Theorem 13.1 *Let* $z = x + iy \in \mathbb{C} \setminus \Lambda(A)$. *Then* $g(x, y)$ *is real analytic in a neighborhood of* (x, y), *if* $\sigma_{\min}((x + iy)I - A)$ *is a simple singular value. The gradient of* $g(x, y)$ *is equal to*

$$\nabla g(x, y) = (\Re(v_{\min}^* u_{\min}), \Im(v_{\min}^* u_{\min})) = v_{\min}^* u_{\min},$$

where $u_{\min}$ and $v_{\min}$ denote the left and right singular vectors corresponding to $\sigma_{\min}$.

From the value of the gradient at any point of the curve, one can make a small prediction step along the tangent followed by a correction step to return to a subsequent point on the curve. Initially, the method needs to locate one point of each non-intersecting boundary $\partial \Lambda_\varepsilon(A)$. The path following approach was proposed in [7]. We denote it by PF and list it as Algorithm 13.4, incorporating the initial reduction to Hessenberg form.

Algorithm 13.4 PF: computing $\partial \Lambda_\varepsilon(A)$ by predictor-corrector path following [7].

Input: $A \in \mathbb{R}^{n \times n}$, value $\varepsilon > 0$.
Output: plot of the $\partial \Lambda_\varepsilon(A)$
1: Transform A to upper Hessenberg form and set $k = 0$.
2: Find initial point $z_0 \in \partial \Lambda_\varepsilon(A)$
3: **repeat**
4:    Determine $r_k \in \mathbb{C}$, $|r_k| = 1$, steplength τ_k and set $\tilde{z}_k = z_{k-1} + \tau_k r_k$.
5:    Correct along $d_k \in \mathbb{C}$, $|d_k| = 1$ by setting $z_k = \tilde{z}_k + \theta_k d_k$ where θ_k is some steplength.
6: **until** termination

Unlike typical prediction-correction, the prediction direction r_k in PF is taken orthogonal to the previous correction direction d_{k-1}; this is in order to avoid an additional (expensive) triplet evaluation. Each Newton iteration at a point z requires the computation of the singular triplet $(\sigma_{\min}, u_{\min}, v_{\min})$ of $zI - A$. This is the dominant cost per step and determines the total cost of the algorithm. Computational experience indicates that a single Newton step is sufficient to obtain an adequate correction. What makes PF so appealing in terms of cost and justifies our previous characterization of the dimensionality reduction as dramatic, is that by tracing $\partial \Lambda_\varepsilon(A)$, a computation over a predefined two dimensional grid Ω_h is replaced with a computation over points on the pseudospectrum boundary. On the other hand, a difficulty of PF is that when the pseudospectrum boundary contains discontinuities of direction or folds that bring sections of the curve close by, path following might get disoriented unless the step size is very small. Moreover, unlike GRID, the only opportunities for parallelism are in the computation of the singular triplets and in the computation of multiple boundaries.

It turns out that it is actually possible to improve performance and address the aforementioned numerical weaknesses by constructing a parallel path following algorithm. We note in passing that this is another one of these cases that we like to underline in this book, where parallelism actually improves the numerical properties of an algorithm. The following two observations are key. First, that when the boundary is smooth, the prediction-correction can be applied to discover several new points further along the curve. This entails the computation of several independent singular triplets. Second, that by attempting to advance using several points at a time, there is more information available for the curve that can be used to prevent the algorithm from straying off the correct path. Algorithm 13.5 is constructed based on these ideas.

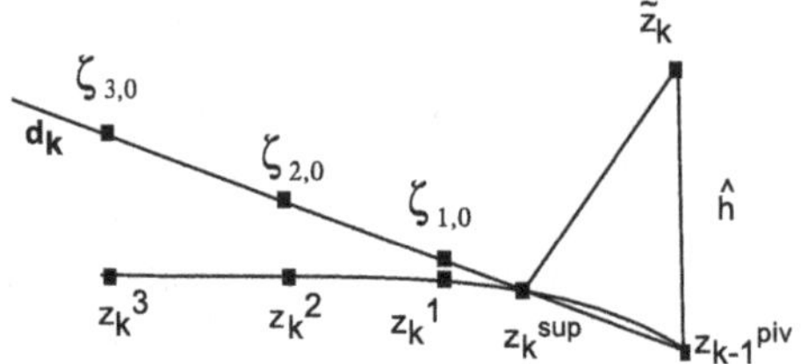

Fig. 13.4 COBRA: Position of pivot (z_{k-1}^{piv}), initial prediction ($\tilde{z}_k$), support (z_k^{sup}), first order predictors ($\zeta_{j,0}$) and corrected points (z_k^j). (A proper scale would show that $h \ll \mathbb{H}$)

It lends itself for parallel processing while being more robust than the original path following. The name used is COBRA in order to evoke that snake's spread neck.

Iteration k is illustrated in Fig. 13.4. Upon entering the repeat loop in line 4, it is assumed that there is a point, z_{k-1}^{piv}, available that is positioned close to the curve being traced. The loop consists of three steps. In the *first prediction-correction step* (line 4–5), just like Algorithm PF, a support point, z_k^{sup}, is computed first using a small stepsize h. In the *second prediction-correction step* (lines 7–8) z_{k-1}^{piv}, z_k^{sup} determine a prediction direction d_k and m equidistant points, $\zeta_{i,0} \in d_k, i = 1, \ldots m$, are selected in the direction of d_k. Then $h = |\tilde{z}_k - z_{k-1}^{\mathrm{piv}}|$ is the stepsize of the method; note that we can interpret $\mathbb{H} = |z_{k-1}^{\mathrm{piv}} - \zeta_{m,0}|$ as the length of the "neck" of COBRA. Then each $\zeta_{i,0} \in d_k, i = 1, \ldots m$ is corrected to obtain $z_k^i \in \partial \Lambda_\varepsilon(A), i = 1, \ldots, m$. This is implemented using only one step of Newton iteration on $\sigma_{\min}(A - zI) - \varepsilon = 0$. All corrections are independent and can be performed in parallel. In the third step (line 10), the next pivot, z_k^{piv}, is selected using some suitable criterion.

The correction phase is implemented with Newton iteration along the direction of steepest ascent $d_k = \nabla g(\tilde{x}_k, \tilde{y}_k)$, where $\tilde{x}_k, \tilde{y}_k$ are the real and imaginary parts of $\tilde{z}_k$. Theorem 13.1 provides the formula for the computation of the gradient of $s(z) - \varepsilon$. The various parameters needed to implement the main steps of the procedure can be found in [7].

In the correction phase of and other path following schemes, Newton's method is applied to solve the nonlinear equation $g(x, y) = \varepsilon$ for some $\varepsilon > 0$ (cf. (13.5)). Therefore, we need to compute $\nabla g(x, y)$. From Theorem 13.1, this costs only 1 DOT operation.

COBRA offers large-grain parallelism in its second step. The number of gridpoints on the cobra "neck" determines the amount of available large-grain parallelism available that is entirely due to path following. If the number of points m on the "neck" is equal to or a multiple of the number of processors, then the maximum speedup due to path following of COBRA over PF is expected to be $m/2$. A large number of processors would favor making m large and thus it is well suited for obtaining $\partial \Lambda_\varepsilon(A)$ at very fine resolutions. The "neck" length H, of course, cannot be arbitrarily large and once there are enough points to adequately represent $\partial \Lambda_\varepsilon(A)$, any further

---

Algorithm 13.5 COBRA: algorithm to compute $\partial \Lambda_\varepsilon(A)$ using parallel path following [48]. After initialization (line 1), it consists of three stages: i) prediction-correction (lines 4-5), ii) prediction-correction (lines 7-8), and iii) pivot selection (line 10).

Input: $A \in \mathbb{R}^{n \times n}$, value $\varepsilon > 0$, number m of gridpoints for parallel path following.
Output: pseudospectrum boundary $\partial \Lambda_\varepsilon(A)$
1: Transform A to upper Hessenberg form and set $k = 0$.
2: Find initial point $z_0 \in \partial \Lambda_\varepsilon(A)$
3: **repeat**
4:     Set $k = k + 1$ and predict $\tilde{z}_k$.
5:     Correct using Newton and compute z_k^{sup}.
6:     **doall** $j = 1, ..., m$
7:         Predict $\zeta_{j,0}$.
8:         Correct using Newton and compute z_k^j.
9:     **end**
10:    Determine next pivot z_k^{piv}.
11: **until** termination

---

subdivisions of H would be useless. This limits the amount of large-grain parallelism that can be obtained in the second phase of COBRA. It is then reasonable to exploit the parallelism available in the calculation of each singular triplet, assuming that the objective is to compute the pseudospectrum of matrices that are large. Another source of parallelism would be to compute several starting points on the curve and initiate COBRA on each of them in parallel.

We note some additional advantages of the three-step approach defined above. The first step is used to obtain the chordal direction linking the pivot and support points. For this we use a stepsize $\hat{h}$ which can be very small in order to capture the necessary detail of the curve and also be close to the tangential direction. Nevertheless, the chordal direction is often better for our purpose: For instance, if for points in the neighborhood of the current pivot $\partial \Lambda_\varepsilon(A)$ is downwards convex and lies below the tangent then the predicted points $\{\zeta_{j,0}\}_{j=1}^m$ lie closer to $\partial \Lambda_\varepsilon(A)$ than if we had chosen them along the tangent. We also note that unlike PF, the runtime of COBRA is not adversely affected if the size of $\hat{h}$ is small. Instead, it is the "neck" length H that is the effective stepsize since it determines how fast we are marching along the curve. Furthermore, the Newton correction steps that we need to apply on the m grid points $\zeta_{j,0}$ are done in parallel. Finally, the selection procedure gives us the opportunity to reject one or more points. In the rare cases that all parallel corrections fail, only the support or pivot point can be returned, which would be equivalent to performing one step of PF with (small) stepsize $\hat{h}$. These characteristics make COBRA more robust and more effective than PF.

13.2.2 Path Following by Triangles

Another parallel method for circumventing the difficulties encountered by PF at segments where $\partial \Lambda_\varepsilon(A)$ has steep turns and angular shape is by employing overlapping triangles rather than the "cobra" structure. We describe this idea (originally from [9]) and its incorporation into an algorithm called PAT.

Definition 13.1 Two complex points a and b are said to be ε-separated when one is interior and the other exterior to $\Lambda_\varepsilon(A)$.

Given two distinct complex numbers z_0 and z_1, we consider the regular lattice

$$S(z_0, z_1) = \{S_{kl} = z_0 + k\,(z_1 - z_0) + l\,(z_1 - z_0)e^{i\frac{\pi}{3}}, \ (k, l) \in \mathbb{Z}^2\}.$$

The nodes are regularly spaced:

$$|S_{k,l+1} - S_{k,l}| = |S_{k+1,l} - S_{k,l}| = \tau,$$

where $\tau = |z_0 - z_1|$. This mesh defines the set $\Omega(z_0, z_1)$ of equilateral triangles of mesh size τ (see Fig. 13.5).

Denote by $\mathscr{O}_L$ the subset of $\Omega(z_0, z_1)$ where $T \in \mathscr{O}_L$ if, and only if, at least two vertices of T are ε-separated. It is easy to prove that, for all $z_0 \neq z_1$, the set $\mathscr{O}_L$ is a finite set.

Definition 13.2 When $T \in \mathscr{O}_L$, two vertices are interior and one vertex is exterior or the reverse. The vertex alone in its class is called the *pivot of T* and is denoted $p(T)$.

Let us define a transformation F which maps any triangle of $\mathscr{O}_L$ into another triangle of the same set: it is defined by

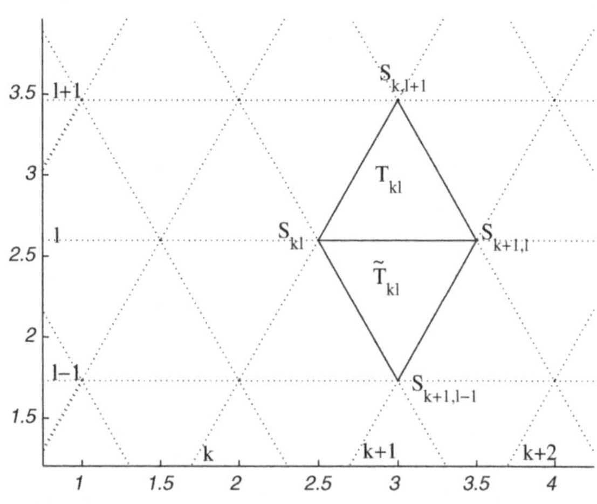

Fig. 13.5 PAT: The lattice and the equilateral triangles

$$
\begin{cases}
F(T) = R(p(T), \theta), \\
\text{with } \theta = \frac{\pi}{3} \text{ if } p(T) \text{ is interior, else } \theta = -\frac{\pi}{3},
\end{cases}
\tag{13.6}
$$

where $R(z, \theta)$ denotes the rotation centered at $z \in \mathbb{C}$ with angle θ.

Proposition 13.3 *F is a bijection from $\mathcal{O}_L$ onto $\mathcal{O}_L$.*

Proof See [9].

In Fig. 13.6, four situations are listed, depending on the type of the two points $p(T)$ and $p(F(T))$.

Definition 13.3 For any given $T \in \mathcal{O}_L$ we define the $F-$orbit of T to be the set $O(T) = \{T_n, n \in \mathbb{Z}\} \subset \mathcal{O}_L$, where $T_n \equiv F^n(T)$.

Proposition 13.4 *Let $O(T)$ be the $F-$orbit of a given triangle T;*

1. *If $N = \mathrm{card}(O(T))$, is the cardinality of $O(T)$, then N is even and is the smallest positive integer such that $F^N(T) = T$.*
2. *$\sum_{i=0}^{N-1} \theta_i = 0 \ (2\pi)$ where θ_i is the angle of the rotation $F(.)$ when applied to T_i.*

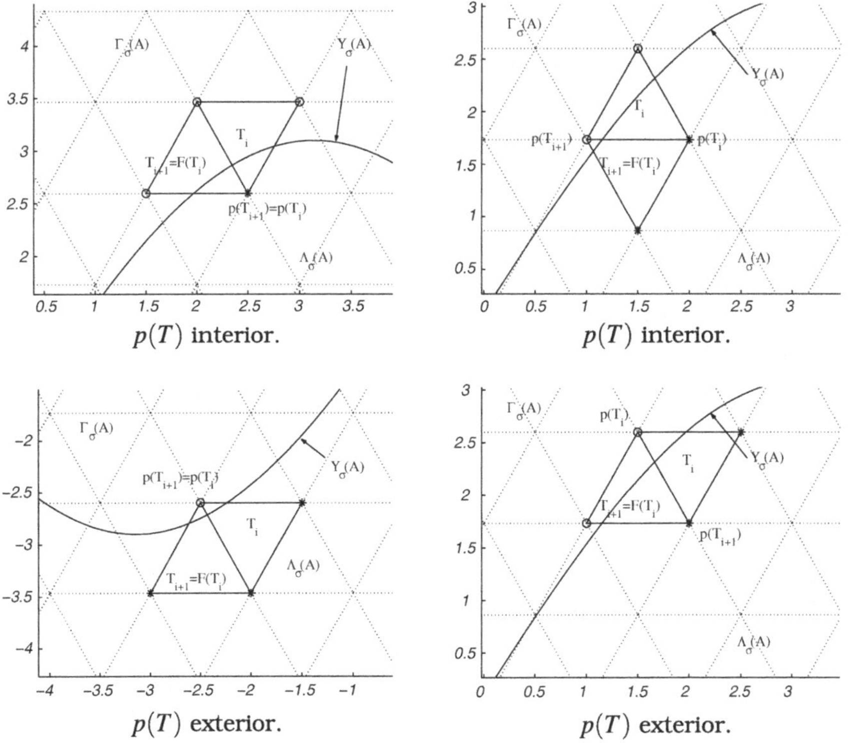

Fig. 13.6 PAT: Illustrations of four situations for transformation F

Algorithm 13.6 PAT: path-following method to determine $\Lambda_\varepsilon(A)$ with triangles.

Input: $A \in \mathbb{R}^{n \times n}, \varepsilon > 0, \tau > 0$ mesh size, θ lattice inclination, z_0 interior point ($\sigma_{\min}(A - z_0 I) \leq \varepsilon$).

Output: An initial triangle $T_0 = (z^i, z^e, z') \in \mathcal{O}_L$, the orbit $O(T_0)$, a polygonal line of N vertices ($N = \text{card}(O(T_0))$) that belong to $\partial \Lambda_\varepsilon(A)$.

1: $h = \tau e^{i\theta}$; $Z = z_0 + h$; $I_t = 0$; //Looking for an exterior point
2: **while** $\sigma_{\min}(A - z_0 I) \leq \varepsilon$,
3:     $I_t = I_t + 1$; $h = 2h$; $Z = z_0 + h$;
4: **end while**
5: $z^i = z_0$; $z^e = Z$;   //STEP I: Defining the initial triangle
6: **do** $k = 1 : I_t$,
7:     $z = (z^e + z^i)/2$;
8:     **if** $\sigma_{\min}(A - zI) \leq \varepsilon$, **then**
9:       $z^i = z$;
10:    **else**
11:      $z^e = z$;
12:    **end if**
13: **end**
14: $T_1 = F(T_0)$; $i = 1$; $\mathscr{I} = \{[z^i, z^e], [z', p(T_1)]\}$;   //STEP II: Building the F−orbit of T_0
15: **while** $T_i \neq T_0$,
16:    $T_{i+1} = F(T_i)$; $i = i + 1$;
17:    From T_i insert in $\mathscr{I}$ a segment which intersets $\partial \Lambda_\varepsilon(A)$;
18: **end while**
    //STEP III: Extraction of points of $\partial \Lambda_\varepsilon(A)$
19: **doall** $I \in \mathscr{I}$,
20:    Extract by bisection $z_I \in \partial \Lambda_\varepsilon(A) \cap I$;
21: **end**

Proof See [9].

The method is called PAT and is listed as Algorithm 13.6. It includes three steps:

1. *STEP I: Construction of the initial triangle.* The algorithm assumes that an interior point z_0 is already known. Typically, such a point is provided as approximation of an eigenvalue of A lying in the neighborhood of a given complex number. The mesh size τ and the inclination θ of the lattice must also be provided by the user. The first task consists in finding a point Z of the lattice that is exterior to $\Lambda_\varepsilon(A)$. The technique of iteratively doubling the size of the interval is guaranteed to stop since $\lim_{|z| \to \infty} \sigma_{\min}(A - zI) = \infty$. Then, an interval of prescribed size τ and intersecting $\partial \Lambda_\varepsilon(A)$ is obtained by bisection. Observe that all the interval endpoints that appear in lines 1–13 are nodes of the lattice.

2. *STEP II: Construction of the chain of triangles.* Once the initial triangle is known, the transformation F is iteratively applied. At each new triangle a new interval intersecting $\partial \Lambda_\varepsilon(A)$ is defined: one endpoint is the pivot of the new triangle and the other endpoint is the new point which has been determined with the new triangle. Since every point T_{kl} of the lattice is exactly characterized by its integer coordinates k and l, the test of equality between triangles is exact.

3. *STEP III: Definition of the polygonal line supported by $\partial \Lambda_\varepsilon(A)$.* This step is performed by a classical bisection process.

The minimum number of triangles in an orbit is 6. This corresponds to the situation where the pivot is the same for all the triangles of the orbit and therefore the angles of all the rotations involved are identical. Such an orbit corresponds to a large τ when compared to the length of $\partial \Lambda_\varepsilon(A)$.

Proposition 13.5 *Assume that Algorithm 13.6 determines an orbit $O(T_0)$ that includes N triangles and that $N > 6$. Let ℓ be the length of the curve $\partial \Lambda_\varepsilon(A)$ and η be the required absolute accuracy for the vertices of the polygonal line. The complexity of the computation is given by the number $\mathcal{N}_\varepsilon$ of points where $s(z)$ is computed. This number is such that*

$$\mathcal{N}_\varepsilon \leq (N+1)\lceil \log \frac{\tau}{\eta} \rceil \; and \; N = O\left(\frac{\ell}{\tau}\right).$$

Proof See [9].

Remark 13.1 When the matrix A and the interior point z_0 are real, and when the inclination θ is null, the orbit is symmetric with respect to the real axis. The computation of the orbit can be stopped as soon as a new interval is real. The entire orbit is obtained by symmetry. This halves the effort of computation.

Remark 13.2 In Algorithm 13.6, one may stop at Step 19, when the orbit is determined. From the chain of triangles, a polygonal line of exterior points is therefore obtained which follows closely the polygonal line which would have been built with the full computation. This reduces the number of computations significantly.

To illustrate the result obtained by PAT, consider matrix `grcar(100)` and two experiments. In the two cases, the initial interior point is automatically determined in the neighborhood of the point $z = 0.8 - 1.5i$. The results are expressed in Table 13.1 and in Fig. 13.7.

Parallel Tasks in PAT

Parallelism in PAT, as in the previous algorithms, arises at two levels. (i) within the kernel computing $s(z)$; (ii) across computations of $s(z)$ at different values $z = z_1, z_2, \ldots$. The main part of the computation is done in the last loop (line 19) of Algorithm 13.6 that is devoted to the computation of points of the curve $\partial \Lambda_\varepsilon(A)$ from a list of intervals that intersect the curve. As indicated by the `doall` instruction, the

Table 13.1 PAT: number of triangles in two orbits for matrix `grcar(100)`

ε	τ	N
10^{-12}	0.05	146
10^{-6}	0.2	174

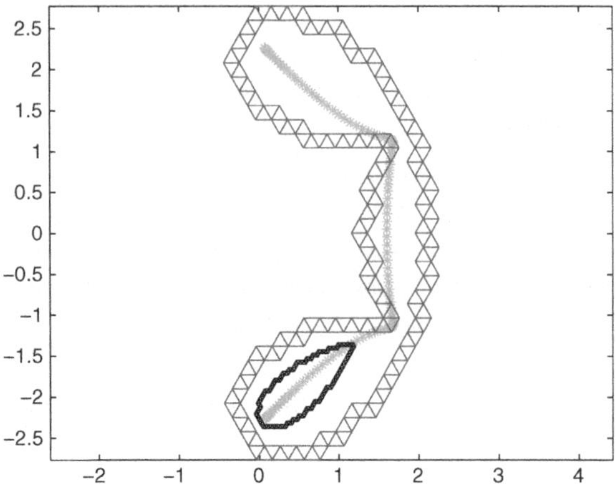

Fig. 13.7 PAT: Two orbits for the matrix `grcar(100)`; eigenvalues are plotted by (*red*) '*', the two chains of *triangles* are drawn in *black* ($\varepsilon = 10^{-12}$) and *blue* ($\varepsilon = 10^{-6}$)

iterations are independent. For a parallel version of the method, two processes can be created: P_1 dedicated to STEP I and STEP II, and P_2 to STEP III. Process P_1 produces intervals which are sent to P_2. Process P_2 consumes the received intervals, each interval corresponding to an independent task. This approach allows to start STEP III before STEP II terminates.

The parallel version of PAT, termed PPAT, applies this approach (see [10] for details). To increase the initial production of intervals intersecting the curve, a technique for starting the orbit simultaneously in several places is implemented (as noted earlier, COBRA could also benefit from a similar technique). When A is large and sparse, PPAT computes $s(z)$ by means of a Lanczos procedure applied to $\begin{pmatrix} O & R^{-1} \\ R^{-*} & 0 \end{pmatrix}$ where R is the upper triangular factor of the sparse QR factorization of the matrix $(A - zI)$. The two procedures, namely the QR factorization and the triangular system solves at each iteration, are also parallelized.

13.2.3 Descending the Pseudospectrum

PF, COBRA, PAT and PPAT construct $\partial \Lambda_\varepsilon(A)$ for any given ε. It turns out that once an initial boundary has been constructed, a path following approach can be applied, yielding an effective way for computing additional boundaries for smaller values of ε that is rich in large grain parallelism.

The idea is instead of tracing the curve based on tangent directions, to follow the normals along directions that $s(z)$ decreases. This was implemented as an algorithm called Pseudospectrum Descent Method (PsDM for short). Essentially, given enough gridpoints on the outermost pseudospectrum curve, the algorithm creates a front along the normals and towards the direction of decrease of $s(z)$ at each point steps until the next sought inner boundary is reached. The process can be repeated to compute

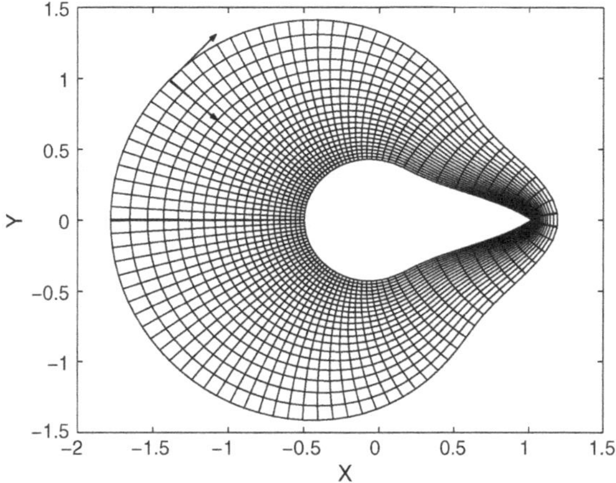

Fig. 13.8 Pseudospectrum contours and trajectories of points computed by PsDM for *pseps*, $\varepsilon = 10^{-1}, ..., 10^{-3}$ for matrix kahan(100). *Arrows* show the directions used in preparing the outermost curve with path following and the directions used in marching from the outer to the inner curves with PsDM

pseudospectral contours. Figure 13.8 shows the results from the application of PsDM to matrix kahan(100) (available via the MATLAB gallery; see also [11]). The plot shows (i) the trajectories of the points undergoing the steepest descent and (ii) the corresponding level curves. The intersections are the actual points computed byPsDM.

To be more specific, assume that an initial contour $\partial \Lambda_\varepsilon(A)$ is available in the form of some approximation (e.g. piecewise linear) based on N points z_k previously computed using COBRA (with only slight modifications, the same idea applies if PAT or PPAT are used to approximate the first contour). In order to obtain a new set of points that define an inner level curve we proceed in 2 steps:

Step 1: Start from z_k and compute an intermediate point $\tilde{w}_k$ by a single modified Newton step towards a steepest descent direction d_k obtained earlier.

Step 2: Correct $\tilde{w}_k$ to w_k using a Newton step along the direction l_k of steepest descent at $\tilde{w}_k$.

Figure 13.9 (from [12]) illustrates the basic idea using only one initial point. Applying one Newton step at z_k requires $\nabla s(z_k)$. To avoid this extra cost, we apply the same idea as in PF and COBRA and use the fact that

$$\nabla s(\tilde{x}_k + i\tilde{y}_k) = (\Re g_{\min}^* q_{\min}, \Im g_{\min}^* q_{\min})$$

where $g_{\min}$ and $q_{\min}$ are the corresponding left and right singular vectors. This quantity is available from the original path following procedure. In essence we approximated the gradient based at z_k with the gradient based at $\tilde{z}_k$, so that

$$z_k = \tilde{z}_k - \frac{s(\tilde{x}_k + i\tilde{y}_k) - \varepsilon}{q_{\min}^* g_{\min}}, \tag{13.7}$$

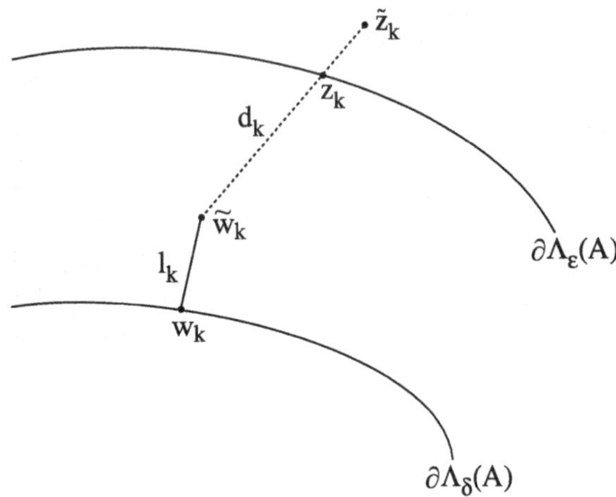

Fig. 13.9 Computing $\partial\Lambda_\delta(A)$, $\delta < \varepsilon$

Applying correction based on gradient information from $\tilde{z}_k$ as in (13.7) it follows that

$$\tilde{w}_k = z_k - \frac{\varepsilon - \delta}{q^*_{\min} g_{\min}}, \tag{13.8}$$

where $\delta < \varepsilon$ and $\partial\Lambda_\delta(A)$ is the new pseudospectrum boundary we are seeking. Once $\tilde{w}_k$ is available, we perform a second Newton step that yields w_k:

$$w_k = \tilde{w}_k - \frac{s(\tilde{w}_k) - \delta}{u^*_{\min} v_{\min}}, \tag{13.9}$$

where the triplet used is associated with $\tilde{w}_k$. These steps can be applied independently to all N points. This is one sweep of PsDM; we list it as Algorithm 13.7 (PsDM_sweep).

Algorithm 13.7 PsDM_sweep: single sweep of method PsDM.

Input: N points z_k of the initial contour $\partial\Lambda_\varepsilon(A)$.
Output: N points w_k on the target contour $\partial\Lambda_\delta(A)$, $\delta < \varepsilon$.
1: **doall** $k = 1, \ldots, N$
2:    Compute the intermediate point $\tilde{w}_k$ according to (13.8).
3:    Compute the target point w_k using (13.9).
4: **end**

Assume now that the new points computed after PsDM_sweep define satisfactory approximations of a nearby contour $\partial\Lambda_\delta(A)$, where $\delta < \varepsilon$. We can continue in this manner to approximate boundaries of the pseudospectrum nested in $\partial\Lambda_\delta(A)$. As noted earlier, the application of PsDM_sweep uses $\nabla s(\tilde{x}_k + i\tilde{y}_k)$, i.e. the triplet at $\tilde{z}_k$ that was available from the original computation of $\partial \Lambda_\varepsilon(A)$ with COBRA or PF. Observe now that as the sweep proceeds to compute $\partial \Lambda_{\tilde{\delta}}(A)$ from $\partial \Lambda_\delta(A)$ for $\tilde{\delta} < \delta$, it also computes the corresponding minimum singular triplet at $\tilde{w}_k$. Therefore enough derivative information is available for PsDM_sweep to be reapplied with starting points computed via the previous application of PsDM_sweep and it is not necessary to run PF again. Using this idea repeatedly we obtain the PsDM method listed as Algorithm 13.8 and illustrated in Fig. 13.10.

Algorithm 13.8 PsDM: pseudospectrum descent method

Input: N points z_k approximating a contour $\partial \Lambda_\varepsilon(A)$.
Output: $M \times N$ points approximating M contours $\partial \Lambda_{\delta_i}, \varepsilon > \delta_1 > \ldots > \delta_M$.
1: $\delta_0 = \varepsilon$
2: **do** $i = 1, \ldots, M$
3:     Compute N points of $\partial \Lambda_{\delta_i}$ by PsDM_sweep on the N points of $\partial \Lambda_{\delta_{i-1}}$
4: **end**

Observe that each sweep can be interpreted as a map that takes as input N_{in} points approximating $\Lambda_{\delta_i}(A)$ and produces N_{out} points approximating $\Lambda_{\delta_{i+1}}(A)$ $(\delta_{i+1} < \delta_i)$. In order not to burden the discussion, we restrict ourselves to the case that $N_{\text{in}} = N_{\text{out}}$. Nevertheless, this is not an optimal strategy. Specifically, since for a given collection of ε's, the corresponding pseudospectral boundaries are nested sets on the complex plane, when $\varepsilon > \delta > 0$, the area of $\Lambda_\delta(A)$ is likely to be smaller than the area of $\Lambda_\varepsilon(A)$; similarly, the length of the boundary is likely to change; it would typically become smaller, unless there is separation and creation of disconnected components whose total perimeter exceeds that of the original curve, in which case it might increase.

The cost of computing the intermediate points $\tilde{w}_k$, $k = 1, \ldots, N$ is small since the derivatives at $\tilde{z}_k$, $k = 1, \ldots, N$ have already been computed by PF or a previous application of PsDM_sweep. Furthermore, we have assumed that $\sigma_{\min}(z_k I - A) = \varepsilon$

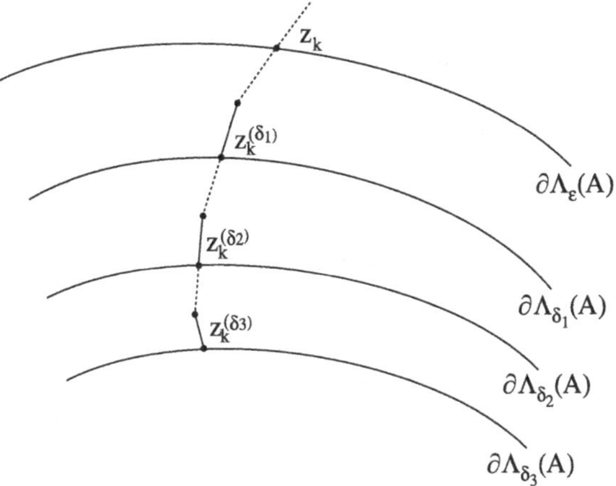

Fig. 13.10 Pseudospectrum descent process for a single point

for all k, since the points z_k approximate $\partial \Lambda_\varepsilon(A)$. On the other hand, computing the final points w_k, $k = 1, \ldots, N$ requires N triplet evaluations. Let $C_{\sigma_{\min}}$ denote the average cost for computing the triplet. We can then approximate the cost of PsDM_sweep by $\mathbf{T}_1 = N C_{\sigma_{\min}}$. The target points can be computed independently. On a system with p processors, we can assign the computation of at most $\lceil N/p \rceil$ target points to each processor; one sweep will then proceed with no need for synchronization and communication and its total cost is approximated by $\mathbf{T}_p = \lceil \frac{N}{p} \rceil C_{\sigma_{\min}}$.

Some additional characteristics of PsDM that are wortht of note (cf. the references in Sect. 13.4) are the following:

- It can be shown that the local error induced by one sweep of PsDM stepping from one contour to the next is bounded by a multiple of the square of the stepsize of the sweep. This factor depends on the analytic and geometric characteristics of the pseudospectrum.
- A good implementation of PsDM must incorporate a scheme to monitor any significant length reduction (more typical) or increase (if there is separation and creation of two or more disconnected components) between boundaries as ε changes and make the number of points computed per boundary by PsDM adapt accordingly.
- PsDM can capture disconnected components of the pseudospectrum lying inside the initial boundary computed via PF.

As the number of points, N, on each curve is expected to be large, PsDM is much more scalable than the other path following methods and offers many more opportunities for parallelism while avoiding the redundancy of GRID. On the other hand, PsDM complements these methods: We expect that one would first run COBRA, PAT or PPAT to obtain the first boundary using the limited parallelism of these methods, and then would apply PsDM that is almost embarrassingly parallel. In particular, each sweep can be split into as many tasks as the number of points it handles and each task can proceed independently, most of its work being triplet computations. Furthermore, when the step does not involve adaptation, no communication is necessary between sweeps.

Regarding load balancing, as in GRID, when iterative methods need to be used, the cost of computing the minimum singular triplets can vary a lot between points on the same boundary and not as much between neighboring points. In the absence of any other a priori information on how the cost of computing the triplets varies between points, numerical experiments (cf. the references) indicate that a cyclic partitioning strategy is more effective than block partitioning.

13.3 Dimensionality Reduction on the Matrix: Methods Based on Projection

When the matrix is very large, the preprocessing by complete reduction of the matrix into a Hessenberg or upper triangular form, recommended in Sect. 13.1.2, is no longer feasible. Instead, we seek some form of dimensionality reduction that provides acceptable approximations to the sought pseudospectra at reasonable cost. Recalling the Arnoldi recurrence (Eq. (9.59) in Chap. 9), an early idea [13] was to apply Krylov methods to approximate $\sigma_{\min}(A - zI)$, from $\sigma_{\min}(H - zI)$, where H was either the square or the rectangular (augmented) upper Hessenberg matrix that is obtained during the Arnoldi iteration and I is the identity matrix or a section thereof to conform in size with H. A practical advantage of this approach is that, whatever the number of gridpoints, it requires only one (expensive) transformation; this time, however, we would be dealing with a smaller matrix. An objection is that until a sufficient number of Arnoldi steps have been performed so that matrix A is duly transformed into upper Hessenberg, there is no guarantee that the minimum singular value of the square Hessenberg matrix will provide acceptable approximations to the smallest one for A. On the other hand, if A is unitarily similar to an upper Hessenberg matrix H, then the ε-pseudospectra of the $(j + 1) \times j$ upper left sections of H for $j = 1, \ldots, n - 1$ are nested (cf. [1]),

$$\Lambda_\varepsilon(H_{2,1}) \subseteq \Lambda_\varepsilon(H_{3,2}) \subseteq \cdots \subseteq \Lambda_\varepsilon(A).$$

Therefore, for large enough m, $\Lambda_\varepsilon(H_{m+1,m})$ could be a good approximation for $\Lambda_\varepsilon(A)$. This idea was further refined in the EIGTOOL package.

13.3.1 An EIGTOOL *Approach for Large Matrices*

EIGTOOL obtains the augmented upper Hessenberg matrix $\tilde{H}_m = H_{m+1,m}$ and ρ Ritz values as approximations of the eigenvalues of interest. A grid Ω_h is selected over the Ritz values and algorithm GRID is applied. Denoting by $\tilde{I}$ the identity matrix I_m augmented by a zero row, the computation of $\sigma_{\min}(\tilde{H}_m - z\tilde{I})$ at every gridpoint can be done economically by exploiting the Hessenberg structure. In particular, the singular values of $\tilde{H}_m - z\tilde{I}$ are the same as those of the square upper triangular factor of its "thin QR" decomposition and these can be computed fast by inverse iteration or inverse Lanczos iteration. A parallel algorithm for approximating the pseudospectrum locally as EIGTOOL can be formulated by modifying slightly the Algorithm in [14, Fig. 2]. We call it LAPSAR and list it as Algorithm 13.9. In addition to the parallelism made explicit in the loop in lines 3–6, we assume that the call in line 1 is a parallel implementation of implicitly restarted Arnoldi.

The scheme approximates ρ eigenvalues, where $\rho < m$ (e.g. the EIGTOOL default is $m = 2\rho$) using implicitly restarted Arnoldi [15] and then uses the additional information obtained from this run to compute the pseudospectrum corresponding to those eigenvalues by using the augmented Hessenberg matrix $\tilde{H}_m$.

It is worth noting that at any step m of the Arnoldi process, the following result holds [16, Lemma 2.1]:

$$(A + \Delta)W_m = W_m H_m, \quad \text{where } \Delta = -h_{m+1,m}w_{m+1}w_m^* \tag{13.10}$$

Algorithm 13.9 LAPSAR: Local approximation of pseudospectrum around Ritz values obtained from ARPACK.

Input: $A \in \mathbb{R}^{n \times n}$, $l \geq 1$ positive values in decreasing order $\mathcal{E} = \{\varepsilon_1, \varepsilon_2, ..., \varepsilon_l\}$; s the number of approximated eigenvalues, OPTS options to use for the approximation, e.g. to specify the types of eigenvalues of interest (largest in magnitude, ...) similar to the SIGMA parameter in the MATLAB eigs function.

Output: plot ε-pseudospectrum in some appropriate neighborhood of ρ approximate eigenvalues of A obtained using ARPACK.

1: [Rvals, $\tilde{H}_m$] = arpack(A, ρ, m, OPTS) //call an ARPACK implementation using subspace size m for $\rho < m$ eigenvalues.
2: Define grid Ω_h over region containing Ritz values Rvals
3: **doall** $z_k \in \Omega_h$
4:    $[Q, R] = \mathrm{qr}(z\tilde{I} - \tilde{H}_m, 0)$ //thin QR exploiting the Hessenberg structure
5:    compute $s(z_k) = \sqrt{\lambda_{\min}(R^*R)}$ //using inverse [Lanczos] iteration and exploiting the structure of R
6: **end**
7: Plot the l contours of $s(z_k)$ for all values of $\mathcal{E}$.

This means that the eigenvalues of H_m are also eigenvalues of a specific perturbation of A. Therefore, whenever $|h_{m+1,m}|$ is smaller than ε, the eigenvalues of H_m also lie in the pseudospectrum of A for this specific perturbation. Of course one must be careful in pushing this approach further. Using the pseudospectrum of H_m in order to approximate the pseudospectrum of A entails a double approximation. Even though when $|h_{m+1,m}| < \varepsilon$ the eigenvalues (that is the 0-pseudospectrum of H_m) is contained in $\Lambda_\varepsilon(A)$, this is not necessarily true for $\Lambda_\varepsilon(H_m)$.

13.3.2 Transfer Function Approach

Based on the matrix resolvent characterization of the pseudospectrum (13.3), we can construct an approximation of the entire pseudospectrum instead of focusing on regions around selected eigenvalues. The key idea is to approximate the norm of the resolvent, $\|R(z)\| = \|(A - zI)^{-1}\|$ based on transfer functions $G_z(A, E, D) = DR(z)E$, where D^*, E are full rank—typically long and thin—matrices, of row dimension n. If $W_m = [w_1, \ldots, w_m]$ denotes the orthonormal basis for the Krylov subspace $\mathcal{K}_m(A, w_1)$ constructed by the Arnoldi iteration, then two options are to use $E = W_{m+1}$ and $D = W_m^*$ or $D = W_{m+1}^*$. If we write

$$W_{m+1} = (W_m, w_{m+1}) \text{ and } H_{m+1,m} = \begin{pmatrix} H_{m,m} \\ h_{m+1,m}e_m^\top \end{pmatrix}$$

and abbreviate $G_{z,m}(A) = G_z(A, W_{m+1}, W_m^*)$ it can be shown that

$$\frac{1}{\sigma_{\min}(\tilde{H}_m - z\tilde{I})} \leq \|G_{z,m}(A)\| \leq \frac{1}{\sigma_{\min}(\tilde{H}_m - z\tilde{I})} + \|G_z(A, W_{m+1}, W_m^*)\tilde{u}\|$$

(13.11)

where $\tilde{u}$ is the $(m+1)$th left singular vector of $\tilde{H}_m - z\tilde{I}$ and that

$$\frac{1}{\sigma_{\min}(\tilde{H}_m - z\tilde{I})} \leq \|G_{z,m}(A)\| \leq \frac{1}{\sigma_{\min}(A - zI)} = \|R(z)\|.$$

Therefore, $\|G_{z,m}(A)\|$ provides an approximation to $\|R(z)\|$ that improves monotonically with m and is at least as good and frequently better than $1/\sigma_{\min}(\tilde{H}_m - z\tilde{I})$ [16, Proposition 2.2]. We thus use it to approximate the pseudospectrum.

One might question the wisdom of this approach on account of its cost, as it appears that evaluating $W_m^*(A - zI)^{-1}W_{m+1}$ at any z requires solving $m+1$ systems of size n. Complexity is reduced significantly by introducing the function $\phi_z = W_m^*(A - zI)^{-1}w_{m+1}$ and then observing that

$$G_{z,m}(A) = ((I - h_{m+1,m}\phi_z e_m^\top)(H_{m,m} - zI)^{-1}, \phi_z)).$$
(13.12)

Therefore, to compute $\|G_{z,m}(A)\|$ for the values of z dictated by the underlying domain method (e.g. all gridpoints in Ω_h for straightforward GRID) we must compute ϕ_z, solve m Hessenberg systems of size m, and then compute the norm of $G_{z,m}(A)$. The last two steps are straightforward. To obtain the term $(A - zI)^{-1}w_{m+1}$ requires solving one linear system per shift z. Moreover, they all have the same right-hand side, w_{m+1}. From the shift invariance property of Krylov subspaces, that is $\mathcal{K}_d(A, r) = \mathcal{K}_d(A - zI, r)$, it follows that if a Krylov subspace method, e.g. GMRES, is utilized then the basis to use to approximate all solutions will be the same and so the Arnoldi process needs to run only once. In particular, if we denote by $\hat{W}_1$ the orthogonal basis for the Krylov subspace $\mathcal{K}_d(A, w_{m+1})$ and by $F_{d+1,d}$ the corresponding $(d+1) \times d$ upper Hessenberg matrix, then

$$(A - zI)\hat{W}_d = \hat{W}_{d+1}(F_{d+1,d} - z\tilde{I}_d),$$

where $\tilde{I}_d = (I_d, 0)^\top$. We refer to Sect. 9.3.2 of Chap. 9 regarding parallel implementations of the Arnoldi iterations and highlight one implementation of the key steps of the parallel transfer function approach, based on GMRES to solve the multiply shifted systems in Algorithm 13.10 that we call TR.

TR consists of two parts, the first of which applies Arnoldi processes to obtain orthogonal bases for the Krylov subspaces used for the transfer function and for the solution of the size n system. One advantage of the method is that after the first part is completed and the basis matrices are computed, we only need to repeat the second part for every gridpoint. TR is organized around row-wise data partition: Each processor is assigned a number of rows of A and the orthogonal bases. At each step, the vector that is to orthogonalized to become the new element of the basis W_m

---

Algorithm 13.10 TR: approximating the ε-pseudospectrum $\Lambda_\varepsilon(A)$ using definition 13.3 on p processors.

Input: $A \in \mathbb{R}^{n \times n}$, $l \geq 1$ positive values in decreasing order $\mathscr{E} = \{\varepsilon_1, \varepsilon_2, ..., \varepsilon_l\}$; vector w_1 with $\|w_1\| = 1$, scalars m, d.

Output: plot of the ε pseudospectrum for all $\varepsilon \in \mathscr{E}$

1: Ω_h as in line 1 of Algorithm 13.1.   //The set $\Omega_h = \{z_i | i = 1, ..., M\}$ is partitioned across processors; processor pr_id is assigned nodes indexed $\mathscr{I}(\text{pr\_id})$, where $\mathscr{I}(1), ..., \mathscr{I}(p)$ is a partition of $\{1, ..., M\}$.

2: **doall** pr_id $= 1, ..., p$

3:     $(W_{m+1}, H_{m+1,m}) \leftarrow \texttt{arnoldi}(A, w_1, m)$

4:     $(\hat{W}_{d+1}, F_{d+1,d}) \leftarrow \texttt{arnoldi}(A, w_{m+1}, d)$

        //Each processor has some rows of W_{m+1} and $\hat{W}_{d+1}$, say $W_{m+1}^{(i)}$ and $\hat{W}_{d+1}^{(i)}$, and all have access to $H_{m+1,m}$ and $F_{d+1,d}$.

5:     **do** $j \in \mathscr{I}(\text{pr\_id})$

6:        compute $y_j = \text{argmin}_y\{\|(F_{d+1,d} - z_j \tilde{I}_d)y - e_1\|\}$ and broadcast y_j to all processors

7:     **end**

8:     Compute $\phi_{z,\text{pr\_id}} = (W_m^{(\text{pr\_id})})^* \hat{W}_d^{(\text{pr\_id})} Y$ where $Y = (y_1, ..., y_M)$

9: **end**

    //The following is a reduction operation

10: Compute $\Phi_z = \sum_{k=1}^p \phi_{z,p}$ in processor 1 and distribute its columns, $(\Phi_z)_k$, $(k = 1, ..., M)$ to all processors

11: **doall** pr_id $= 1, ..., p$

12:     **do** $j \in \mathscr{I}(\text{pr\_id})$

13:        $D_j = (I - h_{m+1,m}(\Phi_z)_j e_m^\top)(H_{m,m} - z_j I)^{-1}$

14:        nm_gz$_j = \|(D_j, (\Phi_z)_j)\|$

15:     **end**

16: **end**

17: Gather all nm_gz$_i$, $i = 1, ..., M$ in processor 1.   //They are the computed values of $1/\|G_{z_i}(A)\|$

18: Plot the contours of $1/\|G_{z_i}(A)\|$, $i = 1, ..., M$ corresponding to $\mathscr{E}$.

---

is assumed to be available in each processor. This requires a broadcast, at some point during the iteration, but allows the matrix-vector multiplication to take place locally on each processor in the sequel.

After completing the Arnoldi iterations (Algorithm 13.10, lines 3–4), a set of rows of W_{m+1} and $\hat{W}_{d+1}$ are distributed among the processors. We denote these rows by $W_{m+1}^{(i)}$ and $\hat{W}_{d+1}^{(i)}$, where $i = 1, ..., p$. In line 6, each node computes its share of the columns of matrix Y. For the next steps, in order for the local ϕ_z to be computed, all processors must access the whole Y, hence all-to-all communication is necessary. In lines 12–15 each processor works on its share of gridpoints and columns of matrix Φ_z. Finally, processor zero gathers the approximations of $\{1/s(z_i)\}_{i=1}^M$. For lines 8–10, note that $\phi_{z,i} = W_i^* \hat{W}_i Y$, where W_i and $\hat{W}_i$, $i = 1, ..., p$ are the subsets of contiguous rows of the basis matrices W and $\hat{W}$ that each node privately owns. Then, $\Phi_z = \sum_{i=1}^p \phi_{z,i}$, which is a reduction operation applied on matrices, so it can be processed in parallel. It is worth noting that the preferred order in applying the two matrix-matrix multiplications of line 8 depends on the dimensions of the local submatrices $W_m^{(i)}$, $\hat{W}_d^{(i)}$ and on the column dimension of Y and could be decided at runtime.

Another observation is that the part of Algorithm 13.10 after line 7 involves only dense matrix computations and limited communication. Remember also that at the end of the first part, each processor has a copy of the upper Hessenberg matrices. Hence, we can use BLAS3 to exploit the memory hierarchy available in each processor. Furthermore, if $M \geq p$, the workload is expected to be evenly divided amongst processors.

Observe that TR can also be applied in stages, to approximate the pseudospectrum for subsets of Ω_h at a time. This might be preferable when the size of the matrix and number of gridpoints is very large, in which case the direct application of TR to the entire grid becomes very demanding in memory and communication. Another possibility, is to combine the transfer function approach with one of the methods we described earlier that utilizes dimensionality reduction on the domain. For very large problems, it is possible that the basis $\hat{W}_{d+1}$ cannot be made large enough to provide an adequate approximation for $(A - zI)^{-1} w_{m+1}$ at one or more values of z. Therefore, restarting is necessary using as starting vectors the residuals corresponding to each shift (other than those systems that have already been approximated at the desired tolerance) will be necessary. Unfortunately, the residuals are different in general and the shift invariance property will be no longer readily applicable. Cures for this problem have been proposed both in the context of GMRES [17] and FOM(Full Orthogonalization Method) [18]. Another possibility is to use short recurrences methods, e.g. QMR [19] or BiCGStab [20]. In FOM for example, it was shown that if FOM is used, then the residuals from all shifted systems will be collinear; cf. [21].

13.4 Notes

The seminal reference on pseudospectra and their computation is [1]. The initial reduction of A to Schur or Hessenberg form prior to computing its pseudospectra and the idea of continuation were proposed in [22]. A first version of a Schur-based GRID algorithm written in MATLAB was proposed in [23]. The EIGTOOL package was initially developed in [24]. The queue approach for the GRID algorithm and its variants was proposed in [25] and further extended in [4] who also showed several examples where imbalance could result from simple static work allocation. The idea of inclusion-exclusion and the MOG algorithm 13.3 were first presented in [2]. The parallel Modified Grid Method was described in [4] and experiments on a cluster of single processor PC's over Ethernet running the Cornell Multitasking Toolbox [26] demonstrated the potential of this parallelization approach. MOG was extended for matrix polynomials in [27] also using the idea of "inclusion disks". These, combined with the concentric exclusion disks we mentioned earlier, provide the possibility for a parallel MOG-like algorithm that constructs the pseudospectrum for several values of ε. The path following techniques for the pseudospectrum originate from early work in [7] where they demonstrated impressive savings over GRID. COBRA was devised and proposed in [8]. Advancing by triangles was proposed in [9, 10]. The Pseudospectrum Descent Method (PsDM) was described in [12]. The programs were written in Fortran-90 and the MPI library and tested on an 8 processor SGI Origin 200 system. The method shares a lot with an original idea described in [28] for the independent computation of eigenvalues.

The transfer function approach and its parallel implementation were described in [16, 29]. Finally, parallel software tools for constructing pseudospectra based on the algorithms described in this chapter can be found in in [30, 31].

References

1. Trefethen, L., Embree, M.: Spectra and Pseudospectra. Princeton University Press, Princeton (2005)
2. Koutis, I., Gallopoulos, E.: Exclusion regions and fast estimation of pseudospectra (2000). Submitted for publication (ETNA)
3. Braconnier, T., McCoy, R., Toumazou, V.: Using the field of values for pseudospectra generation. Technical Report TR/PA/97/28, CERFACS, Toulouse (1997)
4. Bekas, C., Kokiopoulou, E., Koutis, I., Gallopoulos, E.: Towards the effective parallel computation of matrix pseudospectra. In: Proceedings of the 15th ACM International Conference on Supercomputing (ICS'01), pp. 260–269. Sorrento (2001)
5. Reichel, L., Trefethen, L.N.: Eigenvalues and pseudo-eigenvalues of Toeplitz matrices. Linear Algebra Appl. **162–164**, 153–185 (1992)
6. Sun, J.: Eigenvalues and eigenvectors of a matrix dependent on several parameters. J. Comput. Math. **3**(4), 351–364 (1985)
7. Brühl, M.: A curve tracing algorithm for computing the pseudospectrum. BIT **33**(3), 441–445 (1996)
8. Bekas, C., Gallopoulos, E.: Cobra: parallel path following for computing the matrix pseudospectrum. Parallel Comput. **27**(14), 1879–1896 (2001)
9. Mezher, D., Philippe, B.: PAT—a reliable path following algorithm. Numer. Algorithms **1**(29), 131–152 (2002)
10. Mezher, D., Philippe, B.: Parallel computation of the pseudospectrum of large matrices. Parallel Comput. **28**(2), 199–221 (2002)
11. Higham, N.: The Matrix Computation Toolbox. Technical Report, Manchester Centre for Computational Mathematics (2002). http://www.ma.man.uc.uk/~higham/mctoolbox
12. Bekas, C., Gallopoulos, E.: Parallel computation of pseudospectra by fast descent. Parallel Comput. **28**(2), 223–242 (2002)
13. Toh, K.C., Trefethen, L.: Calculation of pseudospectra by the Arnoldi iteration. SIAM J. Sci. Comput. **17**(1), 1–15 (1996)
14. Wright, T., Trefethen, L.N.: Large-scale computation of pseudospectra using ARPACK and Eigs. SIAM J. Sci. Comput. **23**(2), 591–605 (2001)
15. Lehoucq, R., Sorensen, D., Yang, C.: ARPACK User's Guide: Solution of Large-Scale Eigenvalue Problems with Implicitly Restarted Arnoldi Methods. SIAM, Philadelphia (1998)
16. Simoncini, V., Gallopoulos, E.: Transfer functions and resolvent norm approximation of large matrices. Electron. Trans. Numer. Anal. (ETNA) **7**, 190–201 (1998). http://etna.mcs.kent.edu/vol.7.1998/pp190-201.dir/pp190-201.html
17. Frommer, A., Glässner, U.: Restarted GMRES for shifted linear systems. SIAM J. Sci. Comput. **19**(1), 15–26 (1998)

18. Simoncini, V.: Restarted full orthogonalization method for shifted linear systems. BIT Numer. Math. **43**(2), 459–466 (2003)
19. Freund, R.: Solution of shifted linear systems by quasi-minimal residual iterations. In: Reichel, L., Ruttan, A., Varga, R. (eds.) Numerical Linear Algebra, pp. 101–121. W. de Gruyter, Berlin (1993)
20. Frommer, A.: BiCGStab(ℓ) for families of shifted linear systems. Computing **7**(2), 87–109 (2003)
21. Simoncini, V.: Restarted full orthogonalization method for shifted linear systems. BIT Numer. Math. **43**(2), 459–466 (2003)
22. Lui, S.: Computation of pseudospectra with continuation. SIAM J. Sci. Comput. **18**(2), 565–573 (1997)
23. Trefethen, L.: Computation of pseudospectra. Acta Numerica 1999, vol. 8, pp. 247–295. Cambridge University Press, Cambridge (1999)
24. Wright, T.: Eigtool: a graphical tool for nonsymmetric eigenproblems (2002). http://web. comlab.ox.ac.uk/pseudospectra/eigtool. (At the Oxford University Computing Laboratory site)
25. Fraysse, V., Giraud, L., Toumazou, V.: Parallel computation of spectral portraits on the Meiko CS2. In: Liddel, H., et al. (eds.) LNCS: High-Performance Computing and Networking, vol. 1067, pp. 312–318. Springer, Berlin (1996)
26. Zollweg, J., Verma, A.: The Cornell multitask toolbox. http://www.tc.cornell.edu/Services/ Software/CMTM/. Directory Services/Software/CMTM at http://www.tc.cornell.edu
27. Fatouros, S., Psarrakos, P.: An improved grid method for the computation of the pseudospectra of matrix polynomials. Math. Comput. Model. **49**, 55–65 (2009)
28. Koutis, I.: Spectrum through pseudospectrum. http://arxiv.org/abs/math.NA/0701368
29. Bekas, C., Kokiopoulou, E., Gallopoulos, E., Simoncini, E.: Parallel computation of pseudospectra using transfer functions on a MATLAB-MPI cluster platform. In: Recent Advances in Parallel Virtual Machine and Message Passing Interface, Proceedings of the 9th European PVM/MPI Users' Group Meeting. LNCS, vol. 2474. Springer, Berlin (2002)
30. Bekas, C., Kokiopoulou, E., Gallopoulos, E.: The design of a distributed MATLAB-based environment for computing pseudospectra. Future Gener. Comput. Syst. **21**(6), 930–941 (2005)
31. Mezher, D.: A graphical tool for driving the parallel computation of pseudosprectra. In: Proceedings of the 15th ACM International Conference on Supercomputing (ICS'01), pp. 270–276. Sorrento (2001)

Index

A

α-bandwidth, 42, 43

Amdahl's law, 13

APL language, 176

Arnoldi's method, 299–303, 350, 351, 460, 461

   identities, 301

   iteration, 337, 458, 460, 461

   restarted, 355, 459

B

Banded preconditioners, 91, 95, 99

Banded solver, 65, 129, 176

Banded Toeplitz solver, 183–192

Bandweight, 42

Bandwidth (matrix), 38, 40–43, 52, 57, 64, 66–69, 72, 91, 92, 105, 115, 128–130, 141, 145, 157, 176, 180, 191, 352, 363

B2GS, *see* Gram-Schmidt algorithm

BCR, *see* Rapid elliptic solvers

BGS, *see* Gram-Schmidt algorithm

BiCG, 336

BiCGStab, 463

Biconjugate gradient, *see* BiCG

Bidiagonalization

   via Householder reduction, 272

BLAS, 17, 20, 25, 29, 30, 52, 82, 100, 104, 139, 172, 211, 232–235, 243, 264, 272, 384, 410, 426, 439, 462

   _AXPY, 18, 32, 301

   _DOT, 18, 301

   BLAS1, 20, 25, 139, 211, 232, 264, 301

   BLAS2, 19, 25, 82, 100, 172, 232, 234, 243, 264, 394, 395, 426

   BLAS3, 20, 29, 30, 82, 100, 104, 233, 235, 264, 272, 462

   DOT, 448

Block cyclic reduction, 137, 139, 206

Buneman stabilization, *see* Stabilization

C

CFT, *see* Rapid elliptic solvers

CG, 37, 113–115, 297, 298, 312, 318–321, 372–378, 380

   acceleration, 318, 320, 321

   CGNE, 323

   preconditioned, 316

Chebyshev acceleration, 390, 396, 398

Chebyshev-Krylov procedure, 306

Cholesky factorization, 25, 26, 99, 126, 179, 184, 188–191, 237, 251

   incomplete, 399

Cimmino method, 320, 323

Cobra, *see* Path following methods

Complete Fourier transform, *see* CFT

Conjugate gradient, *see* CG

Conjugate gradient of the normal equations, *see* CGNE

CORF, *see* Rapid elliptic solvers

Cramer's rule, 155, 192

CS decomposition, 322, 323

D

Davidson method, 363–370, 381, 389, 398, 399

DD_PPREFIX, *see* Divided differences

DFT, 168, 169, 171

Diagonal boosting, 87, 100

Direct solver, 94, 95, 312, 431

E. Gallopoulos et al., *Parallelism in Matrix Computations*, Scientific Computation, DOI 10.1007/978-94-017-7188-7

The manufacturer's authorised representative in the EU is Springer
Nature Customer Service Centre GmbH, Europaplatz 3, 69115 Heidelberg,
Germany. If you have any concerns regarding our products, please
contact ProductSafety@springernature.com

Printed and bound by CPI Group (UK) Ltd, Croydon, CR0 4YY
08/01/2026
02031474-0008